AF717904

Managing Forest Ecosystems

Volume 37

Series Editors

Mikko Kurttila, *Natural Resources Institute Finland, Finland*
Margarida Tomé, *Instituto Superior de Agronomía, Lisboa, Portugal*
Thomas Seifert, *Faculty of Environment and Natural Resources, University of Freiburg, Freiburg, Germany*

Well-managed forests and woodlands are a renewable resource, producing essential raw material with minimum waste and energy use. Rich in habitat and species diversity, forests may contribute to increased ecosystem stability. They can absorb the effects of unwanted deposition and other disturbances and protect neighbouring ecosystems by maintaining stable nutrient and energy cycles and by preventing soil degradation and erosion. They provide much-needed recreation and their continued existence contributes to stabilizing rural communities.

Forests are managed for timber production and species, habitat and process conservation. A subtle shift from multiple-use management to ecosystems management is being observed and the new ecological perspective of multi-functional forest management is based on the principles of ecosystem diversity, stability and elasticity, and the dynamic equilibrium of primary and secondary production.

Making full use of new technology is one of the challenges facing forest management today. Resource information must be obtained with a limited budget. This requires better timing of resource assessment activities and improved use of multiple data sources. Sound ecosystems management, like any other management activity, relies on effective forecasting and operational control.

The aim of the book series *Managing Forest Ecosystems* is to present state-of-the-art research results relating to the practice of forest management. Contributions are solicited from prominent authors. Each reference book, monograph or proceedings volume will be focused to deal with a specific context. Typical issues of the series are: resource assessment techniques, evaluating sustainability for even-aged and uneven-aged forests, multi-objective management, predicting forest development, optimizing forest management, biodiversity management and monitoring, risk assessment and economic analysis.

More information about this series at http://www.springer.com/series/6247

Klaus von Gadow
Juan Gabriel Álvarez González
Chunyu Zhang • Timo Pukkala • Xiuhai Zhao

Sustaining Forest Ecosystems

Springer

Klaus von Gadow
Georg-August University of Göttingen
Göttingen, Germany

Juan Gabriel Álvarez González
University of Santiago de Compostela
Lugo, Spain

Chunyu Zhang
Beijing Forestry University
Beijing, China

Timo Pukkala
University of Eastern Finland
Joensuu, Finland

Xiuhai Zhao
Beijing Forestry University
Beijing, China

ISSN 1568-1319 ISSN 2352-3956 (electronic)
Managing Forest Ecosystems
ISBN 978-3-030-58713-0 ISBN 978-3-030-58714-7 (eBook)
https://doi.org/10.1007/978-3-030-58714-7

© Springer Nature Switzerland AG 2021
This work is subject to copyright. All rights are reserved by the Publisher, whether the whole or part of the material is concerned, specifically the rights of translation, reprinting, reuse of illustrations, recitation, broadcasting, reproduction on microfilms or in any other physical way, and transmission or information storage and retrieval, electronic adaptation, computer software, or by similar or dissimilar methodology now known or hereafter developed.
The use of general descriptive names, registered names, trademarks, service marks, etc. in this publication does not imply, even in the absence of a specific statement, that such names are exempt from the relevant protective laws and regulations and therefore free for general use.
The publisher, the authors, and the editors are safe to assume that the advice and information in this book are believed to be true and accurate at the date of publication. Neither the publisher nor the authors or the editors give a warranty, expressed or implied, with respect to the material contained herein or for any errors or omissions that may have been made. The publisher remains neutral with regard to jurisdictional claims in published maps and institutional affiliations.

This Springer imprint is published by the registered company Springer Nature Switzerland AG
The registered company address is: Gewerbestrasse 11, 6330 Cham, Switzerland

Changbaishan (长白山) in autumn colours.

Preface

白日依山尽
黄河入海流
欲穷千里目
更上一层楼

Crutzen (2002) assigned the term "Anthropocene" to the current human-dominated geological epoch. He concluded that humankind would remain a major environmental force in the future. Ecologists and engineers should therefore guide society towards environmentally sustainable management. Humans are able to identify the causes and disastrous consequences of neglect and exploitation of the forest ecosystems of the earth. Nowadays, we have both the knowledge and the tools to prevent further damage, and to sustain the world's ecosystems, for our own benefit and for the survival of life on Earth. Humans are capable of managing the environment to provide a range of ecosystem services and benefits to society. For this reason, the study of sustainable ecosystem design has become a central theme of forest ecology.

For a long time, the world of science has been involved in a progressive process of differentiation and specialisation. This process has led to the formation of a great variety of disciplines, each with a specific set of methods, terminologies and publication platforms. "Sustainable management" is an expression that has been, for the past 300 years, the guiding principle of forest science.[1] Biologists use the word "ecosystem" to refer to a community of living organisms within a particular physical environment.[2] The title of this book thus reflects what biologists and foresters want: to ensure the continued productive functioning of a community of trees.

Technological development and rapid communication facilitate global networking and cooperation, which is essential for dealing with complex problems such as long-term global warming. Bastin et al. (2019)[3] estimate that an additional 0.9

[1] Carlowitz, Hannß Carl v., 1713: Sylvicultura oeconomica, oder haußwirthliche Nachricht und Naturmäßige Anweisung zur wilden Baum-Zucht (1713). Available online for free download at https://de.wikisource.org /wiki/Datei:Sylvicultura_oeconomica.pdf.

[2] Egerton, F. N., 2012: Roots of Ecology: Antiquity to Haeckel. University of California Press. 288 p.

[3] Bastin, Jean-Francois; Yelena Finegold, Claude Garcia, Danilo Mollicone, Marcelo Rezende, Devin Routh, Constantin M. Zohner, Thomas W. Crowther, 2019: The global tree restoration potential. Science, Vol. 365, Issue 6448, pp. 76–79. DOI: 10.1126/science.aax0848

billion hectares would be potentially available for afforestation, in addition to existing forest and agricultural land, with a potential to cut the atmospheric carbon pool by about 25%. Such general claims, published in a reputable science journal, are sure to raise high expectations. By sequestering carbon dioxide and storing it in live woody tissues, forests represent a major sink in the global carbon cycle. The magnitude of the sink depends on species-specific growth rates and the particular site conditions, as well as the structure and density of a forest community. We present methods of forest analysis, modelling and design that may be helpful in deriving more specific estimates of forest production and to test such general assumptions.

Forest ecosystems include a great variety of communities of organisms interacting with their physical environment: multi-aged natural forests, even-aged monocultures and secondary forests invaded by foreign species. The challenge is to sustain their ability to function by adapting to changing climates and satisfying a multitude of human needs. The first chapter of this book sets the scene with a discussion about the effects of forest management on ecosystem services. Details about forest observational infrastructures are introduced in the second chapter. The third chapter presents methods of analysing forest density and structure. Models for estimating the shape and growth of individual forest trees are introduced in Chap. 4, and models of forest community production are presented in Chap. 5. Methods and examples of sustainable forest design are covered in Chap. 6.

This book, compiled upon the request of the Beijing Forestry University, and with the support of the Key Project of the National Key Research and Development Plan (2017YFC0504005) of China, is written for an international readership, for students of forest ecology and management. New scientific contributions continue to emerge as we are writing, and the work is therefore never finished. An updated version of this book, modified and extended to suit the needs of teaching and forest research especially in China, will soon appear in the Chinese (and English) language. This will naturally result in a changed order of authors.

We thank Fan Chunyu and her students for technical checking of an earlier draft, and Marga v. Gadow and Thea Brink for language editing, proof reading and many helpful comments.

Göttingen, Germany Klaus von Gadow

Beijing, China Chunyu Zhang [张春雨]

Lugo, Spain Juan Gabriel Álvarez González

Joensuu, Finland Timo Pukkala

Beijing, China Xiuhai Zhao [赵秀海]

Literature

Crutzen PJ (2002) Geology of mankind. Nature 415:23. https://doi.org/10.1038/415023a

Contents

What Is a Forest? 1

Abstract

Policies dealing with a broad range of forestry issues are often based on definitions created for the purpose of assessing global forest resources. Forest definitions emerging from prevailing objectives of land use have diversified, with new ones being added as forest management policies have broadened, focusing not only on timber production but gradually incorporating nontimber forest products, biodiversity conservation, and a variety of ecosystem services (Chazdon RL, Brancalion PHS, Laestadius L, Bennett-Curry A, Buckingham K, Kumar C, Moll-Rocek J, Vieira ICG, Wilson SJ, Conserv Lett 10(1):125–132, 2016. https://doi.org/10.1111/conl.12220). Accordingly, when we search the internet with the question "What is a forest?" we are confronted with a large number of definitions by, among others, the United Nations Food and Agriculture Organization (FAO), the United Nations Framework Convention on Climate Change (UNFCCC), the United Nations Convention on Biological Diversity (UN-CBD), the United Nations Convention to Combat Desertification (UN-CCD 2000), and the International Union of Forest Research Organizations (IUFRO). Keenan et al. (For Ecol Manag 352(2015):9–20, 2015) define "forest" as land of at least 0.5 hectare covered by trees higher than 5 m and with a canopy cover of more than 10%, or by trees able to reach these thresholds, and predominantly being under forest land use (see also FAO, Agricultural Organisation of the United Nations. http://www.fao.org/forest-resources-assessment/en/. Accessed 16 June 2017, 2015). Other definitions are given by Schuck et al. (Compilation of forestry terms and definitions, European Forest Institute, Joensuu, 2002), Bolton et al. (Sci Rep 9(1), 2019. https://doi.org/10.1038/s41598-018-37265-4), and the Convention on Biological Diversity (Definitions: indicative definitions taken from the report of the ad hoc technical expert group on forest biological diversity. https://www.cbd.int/forest/definitions.shtml. Retrieved 29 March 2017, 2017).

© Springer Nature Switzerland AG 2021
K. von Gadow et al., *Sustaining Forest Ecosystems*, Managing Forest Ecosystems 37,
https://doi.org/10.1007/978-3-030-58714-7_1

Picea abies forest in Finland

1.1 Classifications of Plant Communities

Vegetation is often seen as an effective indicator of the ecological functioning of an ecosystem. A classification of vegetation can therefore serve as an important basis for understanding, managing, and conserving ecosystems. Accordingly, Grossman et al. (1998) emphasize the need for a standard classification of ecological communities. The objective of any kind of classification is to facilitate description, comparison, and communication by means of a consistent set of terms for the different categories. Grouping organisms with similar adaptations together removes some of the complexity and can even lead to a better understanding of life on earth. There are numerous classification systems, each resulting in different numbers of units known as ecoregions, ecozones, or biomes. Olson et al. (2001) define *ecoregions* as relatively large units of land containing a distinct assemblage of natural communities and species, with boundaries that approximate the original extent of natural communities prior to major land-use changes. Their map of the terrestrial ecoregions of the world recognizes 867 distinct units. Ecoregion boundaries rarely form abrupt edges; they are bound by ecotones and often include a mosaic of habitats.

1.1.1 Global Classifications

One of the most widely used global systems is the Köppen classification system that was published more than a century ago by the Russian climatologist Wladimir Köppen (Köppen 1884; Geiger 1954; Beck et al. 2018). The Köppen climate classification scheme divides climates into five main groups: A (tropical), B (dry), C (temperate), D (continental), and E (polar). Table 1.1 shows three examples of classification systems that focus on climate, plant community, and vegetation structure.

The Köppen system uses two or three letters to indicate climate types and is shown in the first column in Table 1.1. The second letter in each main group indicates the seasonal precipitation type, while the third letter refers to temperature. The system of plant communities by Schimper and Faber (1935) is listed in the second column of Table 1.1. The system includes 15 global types of vegetation. The third column in Table 1.1 shows Ellenberg's system that is based on vegetation structure and includes six terrestrial categories. These three systems are only three examples among numerous other classification systems. New world maps of biomes are published regularly in an attempt to simplify the huge diversity of environments and plant communities.[1]

[1] at https://askabiologist.asu.edu/explore/biomes (Accessed on 22nd March, 2019)

Table 1.1 Three examples of classification systems emphasizing climate, plant community, and vegetation structure

Climate	Plant community	Structure
Class A	1. Tropical rainforest	I. Closed forests
Tropical rainforest (Af)	2. Subtropical rainforest	II. Woodlands
Tropical monsoon (Am)	3. Monsoon forest	III. Fourrés (shrublands or thickets)
Tropical savanna (Aw, As)	4. Temperate rainforest	IV. Dwarf-scrub and related communities
Class B	5. Summer-green deciduous forest	V. Terrestrial herbaceous communities
Desert (BWh, BWk)	6. Needle-leaf forest	VI. Deserts and other scarcely vegetated areas
Semiarid (BSh, BSk)	7. Evergreen hardwood forest	VII. Aquatic plant formations
Class C	8. Savanna woodland	
Humid subtropical (Cfa, Cwa)	9. Thorn forest and scrub	
Oceanic (Cfb, Cwb, Cfc, Cwc)	10. Savanna	
Mediterranean (Csa, Csb, Csc)	11. Steppe and semidesert	
Class D	12. Heath	
Humid continental (Dfa, Dwa, Dfb, Dwb, Dsa, Dsb)	13. Dry desert	
Subarctic (Dfc, Dwc, Dfd, Dwd, Dsc, Dsd)	14. Tundra and cold woodland	
Class E	15. Cold desert	
Tundra (ET)		
Ice cap (EF)		
Alpine (ET, EF)		
Köppen (1884)	Schimper and Faber (1935)	Ellenberg and Mueller-Dombois (1967)

The specific combinations of climatic variables in the Köppen classification are ecologically relevant. The classification has therefore been widely used to map the geographic distribution of long-term ecological conditions. More recently, the classification has been used as a diagnostic base for identifying climate change and potential changes in vegetation over time. Chen and Chen (2013) used the Köppen classification to quantify climate variation and change during the period 1901–2010, using a dataset of global temperature and precipitation observation. They found that the most significant change between 1901 and 2010 was a distinct increase in the size of the area of dry climates. They also found a significant decrease in the size of the area of the polar climate (E) since the 1980s. The areas of spatially stable climate regions for inter-annual and inter-decadal variations were also identified.

Chan and Wu (2015) examined changes in several Köppen categories, basing their findings on observed data, including uncertainties because of internal climate variability. They found that about 5.7% of the global land area shifted toward warmer and drier climate types between 1950 and 2010, as well as a significant expansion of high-latitude continental climate zones, a shrinkage of polar climates, and an increasing average elevation in tropical and polar climates. They found that these changes of climate types since 1950 cannot be explained as natural variations and that they were most likely driven by anthropogenic factors.

1.1.2 Regional Classifications

A range of environmental conditions exist within each biome. Some organisms may have adapted to a part of the biome, while others are able to thrive within the full range of conditions within the boundaries of a defined biome. Some of that variety is captured in regional classifications. Many regional classification systems were developed by experienced ecologists who were able to see differences and commonalities without getting bogged down by the dazzling variety of structures and patterns that are so typical of the real world. The vegetation in China, a country with a great diversity of climates, includes coniferous forests, deciduous forests, mixed forests, shrubland, deserts, various types of grassland, swamps, and extensive areas of planted forests. South Africa is another example of a region with a great diversity of climate types that include deserts and semideserts (Karoo); a Mediterranean climate with the highly diverse Cape flora locally known as *fynbos*; afromontane forests, and subtropical coastal forests; a particular type of savanna known as *bushveld*; and temperate grasslands. India is also a region with a great diversity of environments, and an old classification system for forest types, dividing their forests into 14 categories, is still in use (Champion and Seth 1968):

Forest Type = wet evergreen forest, semi-evergreen forest, moist deciduous forest, dry deciduous forest, littoral, swamp/mangrove forest, dry evergreen forest, thorn forest, subtropical broad-leaf forest, subtropical pine forest, subtropical dry evergreen forest, montane wet temperate forest.

More recently, Verma et al. (2013) proposed a similar set of forest types for India that included geographical detail. The tropical wet evergreen forest communities of the northeast differ from those of the Western Ghats, and it therefore makes sense to separate the two types. More recently, Reddy et al. (2015) published a map with detailed information about vegetation and land use patterns that exhibit the dramatic changes resulting from the human footprint.

1.1.3 Potential Natural Vegetation

Potential natural vegetation (PNV) is the vegetation that can be expected in specific environmental conditions (climate, geomorphology, and geology) and in the absence of human intervention or major hazard events. Because of the extent to which PNV has been used, the concept can be considered one of the most successful contributions to vegetation science. According to Härdtle (1995), however, applications of PNV have been affected by inconsistent interpretation or incorrect usage. Härdtle found, for example, that incorrect applications resulted from a failure to distinguish between "potential natural vegetation," "reconstructed natural vegetation" and "vegetation developing during succession." In addition, there is considerable uncertainty as to when vegetation should be defined as reversibly or irreversibly altered.

More recently, new methods of machine learning have been used to estimate and map potential natural vegetation on a global scale. Machine learning can be successfully used to model vegetation distribution, and is especially applicable when the training data sets consist of a large number of observations and a large number of covariates. Hengl et al. (2018a, b) presented a map showing the global distribution of natural biomes that was based on a large dataset of terrestrial observations. They found that the biome reconstruction accuracy was lower at ecotonal boundaries and in mountainous areas because of pollen transport issues, particularly the long-distance transport of tree pollen. Many present-day terrestrial biomes show reduced biodiversity, with grasslands being most affected and tundra and boreal forests least affected (Newbold et al. 2016). Extending the coverage of observations of natural and managed vegetation, as well as observations of new ground will allow regular improvement of PNV maps. Despite rapid technical progress, classification systems of plant communities that represent a "garden-of-Eden" view will always be subject to uncertainty.

1.1.4 Novel Forests

Climate changes are affecting biotic components of ecological niches, while social and economic changes, such as the increasing movement of people and goods in a globalized world offer multiple routes for species translocation and dissemination (Ogden et al. 2019). These external drivers facilitate biological invasions of alien species. The number of invasive tree species, areas invaded, types of impacts, and the complexity of management challenges have increased dramatically. Over the past 250 years, humans have moved *Acacia* species native to Australia extensively throughout the world (Richardson et al. 2011a, b), and climate models suggest that about a third of the world's land surface is climatically suitable for Australian acacias. Many *Acacia* species are commercially important and have been extensively planted, but human perceptions of Australian acacias vary in different parts of the world.

From the middle of the nineteenth century, many acacia species were imported from Australia to the Cape region of South Africa to stabilize shifting sand dunes and to serve as wind breaks. Today, they invade the coastal areas from Port Nolloth on the African west coast to Port Elizabeth on the east coast. The following nine invasive species from Australia have been reported to occur along the coastal area from Pringle Bay to Hermanus in South Africa: *Acacia cyclops*, *A. mearnsii*, *A. saligna*, *A. longifolia*, *A. melanoxylon*, *Hakea sericea*, *Paraserianthes lophantha*,

Leptospermum laevigatum, and *Eucalyptus conferruminata*. In Kleinmond, local volunteers, known as the hacking group, have been trying to eradicate the invasive species since the 1980s. Eucalypts are interesting for their modest invasive success despite their wide dissemination. Casuarinas have only recently been widely planted and little is known of their invasion ecology (Richardson et al. 2014). *Acacia mangium*, a fast-growing multipurpose tree used in agroforestry and for restoration of degraded lands, is highly invasive in many regions of the world where it has been introduced outside its native range (Koutika and Richardson 2019).

Humans have consciously or unconsciously generated widespread biological invasions, leading to radical alterations in the functioning of ecosystems. Hui and Richardson (2017) provide substantial evidence that shows how humans have been rearranging the world's biota during and after the era of European colonization, and especially during the period of globalization after the Second World War. Invasions of alien tree species have escalated during the last few decades. The number of invasive tree species, invaded areas, types of impacts, and complexity of management challenges have increased dramatically.

Hui et al. (2011a, b) found that the performance of a species in its native range is correlated with its potential success in new regions. This evidence implies that the movement of a species around the world is not random but favors those species that have particular types of native geographical distributions, and are tied to species invasiveness, functions, and spatial scales. Acacias with large native ranges are more likely to become invasive than those with a narrow range. Species with large native ranges may be more invasive because a large native range can translate into a large niche space and environmental tolerance. Native range size and introduction efforts are significantly positively correlated. Species with high rates of population increase can recover rapidly after perturbation and may have faster spread rates, and so are likely to be observed as invasive more frequently than other species (Sakai et al. 2001).

The ultimate reason for the escalation of tree invasions worldwide is the rapid increase in the human-mediated transport and dissemination of thousands of species for a wide range of purposes, especially in forestry, agroforestry, and ornamental horticulture. Vast plantations of alien trees now dominate many regions of the world. Alien trees are also increasingly conspicuous as ornamental plants and in agroforestry. A number of projects are championing the use of trees, mainly nonnative species, for the provision of goods and services, especially in developing countries (Low 2012). New projects such as *AlienScenarios* are evaluating the range of plausible futures of biological invasions for the twenty-first century (Essl et al. 2019).

1.2 Popular Myths

Naturalness has historically been linked to aesthetic preferences, for example, conditions reminiscent of the deep, dark forests in fairy tales (Bennett 2015). In theory, the degree of "naturalness" of an ecosystem is the difference between its current state and some assumed natural state (Gaston and Blackburn 2000). In reality, ecosystems are subject to continual change even when protected from human use. "Naturalness" is a concept which is widely employed in the management of protected areas, but, because of shifting baselines and differing values, it is not possible to measure and objectively evaluate the degree of what constitutes "naturalness" (Sprugel 1991). The concept therefore continues to be ambiguous, subjective, and contentious.

1.2.1 The Myth of the Climax Community

Frederic Clements claimed that all ecosystems would eventually develop a stable, self-regulating *climax community* in perfect equilibrium (Clements and Weaver 1938). Because of this belief, environmentalist campaigns since the 1960s have sought to create protected areas entirely free from human intervention. It was assumed that fenced-in protected areas would inevitably produce good conservation outcomes (Wapner 2014; Purdy 2015).

Forest ecosystems contribute to cleansing the air and water bodies, and to maintaining a livable biosphere. Forests are important carbon sinks that absorb a large proportion of greenhouse gas emissions and pollutants harmful to humans and other organisms. Another important environmental service of forest ecosystems is the potential to provide a range of habitats for regional biodiversity. Consequently, it is of crucial importance that ecosystem services are included in policies that affect the use of natural resources. Interestingly, managed forests have often been found to have higher biodiversity than protected ones.

Schulze (2017) compared the observations from managed and protected *Fagus sylvatica* and *Picea abies* forests in Romania and Germany. Species richness was highest in the herbaceous layer of the German *Picea* forests, with about twice as many species as in the Romanian *Picea* forest. The higher herbaceous diversity in the German coniferous forests was the result of liming and the associated nitrogen saturation (Reif et al. 2013). The herbaceous diversity in the *Fagus* forests was similar in Germany and Romania (Schmidt 2000), but the diversity of the herbaceous flora in the managed forests was generally higher than in protected forests.

In a recent study, Schulze et al. (2019) found that the changes in forest management over the past 45 years have had a significantly positive effect on the abundance of nonmigratory forest bird species in Central Europe. Climate changes led to longer seasons and milder winters. Forest management that promotes tree diversity was positively and significantly related to the prevalence of nonmigratory forest birds. Nonmigrating forest birds had stable populations with a large variation in species, while the number of birds migrating across continents has declined in recent decades. The results of the study suggest that reports about a decline in bird numbers in forests should distinguish between migratory and nonmigratory bird species. Efforts to mitigate the general decline in bird life should not only focus on forests, but should also include other land-use systems.

A general problem with comparative studies in biodiversity is the fact that there are no truly pristine ecosystems. All natural plant communities have been directly or indirectly modified by humans to provide a particular mix of services. In addition, many forested landscapes are affected by air pollution. Forest management provides essential products and services (Pukkala 2008) and can also have a positive effect on biodiversity (Aravind et al. 2005), on streamflow and water resources (Bosch and Gadow 1990), and on erosion protection (Dragoi and Ciornei 2013). Continuous cover forestry, characterized by selective harvesting, minimal human disturbance, avoidance of clear-felling, and the use of natural regeneration of site-adapted tree species, is believed to be superior to even-aged monocultures regarding a wide range of expectations, especially with those related to biodiversity (Pukkala and Gadow 2012).

1.2.2 The Myth of the Primeval Forest

Labeling a forest as "primeval" is not always accurate. New research provides evidence that extensive areas of the Amazon forest were modified by pre-Columbian societies who cultivated palms, nuts, and other domesticated plants (Ross 2017). The densest human populations were located along the major whitewater rivers (Levis et al. 2012; Erb et al. 2017). A prominent example of a forest that is often referred to as "primeval" is the famous Białowieża forest, a remnant of the forests that once covered much of the central European lowland plains (Falinski 1986). Recent studies have found that humans have used and actively shaped the forest for centuries. Navigable rivers traversing the forest that connect to the Baltic Sea provided a means of transporting forest products (Hedemann 1939; Schama 1995). As early as 1410, the rivers of the Białowieża Forest were noted for their economic importance in transporting timber (Brincken 1826). Figure 1.1 presents a map showing the subdivision of the forest into periodic blocks identified by Roman numerals. This subdivision indicates the use of the classic method of assigning particular forest areas to harvest periods.

Białowieża thus represents an intrinsically altered landscape, the result of a long history of forest management and trade in the Baltic (Sunseri 2012). Vera (2000) has shown that the forest was more open in the past than it is today. Pollen profiles analyzed from peat bogs in the Białowieża National Park show increases in the charcoal to pollen ratio from around 1430 (Mitchell and Cole 1998), which supports evidence that the forest was intensely settled from the sixth to the twelfth century (Gorska 1976; Hedemann 1939). Arable farming and animal husbandry were commonly practiced in the forest by clearing the forest and establishing meadows (Franklin et al. 2002).

Fires were frequent from 1653 to the late 1700s, supporting assumptions of an open, pine-dominated forest with substantial human influence on the fire regime (Niklasson et al. 2010). One of the most significant human impacts was the maintenance of high populations of large herbivores for royal hunting throughout the centuries. The last yew trees, a species known to be particularly vulnerable to browsing, disappeared in Białowieża Forest during the nineteenth century (Wiecko 1972). The protection of large herbivores in the Białowieża National Park has caused a

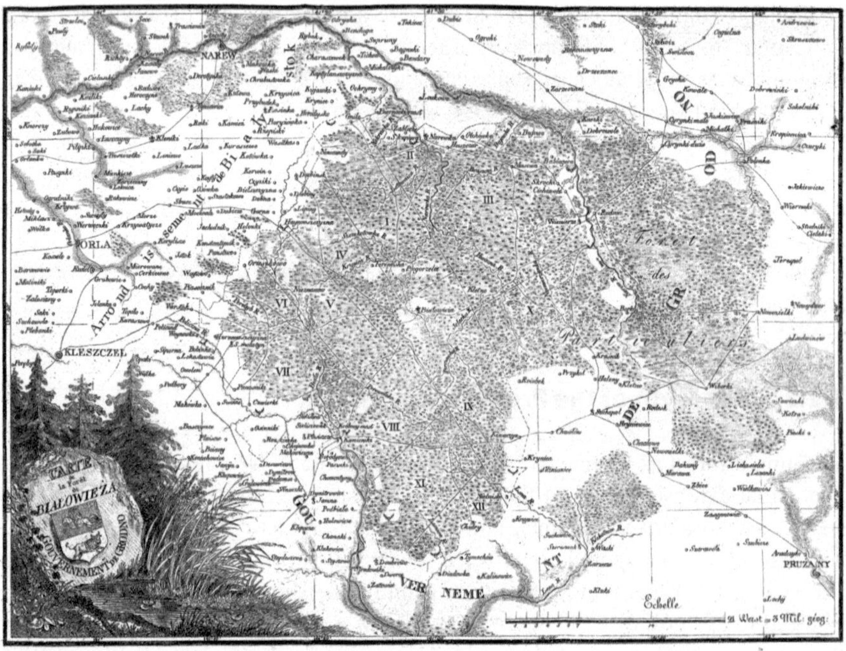

Fig. 1.1 Historical map of 1826 showing Białowieża Forest in Poland and Bialorussia (Brincken 1826). The map shows the subdivision of the forest into periodic blocks – the classic method of assigning particular forest areas to harvest periods

dramatic loss of species diversity. Only a very few species such as *Carpinus betulus* are capable of withstanding browsing pressure (Kuijper et al. 2010).

1.3 Forest Anthromes: Domesticated Ecosystems

Global and regional reports of forest areas use classifications that are based on descriptors of some original biome or climate. Original biomes have been modified, however, and the climate is changing. Most primary forests are believed to have been subject to extensive use which includes indiscriminate logging (Kleinschmit et al. 2016) or sophisticated management based on selective harvesting (Schütz et al. 2012; Puettmann et al. 2015). Forests have also been modified by the human-mediated introduction of organisms to areas well outside their potential range. Most contemporary forest ecosystems are "novel" – they are part of the biosphere and further expansion is likely. Marris (2013) proclaims novel ecosystems to be the "new normal" and suggests that we accept this new reality and "shake off the yoke of history." It makes sense, therefore, to create an empirical basis for observing and understanding the structures and processes of such systems in a more general sense, that is, by accepting that there is no clear distinction between "natural" and "novel."

Hui and Richardson (2017) present a conceptual view of the global extent of contemporary ecosystems which they call anthromes. The difference between particular anthromes is determined by human population density, land use and the integrity of habitat and biotic communities. Forestry is an important land use in the category "semi-natural anthromes." Many habitats in that category are novel and include, for example, nonnative communities which now exist outside their original distributional range and are forming new kinds of systems that are (a) without historical precedent and (b) self-sustaining. Attempts to restore such partially exotic communities to some approximation of a pristine condition are often considered a fruitless effort.

These findings lead us to the conclusion that there appears to be a need for a more realistic classification of forest communities. A first step would be to quantify the compositional differences in species assemblages that are based on structure and biology (Chao et al. 2005; Swenson et al. 2012; Legendre and Cáceres 2013; Hao et al. 2019a, b). Dissimilarity in community composition is one of the most conspicuous features by which different forest ecosystems can be distinguished (Jost et al. 2011). Assessing compositional differences between forest communities is important for several reasons. Dissimilarities between forests can reveal specific habitat effects that shape forest composition and structure (Socolar et al. 2016). Assessing dissimilarities between forest communities is essential for evaluating species invasions, changes caused by selective tree harvesting, or the effects of climate change on species composition. In addition, effective measurements of community dissimilarity can contribute to a more meaningful classification of forest vegetation (Hao et al. 2019a, b).

Jaccard (1900) was probably the first to propose a method for measuring the degree of community similarity and dissimilarity based on the number of species shared by two communities and the number of species unique to each of them. Two additional indices were subsequently proposed to estimate the differences between communities (Sørensen 1948; Bray and Curtis 1957). The Jaccard and Sørensen indices use species presence and species absence data. The Bray-Curtis index, a modified version of the Sørensen index, includes species abundance (Chao et al. 2005). These three indices have become the most widely used measures for assessing community similarity or dissimilarity in community ecology (Anderson et al. 2006). Several other indices based on species incidence or species abundance have also been developed (Chao et al. 2005; Legendre and Legendre 2012). Examples are the Chi-square distance index (Fenelon and Lebart 1971), the Canberra index (Lance and Williams 1967; Stephenson et al. 1972), and the Morisita-Horn index (Magurran 2004).

These indices have been widely used in forest ecology, and they contribute substantially to the understanding of community dissimilarity. However, community dissimilarity is affected not only by differences in species abundance or incidence, but also by the biological heterogeneity among species (Clarke and Warwick 1998, 2001). Accordingly, Hao et al. (2019a, b) introduced a new measure of community dissimilarity that incorporates both the information of species frequencies and the biological heterogeneity among species. The approach is based on a new version of the original Avalanche index proposed by Ganeshaiah et al. (1997) and Ganeshaiah and Shaanker (2000). The discriminating avalanche system introduced in the study of Hao et al. (2019a, b) is used to measure the dissimilarity between tree species, based on taxonomy, phylogeny, and traits. The application is demonstrated by using observations from several stem-mapped forest plots. Compared to traditional approaches, the discriminating avalanche values show a more realistic estimate of community dissimilarities, indicating a greater similarity among communities when species are closely related. The avalanche approach will be explained in greater detail in Chap. 3 of this book.

Community "similarity" or "dissimilarity" is a qualitative human construct which has no precise mathematical definition. Nevertheless, measuring community dissimilarity should rely on some quantitative measurements devised for a specific purpose (Chao et al. 2005). Dissimilarity means "not the same." When we state that two forest communities are the same, what we are really doing is disregarding the particular differences we consider irrelevant. Therefore, although numerous measures have been published, the development and testing of new ones for quantifying community dissimilarity more accurately, more comprehensively and objectively remains a priority in community ecology (Chao et al. 2005).

Besides the avalanche approach for assessing biological distances, the transportation model of linear programming can also be used to evaluate the biological distance between forest communities, the objective being to minimize the "cost" of transforming community i into community j:

$$\min \rightarrow Z = \sum_{i}^{m}\sum_{j}^{n} \text{distance}_{ij} \cdot X_{ij} \tag{1.1}$$

where distance_{ij} is some measure of the biological distance between species i and j. X_{ij} are the numbers (or relative proportions) of the different species in communities i and j. The constraints would be the numbers (or relative proportions) of the different species in communities i and j:

$$\sum_{j}^{n} X_{ij} \leq \text{available}_i \quad \text{and} \quad \sum_{i}^{m} X_{ij} \geq \text{required}_j \tag{1.2}$$

We are not aware of any studies that have used this approach. The ability to measure community dissimilarity is a prerequisite for the meaningful classification of forest ecosystems.

Forests represent a remnant wilderness of high recreational value in the densely populated industrialized countries, a natural resource threatened by elimination or degradation in other regions, and a renewable reservoir of essential raw materials for the wood processing industry. An estimated 2908 million hectare or 23% of the continental surface is classified as productive forest land, and this has been subdivided by Solberg (1996) into five categories of management (Table 1.2).

Table 1.2 Estimated area of productive forest and harvest intensity for five categories of management as defined by Solberg (1996)

Mgmt category	Description	Million ha	Harvest intensity (m^3/ha/year)
M1	Extensively managed native forest in temperate and tropical regions	2401	0.3
M2	Public forests in 16 temperate zone countries, including some exotic plantations (New Zealand) and extensively managed native forests (e.g., Canada)	218	1.4
M3	Privately owned nonindustrial forests in 16 temperate zone countries, including exotic plantations (Australia, Chile, New Zealand) and native plantations (USA, Scandinavia, Japan)	218	1.9
M4	Fast-growing commercial plantations in the tropics, private and public	23	2.8
M5	Forests owned by timber industries in 16 temperate zone countries	43	4.6
Total		2908	0.5

Danneyrolles et al. (2019) assembled a dataset of more than 100,000 tree species list from the nineteenth century across a broad region in temperate eastern Canada and added more recent forest inventory data to test the effects of changes in anthropogenic disturbance, temperature, and moisture on forest dynamics. The specific objective was to evaluate changes in forest composition by quantifying the affinities of co-occurring tree species with temperature, drought, light, and human disturbance. They found that anthropogenic disturbances were the dominant causes of compositional change in these forests. Disturbance-related adaptations of different taxa were captured by two indices. One index focused on well-established shade tolerance scores, is intended to capture responses to severe disturbances (e.g., clear-felling and anthropogenic fire). The other index (CDI) was developed specifically for this study to capture expected responses to a wider range of disturbance severities, such as selective logging. This CDI showed the strongest temporal change, involving increases for a broad spectrum of disturbance-adapted taxa.

When a somewhat simplified classification based on the development of timber volume over age or time is used, two types of sustainable forest management system can be distinguished: a rotation forest management (RFM) system with standard silvicultural treatments and a continuous cover forestry (CCF)[2] system with selective harvesting (Fig. 1.2). The distinction between the two systems is usually the result of decisions relating to the cost of timber harvesting, simplicity of management, or intangible benefits such as ecosystem resilience.

[2] CCF management, also known as "near-natural forest management" or "forest gardening" is characterized by selective harvesting and the use of natural regeneration and site-adapted tree species (Griesel and Gadow 1995).

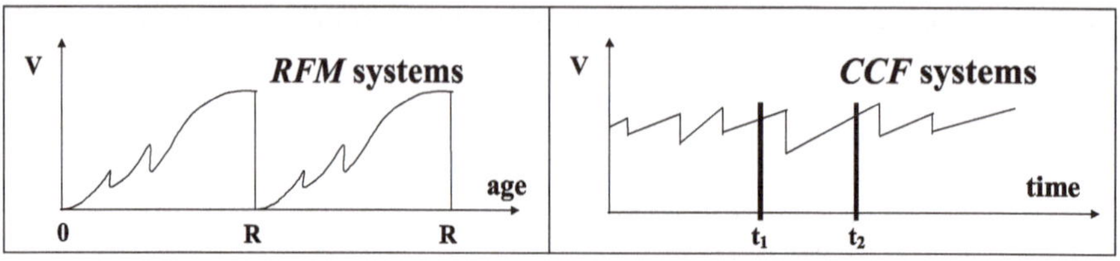

Fig. 1.2 Development of biomass (v) over age or time for two prototype systems of forest management: rotation management system (left) and continuous cover system (right)

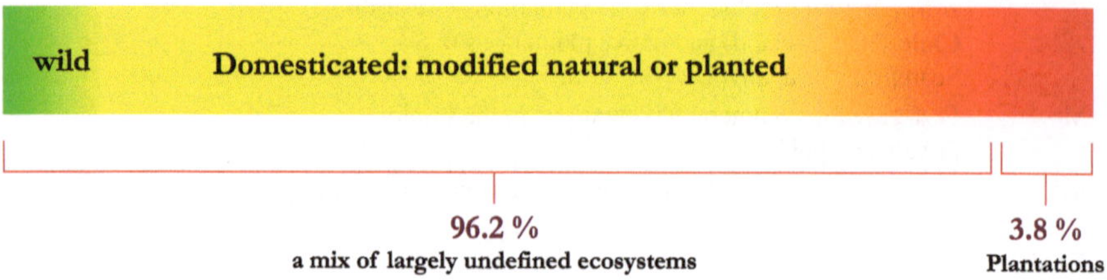

Fig. 1.3 Simplified diagram which attempts to give a realistic impression of the world's forests (the estimated percentages are derived from recent FAO reports)

RFM systems with fast-growing timber species and intensive silviculture are found in the southern hemisphere (Chile, South Africa, Australia, and New Zealand), in the southeastern United States, in many parts of Asia and in the Mediterranean regions. Low intensity RFM systems are usually found in regions controlled by timber processing companies that require regular raw material supplies. In a forest managed under a CCF system, the forest age is undefined. Forest development does not follow a cyclic harvest and regeneration pattern. Instead, it *oscillates* around some "ideal" residual level of growing stock (Gadow et al. 2007, 2013; Puettmann et al. 2015; Messier et al. 2019).

Many forest communities have been domesticated and consist of a mix of species that have adapted to a particular environment and species that were introduced by man. A range of methods for describing these communities will be presented in Chap. 3. A general classification which reflects the reality of continuous change brought about by human use and biological invasion presents a more realistic view than many traditional classification systems. In the absence of a complete and detailed observational coverage of the world's forest ecosystems, the category "Domesticated: modified natural or planted" in Fig. 1.3 is probably a more appropriate label for the category "Modified natural" used in recent FAO reports.

Forest vegetation is dynamic and highly irregular, and is characterized by a unique set of tree species that often includes a considerable proportion of nonnative ones. Forest structures usually represent a high degree of complexity with very specific patterns of density and structural diversity (see Chap. 3). Observational infrastructures are therefore needed for the analysis and description of these complex ecosystems.

1.3.1 To Act or Not to Act?

The fact that humans are part of nature is so obvious that there should be no need to mention it. However, a study by Vining et al. (2008) based on three separate questionnaires has shown that even though the majority of the participants considered themselves part of nature, natural environments were largely described as places absent from any human interference. The authors conclude that gaining an understanding of this apparent contradiction may lead to a better awareness of the importance of people's perceptions. Exactly because of the realization that neglect can result in the demise or even destruction of a healthy ecosystem, a new discipline called "ecological engineering" has emerged. The objective of ecological engineering, as expressed in its own scientific journal, is to design sustainable ecosystems that integrate human society with its natural environment for the benefit of both (Mitsch 2012).

Białowieża, an important forested landscape because of its ecological significance, particular history and large area, has become the site of an intense international debate following a serious bark beetle (*Ips typographus*) outbreak. The controversy centers around two opposing ecological paradigms: (a) to prevent human interference and observe what happens (the *no action* paradigm); (b) to salvage trees infected by *Ips typographus,* and thus reduce the fire risk and CO_2 emissions and prevent further spread of the beetle (the *adaptive action* paradigm). Białowieża is subject to various national acts and regulations as well as a range of international restrictions that are relevant in certain areas, including nature reserves and "reference areas"; protection areas for certain bird species; protection of "nature monuments," "pioneer" and wetland forests. Specific restrictions that apply in particular locations may be irrelevant in others. The complexity of these legal and administrative restrictions complicates planning, because any attempt to solve the conflict must recognize the relevance of the different legal frameworks. Near-natural and ecologically meaningful approaches to managing these ecosystems may provide a solution to resolving conservation conflicts. New methods of managing forests are based on the rationale that since natural forests are able to maintain viable populations of all forest-dwelling species, forest management that follows natural forest dynamics should also be good for maintaining biodiversity.

New methods of ecosystem design present an opportunity for resolving conservation conflicts. Ecosystem design refers to the spatial-temporal organization of a forest and involves the search for a combination of management alternatives (including the management of alternative "protection") that provides a desirable mix of products and services to the landscape. Zoning is sometimes used as a means to offset the impact of increased environmental demands on timber production (Ontario 2000). A zoning approach enforces a spatial segregation of types of land use that is based on specific regulations. A spatial mosaic of national and international regulations and restrictions apply in different parts of the Białowieża landscape. For this reason, small-scale spatial mixing of different land use options would be a more realistic approach than the entrenchment of a simple zonal design. Białowieża could become a continuously evolving landscape where spatial patterns change in response to changing human demands and shifting environmental conditions (Tress and Tress 2001). The basic concept is simple and can be summarized according to a proposal made by Pukkala and Kangas (1993). They defined the utility of a specific design as follows:

$$Maximize\boldsymbol{U} = \sum_{i=1}^{n} w_i \cdot u_i(q_i)$$

where U is the utility of a given combination of alternatives; n refers to the number of services; w_i is the relative weight of service $i \left(0 \le w_i \le 1; \sum_{j=1}^{m} w_i = 1 \right)$; q_i is the realized amount of service i for a given combination of alternatives; and $u_i(q_i)$ is the partial utility function for service i ($0 \le u_i(q_i) \le 1$).

The historical subdivisions of Białowieża reveal a systematic grid of primary units of landscape design (Brincken 1826; Hedemann 1939). Modern GIS technology combined with advanced modeling tools is available to predict the development of each primary spatial unit, based on advanced models of ecosystem dynamics and the preferences of *bona fide* stakeholders (including conscious "no action" decisions or "protect old habitat trees" policies). The book *Designing Green Landscapes* at https://link.springer.com/book/10.1007/978-1-4020-6759-4 contains examples of successful implementation. Alternative formulations and refinements of the basic concept include period weighting or formulations as a mathematical programming model. Applications can be found in Gadow and Pukkala (2008). Additional examples concern the work done in the forests bordering the Indira Gandhi canal in India (Tewari and Gadow 2010), and a forest in the Solling Mountains in Germany (Chen and Gadow 2008). Unpublished applications of this approach include a novel ecosystem dominated by alien invaders, a coastal dune ecosystem, and a mountain ecosystem in South Africa (see Chap. 6 for further details).

Over the past few decades, it has become clear that the separation of humans from nature is a cultural construct of relatively recent origin that has a negative impact on people and can cause significant and unexpected ecological changes (Wapner 2014). "Nature" does not exist apart from humanity; the world we inhabit is the one we have made (Purdy 2015). A significant amount of research has traced the negative social, cultural, and economic impacts of people living near protected areas. A study reviewing developments during the past half century that was based on principles of naturalness found that contrary to expectations, managing exclusively for naturalness did not serve to protect all wilderness values and purposes (Bowman et al. 2011). Active management is therefore often required in protected areas to create and maintain an illusion of naturalness.

Ecological reconstructions show that the world's ecosystems have changed significantly during the past 10,000 years (a period known as the Holocene) as a result of human action, especially those involving fire and agriculture, and environmental conditions such as natural and human-induced climate changes (Parmesan 2006). A comprehensive investigation into whether it is possible to restore the forests of the American northeast to a historic "natural" condition concluded: "There are no static baseline conditions that exist or have existed for comparison with current conditions or for use as a target for restoration activities" (Foster et al. 2008). Accordingly, there is little historical or ecological justification for preventing humans from continuing to shape and influence the existing forest ecosystems. A more diverse landscape that includes mixed hardwoods and softwoods, remnant old-growth, open woodlands and commercial forest will provide the greatest diversity of forest structure with a greater array of ecological services.

The earth's climate has been relatively stable during the Holocene. A very real threat of our time is the prospect of climate change. Numerous studies have shown how human behavior is disrupting many of the earth's interconnected systems. Most severely affected are the atmosphere, water cycles, and soil. Humans are destabilizing the global biosphere, exploiting its natural resources and depositing large quantities of waste. Habitats are changing, and many animal and plant species are becoming extinct (Kolbert 2014; Steffen 2015). The contemporary biosphere differs significantly from that of previous stages of evolution because of many anthropogenic modifications and perturbations. These include the global homogenization of flora and fauna and human appropriation of 25–40% of net primary production (Steffen et al. 2016).

Any deviation from Holocene standards will present human societies with huge challenges. One of the fundamental issues of our time is the need to protect the fragile biosphere from our own destructive power. Innovative approaches are required regarding agricultural production as well as the management of forest ecosystems. Ecological collapse is a new existential threat (Harari 2019). Predicted changes of the climate and the resulting disturbances from wind, fire, and beetles may increasingly change the composition and viability of the world's forest ecosystems. Recent outbreaks of *Ips typographus* are representative of a global trend that has seen an unprecedented proliferation of beetles in central European (Lausch et al. 2013) and North American forests (Raffa et al. 2008). Such hazards require active management aimed at the restoration of ecosystem viability and resilience.

Literature

Anderson MJ, Ellingsen KE, McArdle BH (2006) Multivariate dispersion as a measure of beta diversity. Ecol Lett 9(6):683–693

Aravind NA, Manjunath J, Rao D, Ganeshaiah KN, Shaanker RU, Vanaraj G (2005) Are red-listed species threatened? A comparative analysis of red-listed and non-red-listed plant species in the Western Ghats, India. Curr Sci 88(02):258–265

Beck HE, Zimmermann NE, McVicar TR, Vergopolan N, Berg A, Wood EF (2018) Present and future Köppen-Geiger climate classification maps at 1-km resolution. Sci Data 5:180214

Bennett B (2015) Plantations and protected areas: a global history of forest management. MIT Press, Cambridge, MA

Bosch JM, Gadow K v (1990) Regulating afforestation for water conservation in South Africa. South AfFor J 153:41–54

Bowman DMJS, Balch J, Artaxo P, Bond WJ, Cochrane MA, D'Antonio CM, DeFries R, Johnston FH, Keeley JE, Krawchuk MA, Kull CA, Mack M, Moritz MA, Pyne S, Roos CI, Scott AC, Sodhi NS, Swetnam TW (2011) The human dimension of fire regimes on earth. J Biogeogr 38:2223–2236. https://doi.org/10.1111/j.1365-2699.2011.02595

Bray JR, Curtis JT (1957) An ordination of the upland forest communities of southern Wisconsin. Ecol Monogr 27:325–349

Brincken Jv (1826) Mémoire descriptif sur la forêt impériale de Białowieża en Lithuanie (A Brief Description of the Imperial Białowieża Forest in Lithuania). N. Glücksberg, Imprimeur-Libraire de l'Université Royale. p. 127

Champion HG, Seth SK (1968) A revised survey of the forest types of India. Govt. of India Press, Delhi, 404 p

Chan D, Wu Q (2015) Significant anthropogenic-induced changes of climate classes since 1950. Sci Rep 5(13487):13487. Bibcode:2015NatSR...513487C. https://doi.org/10.1038/srep13487. PMC4551970.PMID26316255

Chao A, Chazdon RL, Colwell RK, Shen TJ (2005) A new statistical approach for assessing similarity of species composition with incidence and abundance data. Ecol Lett 8:148–159

Chen BW, Gadow K (2008) Combining spatial and other objectives in forest design. For Stud|Metsanduslikud Uurimused 48:30–40

Chen L, Wang L, Yeerjiang B, Zhang C, Zhao X, Gadow K (2013) Seed dispersal and seedling recruitment of trees at different successional stages in a temperate forest in northeastern China. J Plant Eco 7:337–346. https://doi.org/10.1093/jpe/rtt024

Clarke KR, Warwick RM (1998) A taxonomic distinctness index and its statistical properties. J Appl Ecol 35:523–531

Clarke KR, Warwick RM (2001) A further biodiversity index applicable to species lists: variation in taxonomic distinctness. Mar Ecol Prog Ser 216:265–278

Clements FE, Weaver JE (1938) Plant ecology, 2nd edn. McGraw-Hill Book Company, New York

Danneyrolles V, Dupuis S, Fortin G, Leroyer M, de Römer A, Terrail R, Vellend M, Boucher Y, Laflamme J, Bergeron Y, Arseneault D (2019) Stronger influence of anthropogenic disturbance than climate change on century-scale compositional changes in northern forests. Nat Commun 10:1265. https://doi.org/10.1038/s41467-019-09265-z

Dragoi M, Ciornei I (2013) A balck-box approach on assessing the opportunity cost of deforestation. Land Use Policy 34:314–320

Ellenberg H, Mueller-Dombois D (1967) Tentative physiognomic-ecological classification of plant formations of the Earth [based on a discussion draft of the UNESCO working group on vegetation classification and mapping.] Berichte des Geobotanischen Institutes der Eidg. Techn. Hochschule, Stiftung Rübel, Zürich 37(1965–1966):21–55

Erb K-H, Kastner T, Plutzar C, Bais ALS, Carvalhais N, Fetzel T, Gingrich S, Haberl H, Lauk C, Niedertscheider M, Pongratz J, Thurner M, Luyssaert S (2017) Unexpectedly large impact of forest management and grazing on global vegetation biomass. Nature 553:73–76. (04 January 2018)

Essl F, Lenznerl B, Courchamp F, Dullinger S, Jeschke JM, Kühn I, Leung B, Moser D, Roura-Pascual N, Seebens H (2019) Introducing AlienScenarios: a project to develop scenarios and models of biological invasions for the 21st century. NeoBiota 45:1–17. https://doi.org/10.3897/neobiota.45.33366

Falinski J (1986) Vegetation dynamics in temperate lowland primeval forests: ecological studies in Bialowieza Forest. Dr W Junk/Kluwer Academic Publishers, Dordrecht, 525 pp

Fenelon JP, Lebart L (1971) Statistique et informatique appliquées. Dunod, Paris

Foster DR, Donahue B, Kittredge DB, Motzkin G, Hall B, Turner BL, Chilton E (2008) New England's forest landscape. In: Redman CL, Foster DR (eds) Ecological legacies and conservation patterns shaped by agrarian history, Agrarian landscapes in transition. Oxford University Press, New York

Franklin JF, Spies TA, Van Pelt R, Carey AB, Thornburgh DA, Berg DR, Lindenmayer DB, Harmon ME, Keeton WS, Shaw DC, Bible K, Chen JQ (2002) Disturbances and structural development of natural forest ecosystems with silvicultural implications, using Douglas-fir forests as an example. For Ecol Manage 155(1-3):399–423

Gadow Kv, Pukkala T (2008) Designing green landscapes, Vol. 15 of the book series "Managing forest ecosystems". Springer, Dordrecht, 322 p

Gadow K, Kurttila M, Leskinen P, Leskinen L, Nuutinen T, Pukkala T (2007) Designing forested landscapes to provide multiple services. CAB Rev 2(038):1–15

Gadow Kv, Zhao XH, Corral Rivas JJ, (2013) Retention strategies for multi-species forests. In: Proceeding of the international symposium for the 50th anniversary of the forestry sector planning in Turkey, 26–28 November 2013, Antalya

Ganeshaiah KN, Chandrashekara K, Kumar ARV (1997) Avalanche index: a new measure of biodiversity based on biological heterogeneity of the communities. Curr Sci 73:128–133

Ganeshalah KN, Shaanker UR (2000) Measuring biological heterogeneity of forest vegetation types: avalanche index as an estimate of biological diversity. Biodivers Conserv 9:953–963

Gaston KJ, Blackburn TM (2000) Pattern and process in macroecology. Blackwell Science, Oxford

Geiger R (1954) Klassifikation der Klimate nach W. Köppen [Classification of climates after W. Köppen]. Landolt-Börnstein – Zahlenwerte und Funktionen aus Physik, Chemie, Astronomie, Geophysik und Technik, alte Serie. Springer, Berlin, 3. pp 603–607

Gorska I (1976) Badania archeologiczne w Puszczy Białowieskiej'. [Archaeological investigations in Bialowieza Forest] Archeologia Polski 21(1)

Griesel F, Gadow Kv (1995) Naturgemäßer – naturnaher – ökologischer Waldbau – Begriffsbestimmungen, Konzepte und Richtlinien, Anwendungsbeispiele. Internal report, Institute of Forsteinrichtung, Univ of Göttingen: 20pp

Grossman DH, Faber-Langendoen D, Weakley AS, Anderson M, Bourgeron P, Crawford R, Goodin K, Landaal S, Metzler K, Patterson KD, Pyne M, Reid M, Sneddon L (1998). International classification of ecological communities: terrestrial vegetation of the United States. Volume I. The National Vegetation Classification System: development, status, and applications. The Nature Conservancy, Arlington, Virginia, USA

Hao M, Ganeshaiah KN, Zhang C, Zhao X, von Gadow K (2019a) Discriminating among forest communities based on taxonomic, phylogenetic T and trait distances. For Ecol Manage 440(2019):40–47

Hao M, Corral-Rivas JJ, González-Elizondo MS, Ganeshaiah KN, Nava-Miranda MG, Zhang C, Zhao X, von Gadow K (2019b) Assessing biological dissimilarities between five forest communities. For Ecosyst 6:30

Harari YN (2019) 21 lessons for the 21st century. Jonathan Cape, London, 352 p

Härdtle W (1995) On the theoretical concept of the potential natural vegetation and proposals for an up-to-date modification. Folia Geobot 30:263. https://doi.org/10.1007/BF02803708

Hedemann O (1939) L'histoire de la foret de Białowieża jusqu'a 1798 (The history of the Forest of Białowieża before 1798). Institut de Recherches des Forets Domaniales Warszawa, Travaux et comptes rendus, Seria A, Nr. 41: 310 pp

Hengl T, Walsh MG, Sanderman J, Wheeler I, Harrison SP, Prentice IC (2018a) Global mapping of potential natural vegetation: an assessment of machine learning algorithms for estimating land potential. PeerJ Preprints 6:e26811v1. https://doi.org/10.7287/peerj.preprints.26811v1

Hengl T, Nussbaum M, Wright MN, Heuvelink GB (2018b) Random forest as a generic frame- work for predictive modeling of spatial and spatio-temporal variables. PeerJ Preprints 6:e26693v1

Hui C, Richardson DM (2017) Invasion dynamics. Oxford University Press, Oxford, 322 p

Hui C, Richardson DM, Robertson MP, Wilson JRU, Yates CJ (2011a) Macroecology meets invasion ecology: linking the native distributions of Australian acacias to invasiveness. Divers Distrib 17:872–883

Hui G, Zhao X, Zhao Z, Gadow K (2011b) Evaluating tree species spatial diversity based on neighborhood relationships. For Sci 57(4):292–300

Jaccard P (1900) Contribution au problème de l'immigration post-glaciaire de la flore alpine. Bulletin de la Société Vaudoise des Sciences Naturelles 36:87–130

Jost L, Chao A, Chazdon RL (2011) Compositional similarity and β (beta) diversity. In: Magurran AE, McGill BJ (eds) Biological diversity: frontiers in measurement and assessment. Oxford University Press, New York, pp 66–84

Kleinschmit D, Mansourian S, Wildburger C, Purret A (eds) (2016) Illegal logging and related timber trade – dimensions, drivers, impacts and responses. A global scienti c rapid response assessment report. IUFRO World Series Volume 35.Vienna. 148 p

Kolbert E (2014) The sixth extinction; an unnatural history. Bloomsbury, London

Köppen W (1884) Translated by Volken, E.; Brönnimann, S. "Die Wärmezonen der Erde, nach der Dauer der heissen, gemässigten und kalten Zeit und nach der Wirkung der Wärme auf die organische Welt betrachtet" [The thermal zones of the earth according to the duration of hot, moderate and cold periods and to the impact of heat on the organic world)]. Meteorologische Zeitschrift (published 2011). 20 (3): 351–360. Bibcode:2011MetZe..20..351K. doi:10.1127/0941-2948/2011/105 – via

Koutika L-S, Richardson DM (2019) Acacia mangium Willd: benefits and threats associated with its increasing use around the world. For Ecosyst 6(2). https://doi.org/10.1186/s40663-019-0159-1

Kuijper DPJ, Jędrzejewska B, Brzeziecki B, Churski M, Jędrzejewski W, Żybura H (2010) Fluctuating ungulate density shapes tree requirement in natural stands of the Białowieża Primeval Forest, Poland. J Veg Sci 21:1082–1098

Lance GN, Williams WT (1967) Mixed-data classificatory programs I – Agglomerative systems. Aust Comput J 1(1):15–20

Lausch A, Heurich M, Fahse L (2013) Spatio-temporal infestation patterns of *Ips typographus* (L.) in the Bavarian Forest National Park, Germany. Ecol Indic 31:73–81

Legendre P, Cáceres MD (2013) Beta diversity as the variance of community data: dissimilarity coefficients and partitioning. Ecol Lett 16(8):951–963

Legendre P, Legendre L (2012) Numerical ecology, 3rd edn. Elsevier, Amsterdam

Levis C, Souza PF, Schietti J, Emilio T, JLPdV P et al (2012) Historical human footprint on modern tree species composition in the Purus-Madeira Interfluve, Central Amazonia. PLoS One 7(11):e48559. https://doi.org/10.1371/journal.pone.0048559

Low T (2012) Australian acacias: weeds or useful trees? Biol Invasions 14:2217–2227

Magurran AE (2004) Measuring biological diversity. Blackwell, Oxford

Marris E (2013) Perspective: coming of age in a trash forest. In: Hobbs et al. Wiley-Blackwell, Chapter 29: 245

Messier C, Bauhus J, Doyon F, Maure F, Sousa-Silva R, Nolet P, Mina M, Aquilue N, Fortin M-J, Puettmann K (2019) The functional complex network approach to foster forest resilience to global changes. For Ecosyst 6:21. https://doi.org/10.1186/s40663-019-0166-2

Mitchell FJG, Cole E (1998) Reconstruction of long-term successional dynamics of temperate woodland in Bialowieza Forest, Poland. J Ecol 86:1042–1059

Mitsch WJ (2012) What is ecological engineering? Ecol Eng 45:5–12

Newbold T, Hudson LN, Hill SLL et al (2016) Global effects of land use on local terrestrial biodiversity. Nature 520:45–50

Niklasson M, Zin ME, Zielonka T, Feijen M, Korczyk AF, Churski M, Samojlik T, Jedrzejewska B, Gutowski JM, Brzeziecki B (2010) A 350-year tree-ring fire record from Białowieza Primeval Forest, Poland: implications for Central European lowland fire history. J Ecol 98:1319–1329

Ogden NH, Wilson JRU, Richardson DM, Hui C, Davies SJ, Kumschick S, Le Roux JJ, Measey J, Saul W-C, Pulliam JRC (2019) Emerging infectious diseases and biological invasions: a call for a one health collaboration in science and management. R Soc Open Sci 6:181577. https://doi.org/10.1098/rsos.181577

Olson DM, Dinerstein, E, Wikramanayake, ED, Burgess ND, Powell GVN, Underwood EC, D'amico JA, Itoua I, Strand HE, Morrison JC, Loucks CJ, Allnutt TF, Ricketts TH, Kura Y, Lamoreux JF, Wettengel WW, Hedao P, Kassem KR (2001) Terrestrial ecoregions of the world: a new map of life on earth. BioScience 51(11), (1 November). https://doi.org/10.1641/0006-3568(2001)051[0933:TEOTWA]2.0.CO;2

Ontario Ministry of Natural Resources (2000) Ontario forest accord. Queen's Printer for Ontario. (http://www.mnr.gov.on.ca/mnr/oll/ofaab/accord.html, as seen on Sept 27, 2000)

Parmesan C (2006) Ecological and evolutionary responses to recent climate change. Annu Rev Ecol Evol Syst:637–669

Puettmann KJ, Wilson SMG, Baker SC, Donoso PJ, Drössler L, Amente G, Harvey BD, Knoke T, Lu Y, Nocentini S, Putz FE, Yoshida T, Bauhus J (2015) Silvicultural alternatives to conventional even-aged forest management – what limits global adoption. For Ecosyst 2:8. https://doi.org/10.1186/s40663-015-0031-x

Pukkala T (2008) Integrating multiple services in the numerical analysis of landscape design. In: Gadow Kv, Pukkala T (eds) Designing green landscapes. Vol 15 of the Springer book series managing forest ecosystems, pp 137–168

Pukkala T, Gadow Kv (eds) (2012) Continuous cover forestry, Book series managing forest ecosystems Vol 24, © Springer

Pukkala T, Kangas J (1993) A heuristic optimization method for forest planning and decision making. Scand J For Res 8:560–570

Purdy J (2015) After nature – a politics for the anthropocene. Harvard University Press, Cambridge

Raffa KF, Aukema BH, Bentz BJ, Carroll AL, Hicke JA, Turner MG, Romme WH (2008) Cross-scale drivers of natural disturbances prone to anthropogenic amplification: the dynamics of bark beetle eruptions. BioScience 58(6):501–517. https://doi.org/10.1641/B580607

Reddy SC Jha CS Diwakar PG Dadhwal VK (2015) Nationwide classification of forest types of India using remote sensing and GIS. Environ Monit Assess 187(12). https://doi.org/10.1007/s10661-015-4990-8. https://www.researchgate.net/publication/285020645_Nationwide_classification_of_forest_types_of_India_using_remote_sensing_and_GIS/figures?lo=1

Reif A, Schulze ED, Ewald J, Rothe A (2013) Waldkalkung – Bodenschutz contra Naturschutz. Waldökologie, Landschaftsforschung und Naturschutz

Richardson DM, Pyšek P, Carlton JT (2011a) A compendium of essential concepts and terminology in invasion ecology. In: Fifty years of invasion ecology – The legacy of Charles Elton. Wiley-Blackwell, Oxford, pp 409–420

Richardson DM, Carruthers J, Hui C, Impson FAC, Miller JT, Robertson MP, Rouget M, Le Roux JJ, Wilson JRU (2011b) Human-mediated introductions of Australian acacias – a global experiment in biogeography. Divers Distrib 17:771–787

Richardson DM, Hui C, Nuñez MA, Pauchard A (2014) Tree invasions – patterns, processes, challenges and opportunities. Biol Invasions 16:473–481. https://doi.org/10.1007/s10530-013-0606-9

Ross E (2017) Amazon rainforest has been shaped by an ancient hunger for fruits and nuts. Nature, https://doi.org/10.1038/nature.2017.21576

Sakai AK, Allendorf FW, Holt JS, Lodge DM, Molofsky J, With KA, Baughman S, Cabin RJ, Cohen JE, Ell-strand NC, McCauley DE, O'Neil P, Parker IM, Thompson JN, Weller SG (2001) The population biology of invasive species. Annu Rev Ecol Syst 32:305–332

Schama S (1995) Landscape and memory. Harper Collins, London, 652 pp

Schimper AFW, von Faber FC (1935) Pflanzengeographie auf physiologischer Grundlage, 3rd edn. Fischer, Jena

Schmidt W (2000) Biotische Mannigfaltigkeit – Indikator für intakte Wälder? Beispiel aus Studien über Bodenvegetation in Naturwaldreservaten. NUA 4:108–127

Schulze ED (2017) Effects of forest management on biodiversity in temperate deciduous forests: an overview based on Central European beech forests. J Nat Conserv 43:213–226

Schulze ED, Craven D, Durso AM, Reif J, Guderle M, Kroiher F, Hennig P, Weiserbs A, Schall P, Ammer C, Eisenhauer N (2019) Positive association between forest management, environmental change, and forest bird abundance. For Ecosyst 6:3

Schütz J-Ph, Pukkala T, Donoso P, Gadow Kv (2012) Historical emergence and current application of CCF. In: Pukkala T, Gadow K.v (eds) Continuous cover forestry. book series managing forest ecosystems. vol 24, © Springer, pp 1–28

Socolar JB, Gilroy JJ, Kunin WE, Edwards DP (2016) How should beta-diversity inform biodiversity conservation? Trends Ecol Evol 31(1):67–80

Solberg B (ed) (1996) Long-term trends and prospects in wood supply and demand for wood and implications for sustainable forest management. EFI research report No. 6, 106 p

Sørensen TA (1948) A method of etablishing groups of equal amplitude in plant sociology based on similarity of species content, and its application to analysis of the vegetation on Danish commons. Biologiske Skrifter Kongelige Danske Videnskabernes Selskab 5:1–34

Sprugel DG (1991) Disturbance, equilibrium, and environmental variability – what is natural vegetation in a changing environment. Biol Conserv 58:1–18

Steffen W (2015) Planetary boundaries: Guiding human development on a changing planet. Science 347:6223

Steffen W, Leinfelder R, Zalasiewicz J, Waters CN, Williams M, Summerhayes C, Barnosky AD, Cearreta A, Crutzen P, Edgeworth M, Ellis EC, Fairchild IJ, Galuszka A, Grinevald J, Haywood A, do Sul JI, Jeandel C, McNeill JR, Odada E, Oreskes N, Revkin A, Richter DB, Syvitski J, Vidas D, Wagreich M, Wing SL, Wolfe AP, Schellnhuber HJ (2016) Stratigraphicand Earth System approaches to defining the Anthropocene. Earth'sFuture 4:324–345. https://doi.org/10.1002/2016EF000379

Stephenson W, Williams WT, Cook SD (1972) Computer analyses of Petersen's original data on bottom communities. Ecol Monogr 42(4):387–415

Sunseri T (2012) Exploiting the Urwald: German post-colonial forestry in Poland and central Africa, 1900–1960. Past Present 214:305–342

Swenson NG, Stegen JC, Davies SJ, Erickson DL, Forero-Montana J, Hurlbert AH, Kress WJ, Thompson J, Uriarte M, Wright SJ, Zimmerman JK (2012) Temporal turnover in the composition of tropical tree communities: functional determinism and phylogenetic stochasticity. Ecology 93:490–499

Tewari VP, Gadow Kv (2010) Domesticated ecosystems – analysis and design – an example involving Indira Gandhi Canal Plantations in Rajasthan, India. Poster presented at the IUFRO World Congress in Seoul/Korea; August 2010

Tress B, Tress G (2001) Capitalising on multiplicity: a trans- disciplinary systems approach to landscape research. Landsc Urban Plann 57:143–157

UN-CCD (2000) United Nations Convention to Combat Desertification. https://www.unccd.int

Vera FWM (2000) Grazing ecology and forest history. CAB International, Wallingford, 506 pp. http://lib.du.ac.ir/documents/10157/60360/Grazing+Ecology+and+Forest+History+-+(2000).pdf

Verma M, Negandhi D, Wahal AK, Kumar R (2013) Revision of rates of NPV applicable for different class/category of forests. Indian Institute of Forest Management. Bhopal, India

Vining J, Merrick MS, Price EA (2008) The distinction between humans and nature: human perceptions of connectedness to nature and elements of the natural and unnatural. Hum Ecol Rev 15(1):1–11

Wapner P (2014) The changing nature of nature: environmental politics in the Anthropocene. Glob Environ Pol 14:36–54

Wiecko E (1972) Puszcza Białowieska. Panstwowe Wydawnictwo Naukowe, Warsaw. 193 pp

Forest Assessment and Observation

2

Abstract

Forest assessment, including remote assessment, is a vast and specialized topic that cannot be covered adequately in a textbook with a focus on sustaining forest ecosystems. Specific issues of forest assessment (a brief reference to the nested plot design and to the sampling of low woody vegetation) are included here, but the main focus of this section is on continuous observation of the biological and structural dynamics of forest ecosystems by using field plots with mapped trees identified by experienced field taxonomists.

The Food and Agricultural Organization of the United Nations (FAO) has monitored the world's forests at 5- to 10-year intervals since 1946. The assessments are based on country reports prepared by national correspondents and remote-sensing data collected by FAO together with regional partners. These reports provide rough estimates about changes in forest areas. Detailed descriptions of the structure and species composition of the world's forest ecosystems that are based on high-resolution sensors and large networks of field plots are also beginning to emerge.

© Springer Nature Switzerland AG 2021

K. von Gadow et al., *Sustaining Forest Ecosystems*, Managing Forest Ecosystems 37,
https://doi.org/10.1007/978-3-030-58714-7_2

Forest Observational Study Chichimoco in Durango, Mexico

2.1 Forest Assessment

Observational infrastructures for collecting information about forest ecosystems include field experiments, continuous observation, and forest sampling (Fig. 2.1). **Field experiments** are established to test a response to specific treatments. This system is actively manipulated. Treatments include different planting densities, planted species mixtures, or fertilizer applications. The timing of field assessment is defined by the experimental protocol. **Continuous observation** provides information that is useful for understanding ecosystem structure and functioning and provides an empirical basis for research and sustainable management. Observations are collected without manipulating the system. To evaluate the response to specific disturbances such as forest fire or harvest events, assessments should immediately follow these disturbances. Observations from global forest observational networks can provide a new basis for a more realistic and meaningful classification of the world's forests, based on biological characteristics and human impact. The focus of **sampling** is on the assessment of the state of the forest resource, usually at regular intervals, and to prepare a report with reference to a specific area and time. Common examples are local or national forest assessments (McRoberts et al. 2007; Tomppo et al. 2010; Zeng et al. 2015; Alekseev et al. 2019). The objective of sampling is to

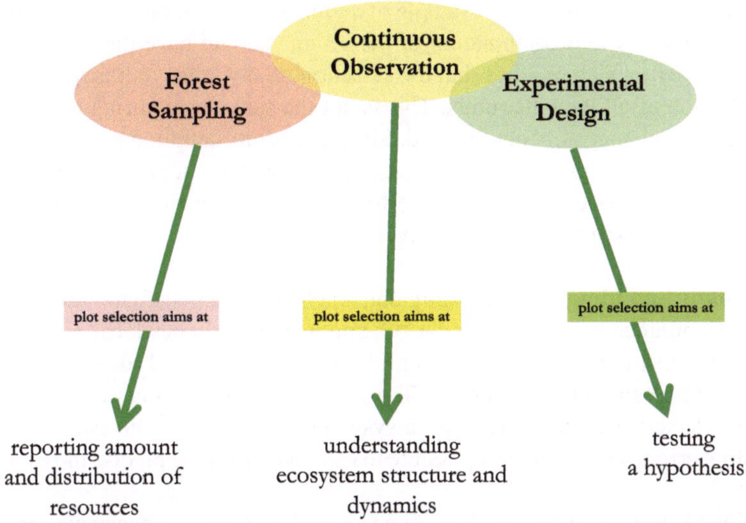

Fig. 2.1 Three strategies for collecting observations about forests: sampling, continuous observation, and field experiments

obtain, at reasonable cost, accurate estimates of specific target variables. Estimates are considered to be accurate if they are close to the true population value. The theory and practice of forest sampling embraces a great variety of methods. Terrestrial sample plots can be permanent or temporary, of different shapes and sizes, and distributed in different ways. The area of a sample plot can be fixed or variable. Effective sampling is characterized by unbiasedness, a small sampling error and consistency. Unbiasedness implies that the mathematical expectation of the sample statistic is identical to the population parameter.

An estimate is consistent if, with increasing sample size, the sample value approaches the true population value. An excellent introduction to forest sampling is the monograph by Freese (Freese 1962). Numerous relevant studies of sampling theory and methods applicable to both timber and nontimber parameters have appeared since (e.g., Kleinn and Köhl 1999; Lorenz et al. 2005; Köhl et al. 2006; McRoberts et al. 2007; Mandallaz 2008; Tomppo et al. 2010; Kershaw Jr et al. 2016).

This section presents a very brief introduction to forest sampling, a critical analysis of the nested plot design, and a new method for sampling forest regeneration. A particular strategy of forest assessment is defined by the type and number of sampling units and by the manner of selecting and distributing these units over the entire area of interest. The most important sampling methods used in forest assessments are the following (standard forest inventory textbooks provide more detail):

Sampling by "Reasoned Choice" (also known as *nonrandom*, *selective*, *opportunistic,* or *convenience sampling*): Sample units thought to be characteristic of the population under study are selected. This technique relies on the researcher's ability to select the sample units. The habit of subjectively locating sample plots is prone to produce biased estimates, and the approach is therefore not very popular.

Simple Random Sampling A sample of *n* units is drawn at random from the population. Every possible combination of the *n* units should have the same chance of being selected, which requires that the selection of any given unit is independent of the selection of all the other units. This is not the same as requiring that every unit has the same chance of being selected. Simple random sampling is used when there is no prior information about the target population. The method is seldom used to assess forest ecosystems, because the effort of selecting and locating units is usually considered too expensive or too involved.

Systematic Sampling The sampling units included in a systematic sample are laid out in accordance with some regular scheme which facilitates locating them in the field. The most common pattern is a grid with equally spaced rows and a constant distance between units within rows. The location of the starting point is often randomly selected. The main problem in systematic sampling is the estimation of the sampling error; it is not always possible to establish just how precise the estimate is.

Stratified Random Sampling It is often practical to set up homogeneous nonoverlapping subpopulations known as strata. Each stratum is sampled separately with the aim of improving precision. The estimates for the individual strata are eventually combined to obtain an overall estimate. Stratified random sampling will produce a more precise overall estimate than simple random sampling if the variation among units within strata is less than the variation among units that are not in the same stratum.

Multistage Sampling It makes sense to concentrate the sampling effort spatially when the cost of locating a sampling unit and traveling between units is high compared to the cost of the measurements. This kind of sampling is usually done in two stages. The first stage involves drawing at random m out of M primary units (PU). At the second stage within each PU, n out of the N possible secondary units (SU) is selected, again at random. Multistage sampling may yield estimates of a given precision at lower cost than simple random sampling.

Cluster Sampling As in multistage sampling, cluster sampling is used when the cost of locating sampling units is high relative to the cost of taking the measurements. The sampling unit is a cluster of individual units. The entire population is divided into clusters and then the clusters are randomly selected. The cluster unit value is the proportion of the individuals in the group having the specified attribute.

Multiphase Sampling (also called *double sampling*): As in stratification, regression estimators can increase the precision of sampling by making use of supplementary information about the population. Consider two variables, x and y, that have to be sampled. A large sample is taken during the primary phase to obtain a good estimate of the mean of x. A secondary subsample is taken on a successive occasion where the y values are also measured to give an estimate of the relation of y to x. The mean of x is then applied to the fitted relationship to obtain an estimate of the mean of y. One of the most popular applications of this method involves successive inventories where only a small fraction of units is remeasured.

Temporary and Permanent Sample Units A distinction is made between a temporary and permanent sample plot. This distinction is partly based on the need for remeasurement, but more specifically on whether the plot location has been marked for remeasurement. Temporary plots are typically used to obtain current estimates of specific ecosystem attributes. A permanent plot is measured on two or more occasions, and the purpose may be to determine growth and mortality rates or to calibrate existing growth models. A "matched plot procedure" is used where only some (randomly selected) plots are remeasured at each subsequent remeasurement occasion. Although not all the original plots are actually remeasured, all are permanent plots because of the fact that their precise locations had to have been recorded in order for them to be candidates for remeasurement. Permanent plots can also be used in an experiment designed to assess responses to specific treatments or environmental changes.

Permanent sample plots are about two to four times more expensive than temporary sample plots (Cunia 1974), but they provide greater precision about the rates of change in tree characteristics in a specific population. The following two formulae for the calculation of the standard deviation of a difference illustrate the superiority of permanent plots in estimating change. The standard error of the difference in the biomass estimate in temporary and permanent plots is:

(a) **Temporary plots**:

$$s_d^2 = s_x^2 + s_y^2 \tag{2.1}$$

where d = y − x = the average difference in biomass, s_d = standard error of the difference in biomass; s_x = standard error of the initial inventory; s_y = standard error of the subsequent inventory.

(b) **Permanent plots**:

$$s_d^2 = s_x^2 + s_y^2 - 2rs_x s_y \tag{2.2}$$

where r is the correlation coefficient between the samples in both inventories. With respect to the linkage of the samples during subsequent remeasuring, three designs can be recognized (see Cunia 1974):

(a) All sample plots are temporary when new samples are selected independently of previous ones during successive enumerations.
(b) All sample plots are permanent when the same samples are remeasured on each successive occasion.
(c) Some sample plots are permanent and some are temporary, that is, a certain proportion of samples are common to both occasions.

Design (c) can be interpreted as the general design, while (a) and (b) are special cases. The accuracy of biomass estimates depends largely on the accuracy of the number of stems estimate. Interplot variability of the stem count yields a variance from which confidence intervals of the estimated stems per hectare can be derived. At an even finer level of detail, the density distribution of trees within plots is used to characterize the patchiness of the stand. This information is essential when one has to select the appropriate sampling intensity and plot size.

2.1.1 Some Selected Issues

The field procedures for a forest inventory system include the type, shape, and size of sample plots; the sampling intensity; the selection of sample plots; and the cruising interval. Several types of sampling unit are used in forest assessments, such as **sample points** based on the angle count method; **row plots** covering a specified number of rows and check rows in planted forests; **fixed area plots** of circular, square, or rectangular shape; and **nested sample plots** (for details, see Gadow and Bredenkamp 1992).

What is the optimum size of a sample plot? The available budget can be spent on a few large plots or on a greater number of small plots. The greater the sample plot area, the smaller the coefficient of variation (Van Laar 1981). The ideal plot size is affected by stand homogeneity and density and is always a compromise between precision and operational convenience. A popular design of forest sampling is the nested plot. Because of its extensive application, this design deserves particular attention.

2.1.1.1 The Nested Plot Design

The objective of the nested design is to obtain unbiased estimates of timber volumes by concentrating the measurement effort on the big trees and reducing the effort on the small trees. A bigger plot for the bigger trees (that contribute most to the estimate of timber volume) and the small "inner" plot for the small trees are treated as separate entities (Fig. 2.2).

The distinction is based on tree size, that is, inclusion probabilities associated with each tree are linked to diameter. Each tree represents a value per hectare based on its plot expansion factor which is equal to the inverse of the plot area in hectare [1/(plot area in hectare)]. For example, the expansion factor for an inner plot with radius 5 m is equal to $EF = 10000/(pi*5^2) = 127.3$; for a bigger plot of 500 m^2, the EF would be equal to $10000/500 = 20$ (for details, see Köhl et al. 2006:155; FIA 2012; Kershaw Jr et al. 2016). Three examples of nested plot designs are presented in Fig. 2.3.

The nested design became popular during the 1960s because it provides unbiased estimates of basal area and timber volume. Other important target variables, such as species richness and forest structure, emerged during the past 30 years. When nested circular plots are used for permanent monitoring, then the plot itself represents the population of interest. It would therefore be of interest to evaluate the effects of nesting on estimates of forest structure and diversity, and on the periodic change of these attributes. Figure 2.4 shows the results of a simulation study in a natural afromontane forest with mapped trees, based on 1000 randomly selected plot positions. The distributions of six structural plot attributes were calculated for the full (blue columns) and nested design (red columns).

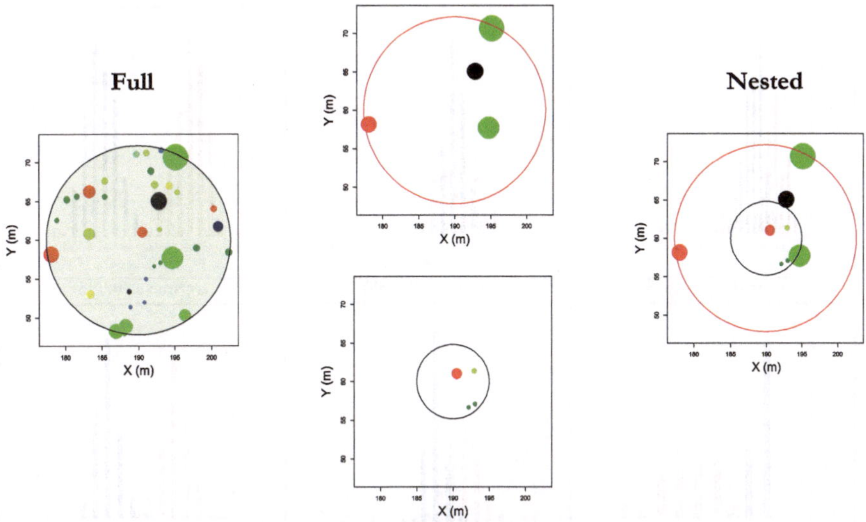

Fig. 2.2 Example of a fully assessed full plot (left) and a partially assessed nested plot (right). The inner circle radius is 5 m, the outer radius is 12.62 m; the threshold DBH is 25 cm. The two parts of the nested plot are shown in the center (Data: Knysna FVC plot)

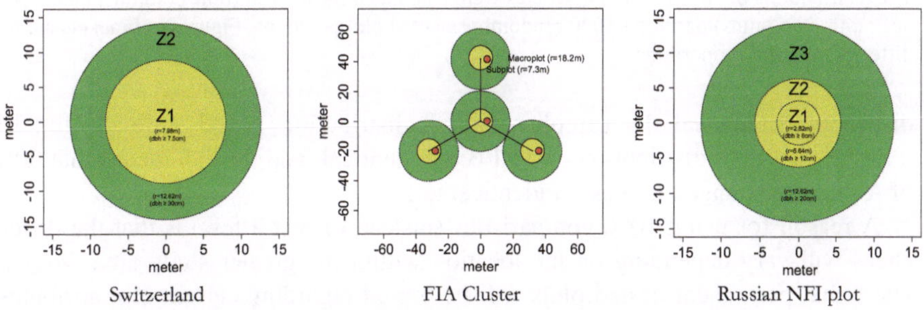

Fig. 2.3 Three examples of nested plot designs used in permanent monitoring

The six structural plot attributes studied were crowding (the number of trees per hectare), density (the basal area per hectare), richness (the number of species), mingling (the number of species as a proportion of the number of trees within the plot area), DBH CV (the DBH coefficient of variation), and Clark & Evans (the Clark & Evans index for evaluating the spatial distribution of tree positions). The two designs were compared by using the absolute discrepancy (d) between the relative frequencies of a given attribute in the nested and the full design; this gives the magnitude of the differences (Gregorius 1974):

$$0 \le d = \frac{1}{2}\sum_{(i=1)}^{n}\left|p_i - q_i\right| \le 1 \tag{2.3}$$

where p_i and q_i are the relative frequencies in the *i*th frequency class. Thus, d gives the proportion of the frequencies that must be exchanged to make the two

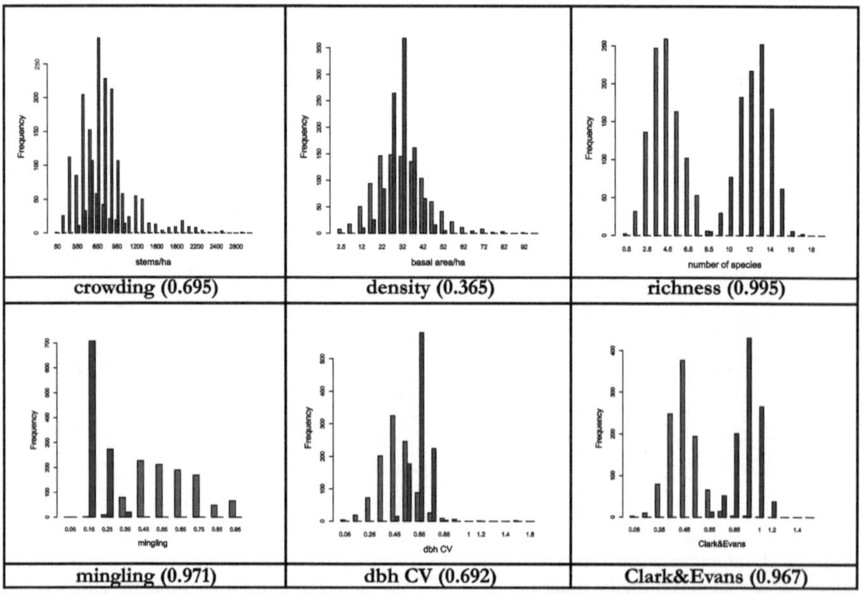

Fig. 2.4 Results of a simulation study in an undisturbed afromontane forest with mapped trees. The distributions of six structural attributes were calculated for the full (blue columns) and nested design (red columns) based on 1000 randomly selected plot positions. Figures in brackets are the differences in the proportions

distributions identical; d = 1 implies that two distributions A and B are disjoint; d = 0 means that both distributions A and B are identical, and d = 0.5 means that 50% of A must be changed to make it identical to B.

A reason for using 1000 comparisons (instead of just a few) is that the differences will vary depending on the location within the greater study area. We can guess that permanent nested plots will be biased regarding certain plot attributes. But it is important to know by how much the nested design affects the estimates of the species richness, spatial pattern, species mingling, diameter coefficient of variation, and other attributes of forest structure. As can be seen in Fig. 2.4, the discrepancies are quite substantial for species richness (0.995), species mingling (0.971), Clark & Evans spatial pattern (0.967), and DBH coefficient of variation (0.692). These results agree with those obtained in a more comprehensive study involving not just one, but a range of nested designs (Ho-Tung Lin et al. 2019).

2.1.1.2 Simulating a Test Area with Mapped Trees

Simulation studies are essential for evaluating the efficiency of different sampling designs (Henttonen and Kangas 2015). Mapped test areas are required with a specific spatial pattern, tree species, and the diameters of the trees. A test area can be a real forest or a simulated one with a defined spatial pattern, a defined number of tree species, and a DBH distribution of each species. The function "Random_forest" can be used to test the efficiency of a particular sampling strategy.

```
# A function that generates a random forest, based on certain assumptions
# Generate positions - uniformly distributed - Poisson forest
# size = length of square plot (m); trees = number of trees in plot
# Nspec = number of species; Sprop = vector of species proportions
Random_forest<-function(size,trees, Nspec,Sprop,mean,sd) {
  x<-runif(trees,0,size); y<-runif(trees,0,size)          # random positions
  sp<-sample(1:Nspec,100,replace=TRUE,prob=Sprop)
  s <-matrix(NA,Nspec, trees)
  for (i in 1:Nspec) s[i,] <- sp==i                      # for each sp a vector TRUE or FALSE
  D <-rep(NA, trees)                                     # set up dbh vector
  for (i in 1:Nspec) {D[s[i,]] <- rnorm(sum(s[i,]), mean[i],sd[i]) }    # generate DBH's
return(cbind(x,y,D,sp))}
# -------------------------------------------------
# Apply the function and extract x,y,D,sp from return vector
size = 200; trees = 400; Nspec = 3; Sprop= c(50,20,30); mean=c(20,40,60); sd=c(3,6,4)
ww<-Random_forest(size,trees, Nspec,Sprop,mean,sd)
x<-ww[,1]; y<-ww[,2];D<- ww[,3];sp<- ww[,4];
# Plot with point size proportional to DBH and col=species
plot(x,y,cex=D/20,pch=19,col=sp)
```

The function requires the plot size, the number of species, and their proportions, as well as DBH means and standard deviations. An example of a four-hectare simulated random forest with three species in proportions of 50%, 20%, and 30% and a DBH mean and standard deviation for each species is shown in Fig. 2.5.

Each run will generate a different random forest with the same species frequencies and diameter distribution, but different spatial mingling of species and different small-scale patterns of tree size variability.

Assignment

Simulate and plot a one-hectare random forest with 1 200 trees and 2 species in the proportion [A = 0.3, B = 0.7]. The DBHs are normally distributed with means [A = 12 cm; B = 35 cm] and standard deviations [A = 2; B = 3.5].

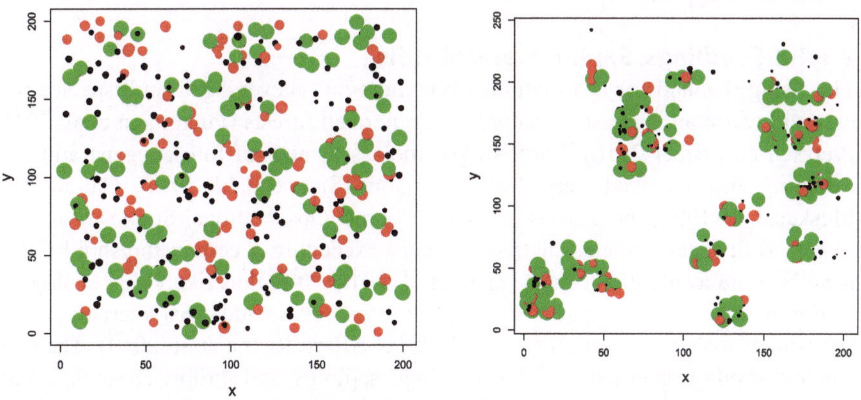

Fig. 2.5 Simulated random forest (left) and a forest with a cluster distribution (right), based on the R code given in this section

The corresponding code for a cluster forest requires a specification of the number of clusters and their centers, the number of trees in each cluster and their distances from the cluster centers, as well as details of species and DBHs generated by using the Weibull or Normal distributions:

```
Nc<-20                                # 1. Specify the number of clusters (Nc)
 xc<-runif(Nc,0,200); yc<-runif(Nc,0,200)  # 2. Generate cluster centers (e.g. uniformly
 lambda<-20                           # 3. Specify the number of trees per cluster (lambda)
 Nt <-rpois(Nc,lambda)                # 4. number of trees in each cluster (Poisson)
 theta<-0.125                         # 5. (x,y) distance from the cluster center (theta)
 # 6. For each tree, generate the x and x distance from its cluster center (e.g.
exponential)
 # and hence the positions of each tree
 N<-sum(Nt)                               # total number of trees
 x<-y<-rep(0,N)                           # create vectors for the positions
 k<-0                                     # tree "number"
 for (i in 1:Nc)                          # loop over clusters
 {for (j in 1: Nt[i])                     # loop over trees within each cluster
    {k<-k+1                               # tree "number"
     x[k]<-xc[i]+rexp(1,theta)            # compute the x-position of this tree
     y[k]<-yc[i]+rexp(1,theta)}}          # compute the y-position of this tree}}
# Generate species - independently in a given proportion 5:2:3
 sp<-sample(1:3,N,replace=TRUE,prob=c(50,20,30))
 # Detemine which trees are of species 1, 2, 3
 ind1 <- sp==1;  ind2 <- sp==2; ind3 <- sp==3
 dbh <- rep(NA,N)                                 # Set up an empty vector for the dbhs
 dbh[ind1] <- rweibull(sum(ind1),2.5,12)          # Fill in the dbhs for species 1 (Weibull)
 dbh[ind2] <- rnorm(sum(ind2),40,6)               # Fill in the dbhs for species 2 (Normal)
 dbh[ind3] <- rnorm(sum(ind3),60,4)               # Fill in the dbhs for species 3 (Normal)
 plot(x,y,cex=dbh/20,pch=19,col=sp)
```

There are many alternatives for generating a forest ecosystem with specific attributes. The R library spatstat has a variety of functions for simulating, comparing, or modeling a forest with specific attributes. The simulated forests can then be used to evaluate alternative sampling designs. It is relatively easy to generate specific point patterns with the R spatstat library. But there are pitfalls. However, misuse or a false interpretation of results can be avoided if the particular functions are studied in detail before application.

2.1.1.3 Seedlings, Saplings, and Shrubs

The rising demand for near-natural silvicultures in uneven-aged multispecies forests will cause an increase of naturally regenerated forests (Puettmann et al. 2015; Pretzsch and Biber 2016). Surprisingly, the assessment of forest regeneration by sampling has received very little attention (Staupendahl 1997; Gadow and Meskauskas (1997). Numerous permanent Forest Observational Studies in natural forests with measurements of trees, usually exceeding a certain threshold DBH, have become available recently (Zhao et al. 2014). It is thus likely that ecological studies in multispecies natural forests will increasingly include characteristics of the subcanopy woody vegetation (Xu et al. 2019; Alfaro Reyna et al. 2019). The subcanopy woody vegetation includes seedlings, saplings, and shrubs. These three categories are defined as follows: **seedlings** include all subcanopy trees with a height exceeding 20 cm and less than 0.5 m; **saplings** are individuals of tree species with a height exceeding 0.5 m and a DBH less than 5 cm; **shrubs** are woody, often multistemmed, plants that are usually not classified as trees because of their low maximum height. Only plants with a height exceeding 20 cm are assumed to qualify as established regeneration.

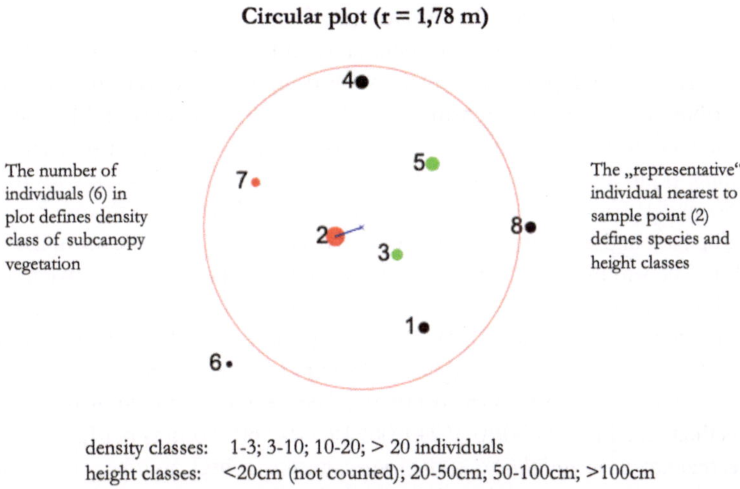

Fig. 2.6 Imaginary sample plot with a radius of 1.78 m (plot area = 10 m^2) for assessing low woody vegetation. The density estimate is 6000 individuals per hectare; the plot representative **species** and its **height** class are defined by sapling No 2, the individual located closest to the plot center

What are the essential characteristics of a particular subcanopy woody vegetation? The total number and the mean height of individuals are of very limited value in heterogenous regeneration areas where groups of high density may occur in close proximity to areas without any regeneration (Staupendahl 1997). The distribution of subcanopy woody plant species and their densities and heights are therefore more suitable target variables for characterizing a particular subcanopy woody vegetation. We will introduce two sampling methods with a potential for assessing low subcanopy woody plants, the *regeneration circular plot* method and the *regeneration neighborhood* method.

2.1.1.4 The Regeneration Circular Plot

Staupendahl (1997) proposed a method based on a sample of 10-m^2 circular plots designed to provide unbiased estimates of density at reasonable cost. There are various practical options for assessing sapling heights and circle perimeters in the field. Staupendahl (1997) proposed to use a measuring stick with 20-, 50-, and 100-cm height markings and a 1.78-m cord attached to the stick for delineating the circle perimeter. One of four density and height classes is assessed, as explained in Fig. 2.6. The class frequencies derived from the representative trees represent area proportions.

The method is very effective in assessing sapling densities, but rare species may not be captured, especially at low sampling intensities. In situations where sapling species and sapling heights are important target variables, an alternative may be to assess more than one individual nearest to the sample point.

2.1.1.5 The Regeneration Neighborhood

Gadow and Meskauskas (1997) presented a sampling method known as the *regeneration neighborhood group*. The method was designed to deliver information about the distributions of species, densities, and heights. The species and heights of the four plants nearest to the sample point are assessed, as well the distances between the sample point and the second and third nearest neighbor. A stick with markings at 20, 50, and 100 cm is used to classify plant heights. The same stick with markings at 20, 50, 100, and 200 cm is used to classify the distances to the second and third neighbor. The sapling density is calculated as the average distance between the second and third neighbors as proposed by Köhler (1951). The number of individuals per ha is then calculated as 10,000 divided by the mean of the squared distances between the second and third neighbors. In Fig. 2.7, the distances to saplings No 3 and 5 are 0.5 m and 1.06 m, respectively. The density estimate at this particular point is thus equal to 10,000/(0.5^2+1.06^2)/2=14560 saplings per ha.

The *regeneration neighborhood group* captures a high number of species and allows an assessment of the diversity (and the spatial mingling) of the different species, which is an advantage if such information is of interest. A disadvantage of the method is the poor performance in estimating plant density, especially in subcanopy communities with an irregular spatial distribution.

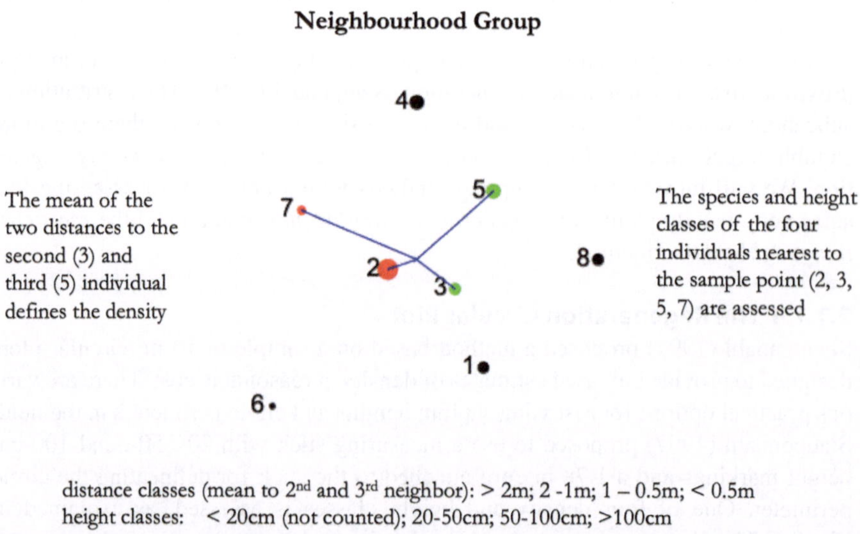

Fig. 2.7 Imaginary sample for assessing low woody vegetation. The species and heights of the four plants nearest to the sample point are assessed, as well the distances between the sample point and the second and third nearest neighbors. The estimate of the number of individuals per ha is then calculated as 10,000 divided by the mean of the squared distances between the second and third neighbors

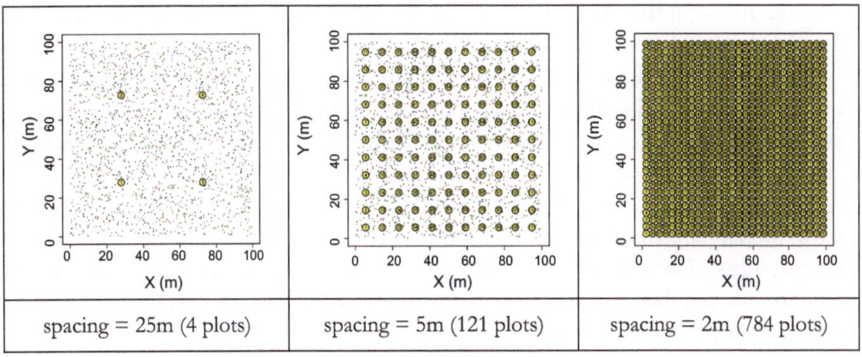

Fig. 2.8 Sample plot locations on a regular grid for three spacings

```
# A function that generates a random forest, with tree locations uniformly distributed
# size = side length of square plot (m); the standard size is 1 ha (size=100)
Random_forest.f<-function(size) {
  area=size^2; mindensity=0.2*area; maxdensity=0.35*area
  trees = sample(mindensity:maxdensity,1)                    # number of trees in plot
  Nspec = sample(2:10,1)                                     # number of species
  Sprop = sample(1:Nspec,Nspec,replace=F); S=Sprop/sum(Sprop)      # sp proportions
  mean=sample(20:80,Nspec,replace=T); sd=sample(3:6,Nspec,replace=T)# distrib pars
  x<-runif(trees,0,size); y<-runif(trees,0,size)             # tree locations
  sp<-sample(1:Nspec,trees,replace=T,prob=S)
  s <-matrix(NA,Nspec, trees)
  for (i in 1:Nspec) {s[i,] <- sp==i}                  # for each sp a vector TRUE or FALSE
  H <-rep(NA, trees)                                   # set up seedling height vector
  for (i in 1:Nspec) {H[s[i,]] <- rnorm(sum(s[i,]), mean[i],sd[i]) }   # heights
return(data.frame(x,y,H,sp))}
# ---------------------------------------------
# Apply the function and extract x,y,H,sp from return vector
ww<-Random_forest(100)
x<-ww[,1]; y<-ww[,2];H<- ww[,3];sp<- ww[,4];
# Plot with point size proportional to height (cm) and col=species
par(mar=c(5,6,2,2) + 0.1)            # plot Margins
plot(x,y,cex=H/400,xlab="X (m)", ylab="Y (m)",col=sp,cex.lab=2)
write.csv(ww, file ="/Users/kg/Desktop/RandomForest.csv")  # for re-use
```

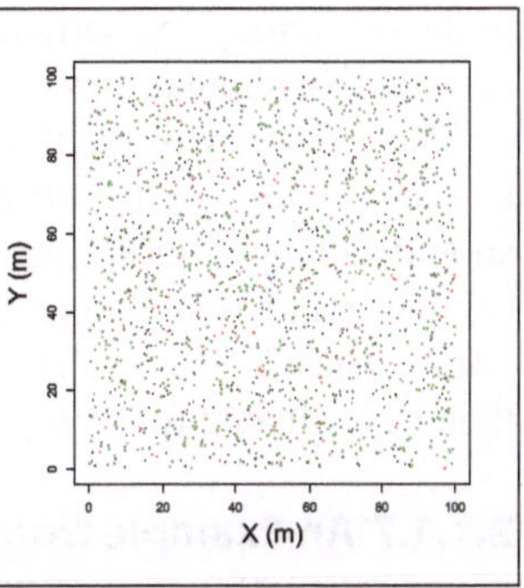

Fig. 2.9 Simulated 1-hectare random forest for three species based on the function Random_forest.f; the code is shown at left

2.1.1.6 Generating a Sampling Scheme on a Regular Grid

The area of each sample circle equals 10 m^2, with radius 1.78 m. The minimum distance of any circle center to the nearest edge of a forest is thus 1.78. All the circular plots are numbered (Fig. 2.8).

The task is to find all the coordinates of the circle centers for a specified number of circles from the total sampling frame using sampling without replacement and to count the total number of individuals, the number of species, and the mean sapling heights in each sample.

We may simulate a population which can be used to evaluate the efficiency of the two methods. There are many alternatives for generating a forest with specific attributes and spatial patterns. A large number of random forests may be generated for each of a certain number of species richness classes (e.g., 2, 4, 6, 8, 10 species). Sapling heights may be normally (or Weibull) distributed with parameters generated at random for each species within certain bounds, for example, means (0.2, 1.3) and variances (0.1, 0.7). An example of such a random forest is presented in Fig. 2.9, using the function Random_forest.f. This function differs from the function

Random_forest in that the number of species, their proportions, and physical attributes are not predefined, but selected at random within certain bounds.

Each run will generate a different random forest with a particular species richness, sapling height distribution, and spatial mingling of species. The corresponding code for a cluster forest requires a specification of the number of clusters and their centers, the number of trees in each cluster, and their distances from the cluster centers, as well as details of species and DBHs based on the Weibull or Normal distributions. The simulated forests will then be used to evaluate alternative sampling designs. The efficiency of the two sampling designs may be evaluated by generating a regular or random grid of sample locations. A regular grid is often preferred for practical reasons.

We may assume that the *regeneration circular plot* method is more efficient for assessing the density of the subcanopy vegetation, whereas the *regeneration neighborhood group* method is especially useful for assessing the biology and structure of the subcanopy vegetation. The assumptions may be evaluated based on the results of extensive simulation studies.

The efficiency of the two methods may be compared using simulated regeneration areas with specific attributes. The attributes of the simulated regeneration areas should include at least (a) the total number of individuals, (b) the number of species, (c) their relative frequencies, and (d) height distributions, and the spatial pattern of points. The efficiency of the two sampling designs in estimating these variables may then be evaluated for a range of sampling intensities.

2.1.1.7 An Example from Mongolia

Kirchhoff (2003) applied the circular plot method to assess forest regeneration in the Khentii Mountains in northern Mongolia, where the Siberian forest belt borders the steppe. The natural regeneration was assessed in three different forest environments. Figure 2.10 gives a brief description of three assessment sites. The three graphs show the main topographic features and occurrence of different tree species such as, among others, *Larix sibirica*, *Betula platyphylla*, *Pinus sibirica*, *Abies sibirica*, and *Picea obovata*.

Covering 15,000 km^2 of forest and grassland, the Khentii Mountains are subdivided into two subprovinces, the western and the eastern Khentii (Savin et al. 1988). In the western Khentii, the lower forest line and the permafrost belt start at a lower altitude than in the eastern Khentii, resulting in different forest types: the *dark taiga mountain forest* which occurs at altitudes between 1200 m and 1600 m where *Pinus sibirica* is dominant; the *larch/birch mountain forest* at altitudes between 800 m and 1200 m where *Larix sibirica* and *Betula platyphylla* are dominant; and the *riparian woodland* (Mühlenberg et al. 2000). Recruitment potential is a key factor of ecosystem dynamics in the southern Taiga forests which are regularly affected by large-scale wildfires. Kirchhoff's assessment of plot height, plot species, and plot density is an example of a rapid and effective regeneration assessment covering a large area.

Burnt forest *Sharlang.* Altitude 1000 m - 1100 m, slope 20° - 30°; regeneration in strips parallel with slope. Two distinct regeneration age classes were found	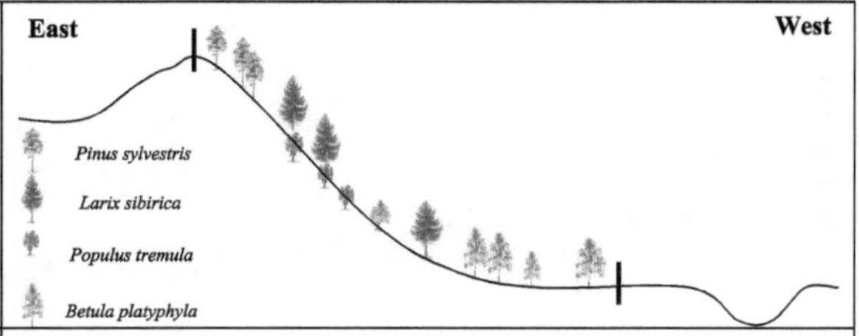
Managed forest *Hausberg.* Altitude 900 m, moderate slope; regeneration in clumps; new regeneration of *Abies sibirica*	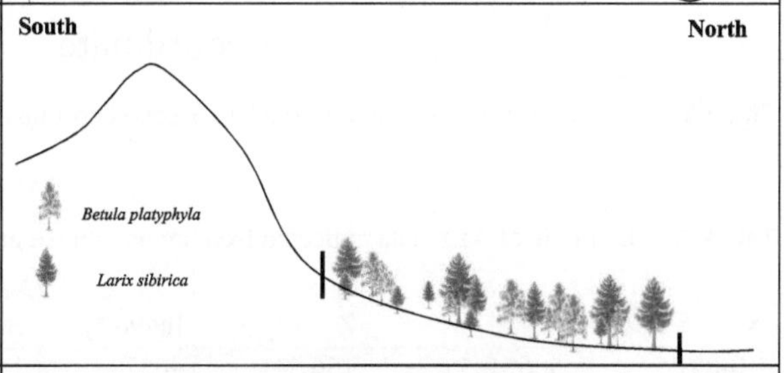
Virgin "dark" taiga forest *Sangstai.* Altitude 1500 m, flat; dense forest with few gaps; regeneration in clumps	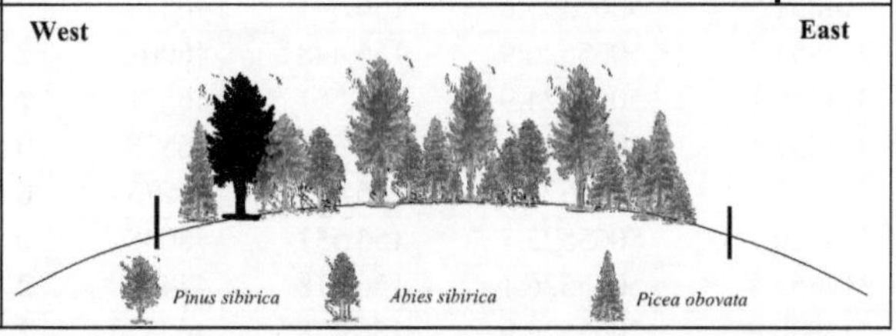

Fig. 2.10 Three examples of typical topographic features in areas where natural regeneration was assessed by Kirchhoff (2003)

2.1.2 Remote Assessments

Remote sensing is increasingly replacing traditional methods of field measurement. Of particular interest among recent developments are new operational applications of the Landsat system, airborne and terrestrial laser scanning, digital aerial photogrammetry, and MODIS applications (Wulder et al. 2019; Goodbody et al. 2019).

2.1.2.1 Use of Airborne Laser Scanning in Forest Inventory

The use of airborne laser scanning (ALS) has increased rapidly in forest inventories. Like radar, ALS is an active remote-sensing method since it measures the return echoes of the laser pulses that the system itself emits. Most traditional remote-sensing methods such as aerial photography are passive since they measure the intensity of reflected sun light or other wavelengths of electromagnetic radiation.

ALS can be conducted from manned airplanes or unmanned aerial vehicles (UAV). The laser beam scans the terrain using swaths perpendicular to the flight direction of the aircraft. Parameters affecting the accuracy of the method and interpretation of the data are flight altitude and speed, scanning angle, and pulse

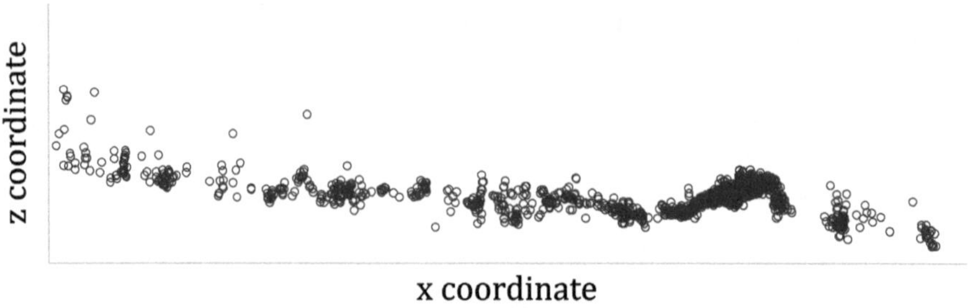

Fig. 2.11 A narrow strip (in y direction) of ALS-echo coordinates

Table 2.1 Example of ALS data collected from an unmanned aerial vehicle

X	Y	Z	Intensity	Vegetation (0) or ground (2)	Return number
314851.5	5065525.6	156.541	46578	2	1
314851.2	5065526.9	156.448	46001	2	1
314850.1	5065524.9	156.661	46571	2	1
314851.5	5065525.6	156.541	46578	0	1
314850.7	5065525.6	156.645	45573	0	1
314850.7	5065525.7	156.557	48048	0	1
314852.4	5065526.1	156.518	44464	2	1
314850.9	5065524.9	156.478	46582	2	1
314851.2	5065526.9	156.448	46001	2	1
314851.0	5065525.6	156.509	46957	2	1
314852.0	5065525.5	156.394	45855	2	1
314850.7	5065524.9	156.563	45731	0	1
314852.0	5065523.8	156.476	44631	2	1
314850.6	5065524.6	156.622	46384	0	1
314850.1	5065524.9	156.661	46571	0	1
314930.7	5065541.0	158.827	37493	2	2
314910.6	5065545.6	158.563	47774	2	1
314895.7	5065550.6	158.493	39174	2	1
314932.1	5065531.2	158.921	40480	2	2

repetition frequency. Together they result in certain pulse density (number of pulses per square meter), which affects the methodology that can be used to interpret forest characteristics from the pulse data. The ALS machinery also includes global positioning system (GPS) and inertia measurement units. They make it possible to calculate the x, y, and z (elevation) coordinates of every return echo (Fig. 2.11).

The same pulse may reflect several times from different objects, for instance, first from the tree crown and then from the ground. Table 2.1 shows an example of data obtained from ALS scanning.

The number of the return echo of a certain laser pulse is also recorded, making it possible to classify them, for instance, to first echoes, last echoes, and other echoes.

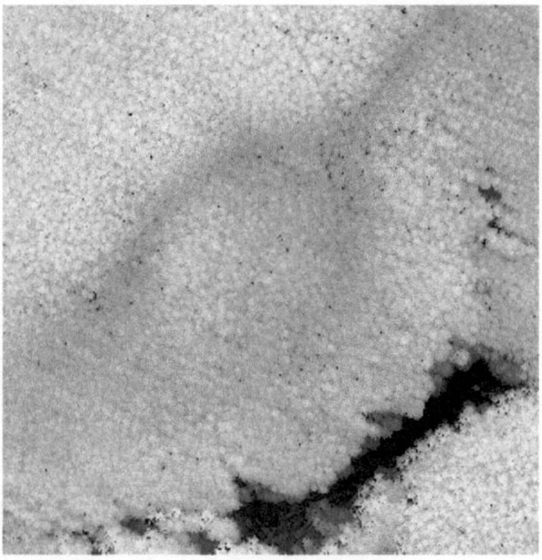

Fig. 2.12 Canopy height model created from ALS pulse data. The height of the canopy has been calculated for 1-m^2 grid cells. Light tone indicates tall trees. Crowns of individual trees can be seen from the visualization

Echoes are also usually classified as vegetation echoes and ground echoes. The intensity of the return pulse is also recorded.

Interpretation of ALS Data

ALS data can be used to interpret the elevation of both the ground surface and the surface of the tree canopy. The last echoes are typically assumed to indicate the elevation of the terrain. Delaunay triangulation is one of the methods that has been used to derive the digital terrain model (DTM) from the last echoes. The elevation of the tree canopy, designated as the digital surface model (DSM), is often calculated from the first echoes. For example, a 1-m square grid is generated and the highest echo within a certain radius is taken as the elevation of the grid point. If there are no echoes within a certain distance from the grid point, the height is interpolated from adjacent grid points.

Subtracting the ground elevation (DTM) from the elevation of the canopy (DSM) gives the canopy height model (Fig. 2.12). The canopy height model (CHM) can be used in forest inventories in different ways. For example, it can provide visual information for manual stand delineation. The delineation can be based on the combined use of aerial photographs, maps, and the CHM. Once the stands have been delineated, the CHM gives accurate information on tree height in different stands.

It is also possible to use the CHM in automated stand delineation or for the interpretation of individual trees. The idea of individual tree detection (ITD) is to find the tree tops from the pulse data. The coordinates of these points give the tree height (z coordinate) and the location (x and y coordinates) of a particular tree. The procedure searches for the local maxima from the data, that is, points that are located higher than their surroundings.

The CHM can be used to identify local maxima, but the exact location and height of the tree can be taken from the original pulse data, by finding the highest pulse from the cell where the local maximum is located. Individual tree detection can be continued to the delineation of tree crowns (Fig. 2.13). It can be assumed that

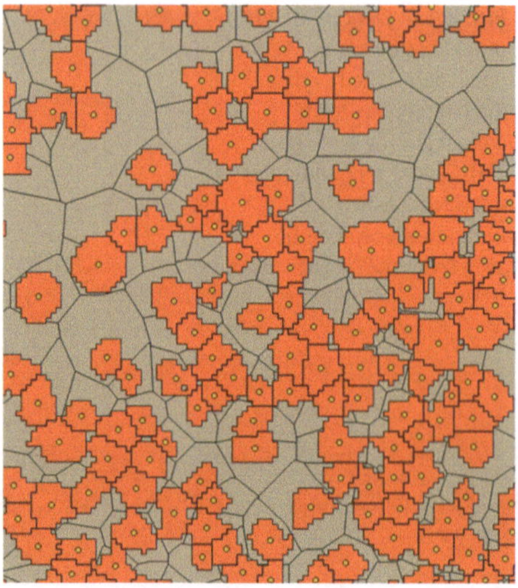

Fig. 2.13 Three locations and crown areas interpreted from ALS data (Packalen et al. 2020). The polygons (straight lines) around trees are a variant of Voronoi polygon. They can be used as initial segments in region merging, or individual tree areas in tree-level forest planning applications

valleys in the elevation data correspond to crown perimeters. Starting from the tree top, the area of individual tree crown is enlarged until a local valley is met (elevation starts to increase again) or there is a gap in the canopy or a maximum distance from tree top is exceeded.

If there are field measurements of crown diameter, tree height, and stem diameter from the same forest, it is possible to fit a model for predicting tree diameter at breast height (DBH) from ALS-interpreted tree height and crown width. Tree heights and crown diameters measured in the field may deviate systematically from the values of the same variables interpreted from pulse data. For example, it has been found that the tree heights, estimated from CHM or from the pulse data, are often smaller than the true tree heights. This is because the laser pulses do not always hit the highest points of the tree tops, which can be very thin (for instance, the latest annual shoot of a fast-growing young tree). Therefore, the model for predicting DBH from ALS data should preferably be fitted between ALS-detected tree height (H_{ALS}) and crown diameter (CW_{ALS}), and field-measured DBH (D):

$$D = f\left(H_{ALS}, CW_{ALS}\right) \tag{2.4}$$

This kind of modeling requires that the same individual trees detected from ALS data are also found and identified in the field. Therefore, the coordinates of the trees measured in the field must be determined accurately. Then, it is examined whether the same tree can be found from the individual trees detected from ALS data. Those trees for which a match is found (e.g., field and ALS coordinates are close enough to each other, and height is of the same magnitude) can be used for modeling.

After this step, each detected tree has a location, height, DBH, and crown diameter. These characteristics can be used to calculate additional growing stock variables such as number of trees per hectare and stand basal area. If the stand is a plantation (one species), tree species can be obtained from plantation records. In case of naturally regenerated forests or mixed plantations, tree species identification is more problematic. Alpha shapes, calculated from the pulse data for individual tree crowns, have been proposed as a means to interpret the species of the detected trees (Vauhkonen et al. 2009). Combined use of digital aerial photograph and ALS data also helps to identify tree species.

Area-Based Interpretation

Individual tree detection produces very detailed information for forest planning. However, the method requires high pulse density, which increases costs. In addition, the method is prone to systematic errors, especially in mixed, multistoried, and naturally regenerated forests. This is because all trees are seldom detected in these forests. Two or a few trees may be very close to each other, and these tree groups may be interpreted as only one tree. Suppressed trees growing below the crowns of dominant trees may also remain undetected.

Due to these reasons, another interpretation line, the area-based approach, is still the most common method in ALS-based forest inventories. In this method, the ALS data, perhaps together with other data sources, are first used to divide the forest into homogeneous areas, called segments (or micro segments). Then, the stand attributes are derived for each segment using modeled relationships between pulse data and stand attributes.

The method requires that a set of field plots is measured for the interpretation area. The number of plots depends on the variability of the inventoried area and may range from few tens (even-aged plantations) to few hundreds (several stand types and species compositions). The exact location of each field plot is determined with GPS. Then, the ALS pulse data are used to calculate several variables for each field plots. Examples of variables calculated from the pulse data are maximum height, average height, number of echoes per area unit (pulse density), and various percentiles of the height distribution of the echoes.

Growing stock and pulse variables calculated for the field plots can be used to fit regression models for predicting stand characteristics from pulse variables. Another possibility is to use the pulse variables of the field plots and the interpretation units (e.g., segments) to find the most similar field plots for every interpretation unit. Then, the measured stand variables of the most similar pots are imputed for the segment. This approach does not need modeling, and it guarantees that different stand attributes of the interpretation unit are logically related to each other.

The interpretation units are not necessarily segments. They can also be grid cells or stands delineated visually on aerial photographs. From the interpretation point of view, the size of the interpretation units should be more or less the same as the area of the field plot. Using grid cells as interpretation units makes this possible. For example, if the area of the field plots is 250 m^2, the interpretation unit might be a grid cell of 16 × 16 m (256 m^2). This is the case for instance, in Finnish ALS-based forest inventory, where the stand variables are interpreted for both existing stands and 256-m^2 grid cells.

Using grid cells as interpretation units leads to the problem of mixed cells. Mixed cells are cells that cross stand boundaries. Pulse variables calculated for mixed cells,

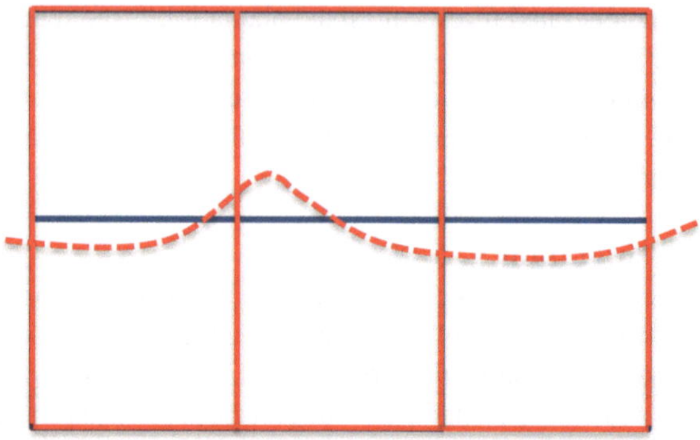

Fig. 2.14 The idea of nanosegment. Segment boundaries (dashed line) are overlaid with a regular square lattice, and the small parts of square cells created by the overlay are joined to an adjacent cell of the same segment. Red lines show the boundaries of nanosegments. Nanosegments have the advantage that the whole segment is located within the same stand (nanosegments do no cross stand boundaries)

such as mean height or percentiles of the height distribution of echoes, are averages of two or more dissimilar stands, and may not characterize any of the stands. The mixed-cell problem can be overcome by creating an overlay of (homogeneous) segments and grid cells (Fig. 2.14). The resulting small subareas of cells are joined to an adjacent cell of the same segment. The cells modified in this way are called nanosegments (Pippuri et al. 2013). Nanosegments are interpretation areas which do not cross stand boundaries and have nearly similar size as the field plots of area-based ALS inventory.

2.1.2.2 Numerical Delineation of Forest Stands

Segmentation

A common step in the area-based interpretation of ALS data is the creation of segments, which are homogeneous subareas of forest. Then, stand attributes are interpreted for the obtained segments. However, segmentation is not a necessary step in the area-based ALS inventory because a regular grid of square-shaped or hexagonal cells can also be used, as well as existing stand boundaries. Segmentation has the advantage that the interpretation units are homogeneous and do not cross stand boundaries. Compared to the use of small cells, segmentation greatly reduces the number of interpretation unis, making all subsequent calculations faster.

A number of methods have been proposed for the segmentation of ALS data. Most methods start from very small initial cells (e.g., 1 × 1 m or 2 × 2 m grid cells). The initial cells are so small that the proportion of mixed cells remains low. Instead of square cells, it is also possible to use hexagon-shaped cells, which usually result in smoother segment boundaries. In addition, in cases where the locations of individual trees have been interpreted from the ALS data, it is possible to create Voronoi polygons around the tree locations (Fig. 2.13) and use these polygons as the starting units of segmentation.

The method most commonly used in forestry is region merging. In this method, the most similar adjacent areas (square cells, hexagons, Voronoi polygons) in terms of the pulse characteristics are merged, resulting in a new larger segment. This process of removing boundaries between two adjacent and similar areas is continued until the size of the segments is suitable, and within-segment variation in pulse characteristics is small compared to between-segment variation. It is also possible to take the shape of the segments into account in segmentation.

Automated Stand Delineation

Although many segmentation methods have been developed and described in forestry literature, the topic has not yet been exhaustively studied. Especially, the use of existing "general-purpose" algorithms such as cellular automata, self-organizing maps, metaheuristics, and various self-learning systems deserves further research.

The same algorithms that are used in the segmentation of ALS pulse data can also be used with other types of data. For example, it is possible to segment digital aerial photographs or the canopy height model calculated from the ALS pulse data. In addition, if the stand variables are interpreted for grid cells, it is possible to use the interpreted stand variables as the basis for segmentation. Moreover, if the segments used in area-based ALS inventory are very small, it is possible to aggregate these micro segments further into larger areas, so as to reduce the computational burden of forest planning calculations.

Segmentation methods can use several different data sources and initial spatial units, and aim at different average sizes of the segments. A common target is to delineate homogeneous subareas of forest that correspond to the concept of stand (sometimes called subcompartment). Therefore, the segmentation methods described in this section are called automated stand delineation methods.

Cellular Automata in Stand Delineation

One of the methods recently tested in stand delineation is cellular automaton (CA). Cellular automata are self-organizing algorithms based on the assumption that the interaction between cells decreases rapidly with increasing distance. Although cellular automaton refers to cells, it can also be used with other types of calculation units, for instance, micro segments. Each unit takes one of a limited number of states, which in case of stand delineation are stand numbers. The purpose of CA is to find the optimal stand number for each cell, by considering the variables of the cell itself as well as the local neighborhood of the cell (Jia et al. 2020). When CAs are used for stand delineation, the purpose is to create a layout where the stands are homogeneous enough and of suitable size and shape. These are rather subjective criteria, but they can be at least partly measured by numerical indicators. For example, the degree of variance explained by the delineation can be used to measure homogeneity (in terms of any variable). The R2 of a variable is calculated from

$$\mathrm{R2} = 1 - SSE / SST \tag{2.5}$$

where *SSE* is variation not explained by the delineation and *SST* is the total variation of the characteristic in the area. *SST* and *SSE* are calculated as follows:

$$SST = \sum_{j=1}^{N} \sum_{i=1}^{n_j} \left(y_{ij} - \overline{y} \right)^2 \tag{2.6}$$

$$SSE = \sum_{j=1}^{N} \sum_{i=1}^{n_j} \left(y_{ij} - \overline{y_j} \right)^2 \tag{2.7}$$

where N is the number of stands, n_j is the number of cells in stand j, y_{ij} is the value of the variable in cell i belonging to stand j, $\overline{y}$ is the overall mean of the variable, and $\overline{y}_j$ is the mean value of the variable among the cells that belong to stand j.

Stand delineation seldom aims at creating stands as large as possible. Very large stands may be problematic in forest planning, for instance, when the aim is to harvest the same volume every year. A low number of large stands may make this impossible unless the stands are partitioned into smaller substands during the planning calculations. However, such partitioning contradicts the purpose of stand formation, because stands are by definition regarded as indivisible spatial units of a forest. Therefore, the usual aim is to avoid both very small and very large stands. This objective can be attained by using a sigmoid type relationship between the stand area and the quality of the segmentation (Fig. 2.15).

The shape of the stands can be measured using the ratio of the border length of the stand and the square root of the number of cells that form the stand. Another measure is the border length divided by the shortest possible border of a stand of the same size (circle). The cellular automata developed so far for stand delineation first divide the forest into square-shaped initial stands (e.g., 2 ha each). Then, the "most suitable" stand number is given to each cell for several iterations (Fig. 2.16).

A cell gets the stand number of one of its adjacent stands, which are eight in number (stand number of the cell to the east, west, north, and south, and four corner cells). Figure 2.17 illustrates the evolution of the stand delineation during a short CA run.

In one application of CA (Pukkala 2019b), the stand number given to a cell (or any other spatial unit) depends on a score that is calculated to each of its adjacent cells (or any other spatial unit):

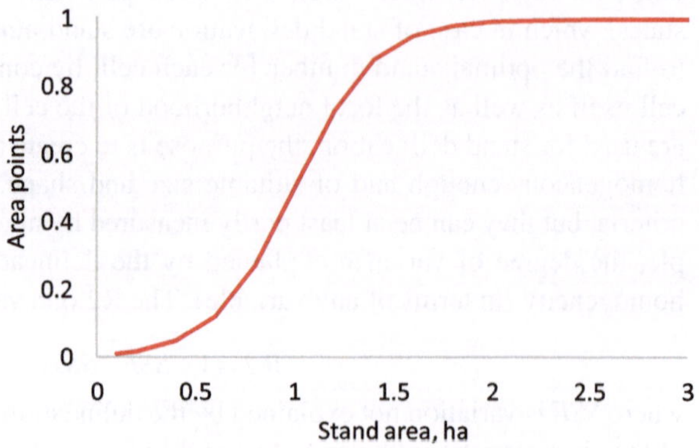

Fig. 2.15 A sigmoid type of relationship between stand area and the stand's area score leads to avoiding both very small and very large stands when area points are maximized simultaneously with the degree of explained variance (R^2)

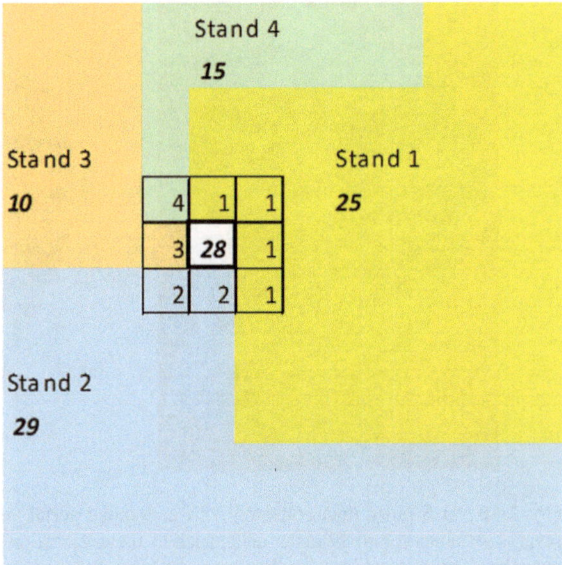

Fig. 2.16 In a cellular automaton, a cell is joined to one of its adjacent stands. The white area is joined to Stand 1 or Stand 2 if the basal area is the only stand variable considered (the boldface italic numbers **10, 15, 25, 28,** and **29** are stand basal areas in m²ha⁻¹). The decision depends on the weights of the common border of the cell and the stand, the similarity of stand characteristics in the cell and the stand, and the area of the stand

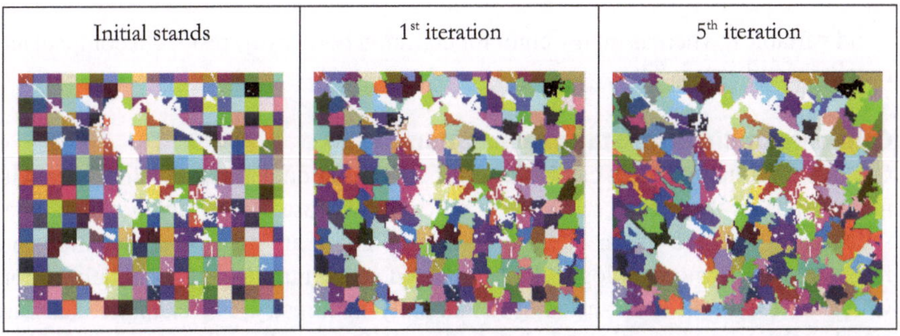

Fig. 2.17 Development of stands during five iterations of the cellular automaton. The size of the initial stands (left) is 3 ha. Each cell is joined to one of its adjacent stands for several iterations using the principles explained in Fig. 2.16

$$S_{ij} = a_1 s_1 \left(B_{ij}\right) + a_2 s_2 \left(A_j\right) + a_3 s_3 \left(D_{ij}\right) \tag{2.8}$$

where S_{ij} is the score of assigning the number of stand j to cell i, B_{ij} is the proportion of common border between cell i and stand j (of the total border length of cell i), A_j is the area of stand j, and D_{ij} is the difference in stand characteristics between cell i and stand j. Increasing stand area and increasing length of common border increase the score, and increasing difference in stand variables between the cell and the stand decreases the score. If the weight of the difference in stand characteristics (a_3) is high, the stands tend to be small and irregular with little within-stand variation in

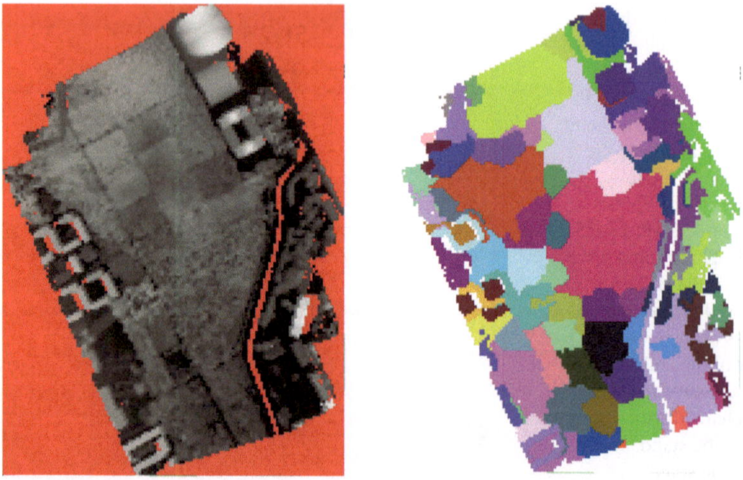

Fig. 2.18 ALS pulse data collected with unmanned aerial vehicle have been used to calculate several variables for small raster cells, such as mean elevation of the echoes (left) and maximum elevation. These variables have then been used in cellular automaton to divide the area into homogeneous micro segments (right, different colors indicate different micro segments)

stand variables, whereas high weight for common border (a_1) results in compact and round-shaped stands.

Other Possibilities for Stand Delineation

CAs are flexible tools that can be used for segmentation and stand delineation with many types of data (Fig. 2.18). They are multipurpose algorithms not developed specifically for forestry or stand delineation. There are also other potential, not forestry-specific methodologies for automated segmentation and stand delineation, which have not yet been tested systematically in forestry. For example, standard metaheuristics such as simulated annealing, tabu search, and genetic algorithms could be used for stand delineation more or less in the same was as they have been used in spatial optimization in the context of forest management planning. In this case, the optimization problem consists of finding the optimal combination of stand numbers for the cells, aiming at continuous and large enough stands, small within-stand variation in stand attributes, and large between-stand variation.

Another interesting methodology are self-organizing maps (SOM), which are also called Kohonen maps according to the Finnish academic Teuvo Kohonen who developed the method during the 1980s. Self-organizing maps can be used to cluster data points (e.g., grid cells) into homogeneous stands (Fig. 2.19).

The clusters (stands) do not automatically form spatially continuous areas, but it might be possible to modify the basic SOM algorithm so that the geographical location of the data points plays an important role in clustering

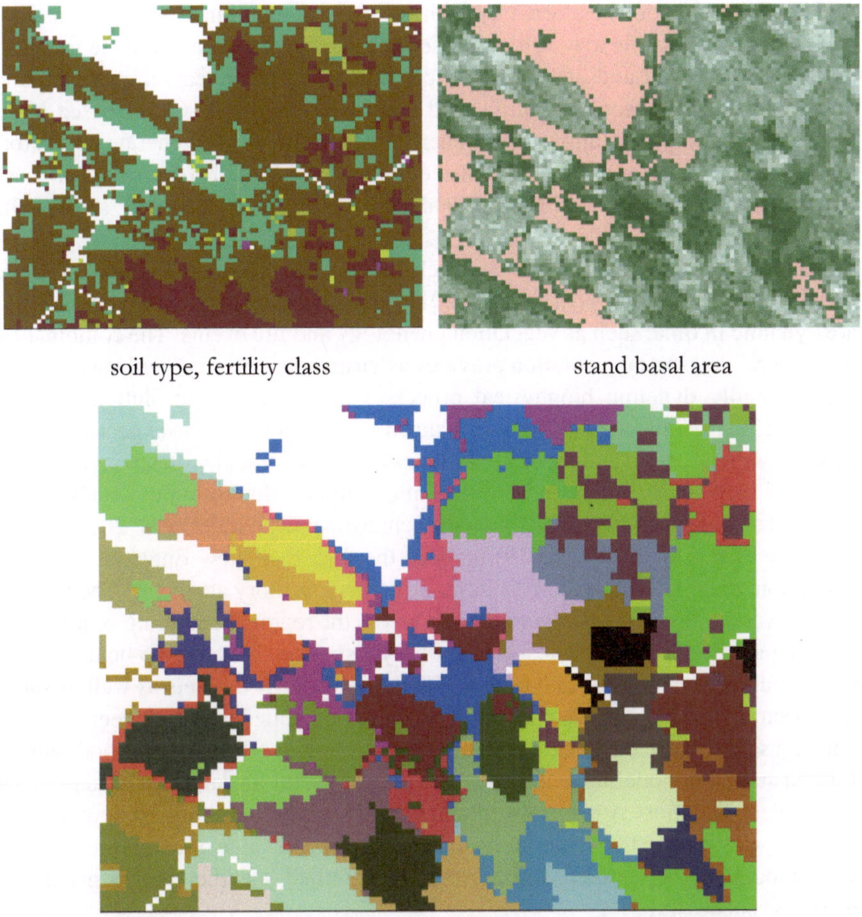

Fig. 2.19 Segmentation (bottom) created by self-organizing map based on soil type, fertility class (top left), stand basal area (top right), mean diameter, and mean height. White cells are land uses other than productive forest. Pink color in the basal area map (top right) indicates cells where basal area is zero. The cell size of the raster is 16 m by 16 m

2.1.2.3 Landsat

Formal planning of the Landsat program, managed by the United States Geological Survey and the National Aeronautics and Space Administration (NASA), began in 1967. The development of the Landsat program and its ability to meet increasing national and international information needs have been described in several recent reviews. A comprehensive history of the Landsat program was compiled by Goward et al. (2017). Wulder et al. (2019) described major trends in Landsat research during the past 50 years. There have been dramatic changes in the Landsat program, especially since 2008, when the Landsat data policy changed and the Landsat archive became freely accessible, that is, open access for research.

The historical context offered by archival imagery combined with new assessments makes it possible to develop time series algorithms that can produce information on trends. The Landsat system comprises more than 50 years of planning and operations and has become multifaceted and highly complex. Cloud-based high-performance computing allows researchers and practitioners to generate new information over large areas. An extensive and sophisticated Landsat user base implements advanced methods of analyses on an unprecedented scale (Zhe et al. 2019).

Landsat-8, Sentinel-2A, and Sentinel-2B together provide a global average revisit interval of 2.9 days (Li and Roy 2017). This allows mapping of processes that are dynamic in time, such as vegetation phenology and fire events. The combination of temporal and spatial resolution provides a "virtual observatory" with an ability to extract highly dynamic biophysical processes at subhectare resolution (Wulder et al. 2015). The launch of Landsat-9 planned for the late 2020s is expected to have a two-day revisit cycle and further continuity for reassessment with an ongoing data stream. Many years of data and the resulting temporal dimension of Landsat have emerged as one of the program's greatest achievements (Wulder et al. 2019).

Ayrey et al. (Ayrey et al. 2019) mapped the forests of New England at a 10-m resolution, making estimates of 14 different forest inventory attributes. They aggregated several public LiDAR datasets from across the region, combining acquisitions of varying pulse density and seasonality. They also assembled 24,606 field plots for model training, validation, and testing. They used LiDAR datasets as well as satellite spectral, phenological, and disturbance data and applied a form of deep learning known as "three-dimensional convolutional neural networks." The method outperformed standard modeling techniques and proved useful for large-scale mapping at fine scales. The most successful estimates in this study were those that quantified tree size, while those that quantified tree density or percentage of coniferous species were moderately successful and those that quantified specific species groupings were less successful.

Forest inventories rely increasingly on airborne laser scanning (ALS). In Finland, the multisource national forest inventory data in combination with airborne laser scanning makes large-scale mapping at fine scales possible. Detailed forest information is available for 16 × 16 m grid cells. The site variables include land use category (e.g., productive forest, stunted forest, and agricultural land), soil group (mineral soil, spruce mire, pine bog, and open bog), and fertility class (mesotrophic herb-rich, herb-rich, mesic, subxeric, xeric, and barren heath). The growing stock variables include basal area, number of trees per hectare, stand age, mean tree diameter, and mean tree height for pine, spruce, and broadleaved species. These data are available for all the private forests in Finland and are freely available for forest planning and research. Pukkala (2019a) describes an application that compares the use of the small raster cells as calculation units or aggregates the cells into segments for spatial planning. The use of cells made it necessary to use spatial optimization to aggregate harvest activities and other treatments into blocks that were large enough for the practical implementation of his plan.

2.1.2.4 Terrestrial Laser Scanning

Terrestrial laser scanning (TLS) is technology for measuring surrounding objects using active Light Detection and Ranging (LiDAR). "Active" means that the system itself generates light to measure objects on the ground. Light emitted from a rapidly firing laser is reflected from tree stems and branches. The reflected light then returns to the LiDAR sensor where it is recorded for range and precise angular measurements to capture 3D point observations from the object surfaces (Liang et al. 2016a, b). The scanning mechanisms enable TLS to capture very dense measurements in a short period. In a typical TLS instrument, the scanner measures the surrounding environment stepwise, using a fast vertical mirror rotation and a slower horizontal instrument rotation. In the vertical direction, the laser beam starts, for example, from the scanner zenith and rotates to the lowest scanning position below the horizontal plane of the instrument. The laser beam then continues to the scanner zenith on the other side of the instrument. In the horizontal direction, the scanner turns 180 degrees and can scan both sides of the instrument simultaneously.

After a decade of active research, TLS has not yet been generally accepted as an operational tool in forest inventories. Its application is hampered mainly by difficulties in the automation of the point cloud processing that provides convincing measurement results of the most important forest inventory parameters. Reliable methods for detecting important tree attributes, such as tree species, are still lacking. Other important factors that limit the general use of this technology include the relatively high cost of the instrument, the limited software, and the lack of trained personnel. Consequently, it may take some time before TLS systems will be used operationally.

In spite of these difficulties, however, TLS research is high on the scientific agenda of many research teams, and new results are published continuously. For example, stem diameters at different tree heights are assessed by fitting cylindrical or circular shapes into the TLS point cloud. Pitkänen et al. (2019) propose an automatized processing chain that consists of two fitting phases and an additional taper curve calculation to refine the final diameter measurements. The method was tested with a set of 505 trees, and the results showed that the proposed processing chain was capable of producing accurate diameter measurements.

Giannetti et al. (2018) integrated airborne laser scanning (ALS) and TLS data in a Mediterranean forest to assess five single-tree attributes: tree position, stem diameter at breast height, tree height, height to crown base, and crown projection area. They obtained point clouds by using a handheld mobile laser scanner as well as a static terrestrial laser scanner, both singly and in combination with ALS. The precision of single-tree predictions, in terms of bias and root mean square error, was evaluated against measurements obtained with traditional instruments. They found that both static TLS and the handheld mobile laser scanner produced accurate results in estimating tree position, stem diameter at breast height, and crown projection area. Height to crown base was the most difficult attribute to assess with sufficient accuracy. The integration with ALS increased the accuracy of the measurements of tree heights.

2.1.2.5 Estimating NPP Based on MODIS

Productivity estimates are important data with which the mass budget of a forest ecosystem can be characterized. Net primary production (NPP), which describes the carbon uptake by vegetation through photosynthesis, is a measure for carbon flux. Large-scale carbon assessments are traditionally provided by national forest inventories and occasionally also by flux towers. Remote-sensing methods such as MODIS (MODerate resolution Imaging Spectroradiometer) are often used to estimate local leaf area index and NPP (Wang et al. 2004; Zhao et al. 2006). MOD17A3 is one of the most widely used data sources for studies of the global carbon cycle (Gulbeyaz et al. 2018; Huang et al. 2018).

Gulbeyaz et al. (2018) found that the MODIS NPP estimates showed some limited and variable agreement with local measurements in Turkey. Neumann et al. (2015) suggested that differences in stand density should be considered to explain observed discrepancies between "space-based" MODIS estimates and "ground-based" NPP estimates. Two forests of the same type and development stage, and on the same site, may differ in density, and this can significantly affect their NPP values. Thus, coarse scale MODIS products should be used with caution and validated with field observations. Neumann et al. (2015) also found that using local, instead of global, climate data produced better estimates of the MODIS NPP predictions.

A recent example of a study involving MODIS NPP predictions investigated the spatiotemporal pattern of net primary productivity, evapotranspiration (ET) and water use efficiency, and their relations with the local climatic and forest attributes of a landscape in northeastern China (Li et al. 2019). The time series of spatial data on NPP and evapotranspiration were extracted from the global datasets provided by MOD16A3 and MOD17A3 for the period 2000–2014. The time series of spatial patterns of annual precipitation and annual mean temperature were obtained as grid maps for regional meteorological variables. Forest patches were classified as coniferous, broadleaved, and mixed species, and as natural and planted. In addition, five age-related classes (young, mid-age, near-mature, mature, and old-growth) were identified by means of the forest inventory datasets of China. Across the forested landscape, the mean annual NPP varied between about 200 gC and 640 gC•per m^2 per annum, and evapotranspiration varied between 441.5 mm and 784.0 mm per annum. The spatial variations in NPP and ET were generally attributable to forest type, stand age and density, establishment mode, and temperature variables. Biomass carbon sequestration is one of the key ecosystem services of forests, as it mitigates the effects of greenhouse gas emissions and the threat of climate change. The natural forest patches had significantly greater NPP than planted forests.

2.1.3 Cost–Benefit Issues

One of the issues in forest sampling is the relation between cost and precision. If we define the information level as the quotient between information required and information available, then the optimum sampling intensity is somewhere between total ignorance and complete information. The variables benefit, cost, and information level are usually not precisely quantifiable, and it is therefore not possible to perform an explicit cost–benefit analysis. However, surrogate or partial measures of benefit, such as variance and the cost associated with achieving increasing precision, can be used. As one generally begins with total ignorance, any initial increase in the information level produces large benefits at small cost. Thereafter, the incremental improvement in benefit rapidly diminishes until it becomes prohibitively expensive to obtain additional information.

Systematic sampling is widely used in forest assessment as it facilitates locating plots in the field. An element of randomness is often introduced by choosing a random starting point that then defines the location of the remaining n-1 plots. There is no harm in doing so, provided the population of sampling units is in "random order," that is, not aligned with natural strata such as soil types or slopes. If the systematic design does not coincide with any natural patterns in the field, the results of systematic and random sampling are equivalent. This is important as most calculations are based on the assumption of randomness (Tomppo et al. 2010).

When we wish to determine the sampling intensity, it is not only the sampling fraction (the proportion of the forest area that needs to be sampled) that is important but also the number of sample units needed to produce an estimate with the required precision. The sampling intensity is influenced by the variability of the target variable, the desired precision, and the cost constraints. It is generally desirable to minimize the sampling intensity subject to a required degree of precision (Heikkinen 2006).

2.2 Field Experiments

Field experiments for testing the system response to particular treatments include spacing experiments, short-rotation coppice experiments, and ecosystem functioning and biodiversity experiments.

2.2.1 Spacing Experiments for Planted Ecosystems

A famous example of a manipulated field experiment for planted forests is the *Pinus radiata* study at Glencoe Hill in southern Australia. The experiment was established in 1985 and included three thinning and four fertilizer treatments with four replicates each, distributed over 144 plots in a factorial design (O'Hehir 2001). Because of its ambitious design, the experiment necessitated finding a large site of approximately 70 ha that was relatively homogeneous in terms of soil type, age, and planting density. The choice of sites that *met all* these criteria was extremely limited, which seems to be a typical problem of field experiments involving trees (Kleinn and Köhl 1999). Some forest field experiments have been remeasured for extended periods of time. An example of such a long-term field study is the so-called *CCT* spacing experiment established in South Africa during the 1930s by A.J. O'Connor. Figure 2.20 shows a map with the geographical distribution of the most important South African CCT spacing experiments.

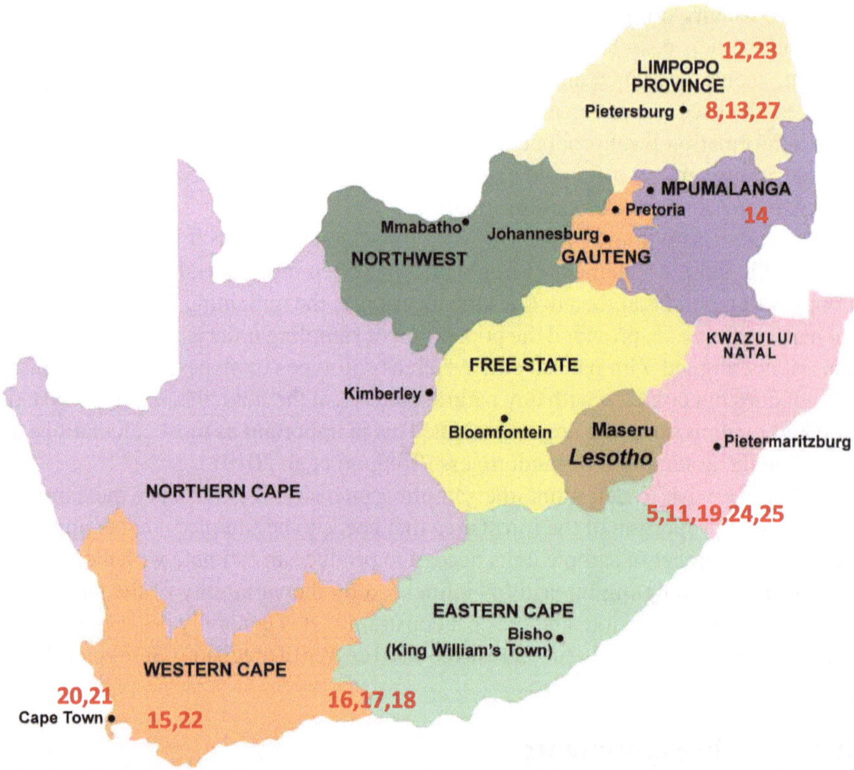

Fig. 2.20 Geographical distribution of the most important South African CCT spacing experiments (basic series)

The CCT experiment features a wide range of densities, often with several replications, and plots have been continuously remeasured until recently (Gadow and Bredenkamp 1992:55; Bredenkamp et al. 2000). Most of the *CCT* experiments consisted of up to 18 plots of 0.04 ha each, and sometimes with up to four replicates. The basic series involves nine plots representing a range from extremely low to very high densities; it therefore provides answers to such fundamental questions as the maximum density that can be expected on a given site, or the relation between stand density and production.

The basic series of eight densities was established by planting all the plots at the highest densities and then thinning the plots progressively "ahead of competition." Two stands representing different spacings were compared at a time. Initially, one of these was plot 8, the lowest stand density tested. As soon as the mean DBH in the more densely planted plot was 2.5 mm (1/10 inch) less than the mean DBH in the "suppressed" plot, it was assumed that competition had started and the "free-growing" plot and all the plots not yet at their final density were thinned. This rule of "thinning ahead of competition" was applied almost monthly in the early stages. The thinning operations were usually completed when the trees were about 8 years

Table 2.2 Stem number reduction scheme for reaching the nominal densities in the basic CCT series (example Nelshoogte experiment)

Stems per ha at age									
Plot	0	2	4	4	5	6	7	8	Stems/ha at age 20
1	2965	2965	2965	2965	2965	2965	2965	2965	2250
2	2965	1483	1483	1483	1483	1483	1483	1483	1361
3	2965	1483	988	988	988	988	988	988	941
4	2965	1483	988	741	741	741	741	741	721
5	2965	1483	988	741	494	494	494	494	482
6	2965	1483	988	741	494	371	371	371	371
7	2965	1483	988	741	494	371	247	247	247
8	2965	1483	988	741	494	371	247	124	124

old. The nominal stand densities for the Nelshoogte experiment, established by the thinning process to represent particular spacings, are shown in Table 2.2.

In addition to the basic series of eight plots, many spacing experiments included a series of plots in which a variety of thinning treatments were applied. The difference was the age at which the thinning operations commenced and the degree of release in terms of stem number and volume of timber removed.

2.2.2 Short-Rotation Coppice Experiments

The upsurge in the demand for timber as a source of energy requires specific experimental designs for developing productive bioenergy plantations, and typically involves willow, poplar, and other fast-growing tree species. Food, energy, and biodiversity conservation compete on a limited terrestrial surface. The use of plant products for energy must therefore be placed in the context of multiple competing uses (Bauhus et al. 2010).

Different tree species, different environmental conditions, and silvicultural treatments can produce diverse volume-age relations, which is an indication of the uncertainties that still exist despite the great number of short-rotation yield studies that have been conducted. For this reason, new field trials are often essential (Liesebach et al. 1999; Heaton 2000; Hofmann 2005). The methods adopted in these studies are similar and usually include a full count of shoots in the research plot, measurements of shoot lengths and diameters at a convenient reference height, and destructive samples of biomass. These measurements can be processed to derive estimates of numbers of stools and shoots, the distribution of shoot basal diameters, lengths and biomass, as well as total standing biomass for the area of interest. It is important to note that, in contrast to traditional yield studies, both the shoot age and the root age determine production rates in coppice plantations. This requires particular experimental designs.

Szczukowski et al. (2002) reported an increase in the average annual biomass production in their Kwidzyn experiment in northeastern Poland, from 14.1 t dry mass per hectare in the first year to 16.1 t dry mass per hectare in the second year

Table 2.3 Yields (in ton dry mass per hectare) for *Salix viminalis x gigantea* clone 1047, at different root and shoot ages, according to Szczukowski et al. (2002) and Walotek et al. (2006). Estimates (brown) are based on results for experiments Kwidzyn (blue), Zamosc (green), and Jelenia Gora (yellow)

shoot	root age						
age	1	2	3	4	5	6	7
1	15.7	13.5	11.3	15.1	18.9	20.4	21.9
2		14.8	12.2	20.3	25.4	27.4	29.4
3			19.0	25.5	31.9	34.4	36.9
4				43.2	54.1	58.4	62.6

and 21.6 t dry mass per hectare in the third year. The highest average annual increment was 26.4 t dry mass per hectare for *Salix viminalis*. The live root system stores carbohydrates and nutrients and makes deeper soil regions accessible; it is therefore vital for shoot growth. Table 2.3 shows where information about combinations of root age and shoot age yield was lacking for the Kwidzyn, Zamosc, and Jelenia Gora experiments established for the *Salix viminalis x gigantea* clone 1047 in Poland.

The growth of coppice shoots is generally much more vigorous than the growth of planted individuals of a given species at the same age and on the same site. To obtain realistic and useful information about the production potential of coppice plantations, it is therefore necessary to assess all possible combinations of root age and shoot age for one clone, one site, and one planting density. This approach facilitates estimation of aboveground biomass production rates, but requires large experimental areas and extended observation periods.

2.2.3 Ecosystem Functioning and Biodiversity Experiments

Monospecific plantations are known to be vulnerable to abiotic and biotic disturbances. For this reason, experiments are designed to evaluate planted forests that are diverse in genotypes, species, and structure (Bauhus et al. 2010). Accordingly, several new planted experiments were established during recent years with a focus on evaluating the functional significance of tree species diversity.

Forest biodiversity/ecosystem functioning (BEF) experiments that are based on concepts developed for grassland ecosystems can evaluate the response to tree species loss. Plots of high species richness are compared to plots from which species have gradually been eliminated. Bruelheide et al. (2014) describe an experiment that started in 2009 at the subtropical forest site Xingangshan in Jiangxi province in China. The research focus was on measuring carbon and nitrogen storage, nutrient cycling, and prevention of soil erosion. The main experiment was established at two sites, covered 50 ha in total, involved 566 plots of 400 trees each, and ranged in diversity from monocultures to mixtures of 24 species. The experiment used 42 native tree species and 10 shrub species. More than 400,000 tree and shrub saplings were planted.[1] The design of the biodiversity/ecosystem functioning experiment favored many small plots rather than a few large plots.

One criticism referred to the fact that, because of unavoidable biological correlations which can be seen as "hidden treatments," it is usually not possible to design studies where diversity is completely orthogonal from other variables (Huston 1997). Furthermore, the sampling effect implies that more diverse communities are more likely to contain high-performing species, which can generate a positive relation between biodiversity and ecosystem functioning if such species are dominant within the community (Tilman et al. 1997).

Different species combinations are planted in Europe to study the effects of species richness on long-term C-sequestration and nutrient cycling. The "functional significance" of tree species diversity is evaluated by mixing specific traits, such as leaf type, light requirement, growth rate, and litter C/N ratio (Bauhus et al. 2010; Verheyen et al. 2013). The BIOTREE experiment, initiated in 2004 in Thüringen, Germany, consists of three sites with groups of plots on former grasslands with contrasting soil types and different combinations of tree species (Scherer-Lorenzen et al. 2007). The calcareous site *Mehrsted* includes 40 plots of 40 ha, planted with 1, 2, ... 6 species, and the acidic site Kaltenborn has 16 plots on 23 ha planted with 1, 2, ... 4 species. The Bechstedt site has 25 hexagonal plots on 11 ha and a 16 species pool. Individual plots cover 0.17 ha (Bechstedt), 0.6 ha (Kaltenborn), and 1.0 ha (Mehrstedt). Another example of a new diversity experiment is FORBIO, established at three sites in Belgium with a diversity per plot ranging from one to four tree species. Trees were planted in small monospecific patches. These patches were arranged in a checkerboard pattern in two species mixtures and randomly attributed to the species in three and four species mixtures (see Verheyen et al. 2013 for details).

A rigorous experimental design, based on very specific manipulations that are not found in the natural environment, is required to address a specific hypothesis that cannot be tested by merely using available observations. An example is the Roof Project established in the Solling Hills of northern Germany. The aim of that project was to evaluate the effects of changes in environmental conditions in a Norwegian spruce forest by manipulating the input of nutrient and water supply (Dohrenbusch et al. 2003). The response included effects on soil water chemistry and root growth. It was shown that atmospheric element input largely controlled soil chemistry in the 60-year-old spruce forest on acid soil. The particular forest exhibited a time-lagged response to a changing chemical environment, but rapid responses to drought stress.

Another example is the Höglwald experiment in Bavaria, established in a forest originally dominated by broadleaved species which had been largely replaced by spruce plantations. Three small *Fagus sylvatica* plots (aged 160, 90, and 72 years) were compared with the surrounding *Picea abies* plantations aged 90, 85, and 58 years (Kreutzer 1993; Kreutzer and Weis 1998). While the basic soil processes were

[1] http://www.bef-china.de/index.php/en/sites/home-1, accessed on October 20, 2015.

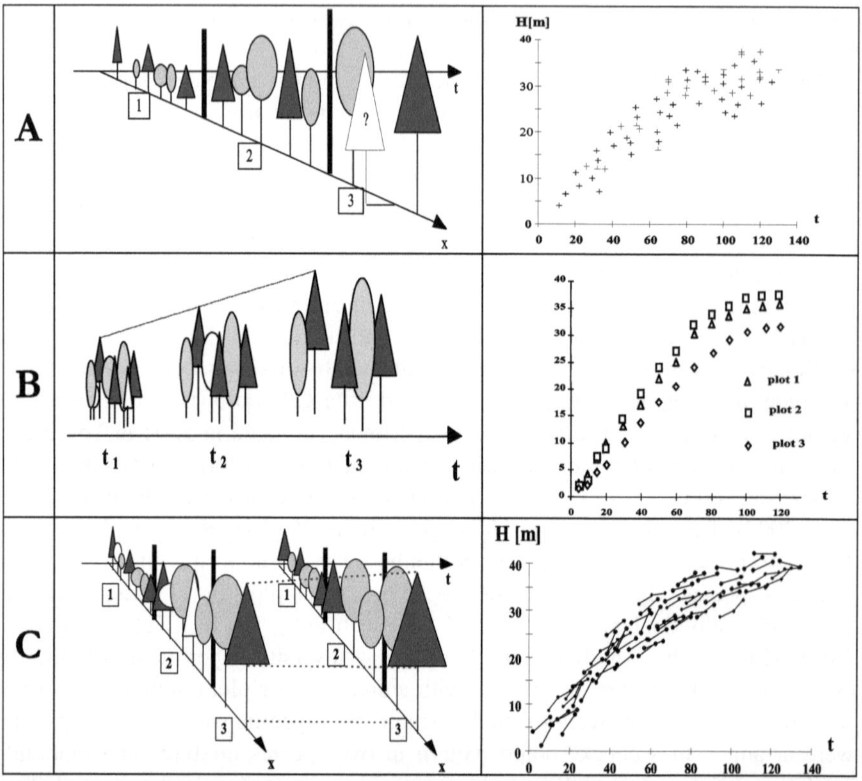

Fig. 2.21 Schematic representation of (**a**) cross-sectional study with temporary plots; (**b**) longitudinal study with permanent plots; (**c**) interval study with at least one remeasurement. Right: Hypothetical data series derived from the different plot types. White trees are removed during a thinning operation. The x-axis signifies the tree position; the symbol t indicates the time axis (see Gadow and Hui 1999; Burkhart and Tomé 2012:341)

found to be similar for both species, the flux rates were different. Nitrogen leaching was negligible in the *Fagus sylvatica* plots but important in the *Picea abies* plots. Throughfall, deposition, and soil leaching of sulfate, nitrogen compounds, and acidity were found to be significantly higher in *Picea abies* plots than in the *Fagus sylvatica* plots. The difference between the two species increased with the deposition load and was negligible in remote, relatively unpolluted areas.

2.2.4 Longitudinal, Cross-sectional, and Interval Studies

Manipulated experiments are often designed as *long-term* or *longitudinal studies*. Longitudinal studies are commonly used to describe and explain trends (Crawley 2005:180). One disadvantage of a long-term or "permanent" experiment is the high maintenance cost of the research infrastructure and the long wait for results (Fig. 2.21).

In a *cross-sectional study* (also known as a *chronosequence*), measurements from a set of temporary field plots which may cover a wide range of ages and environmental conditions are all collected at one point in time. Plots are assessed only once and results are available relatively quickly. The sequence of remeasurements in time of a longitudinal study is therefore substituted by simultaneous measurements in space. This method was used extensively during the nineteenth century (Kramer 1988, p. 97). A *cross-sectional study* provides one-time measurements on many individuals, but no growth rates. A *longitudinal study* does assess change, but because of the high cost of maintaining long-term studies, it is usually only done on relatively few plots.

A practical compromise is the *Interval Study:* plots are measured at least twice to assess rates of change between the two assessments (as in a *longitudinal study*), and at the same time, a variety of sites are assessed, as in a *cross-sectional study* (Zhao et al. 2014). *Interval studies* can therefore exploit the primary advantages of *longitudinal studies* (determining rates of change) and *cross-sectional studies* (broad coverage of site conditions and minimum wait for data). An important question concerns the balance between the number of growth intervals that should be assessed in one particular location and the range of environmental and treatment conditions that needs to be covered. In the ideal case, one growth interval provides sufficient results, and the funds that would have been required for assessing a second interval at that particular location can be spent on collecting a new set of interval data in another locality. Obviously, interval plots continue to be studied for multiple growth periods for as long as new results are expected, but ongoing critical assessment is required regarding the trade-off between abandoning a current interval plot and replacing it with a new plot, at another location, to increase the variety of initial states for which a response is required (Tewari and Gadow 2008).

2.3 Continuous Forest Observation

Observations collected over long periods of time in the same location are essential for both management and research (Sagarin and Pauchard 2012). Sustainable management requires evidence to estimate harvest potential, such as information about tree growth and the productive potential of a forest in response to changing environmental conditions. Forest ecological research depends on remeasured field plots, preferably remeasured over long periods of time and covering sufficiently large areas to assess effects of scale (Condit et al. 2013; Kenfack et al. 2014; Gadow et al. 2016a, b). Such installations represent an important research infrastructure for addressing a range of topics, including community ecology, effects of climate change, and human disturbance.

2.3.1 Early Observational Infrastructures

The first forest field plots, initiated in Europe during the nineteenth century, were established to measure tree growth on different sites in response to specific thinning treatments. Seckendorf (1881) was among the early scientists who developed many

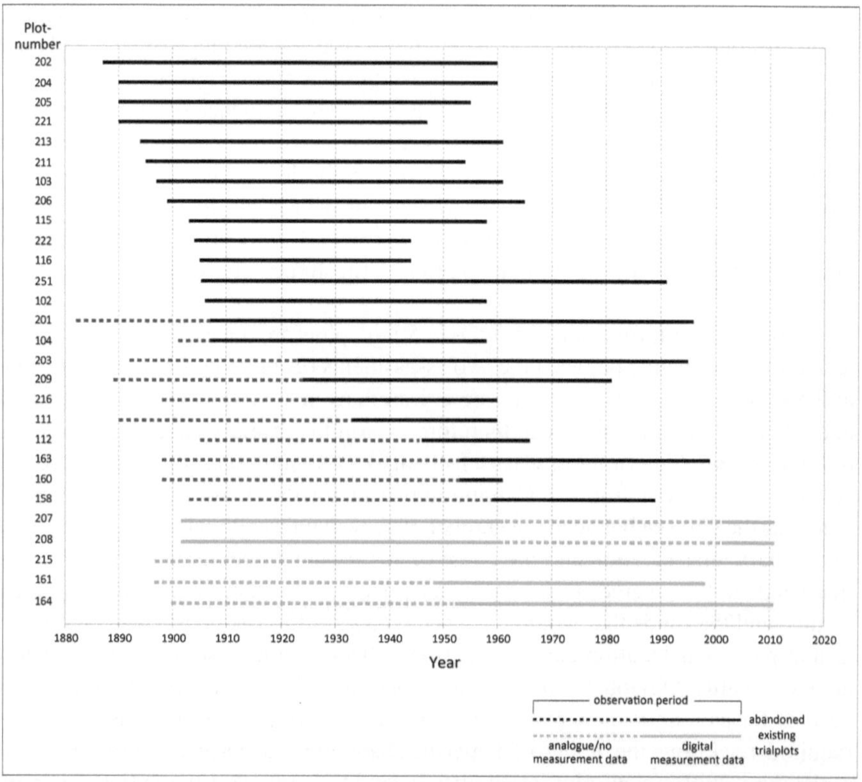

Fig. 2.22 History of the field trials of the Austrian Forest Research Center (see Kristöfel 2017). Black lines show abandoned trials, and dotted lines indicate years during which observations were either not available at all or only available on paper

of the basic concepts relating to the establishment and maintenance of permanent field plots. An overview of the field trials of the Austrian Forest Research Institute was presented by Kristöfel (2017; see Fig. 2.22).

Some of these old field plots have been remeasured for over a century, providing valuable information on long-term response to variations in climate (Assmann 1961; Spellmann et al. 1996; Pretzsch 2009). The old growth and yield trials were complemented during the twentieth century with *manipulated* field experiments that were increasingly based on statistical principles.

2.3.2 New Forest Observational Studies

Natural forests exhibit complex structures and changing spatial patterns. A designed planting of species mixtures cannot mimic the complexity resulting from local variations in tree species, tree ages, tree dimensions, and the varying spatial arrangement of these attributes that is so characteristic of a natural forest. Experimental plots are often too small to capture scale-dependent phenomena such as scale effects on community structure (Zhang et al. 2010a, b) or species-habitat associations and demographic variation (Zhang et al. 2009, 2012; Kenfack et al. 2014). It is not surprising, therefore, that forest observational studies, also known as *quasi experiments*, have become increasingly popular during recent years (Gadow et al. 2016a, b). Extensive new plot networks have been established in China and Mexico, while specific initiatives like the Global Forest Biodiversity Initiative have begun to aggregate forest observational data on a global scale.

2.3.2.1 China

A forest observational network was initiated in China in 2005 by the Beijing Forestry University (Zhao et al. 2014). After the establishment of a few plots in the Changbai Mountains, the network expanded rapidly; it currently includes observations from almost all the main natural forest ecosystems in China. Only 3 of the 19 plots of the network are smaller than 4 ha, while 11 plots have areas between 10 ha and 30 ha. All the trees and shrubs exceeding 1 cm in breast-height diameter are tagged and mapped in each field plot (Zhao et al. 2012). Large plots provide a rich set of observations, with new insights emerging, such as unexpected competition effects (Seifert et al. 2014), patterns of density and various effects of scale on community structure and diversity (Zhang et al. 2015), and new findings related to the productivity/diversity relation (Hao et al. 2018; Cheng et al. 2018).

Observations are not limited to tree measurements but include a variety of particular assessments, usually extending over several growing seasons. Examples are studies of seedling survival within specific neighborhoods (Wang et al. 2013a, b; Chen et al. 2013; Yan et al. 2015) and repeated assessment of the flowers and fruit of dioecious trees and shrub species to study their reproductive allocation (Wang et al. 2014). Soon after the first installations and especially after remeasurements became available, new initiatives were started, including a study of spatial associations of dominant tree species in the Changbai Mountains (Zhang et al. 2012); neighboring competition and habitat effects of a dioecious tree species (Zhang et al. 2009); new methods of quantifying ecosystem structure and diversity (Ni et al. 2014); and studies involving species–habitat associations (Zhang et al. 2012).

Another famous observational network is CForBio, a network for monitoring forest biodiversity in China (Ma 2015). The Chinese Forest Biodiversity Network was started to study the maintenance of biodiversity in China's forest ecosystems. Feng et al. (2016) summarized the research progress in CForBio in their review of the network and also made suggestions for future research. The research activities based on CForBio focused mainly on ecological factors, such as environmental filtering, biotic interactions, small-scale dispersal limitation, and the effects of climate change on forest community composition. China has tropical, subtropical, temperate, and boreal forests. Temperate forests in China, Europe, and North America have a similar species composition and community structure. However, temperate forests in China have a higher taxonomic diversity and contain more unique paleo-endemics than the forests in Europe and North America. All the biodiversity hotspots in China for both plants and animals, as well as their endemic and threatened components, are located in mountainous areas (Tang et al. 2006).

2.3.2.2 South Africa

The unique afro-temperate ecosystems, which are part of the afromontane group of forests (White 1978), are found on the coastal platform and foothill zone of the Outeniqua and Tsitsikamma Mountains in South Africa, at elevations between 190 m a.s.l. and 520 m a.s.l. (Seydack et al. 2012). Breitenbach's research (1974) was based on earlier studies by Phillips (1931) and Laughton (1937) and recognized 98 tree species in six forest types. Observational study sites were established in 1972 in the Diepwalle research area in medium-moist and moist high forest types by officers of the Saasveld Forest Research Centre of the then Department of Forestry. The plots were remeasured in 1978. Further remeasurements, including mapping of tree positions, took place in 1987, 1997, and 2007 under the initiative and jurisdiction of the Scientific Services Section of the Department of Water Affairs and Forestry and, since 2006, of the South African National Parks. Repeated DBH measurements and mapped tree coordinates are therefore available for a period of 43 years.

All tree species were identified in the Diepwalle research area. Fourteen species which had been observed to reach the forest canopy in an undisturbed forest were designated as "canopy species." One of these, *Podocarpus falcatus*, sometimes occurs as a very tall individual, a so-called emergent. Eleven species, identified as subcanopy species, never reach the canopy in an undisturbed forest. Five of these are "shrublike" subcanopy species that are difficult to distinguish from shrubs (Seifert et al. 2014). A way to simplify a species-rich ecosystem is to develop functional groups, for example, by height-growth ordination, by differentiating between canopy and permanent subcanopy species, and by subdividing the canopy group into mature and immature individuals. A specific cluster process which involves a segregation in the relations between diameters and heights was applied by Lujan-Soto et al. (2015) to identify species-specific DBHs separating mature and immature canopy trees.

2.3.2.3 India

Biogeographically, India is situated at the tri-junction of three realms, namely the Afro-Tropical, Indo-Malayan, and Paleo-Arctic realms, and therefore has characteristic elements of each of these realms. This location makes the country rich and unique in biological diversity. India's natural ecosystems include evergreen tropical rain forests on the Andaman and Nicobar Islands, the Western Ghats, and the northeastern states; dry alpine scrub forests in the northern Himalayas; semi-evergreen rain forests, deciduous monsoon forests, thorn forests, and subtropical pine forests in the lower montane zone; and temperate montane forests (Champion and Seth 1968; Lal 1989).

Forest observational studies in India are known as *Linear Tree Increment Plots*, *Linear Sample Plots*, and *Permanent Preservation Plots*. A common characteristic of these installations is the fact that tree diameters are measured at periodic intervals. The *Hopea parviflora* forests in the south and the *Shorea robusta* forests in the north seem to have been the first locations where observational studies were started during the early years of the twentieth century (for details, see Tewari et al. 2014c). The oldest Preservation Plot was established in 1905 in the *Shorea robusta* forests of Bihar (Kaul et al. 1975). The available records show that there were 309 Preservation Plots in the country, of which 187 were located in natural forests and 122 in plantations. It is not always known what happened to these old plots. Many were abandoned and are very difficult to relocate. However, the Indian Council of Forestry Research and Education (ICFRE) intends to "revive" these early field plots for permanent observation.

In recent years, several new observational studies were established by staff of the Indian Council of Forest Research and Education: two in Bangalore within the grounds of the Institute of Wood Science and Technology, two in Kaylana near the Arid Forest Research Institute in Jodhpur, one by the staff of the Rain Forest Research Institute in the Gibbon Reserve near Jorhat in Assam, and another one at the Himalayan Forest Research Institute in a coniferous forest in Shimla. These plots represent pilot studies and demonstrate different methods of field assessment and quantitative analysis. The Forest Survey of India also established several new and permanent field plots in 2017 (Gadow 2017).

2.3.2.4 Mexico

The forests of the Sierra Madre Occidental in Durango, Mexico, cover an area of approximately 5 million hectares in total. These forests are unique. They contain 27 coniferous tree species, among them 20 different species of *Pinus* and 43 species of *Quercus* (González et al. 2007). Unevenly aged and selectively harvested pine-oak communities are predominant (Aguirre et al. 2003; Wehenkel et al. 2011; Corral-Rivas et al. 2012). The sustainable use of this socio-ecological system requires effective monitoring to identify optimum levels of stocking. Accordingly, an extensive network of observational field plots was established between 2007 and 2018 on the Sierra Madre Occidental. Each of these study sites covers an area of a quarter hectare (50 × 50 m). More than 1400 plots have already been remeasured at least once.

The unevenly aged communal forests of the Mexican Sierra Madre Occidental have been subject to Continuous Cover Forestry (CCF) with selective harvesting for more than a century. To ensure that harvests are sustainable, specific monitoring systems were developed which are based on those developed for the Swiss Plenter Forests. Sustainability in CCF management systems implies that the harvested timber volume is controlled. It also implies that the structural disturbance caused by harvesting events is minimized. Structure is central to CCF silviculture because it determines the current output of products and services as well as the resilience and future dynamics of the managed ecosystem. Pioneering studies were done a few years after the first installation of the FOS in Durango, for example, the development of indices to describe forest spatial structure (Corral-Rivas et al. 2010); new methods of quantifying forest structure differences between ecosystems (Wehenkel et al. 2014); the development of local and generalized height–diameter relations in mixed unevenly aged forests (Crecente-Campo et al. 2014); studies involving the grouping of tree species (Lujan-Soto et al. 2015); modeling diameter distributions of natural forests (Corral-Rivas et al. 2016); estimation of the biomass of mixed and unevenly aged forests by using remote-sensing data (López-Serrano et al. 2015); and the effect of density and structure on production in the communal forests of the Mexican Sierra Madre Occidental (Corral-Rivas et al. 2018).

2.3.2.5 Global Forest Observation

Many of the existing forest field plots represent major investments and have become part of international and multidisciplinary platforms involving scientists with different backgrounds and traditions (Peterson and Monserud 2002; Mårell and Leitgeb 2004). The following list includes some prominent examples of North American, European, and international forest research networks:

- MAB (Man and Biosphere Reserves), a global network which includes 651 biosphere reserves in 120 countries (http://www.unesco.org/new/en/natural-sciences/environment/ecological-sciences/man-and-biosphere-programme/)
- LTER (Long-term Ecosystem Research and Monitoring), a platform established in 1980 in the United States for long-term ecosystem research and monitoring with linkages to national as well as global research activities (www.lter.edu)
- ILTER, an International Long-Term Ecological Research network which coordinates the maintenance of a large number of long-term datasets (www.ilternet.edu/)
- EuMIXFOR, a research network supported by the COST framework
- TreeDivNet, a platform for tree diversity experiments worldwide (Verheyen et al. 2013)
- FunDivEUROPE, a network of forest plots along tree species diversity gradients in six major European forest types (Baeten et al. 2013)

The Global Forest Biodiversity Initiative (GFBI)[2] is a relatively new nongovernment multistakeholder research collaborative that provides a platform for consolidating observations from permanent forest field plots across the world. Table 2.4 presents some examples of forest observational networks that have become part of GFBI.

According to Kafatos and Eisner (2004), many of the great "research triumphs" are products of interdisciplinary inquiry and collaboration. Attracting scholars from a variety of disciplines is therefore an important long-term objective of many forest observational studies. The potential for interdisciplinary research can be facilitated by integrating methods, concepts, and data from several disciplines (Ausburg 2006). Such efforts can advance fundamental understanding or help to solve problems where solutions are beyond the scope of a single discipline.

2.3.2.6 Machine Learning

Artificial intelligence (AI) is a general term that refers to the creation of human-like machine intelligence that can learn, reason, plan, perceive, and even process natural language. The ultimate aim in developing AI applications is to solve problems in a better and faster way than humans do. Rule-based reasoning, decision tree learning, expert systems, and Bayesian networks were popular approaches associated with the first applications of artificial intelligence (Winston 1977; Gadow 1991; Reynolds and Schmoldt 2006). The development of neural networks, which are designed to work by classifying information in the same way a human brain does, allowed for a real breakthrough. AI is now omnipresent and performs all kinds of tasks, such as moving

[2] Refer to http://www.gfbinitiative.org (accessed on January 10, 2019)

Table 2.4 Characteristics and references of some existing forest inventory networks that were established before 2000 and deposited at the GFBI global database

Region	Year	No of plots	References
North, Central, and South America			
Alaska, USA	1994–2015	2,024	Malone & Liang (2009)
NFI, USA	1968–2014	591,429	O'Connell et al. (2014)
NFI, Canada	1992–2007	979	Gillis et al. (2005)
Canada	1970–2014	338,823	Paquette & Messier (2011)[4]
Mexico	1984	675	Bongers et al. (1988)
Canada	1958–2013	9,085	Chen & Luo (2015)
Chile	1981–2014	498	Salas et al. (2006)
French Guyana	1984–2015	1,456	Hérault & Pétronelli (2017)
Bolivia	1998–2010	402	de-Miguel et al. (2014)
Eurasia			
Russia	1970, 1973	1,682	Tchebakova et al. (2016)[31]
NFI, Spain	1980–2008	64,654	DGCN (2008)
Estonia	2012–2016	300	Kiviste et al. (2015)
China (big plots)	2007–2015	33	Zhang et al. (2015)
Africa			
South Africa	1971–2007	7	Gadow et al. (Gadow et al. 2016a, b)
Oceania			
Australia (QLD)	1974–2014	541	Muhairwe (2006)
Australia (NSW)	1977–2009	2,236	Unpublished

around, speaking, recognizing objects and sounds, and detecting patterns and similarities in unlabeled data (Hagan et al. 2014). Google uses machine learning to filter out spam messages from Gmail. Computers are trained to identify human faces. Deep learning is used by Netflix and Amazon to decide what people want to watch or buy.

Over the course of just a few years, machine learning and especially deep learning revolutionized several research fields, such as bioinformatics or medicine. These tools are increasingly applied in ecology because of the ever-growing size and complexity of ecological datasets. Christin et al. (2018) present a literature review of the implementation of deep learning in ecology. Hu et al. (2018) used a neural network based on Landsat-8 Operational Land Imager imagery. This network was based on a hierarchical sampling strategy to realize an automatic construction of training datasets and performs automatic classifications of remote-sensing images from Qinhuangdao City, Hebei Province. Their "deep convolutional neural network" provided better classification results than traditional maximum likelihood methods.

Machine learning is an area with a great potential regarding forest assessment. Rather than programming computers with a specific set of instructions, the machine is given a large number of training data, such as a combination of remote data (LIDAR, free Landsat, or sentinel-2) and field measurements. Machines can be taught to learn with or without supervision and to detect, predict, and identify patterns. For deep learning, most codes are available in Python, but it is not easy to use R for deep learning. Of particular interest regarding forest ecosystems is automatic identification of tree species (Zhang et al. 2008). Identification of tree species in species-rich permanent field plots by remote sensing is a relevant example. By being

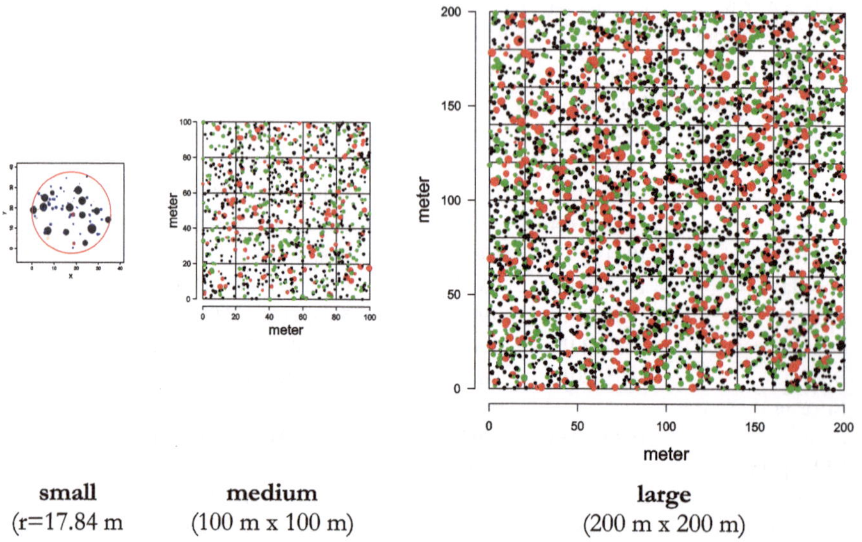

Fig. 2.23 Three standard POP designs, see main text for details

shown many labeled images with instances of the crown structure, bark, or the leaves of a specific tree species, the machine can be made to shape its own rules for deciding whether a particular set of digital pixels signifies that species. To our knowledge, this potential has not been explored yet but might become very relevant, especially when the deplorable shortage of experienced field taxonomists is taken into account.

2.3.3 Establishing an Observational Field Plot

Three standard field plot designs were recommended to be used as permanent observational plot installations in India (Gadow 2017; Fig. 2.23).

Small circular plots of 0.1 ha are useful in broken terrain. Small plots are conveniently established in different forest types to capture gradients, for example, variations in vegetation in response to altitude, soil conditions, and human disturbance; "medium-sized" monitoring and research plots of 1 ha allocated to different forest types to capture detailed forest structures and dynamics and to observe changes through time; and "big" monitoring and research plots of 4 ha each allocated to major forest types for comprehensive ecological analysis of scale-dependent phenomena (Cheng et al. 2018; Condit et al. 2013).

It is estimated that the initial establishment of one small, medium, or large plot by an experienced field crew requires, roughly and on average, a day, a week, and a month, respectively. The observations provide a unique basis for comprehensive ecological analysis, preferably in cooperation with research groups specializing in the biological and physical sciences. Each permanent field plot is assigned a unique identifier (e.g., two digits for State plus two digits for district plus a consecutive number). The GPS coordinates and altitude above sea level of the plot reference corner (or center) are recorded. If the inclination between two quad corner posts is nonzero, a distance correction is necessary, because the standard 20 m distance refers to horizontal distance. When measurements of a circular plot established on a slope with an angle of α degrees are projected on a horizontal plane, they generate an ellipse with its longest and shortest axis being r_s and $r_h=r_s*\cos(\alpha)$. The area of the ellipse is:

$$a_h = \pi \cdot r_s^2 \cdot \cos(\alpha) = a_s \cdot \cos(\alpha) \quad (2.9)$$

where a_s is ground area on the slope; a_h is the projected area on the horizontal plane; r_s is the radius of the circular plot measured on the slope and the longest axis of the ellipse (Van Laar and Akça 2007). When slope is measured as a percentage, conversion to degrees is required. For example, when a circular plot with area 0.05 ha is established on a slope with an angle of 45%, the angle α is $\tan^{-1}(0.45)=24.23°$, and the correction factor is 0.912. The application of a single correction factor, based on average slope, may give biased estimates of area-based variables. It is therefore more practical to lay out a circular plot using a radius that has been corrected for slope.

Several digital photographs should be taken at positions which are marked for easy relocation. A map which facilitates the relocation of the plot reference point should then be drawn. In addition, certain qualitative parameters may be assessed in the vicinity of the plot reference point, for example, land use, legal status, slope, aspect, soil type, regeneration, and human disturbance. The accuracy of area-based variables depends especially on the exact measurement of the plot area. Establishing field plots therefore requires great care.

The decision on plot size is crucial because many relations are scale-dependent. An example is the study by Chisholm et al. (2013) which shows that the relation between tree diversity and ecosystem functions is affected by plot size. Scale dependence, involving innovative methods such as principal coordinates of neighboring matrices (Borcard and Legendre 2002; Griffith and Peres-Neto 2006; Zhang et al. 2010c), requires large plots that can be subdivided to test effects of cell size. For this reason, a small number of large POP should ideally be included into the new design. A forest observational study differs from a field experiment in that it can be compared to a warehouse which holds a vast amount of information about complex forest communities. The aim of an observational study is to study the contents of the "warehouse" without manipulating the system in any way. An ecological community is defined by species, structure, and density, all of which provide specific ecosystem functions and services (Costanza et al. 1997; Daily 1997). Change over time can be monitored more effectively in communities than in individual species, as composition, including species' relative abundance and the proportion of invaders, can be assessed in a community. The structure and diversity of a community is an important indicator of its productivity (Noss 1990).

2.3.4 Assessment Protocols

This section introduces an assessment protocol for a standard 1-hectare square plot. Details for circular plot designs have already been presented by Gadow (2017) and are not repeated here. Each permanent Forest Observational Plot is identified by a

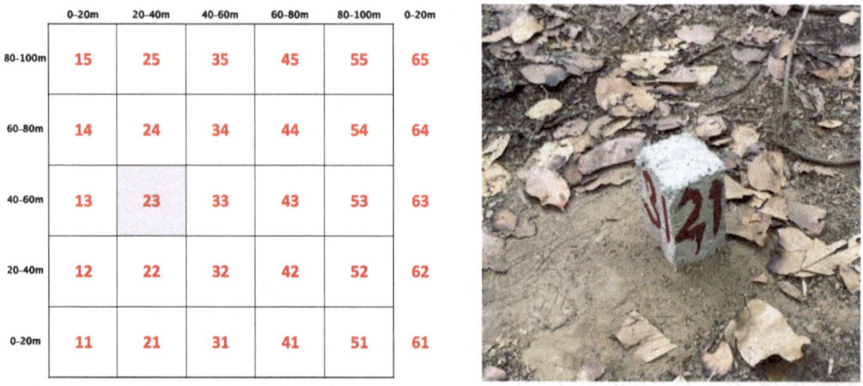

Fig. 2.24 Left: design of a one-hectare plot subdivided into 25 quadrats. Each quadrat measures 20 × 20 m. Quadrat numbering starts at the (south-west) origin; the first digit refers to the column (x-value) and the second digit to the row (y-value). Cell number 23, for example, is located in column 2 and row 3. The origin of quadrat 23 has x-value=20 m and y-value=40 m. Right: Example of an 80-cm pillar used to mark a quadrat corner. The painted numbers refer to adjoining quadrats

geographical reference. The plot center defines the reference coordinates for a circle and the southwestern corner those for a rectangular plot. The plot description includes a number of the attributes of the plot reference point:

Plot name (a unique identifier, e.g., referring to region, plot type, and plot number).
The GPS coordinates of the plot reference point.
The altitude above sea level of the plot reference point (m).
Three digital photos taken from three positions that are marked for easy relocation.
A map to facilitate the relocation of the plot reference point.

In addition, certain qualitative parameters can be assessed in the vicinity a geographical reference point of the plot, for example, land use, legal status, slope, aspect, soil type, regeneration, and human disturbance. The accuracy of all area-based variables depends on the exact measurement of the plot area. As mentioned before, the establishment of field plots therefore requires great care.

To facilitate and standardize data analysis, it is convenient to orientate the boundaries of a rectangular plot on the north–south axis. The GPS coordinates of the plot origin are recorded and that point is permanently marked with a concrete or stone pillar.[3] It is convenient to subdivide the standard 1-hectare square plot into a regular grid of 20 × 20 m quadrats (Fig. 2.24).

This requires a device which measures distance and direction, preferably using a tripod for stability. The twenty-five quadrats (*qads*) are numbered. A convenient numbering scheme for a hypothetical one-hectare plot and its 25 quadrats is

[3] In India, where wild elephants can be found in some forests, a 10 cm x 10 cm x 80 cm concrete or stone pillar is recommended (Gadow 2017). The pillar should be buried 60 cm in the ground with only 20 cm visible above ground level to make it "elephant proof." PVC pipes melt in a bush fire and are not recommended.

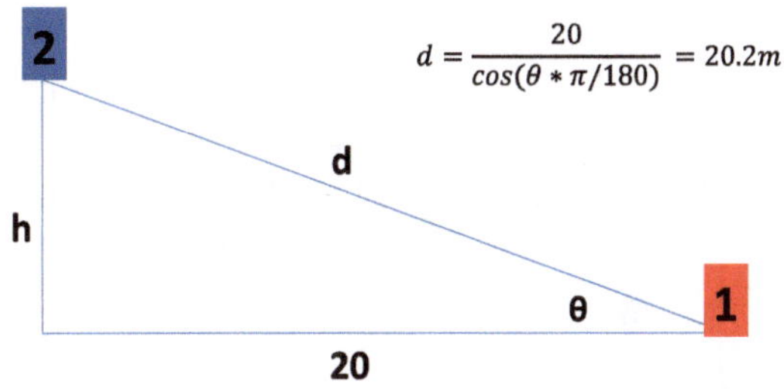

Fig. 2.25 Hypothetical example to demonstrate the calculation of the distance between two quad corners if the inclination is greater than 0 degree, and the resulting difference in the altitude between quad corner 1 and quad corner 2

presented in Fig. 2.24. The altitude above sea level is assessed for the plot reference point with an altimeter or an appropriate app installed on a cell phone. The plot location can be abandoned if there is no feasible alternative within a 100-m radius. A site should, however, not be avoided because of difficult vegetation. Intermediate checking of distances between lines ensures that all south–north lines are exactly parallel. The grid of 20 × 20 m quadrats is a permanent installation, and it should therefore be easy to relocate the quadrats. Radiofrequency identification (RFID) tags, protected within weatherproof capsules, can also be used to mark quadrat corners. A tape and compass (or preferably a Theodolite or "Total" device) are used to measure directions and distances. The standard 20-m distance between quad corners refers to horizontal distance. If the inclination between two quad corner posts is nonzero, a distance correction is therefore necessary (Fig. 2.25).

In Fig. 2.25, the distance d between the two quad corners 1 and 2, with inclination θ=8°, is equal to 20.2 m, which allows the calculation of the difference in altitude h (=2.81 m) between quad corner 1 and quad corner 2.

Two different methods can be used to determine the x–y coordinates of all trees in a 20 × 20 m quadrat. In areas where the view is obstructed by dense vegetation, the quadrat can be subdivided into 16 square cells of 5 m side length. Cell corners are usually not permanent. A 5-m light bamboo stick with markings 1 m apart can be used to measure tree coordinates in the 5-m cells (5-m cell method; see Fig. 2.26). It is important to make sure that the stakes are distinguishable from the permanent quadrat corner stakes.

When all the trees in a quadrat are visible from the origin corner, as can be the case in a forest with big trees and little undergrowth, the distances from the origin should be measured with a tape or rangefinder; the angles can usually be measured with a compass or cell phone app (or a Theodolite). The distance and angle measurements can be converted to x–y coordinates (refer to relevant R code).

Only five variables are assessed for each tree (with DBH >=5 cm) in a quadrat: tree species; x-coordinate; y-coordinate; tree DBH or girth at breast height; and total height. A very important variable is the species of each individual tree. It is often necessary to ask an experienced taxonomist to identify all the tree species in the field. Measuring tree diameters and heights is trivial when compared with the critical importance of correct species identification. The following variables are assessed in each quadrat (recommended variable names are written in brackets) (Table 2.5):

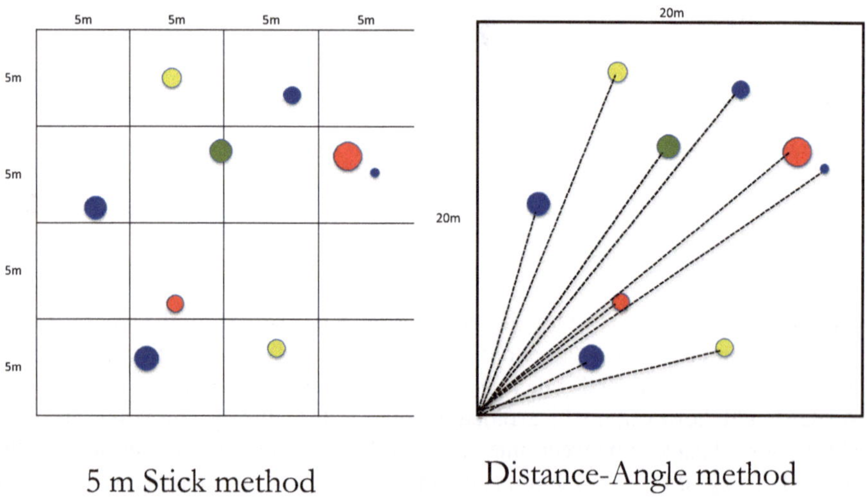

Fig. 2.26 Two methods for determining the x–y coordinates of all the trees in a hypothetical 20 × 20 m quadrat. Left: In dense forest, each quadrat is subdivided into 16 square cells of 5 m side length, and a 5-m light bamboo stick with 1 m markings can be used for easy measurement of the tree coordinates in cells (5-m cell method). Right: Measurement of distances and angles when all the trees can be seen from the quadrat origin (distance-angle method)

Table 2.5 Variables for trees assessed in two quadrats

5-m stick method	Distance-angle method
Quad origin x-value (Quadx0)	Quad origin x-value (Quadx0)
Quad origin y-value (Quady0)	Quad origin y-value (Quady0)
Quad ref. altitude (m) (Quadasl)	Quad ref. altitude (m) (Quadasl)
Cell origin x-value (cell_x0)	Angle from north (theta)
Cell origin y-value (cell_y0)	Distance from quad origin (distance)
Tree number (tree_No)	Tree number (tree_No)
Tree species (species)	Tree species (species)
Azimuth, angle from north (angle_degr)	Azimuth, angle from north (angle_degr)
Distance from plot center (distance_m)	Distance from plot center (distance_m)
Tree girth at breast height (girth_cm)	Tree girth at breast height (girth_cm)
Tree height (height_m)	Tree height (height_m)
Remarks (remarks)	Remarks (remarks)

The data are stored in an Excel table and should be carefully checked for errors. The table is then saved as a .csv or .txt file, which can be read into an R workspace. If available, portable electronic data recorders should be used to avoid errors caused by data transfer from the field sheet to a digital device.

One way to check data accuracy is to perform a systematic remeasurement of randomly selected quadrats from within the completed plot. A remeasurement team remeasures all trees in those quadrats without knowing the first measurements. The remeasurement team should not be the same as the team which carried out the first measurements (Gasparini and Tabacchi 2009). No team should know in advance which quadrats will be remeasured. The simplest approach for choosing remeasurement quadrats is to draw five quadrats at random in each one-hectare plot (20 quadrats in each four-hectare plot) after the first tree measurements have been completed.

2.3.5 Presentation Standards

Some of the advantages of standardization are improvement in technical communication and simplification of work flows. These can include standardized summaries and visualization standards, with standard species codes and colors, and standard maps that indicate the plot location in the vicinity of roads or other prominent landscape features.

2.3.5.1 Species Codes and Colors

It is convenient to establish standard color codes for color graphics of a multispecies forest. Table 2.6 is an example of a color scheme for the Diepwalle observational study in an afromontane forest in South Africa.

Each tree species is associated with a standard species code and three species names (English, Afrikaans, and Latin). The attribute "layer" refers to the vertical layer which a species usually occupies at maturity. The species color is the standard color used in all plot maps and other graphical representations of the study area.

Colors in R can be specified either by name (e.g., col = "red") or as a hexadecimal RGB triplet (such as col = "#FFCC00"). The list in Table 2.6 includes common local names (English and Afrikaans), current botanical names, and color codes for the FVC observational study "Diepwalle" in South Africa. The following R-function *mycolors* assigns a hexadecimal color code to each tree species code (the first column in Table 2.6):

```
mycolors <- function (pl) {plx <- factor(pl)
varvid <- list("#FF7215"="Ass","#C521C5"="BB", "#C521C5"="BH",
"#00FAFA"="BYH","#141414"="Btol",
"#00008B"="FYH","#00008B"="Geel", "#800000" = "HP", "#414141" = "KE", "#5AAABE" = "KH",
"#0000FF" = "Kal","#00FF00" = "Kam","#8055FF" = "Kers","#FF8080" = "Kw",
"#00CD00" = "N","#BEBEBE"="RH","#EE82EE" = "SH","#526600"="Saf","#FF800066"= "Sw",
"#FF000033"="Sy","#FFD700"="V","#414141"="VRH","#00EB00"="WE","#414141" = "WG",
"#FF0000" = "WH","#B22222" = "WP","#414141"="WKP","#660080" = "YH","#CDCD89"= NA)
levels(plx) <- varvid; if (length(plx[is.na(plx)]) > 0) {plx[is.na(plx)] <- "#CDCD89"};
return(as.vector(plx))}
```

Example: the statement mycolors("Kam") returns "#00FF00". Adjectives like "blue" or "lightblue3" can also be used instead of the hexadecimal codes. The R function colors() will show all the 657 built-in color names.[4]

2.3.5.2 Plot Maps

A plot map provides a visual impression of an observational study that reveals possible inconsistencies and obvious errors. Figure 2.27 shows an example of the

[4]Useful references are https://www.nceas.ucsb.edu/~frazier/RSpatialGuides/colorPaletteCheatsheet.pdf and the palette on http://www.stat.columbia.edu/~tzheng/files/Rcolor.pdf.

Table 2.6 Standard species codes, species names, layer specialists, and species colors for the FVC observational study Diepwalle in South Africa

Code	Afrikaans	English	Latin	layer	color	hex
Ass	Assegaai	Assegai	*Curtisia dentata*	canopy		#FF7215
BH	Kaapse Boekenhout	Cape beech	*Rapanea melanophloeos*	canopy		#C521C5
Btol	Swartbas	Bladder-nut	*Diospyros whyteana*	subsmall		#141414
BYH	Duine-olienhout	Dune olive	*Olea exasperata*	sublarge		#00FAFA
Geel	Opregte geelhout	Real yellowwod	*Podocarpus latifolius*	canopy		#00008B
HP	Hardepeer	Hard pear	*Olinia ventosa*	canopy		#800000
Kal	Outeniekwageelhout	Outeniqua yellowwod	*Podocarpus falcatus*	canopy		#0000FF
Kam	Kamassi	Kamassi	*Gonioma kamassi*	sublarge		#00FF00
KE	Klipels	Rock alder	*Canthium mundianum*	subsmall		#414141
Kers	Kershout	Candlewood	*Pterocelastrus tricuspidatus*	canopy		#8055FF
KH	Koolhout	Coalwood	*Lachnostylis hirta*	sublarge		#5AAABE
Kw	Kwar	Quar	*Psydrax obovata*	canopy		#FF8080
N	Notsung	Tree Fuchsia	*Halleria lucida*	sublarge		#00CD00
RH	Kaapse Rooihout	Cape plane	*Ochna arborea*	subsmall		#BEBEBE
Saf	Gewone Saffraan	Common saffron	*Eleodendron croceum*	sublarge		#526600
SH	Stinkhout	Stinkwood	*Ocotea bullata*	canopy		#EE82EE
Sw	Kaapse Swarthout	Cape blackwood	*Maytenus peduncularis*	canopy		#FF800066
Sy	Witsybas	White sily bark	*Cassine eucleiformis*	canopy		#FF000033
V	Bosvlier	Forest elder	*Nuxia floribunda*	canopy		#FFD700
VRH	Vaderlandsrooihout	Wild peach	*Kiggelaria africana*	subsmall		#414141
WE	Witels	White alder	*Platylophus trifoliatus*	sublarge		#00EB00
WG	Wildegranaat	Wild pomegranate	*Burchellia bubalina*	subsmall		#414145
WH	Without	Cape Holly	*Ilex mitis*	canopy		#FF0000
WP	Witpeer	White pear	*Apodytes dimidiata*	canopy		#B22222
YH	Ysterhout	Ironwood	*Olea capensis macrocarpa*	canopy		#660080

FVC789, a combination of three subplots (called vireés) of the FVC observational study Diepwalle in South Africa (a detailed description of the study area can be found in Gadow et al. 2016a, b). FVC789 was subdivided into 3 × 20 = 60 quadrats, each quadrat measuring 20 × 20 m. The aim was to identify the local effects of density and species richness on production. The plot maps in Fig. 2.27 show the first and the last of the five enumerations between 1972 and 2007 in FVC789.

The following R code was used to generate Fig. 2.27 with the 20 × 20 m quadrats:

```
hx<-400; hy<-60; plotarea<-hx * hy                # plot dimensions and total area
nx<-20; ny<-3; cellx<-hx/nx; celly<-hy/ny         # number of cell row and column partitions
cellarea<-hy/ny * hx/nx
dev.new(width=12, height=4)                       # make sure plot window is of this size
plot(0,0,type="n",axes=F,xlim=c(0,nx),ylim=c(0,ny),xlab="X 1987",ylab="Y 1987")
for (iy in 1:ny){for (ix in 1:nx){polygon(c(ix-1,ix,ix,ix-1),c(iy-1,iy-1,iy,iy),border="red", ,density=NA,col=0)}}
points(FVC789$x/cellx, FVC789$y/celly,cex=FVC789$D72*0.03, col=mycolors(FVC789$SANPARKS_Code),) #plot the trees
```

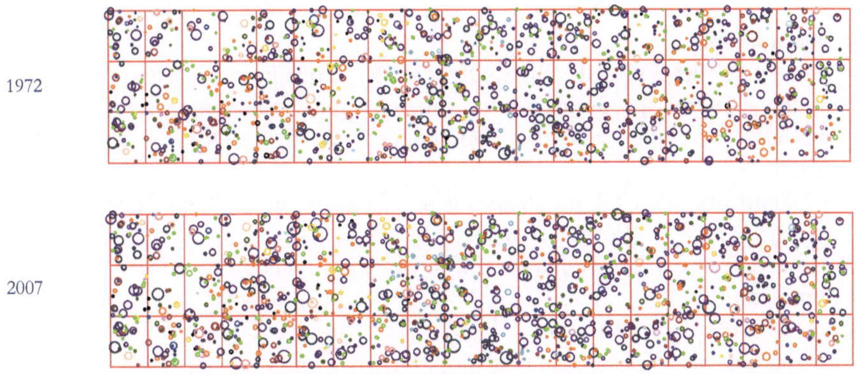

Fig. 2.27 Two plot maps showing the tree locations of the contiguous observational study FVC789 for two of the five available remeasurements (**1972**, 1978, 1992, 1997, **2007**). Colors indicate tree species, and circle radii are proportional to tree DBHs

Plot maps provide a useful first impression of a field plot. Obvious errors are often spotted on such a map.

2.3.5.3 Plot Summaries

This section is a brief introduction to the analysis of the observations collected in a standard one-hectare plot with mapped trees. The following two lines are required for clearing the working environment of a new R-workspace and to import a one-hectare field plot called OGF.

```
rm(list = ls())                                          # clear the working environment
dat<-read.csv("/Users/kg/Desktop/OGF_1ha.csv", header=TRUE) # import data: 1-ha plot
```

The statement length (unique(dat$species)) returns 28. For each of the 28 species, a standard plotting color must be defined. The function *mycolors* implements this requirement by assigning a unique color to each species in OGF.

```
mycolors <- function (pl) {plx <- factor(pl)
varvid <- list("#FF1493" = "Juglans mandshurica", "#C521C5" = "Acer pictum", "#A0522D"
= "Acer mandshuricum", "#FF0000" = "Abies holophylla", "#FFFF00"="Carpinus cordata",
"#DDA0DD" = "Pinus koraiensis", "#EE82EE" = "Corylus sieboldiana", "#8A2BE2" = "Betula
platyphylla", "#4B0082"="Acer barbinerve", "#6A5ACD" = "Ulmus laciniata",
"#98FB98"="Syringa reticulata", "#32CD32" = "Sorbus alnifolia", "#2E8B57" = "Betula
costata", "#006400"="Acer tegmentosum", "#808000" = "Tilia amurensis",
"#00FFFF"="Phellodendron amurense", "#5F9EA0"= "Euonymus verrucosus", "#0000FF"="Prunus
padus", "#0000CD"="Euonymus phellomana", "#778899"="Tilia mandshurica",
"#DCDCDC"="Deutzia parviflora", "#FF0000"="Lonicera chrysantha", "#B8860B"="Fraxinus
mandshurica", "#00FF00" = "Rhamnus davurica", "#87CEEB"="Populus koreana",
"#000000"="Philadelphus schrenkii", "#FFF8DC"="Acer caudatum", "#CDCD89"= NA)
levels(plx) <- varvid; if (length(plx[is.na(plx)]) > 0) {plx[is.na(plx)] <- "#CDCD89"};
return(as.vector(plx))}
```

Assignment

Subdivide the plot OGF into 25 20 cm × 20 m quadrats. Present a plot map with the 28 tree species identified by their assigned colors; the DBHs are represented by circles with the radii proportional to the DBHs measured in the field.

Solution: The OGF plot map is generated using the following code:

hx<-100; hy<-100; plotarea<-hx * hy # plot dimensions and total area

nx<-5; ny<-5; cellx<-hx/nx; celly<-hy/ny # number of cell row and column partitions

cellarea<-hy/ny * hx/nx

par(mar=c(4,4,1,1) + 0.1) # plot Margins

plot(0,0,type="n",axes=F,xlim=c(0,nx),ylim=c(0,ny),xlab="X (m)",ylab="Y (m)", line=1,cex.lab=1.5)

for (iy in 1:ny){for (ix in 1:nx){polygon(c(ix-1,ix,ix,ix-1),c(iy-1,iy-1,iy,iy),border="red", ,density=NA,col=0)}}

points(dat$x/cellx, dat$y/celly,cex=dat$DBH*0.08, pch=19, col=mycolors(dat$species))

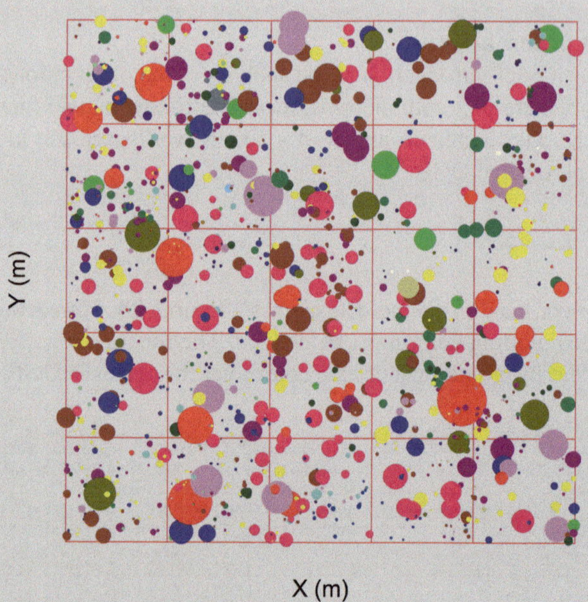

When the plot map has been created, it is also useful to be able to summarize information about each of the 28 species occurring in the OGF plot. The following R code generates a table with standard summaries for each species (Table 2.7):

Table 2.7 Standard summary information about each of the 28 species occurring in the OGF plot

Species	Mean DBH (cm)	Mean height (m)	N trees per ha	BA (m^2/ha)
Abies holophylla	15.9	9.0	64	3.4
Acer barbinerve	3.1	4.3	117	0.1
Acer caudatum	5.7	5.6	7	0.0
Acer mandshuricum	11.4	9.5	188	3.7
Acer pictum	8.2	8.7	261	2.7
Acer tegmentosum	9.5	10.3	70	0.6
Betula costata	14.1	14.9	40	0.8
Betula platyphylla	14.5	9.0	1	0.0
Carpinus cordata	7.0	7.1	265	1.8
Corylus sieboldiana	2.6	4.2	12	0.0
Deutzia parviflora	1.5	3.3	1	0.0
Euonymus phellomana	3.1	3.9	14	0.0
Euonymus verrucosus	2.6	3.5	19	0.0
Fraxinus mandshurica	5.0	6.7	1	0.0
Juglans mandshurica	24.2	17.1	88	4.8
Lonicera chrysantha	1.7	1.7	1	0.0
Phellodendron amurense	10.3	11.4	18	0.2
Philadelphus schrenkii	1.6	1.7	1	0.0
Pinus koraiensis	12.4	7.2	83	2.7
Populus koreana	14.0	16.5	1	0.0
Prunus padus	3.9	4.6	35	0.0
Rhamnus davurica	4.3	4.8	2	0.0
Sorbus alnifolia	16.0	9.6	23	0.9
Syringa reticulata	4.1	5.1	16	0.0
Tilia amurensis	24.0	14.4	27	1.9
Tilia mandshurica	13.0	8.9	5	0.1
Ulmus laciniata	10.2	9.4	140	2.3
Ulmus macrocarpa	37.0	14.8	1	0.1
		Total	**1501**	**26.1**

```
dat$ba<-pi/4*dat$dbh^2                                    # calculate tree basal areas
summary<-data.frame(sp=sort(unique(dat$species)), meanD=tapply(dat$dbh, dat$species,mean),
meanH=tapply(dat$height, dat$species,mean), ba= tapply(dat$ba,dat$species,sum), spsum=
tapply(dat$dbh,dat$species,sum))
summary$N<- summary$spsum/summary$meanD; summary$BA= summary$ba/10000
sumsp<- subset(summary, select=c(1,2,3,6,7)); print(sumsp, row.names=FALSE)
write.csv(sumsp, file ="/Users/kg/Desktop/OGFTable.csv")        # Excel for easier formating
```

The OGF plot is part of a large observational study established in the Jiaohe Forest in northeastern China. The summary table provides a useful first impression of the particular size characteristics of the species and the contribution of each species to total density (BA) and crowding (N).

The R library spatstat (Baddeley 2008) includes a number of useful functions for simulating spatial point patterns. These functions can be used to establish standard forests that can be used to compare an observed spatial structure with a random one and to test the difference, or to fit a spatial model to an observed tree map (Legendre and Legendre 2012).

Literature

Aguirre O, Hui GY, Gadow KV, Jimenez J (2003) Comparative analysis of natural forest sites in Durango, Mexico. For Ecol Manage 183:137–145

Alekseev A, Tomppo E, McRoberts RE, Gadow K v (2019) A constructive review of the State Forest Inventory in the Russian Federation. For Ecosyst 2019:6

Alfaro Reyna T, Martínez-Vilalta J, Retana J (2019) Regeneration patterns in Mexican pine-oak forests. For Ecosyst 6:50. https://doi.org/10.1186/s40663-019-0209-8

Assmann E (1961) Waldertragskunde. Organische Produktion, Struktur, Zuwachs und Ertrag von Waldbeständen. BLV Verlagsgesellschaft, München, Bonn, Wien, p 490

Ausburg T (2006) Becoming interdisciplinary – an introduction to interdisciplinary studies, 2nd edn. Kendall/Hunt Publishing, New York

Ayrey E, Hayesa DJ, Kilbrideb JB, Weiskittel AR (2019) Synthesizing disparate LiDAR and satellite datasets through deep learning to generate wall-to-wall forest inventories of New England. Posted online Mar. 17, 2019. https://doi.org/10.1101/580514

Baddeley A (2008) Analysing spatial point patterns in R. Workshop Notes February 2008. CSIRO

Baeten L, Verheyen K, Wirth C, Bruelheide H, Bussotti F, Finér L, Jaroszewicz B, Selvi F, Valladares F, Allan E, Ampoorter E, Auge H, Avacariei D, Barbaro L, Barnoaiea I, Bastias CC, Bauhus J, Beinhoff C, Benavides R, Benneter A, Berger S, Berthold F, Boberg J, Bonal D, Brüggemann W, Carnol M, Castagneyrol B, Charbonnier Y, Chécko E, Coomes D, Coppi A, Dalmaris E, Dañila G, Dawud SM, Wd V, Wandeler HD, Deconchat M, Domisch T, Duduman G, Fischer M, Fotelli M, Gessler A, Gimeno TE, Granier A, Grossiord C, Guyot V, Hantsch L, Hättenschwiler S, Hector A, Hermy M, Holland V, Jactel H, Joly FX, Juckers T, Kolb S, Koricheva J, Lexer MJ, Liebergesell M, Milligan H, Müller S, Muys B, Nguyen D, Nichiforel L, Pollastrini M, Proulx R, Rabasa S, Radoglou K, Ratcliffe S, Raulund-Rasmussen K, Seiferling I, Stenlid J, Vesterdal L, Kv W, Zavala MA, Zielinski D, Scherer-Lorenzen M (2013) A novel comparative research platform designed to determine the functional significance of tree species diversity in European forests. Persp Plant Ecol Evol Syst 15:281–291

Bauhus J, Pokorny B, Van der Meer P, Kanowski PJ, Kanninen M (2010) Ecosystem goods and services – the key for sustainable plantations. In: Bauhus J, Van der Meer P, Kanninen M (eds) Ecosystem goods and services from plantation forests. Earthscan, London, pp 205–227

Borcard D, Legendre P (2002) All-scale spatial analysis of ecological data by means of principal coordinates of neighbour matrices. Ecol Model 153:51e68

Bredenkamp BV, Kotze H, Kassier H (2000) The langepan CCT. A guide to a correlated curve trend spacing experiment in *Eucalyptus grandis* at Kwambonambi, Kwazulu-Natal, South Africa. 32 pp

Bruelheide H, Nadrowski K, Assmann T, Bauhus J, Both S, Buscot F, Chen XY, Ding B, Durka W, Erfmeier A, Gutknecht JLM, Guo D, Guo LD, Härdtle W, He JS, Klein AM, Kühn P, Liang Y, Liu X, Michalski S, Niklaus PA, Pei K, Scherer-Lorenzen M, Scholten T, Schuldt A, Seidler G, Trogisch S, Gv O, Welk E, Wirth C, Wubet T, Yang X, Yu M, Zhang S, Zhou H, Fischer M, Ma K, Schmid B (2014) Designing forest biodiversity experiments: general considerations illustrated by a new large experiment in subtropical China. Method Ecol Evol 5:74–89

Burkhart HE, Tomé M (2012) Modeling forest trees and stands. Springer, Dordrecht. 457 p

Champion HG, Seth SK (1968) A revised survey of the forest types of India. Govt. of India Press, Delhi, 404 p

Chen L, Wang L, Yeerjiang B, Zhang C, Zhao X, Gadow K (2013) Seed dispersal and seedling recruitment of trees at different successional stages in a temperate forest in northeastern China. J Plant Eco 7:337–346. https://doi.org/10.1093/jpe/rtt024

Cheng Y, Zhang C, Zhao X, Gadow K (2018) Biomass-dominant species shape the productivity-diversity relationship in two temperate forests. Ann For Sci 75:97. https://doi.org/10.1007/s13595-018-0780-0

Christin S, Hervet É, Lecomte N (2018) Applications for deep learning in ecology. bioRxiv preprint first posted online May. 30, 2018; https://doi.org/10.1101/334854

Condit R, Engelbrecht BM, Pino D, Pérez R, Turner BL (2013) Species distributions in response to individual soil nutrients and seasonal drought across a community of tropical trees. Proc Natl Acad Sci U S A 110:5064–5068

Corral-Rivas JJ, Wehenkel C, Castellanos BH, Vargas LB, Diéguez-Aranda U (2010) A permutation test of spatial randomness: application to nearest neighbour indices in forest stands. J For Res 15:218–225

Corral-Rivas JJ, Reyes RI, Wehenkel C, Aguirre-Calderón OA, Gadow Kv (2012) A network of forest observational studies in Durango (Mexico). In: Zhao XiuHai, Zhang ChunYu, Gadow Kv (eds) Forest observational studies. Proceedings of an international workshop at Beijing Forestry University, which convened on 20/21 September 2012: 125–138

Corral-Rivas JS, Torres Rojo JM, Lujan Soto JE, Nava Miranda MG, Aguirre Calderón OA, Gadow Kv (2016) Density and production in the natural forests of Durango/Mexico. Allgemeine Forst- und Jagdzeitung 187(5/6):93–103

Corral-Rivas JJ, González-Elizondo MS, Lujan-Soto JE, von Gadow K (2018) Effects of density and structure on production in the communal forests of the Mexican Sierra Madre Occidental. Southern Forests: a Journal of Forest Science. https://doi.org/10.2989/20702620.2018.1463152

Costanza R, d'Arge R, de Groot R, Farber S, Grasso M, Hannon B, Limburg K, Naeem S, O'Neill RV, Paruelo J, Raskin RG, Sutton P, van den Belt M (1997) The value of the world's ecosystem services and natural capital. Nature 387:253–260

Crawley MJ (2005) Statistics – an introduction using R. Wiley, West Sussex, 327 p

Crecente-Campo F, Corral-Rivas JJ, Vargas-Larreta B, Wehenkel C (2014) Can random components explain differences in the height-diameter relationship in mixed uneven-aged stands? Ann For Sci 71:51–70

Cunia T (1974) On the objectives and methodology of national forest inventories. In: Forest resource inventory. Joint Meeting of IUFRO Groups S4.02 and S4.04, 18–26 June 1978, Institutul de Cerctari si Amenageri Silvice, XI-XXIX. Bucharest

Daily GC (1997) Nature's services: societal dependence on natural ecosystems. Island Press, Washington

Dohrenbusch A, Bredemeier M, Lamersdorf N (2003) Manipulation of nutrient and water input of a Norway Spruce Ecosystem. Springer, Berlin

Feng G, Mi X, Yan H, Li FY, Svenning J-C, Ma K (2016) CForBio: a network monitoring Chinese forest biodiversity. Sci Bull. https://doi.org/10.1007/s11434-016-1132-9

FIA (2012) Forest inventory and analysis. National core field guide. Volume I: Field data collection procedures for phase 2 plots. Version 6.0.: USDA, Forest Service

Freese F (1962) Testing accuracy. For Sci 6:139–145

Gadow K (1991) The structure and derivation of a rule-based model for pruning decisions. Biometrie u Informatik in Medizin u Biologie 23(3):157–166

Gadow Kv (2017) Permanent forest observational plots – assessment and analysis. Report prepared for FAO, October, 2017:. 38pp

Gadow Kv, Bredenkamp BV (1992) Forest management. Academica Press, Pretoria, 151 p

Gadow Kv, Hui GY (1999) Modelling forest development. Kluwer Academic Publishers, Dordrecht, 213 pp. ISBN 0-7923-5488-5

Gadow K, Meskauskas E (1997) Stichprobenverfahren zur Erfassung von Naturverjüngung. AFZ/Der Wald 52(5):247–248

Gadow K, Zhang GQ, Durrheim G, Drew D, Seydack A (2016a) Diversity and production in an afromontane forest. For Ecosyst 3:15. (http://forestecosyst.springeropen.com/articles/10.1186/s40663-016-0074-7)

Gadow K, Zhao XH, Tewari VP, Zhang CY, Kumar A, Rivas JJC, Kumar R (2016b) Forest observational studies: an alternative to designed experiments. Eur J For Res 135(3):417–431

Gasparini P, Tabacchi G (2009) L'Inventario Nazionale delle Foreste e dei serbatoi forestali di Carbonio INFC[page12image168413440]2005. Secondo inventario forestale nazionale italiano. Metodi e risultati. Ministero delle Politiche Agricole, Alimentari e Forestali. 653 (Corpo Forestale dello Stato. Consiglio per la Ricerca e la Sperimentazione in[page12image168419776]Agricoltura, Unità di ricerca per il Monitoraggio e la Pianificazione Forestale, Bologna, Italy, 2011)

Giannetti F, Puletti N, Quatrini V, Travaglini D, Bottalico F, Corona P, Chirici G (2018) Integrating terrestrial and airborne laser scanning for the assessment of single-tree attributes in Mediterranean forest stands. Eur J Rem Sens 51(1):795–807

González EMS, González EM, Márquez LMA (2007) Vegetación y Ecoregiones de Durango. CIIDIR-IPN. Plaza y Valdés, S.A. de C.V. México, D.F. 219 p

Goodbody TRH, Coops NC, White JC (2019) Digital aerial photogrammetry for updating area-based forest inventories: a review of opportunities, challenges, and future directions. Curr For Rep. https://doi.org/10.1007/s40725-019-00087-2

Goward SN, Williams DL, Arvidson T, Rocchio LE, Irons JR, Russell CA, Johnston SS (2017) Landsat's enduring legacy: pioneering global land observations from space. American Society for Photogrammetry and Remote Sensing, Bethesda. (586 p)

Gregorius H (1974) Genetischer Abstand zwischen Populationen – Zur Konzeption der genetischer Abstandsmessung. Silvae Genetica 23:22–27

Griffith DA, Peres-Neto PR (2006) Spatial modelling in ecology: the flexibility of eigenfunction spatial analyses. Ecology 87(10):2603–2613

Gulbeyaz O, Bond-Lamberty B, Akyurek Z, West TO (2018) A new approach to evaluate the MODIS annual NPP product (MOD17A3) using forest field data from Turkey. Int J Remote Sens 39(8):2560–2578. https://doi.org/10.1080/01431161.2018.1430913

Hagan MT, Demuth HB, Beale MH, Jesús OD (2014) Neural network design, 2nd edn. Amazon, Orlando

Hao M, Zhang C, Zhao X, Gadow KV (2018) Functional and phylogenetic diversity determine woody productivity in a temperate forest. Ecol Evol 8:2395–2406

Heaton RJ (2000) The silvicuture, nutrition and economics of short rotation coppice in the uplands of mid-Wales. PhD thesis, Cardiff University

Heikkinen J (2006) Assessment of uncertainty in spatially systematic sampling. In: Kangas A, Maltamo M (eds) Forest inventory: methodology and applications, Managing forest ecosystems, vol 10. Springer, Dordrecht, pp 155–176

Henttonen HM, Kangas A (2015) Optimal plot design in a multipurpose forest inventory. For Ecosyst 2:31

Hofmann M (2005) Pappeln als nachwachsender Rohstoff auf Ackerstandorten – Kulturverfahren, Ökologie und Wachstum unter dem Aspekt der Sortenwahl. Doctoral Dissertation; University of Göttingen

Ho-Tung Lin, Tzeng Yih Lam, Gadow K v, Kershaw JA Jr (2019) Effects of nested plot designs on assessing stand attributes, species diversity and spatial forest structures. For Ecol Manage 457:117658

Huang X, Ma M, Wang X, Tang X, Yang H (2018) The uncertainty analysis of the MODIS GPP product in global maize croplands. Front Earth Sci 12(2). https://doi.org/10.1007/s11707-018-0716-x

Huston MA (1997) Hidden treatments in ecological experiments: re-evaluating the ecosystem function of biodiversity. Oecologia 110:449–460

Jia W, Sun Y, Pukkala T, Jin X (2020) Improved cellular automaton for stand delineation. Forests 11(1):37. https://doi.org/10.3390/f11010037

Kaul ON, Gupta AC, Sharma DC (1975) Preservation plots in India. Indian Forest Bulletin No. 271. FRI, Dehradun

Kenfack D, Chuyong GB, Condit R, Russo SE, Thomas DW (2014) Demographic variation and habitat specialization of tree species in a diverse tropical forest of Cameroon. For Ecosyst 1:22

Kershaw JA Jr, Ducey MJ, Beers TW, Husch B (2016) Forest mensuration, 5th edn. © 2017 Wiley. https://doi.org/10.1002/9781118902028

Kirchhoff B (2003) Erfassung und Beschreibung ausgewählter Waldflächen der Nordmongolei. Master Thesis, University Göttingen: 84 p

Kleinn C, Köhl M (1999) Long term observations and research in forestry. In: Proceedings volume., International IUFRO Symposium held in Costa Rica, February 23–27, 1999

Köhler A (1951) Vorratsermittlung in Buchenbeständen nach Stammdurchmesser und Baumabstand. Allgemeine Forst und Jagdzeitung (AFJZ) 69-74

Köhl M, Magnussen S, Marchetti M (2006) Sampling methods, remote sensing and GIS multiresource forest inventory. Springer, Berlin

Kramer H (1988) Waldwachstumslehre. Verlag Paul Parey, Hamburg/Berlin, 374 p

Kreutzer K (1993) Changes in the role of nitrogen in Central European forests. In: Huettl RF, Mueller-Dombois D (eds) Forest decline in the Atlantic and Pacific region. Springer, Berlin/Heidelberg, pp 82–96

Kreutzer K, Weis T (1998) The Höglwald field experiments – aims, concept and basic data. Plant and Soil 199(1):1–10

Kristöfel F (2017) Das Forstliche Versuchswesen in Österreich. In: Ledermann T (ed) Dauerversuche – Wertvolles Erbe oder Schwere Last? BFW-Berichte 153/2017. Bundesforschungszentrum für Wald, Vienna, pp 57–68

Lal JB (1989) India's forests: myth and reality. Natraj Publishers, New Delhi, 304 p

Laughton FS (1937) The silviculture of the indigenous forests of the Union of South Africa with special reference to the forests of the Knysna region. Sci. Bull. 157, Forestry Series 7, Government Printer, Pretoria: 169 p

Legendre P, Legendre L (2012) Numerical ecology, 3rd edn. Elsevier, Amsterdam

Li J, Roy DP (2017) A global analysis of sentinel-2A, sentinel-2B and Landsat-8 data revisit intervals and implications for terrestrial monitoring. Remote Sens (Basel) 9(9):902

Li X, Farooqi TJA, Jiang C, Liu S, Sun OJ (2019) Spatiotemporal variations in productivity and water use efficiency across a temperate forest landscape of Northeast China. For Ecosyst 6:22. https://doi.org/10.1186/s40663-019-0179-x

Liang JJ, Crowther TW, Picard N, Wiser S, Zhou M, Alberti G, Schulze E-D, McGuire AD, Bozzato F, Pretzsch H, de Miguel S, Paquette A, Hérault B, Scherer-Lorenzen M, Barrett CB, Glick HB, Hengeveld GM, Nabuurs G-J, Pfautsch S, Viana H, Vibrans AC, Ammer C, Schall P, Verbyla D, Tchebakova N, Fischer M, Watson JV, Chen HYH, Lei X, Schelhaas M-J, Lu H, Gianelle D, Parfenova EI, Salas C, Lee E, Lee B, Kim HS, Bruelheide H, Coomes DA, Piotto D, Sunderland T, Schmid B, Gourlet-Fleury S, Sonké B, Tavani R, Zhu J, Brandl S, Vayreda J, Kitahara F, Searle EB, Neldner VJ, Ngugi MR, Baraloto C, Frizzera L, Bałazy R, Oleksyn J, Zawiła-Niedźwiecki T, Bouriaud O, Bussotti F, Finér L, Jaroszewicz B, Jucker T, Valladares F, Jagodzinski AM, Peri PL, Gonmadje C, Marthy W, O'Brien T, Martin EH, Marshall AR, Rovero F, Bitariho R, Niklaus PA, Alvarez-Loayza P, Chamuya N, Valencia R, Mortier F, Wortel V, Engone-Obiang NL, Ferreira LV, Odeke DE, Vasquez RM, Lewis SL, Reich PB (2016a) Positive biodiversity-productivity relationship predominant in global forests. Science 354. https://doi.org/10.1126/science.aaf8957

Liang X, Kankare V, Hyyppä J, Wang Y, Kukko A, Haggrén H, Yu X, Kaartinen H, Jaakkola A, Guan F, Holopainen M, Vastaranta M (2016b) Terrestrial laser scanning in forest inventories. ISPRS J Photogramm Remote Sens 115:63–77. https://doi.org/10.1016/j.isprsjprs.2016.01.006

Liesebach M, Wuehlisch G, Muhs HJ (1999) Aspen for short-rotation coppice plantations on agricultural sites in Germany: effects of spacing and rotation time on growth and biomass production of aspen progenies. For Eco Manag 121:25–39

López-Serrano PM, López-Sánchez CA, Díaz-Varela RA, Corral-Rivas JJ, Solís-Moreno R, Vargas-Larreta B, Álvarez-González JG (2015) Estimating biomass of mixed and uneven-aged forests using spectral data and a hybrid model combining regression trees and linear models. iForest. (early view). 10.3832/ifor1504-008 [online 2015-09-21]

Lorenz M, Varjo J, Bahamondez C, Brack C, Clarke M, Gillis M, Hirvonen H, Kleinn C, Riebau A, Sase H, Totsuka T (2005) Forest assessment for changing information needs. Forests in the global balance – changing paradigms. IUFRO World Series. Volume 17. pp 139–150

Lujan-Soto JE, Corral-Rivas JJ, Aguirre-Calderón OA, Gadow K (2015) Grouping forest tree species on the Sierra Madre Occidental, Mexico. Allgemeine Forst und Jagdzeitung 186(3-4):63–71

Ma KP (2015) Biodiversity monitoring in China: from CForBio to Sino BON. Biodivers Sci 23:1–2. (In Chinese)

Mandallaz D (2008) Sampling techniques for forest inventories. Chapman Hall/CRC, Boca Raton

Mårell A, Leitgeb E (2004) European long-term research for sustainable forestry. www.enfors.org, Appendix MC4-8a

McRoberts RE, Tomppo EO, Næsset E (2007) Advances and emerging issues in national forest inventories. Scand J For Res 25:368–381

Mühlenberg M, Hondong H, Dulamsuren CH, Gadow Kv (2000) Large-scale biodiversity research in the southern Taiga, Northern Mongolia. In: Szaro RC, Peterson CE, Gadow Kv, Kräuchi N (eds) Creat- ing a legacy for sustainable science-based forest management: lessons learned from field experiments. Forest Snow and Landscape Research. 78 (1). Swiss Federal Research Institute WSL, Birmensdorf, Switzerland, pp 93–118

Neumann M, Zhao M, Kindermann G, Hasenauer H (2015) Comparing MODIS net primary production estimates with terrestrial national forest inventory data in Austria. Remote Sens (Basel) 2015(7):3878–3906. https://doi.org/10.3390/rs70403878

Ni R, Baiketuerhan Y, Zhang C, Zhao XH, Gadow Kv (2014) Analysing structural diversity in two temperate forests in northeastern China. For Ecol Manage 316:139–148

Noss RF (1990) Indicators of monitoring biodiversity: a hierarchical approach. Conserv Biol 4:355–364

O'Hehir J (2001) Growth and yield models for South Australian Radiata Pine Plantations incorporating Fertilising and thinning. PhD dissertation, Institute of Land and Food Resources, University of Melbourne

Packalen P, Pukkala T, Pascual A (2020) Combining spatial and economic criteria in tree-level harvest planning. For Eco 7:18

Peterson CE, Monserud RA (2002) Compatibility between wood production and other values and uses on forested lands – a problem analysis. Gen. Tech. Rep. PNW-GTR-564. Portland, OR: US Dept of Agriculture, Forest Service, Pacific Northwest Research Station, 51 p

Phillips JFV (1931) Forest succession and ecology in the Knysna Region. In: Schulze E-D, Caldwell MM (eds) Memoirs of the botanical survey of South Africa 14. The Government Printer, Pretoria

Pippuri I, Maltamo M, Packalen P, Mäkitalo J (2013) Predicting species-specific basal areas in urban forests using airborne laser scanning and existing stand register data. Eur J For Res 132:999–1012

Pitkänen TP, Raumonen P, Kangas A (2019) Measuring stem diameters with TLS in boreal forests by complementary fitting procedure. ISPRS J Photogramm Remote Sens 147:294–306

Pretzsch H (2009) Forest dynamics, growth and yield. Springer, Berlin, p 664

Pretzsch H, Biber P (2016) Tree species mixing can increase maximum stand density. Can J For Res 46:1179–1193

Puettmann KJ, Wilson SMG, Baker SC, Donoso PJ, Drössler L, Amente G, Harvey BD, Knoke T, Lu Y, Nocentini S, Putz FE, Yoshida T, Bauhus J (2015) Silvicultural alternatives to conventional even-aged forest management – what limits global adoption. For Ecosyst 2:8. https://doi.org/10.1186/s40663-015-0031-x

Pukkala T (2019a) Using ALS raster data in forest planning. J For Res. https://doi.org/10.1007/s11676-019-00937-6

Pukkala T (2019b) Optimized cellular automaton for stand delineation. J For Res 30(1):107–119

Reynolds KM, Schmoldt DL (2006) Computer-aided decision-making. In: Shao GF, Reynolds KM (eds) Computer applications in sustainable forest management, Kluwer/Springer book series "Managing forest ecosystems", vol 11, pp 143–169

Sagarin R, Pauchard A (2012) Observation and ecology. Island Press, 213 pp

Savin EN, Milyutin LI, Krasnoshhekov JN, Korotkov IA, Suncov AV, Dugarzhav C, Cogoo Z, Dorzhsuren C, Zhamjansurjen S, Gombosuren N (1988) Forests of the Mongolian People's

Republic (Larch forests of the Eastern Khentey). Soviet-Mongolian Expedition. Nauka, Siberian Department: Novosibirsk, Russia. p. 176. (In Russian)

Scherer-Lorenzen M, Schulze ED, Don A, Schumacher J, Weller E (2007) Exploring the functional significance of forest diversity: a new long-term experiment with temperate tree species (BIOTREE). PerspPlant Ecol Evol Syst 9:53–70

Seckendorf A v (1881) Das Forstliche Versuchswesen, insbesondere dessen Zweck und wirthschaftliche Bedeutung. Faesy & Frick, Vienna, 132p

Seifert T, Seifert S, Seydack A, Durrheim G, Gadow Kv (2014) Competition effects in an afrotemperate forest. For Ecosyst 1:13

Seydack AHW, Durrheim G, Louw JH (2012) Forest structure in selected South African forests: edaphoclimatic environment, phase and disturbance. Eur J For Res 131:261–281

Spellmann H, Wagner S, Nagel J, Guericke M, Griese F (1996) In der Tradition stehend, neue Wege beschreitend. Forst u Holz 51(11):363–368

Staupendahl K (1997) Ein neues Stichprobenverfahren zur Erfassung und Beschreibung von Naturverjüngung. Vortrag anlässlich der Jahrestagung der Sektion Forstliche Biometrie und Informatik des Deutschen Verbandes Forstlicher Versuchsanstalten am 24. 9. 97 in Freiburg

Szczukowski S, Tworkowski J, Klasa A, Stolarski M (2002) Productivity and chemical composition of wood tissues of short rotation willow coppice cultivated on arable land. University of Warmia and Mazury, Olsztyn

Tang ZY, Wang ZH, Zheng CY et al (2006) Biodiversity in China's mountains. Front Ecol Environ 4:347–352

Tewari VP, Gadow Kv (2008) Modelling potential density limiting survival, stand density and basal area growth for pure even-aged *Dalbergia sissoo* stands in a hot arid region of India. For Trees Livelihoods 18:133–150

Tewari VP, Alvarez-Gonzalez JG, Gadow Kv (2014) Dynamic base-age invariant site index models for Tectona grandis in Peninsular India. Southern Forests 76(1):21–27

Tilman D, Lehman CL, Thomson KT (1997) Plant diversity and ecosystem productivity: theoretical considerations. Proc Natl Acad Sci U S A 94:1857–1861

Tomppo E, Gschwantner T, Lawrence M, McRoberts RE (eds) (2010) National forest inventories. Pathways for common reporting, vol XXVI. Springer, 612 p

Van Laar A (1981) The response of Pinus radiata to initial spacing. S Afr For J 121:52–63

Van Laar A, Akça A (2007) Forest mensuration. Vol 13 of the Springer Book Series managing forest ecosystems, 418 p

Vauhkonen J, Tokola T, Packalen P, Maltamo M (2009) Identification of single-tree attributes using airborne laser scanning-based height, intensity and alpha shape metrics. For Sci 55:37–47

Verheyen K, Ceunen K, Ampoorter E, Baeten L, Bosman B, Branquart E, Carnol M, De Wandeler H, Grégoire JC, Lhoir P, Muys B, Setiawan NN, Vanhelle-Mont M, Ponette Q (2013) Assessment of the functional role of tree diversity, the example of the multi-site FORBIO-experiment. Plant Ecol Evol 146:1–10

Walotek P, Murach D, Gadow Kv (2006) Die Korbweide (Salix viminalis) – Ertragschätzung und Nutzungsmodell. Contribution to the German meeting of yield scientists (Ertragskunde-Tagung) on 29/30 May 2006

Wang Y, Woodcock CE, Buermann W, Stenberg P, Voipio P, Smolander H, Hame T, Tian Y, Hu J, Knyazikhin Y, Myneni RB (2004) Evaluation of the MODIS LAI algorithm at a coniferous forest site in Finland. Remote Sens Environ 2004(91):114–127

Wang J, Wu L, Zhao X, Fan J, Zhang C, Gadow K (2013a) Influence of ground flora on Fraxinus mandshurica seedling growth on abandoned land and beneath forest canopy. Eur J For Res 132(2):313–324

Wang J, Shen J, Wu Y, Tu C, Soininen J, Stegen JC, He J, Liu X, Zhang L, Zhang E (2013b) Phylogenetic beta diversity in bacterial assemblages across ecosystems: deterministic versus stochastic processes. ISME J 7:1310–1321

Wang J, Zhang C, Zhao X, Gadow Kv (2014) Reproductive allocation of two dioecious Rhamnus species in temperate forests of Northeast China. iForest 7:25–32. http://www.sisef.it/iforest/contents/?id=ifor1020-007

Wehenkel C, Corral-Rivas JJ, Hernandez-Díaz JC, Gadow Kv (2011) Estimating balanced structure areas in multi-species forests on the Sierra Madre Occidental, Mexico. Ann For Sci 68. https://doi.org/10.1007/s13595-011-0027-9

Wehenkel C, Corral-Rivas JJ, Gadow KV (2014) Quantifying differences between ecosystems with particular reference to selection forests in Durango/Mexico. For Ecol Manage 316:117–124

WeiSheng Z, Tomppo E, Healey SP, Gadow K (2015) The national forest inventory in China: history – results – international context. For Ecosyst 2:23

White F (1978) The afromontane region. In: Werger MJA (ed) Biogeography and ecology of Southern Africa. Dr. W. Junk, The Hague, pp 464–513

Winston PH (1977) Artificial intelligence. Addison-Wesley, Reading

Wulder MA, Hilker T, White JC, Coops NC, Masek JG, Pflugmacher D, Crevier Y (2015) Virtual constellations for global terrestrial monitoring. Remote Sens Environ 170:62–76

Wulder MA, Loveland TR, Roy DP, Crawford CJ, Masek JG, Woodcock CE, Allen RG, Anderson MC, Belward AS, Cohen WB, Dwyer J, Erb A, Gao F, Griffiths P, Helder D, Hermosilla T, Hipple JD, Patrick H, Joseph Hughes M, Huntington J, Johnson DM, Kennedy R, Kilic A, Li Z, Lymburner L, McCorkel J, Pahlevan N, Scambos TA, Schaaf C, Schott JR, Sheng Y, Storey J, Vermote E, Vogelmann J, White JC, Wynne RH, Zhu Z (2019) Current status of landsat program, science, and applications. Remote Sens Environ 225:127–147

Xu W, Luo W, Zhang C, Zhao X, von Gadow K, Zhang Z (2019) Biodiversity-ecosystem functioning relationships of overstorey versus understorey trees in an old-growth temperate forest. Annals of Forest Science

Yan Y, Zhang C, Wang Y, Zhao X, Gadow K (2015) Drivers of seedling survival in a temperate forest and their relative importance at three stages of succession. Ecol Evol:1–13. https://doi.org/10.1002/ece3.1688

Zhang L, Kong J, Zeng X, Ren J (2008) Plant species identification based on neural network. In: 2008 fourth international conference on natural computation, Jinan, 2008, pp 90–94. https://doi.org/10.1109/ICNC.2008.253

Zhang CY, Zhao XH, Gadow K (2009) Gender, neighboring competition and habitat effects on the stem growth of dioecious Fraxinus mandshurica trees in a northern temperate forest. Ann For Sci 66:812–821

Zhang X, Lei Y, Cao QV (2010a) Compatibility of stand basal area predictions based on forecast combination. For Sci 56(6):552–557

Zhang CY, Petráš R, Zhao X, Gadow K (2010b) Estimating beech growth and survival. A study based on longterm experiments in Slovakia. Allgemeine Forst- und Jagdzeitung 181(3/4):44–53

Zhang CY, Zhao XH, Gadow K (2010c) Partitioning temperate plant community structure at different scales. Acta Oecologica 36(2010):306–313

Zhang CY, Zhao Y, Zhao X, Gadow K (2012) Species-habitat associations in a northern temperate forest in China. Silva Fennica 46(4):501–519

Zhang CY, Zhao XH, Gadow K (2015) Maximum density patterns in two natural forests: an analysis based on large observational field studies in China. For Ecol Manage 346:98–105

Zhao M, Running SW, Nemani RR (2006) Sensitivity of moderate resolution imaging spectroradiometer (MODIS) terrestrial primary production to the accuracy of meteorological reanalyses. J Geophys Res 111:1–13

Zhao XH, Zhang CY, Gadow Kv (2012) Forest observational networks. In: Proceedings of the IUFRO-EFI-ICFFI conference ecosystem design for multiple services-with an emphasis on Eurasian Boreal Forests, St. Petersburg Forest Technical University, St. Petersburg/Russia, 9–11 November 2011, pp 34–47

Zhao XH, Corral-Rivas JJ, Zhang CY, Temesgen H, Gadow K (2014) Forest observational studies-an essential infrastructure for sustainable use of natural resources. For Ecosyst 2014:1–8

Zhe Z, Wulder MA, Roy DP, Woodcock CE, Hansen MC, Radeloff VC, Healey SP, Schaaf C, Hostert P, Strobl P, Pekel J-F, Lymburner L, Pahlevan N, Scambos TA (2019) Benefits of the free and open landsat data policy. Remote Sens Environ. https://doi.org/10.1016/j.rse.2019.02.016

Analyzing Forest Ecosystems

3

Abstract

A detailed examination of observations collected in the field, or remotely, is essential for understanding the basic features and dynamics of a forest community and to identify dissimilarities between different communities. Such an analysis typically includes an assessment of forest density and competition, structure and heterogeneity, and specific disturbances caused by a harvest event.

Emergent *Podocarpus falcatus* in an Afromontane Forest

© Springer Nature Switzerland AG 2021

K. von Gadow et al., *Sustaining Forest Ecosystems*, Managing Forest Ecosystems 37,
https://doi.org/10.1007/978-3-030-58714-7_3

3.1 Forest Density

Forest communities are unique in many respects. Trees are the tallest organisms on earth. They also exceed other plants in longevity, intensity of competition, biomass per unit area, and influence on the environment (Zeide 2005). Tree size and longevity both magnify the effect of interaction among individuals, a fact which reveals the inadequacy of density measures employed for other ecological communities (Begon et al. 1996: 569). Forest "density" is a simplified abstraction of the real world. Two forests with the same total biomass per unit area may be very different in terms of structure, number of trees, and species. Nevertheless, the level of site occupancy is an important variable, which significantly affects the productivity of a forest ecosystem. Individual tree growth and survival and even the shape of individual tree stems and crowns can be affected by density.

3.1.1 Measuring Forest Density

Populations of trees usually differ regarding their average size, and the number of individuals is therefore unsuitable as a measure of density (Zeide 1995). A forest with 500 saplings per hectare is obviously less dense than a forest with 500 big trees per hectare. The number of plants is as useful for density estimation as length alone is useful for estimating area. Just as area depends on length and width, density is a function of at least two variables: number of trees and average tree size. Consequently forest scientists pioneered more sophisticated approaches to characterize density than the simple number of organisms that is commonly used in other branches of ecology. West (1982) compared 17 different measures of site occupancy by trees. The most common ones are described in this section. They are *basal area, leaf area index, stand density index, Nilson's sparsity, relative spacing*, and *crown competition factor*.

3.1.1.1 Basal Area

A common measure of density is the sum of the cross-sectional areas at breast height of all the trees in a forest community (Assmann 1961; Kramer 1988). If the diameter at breast height (DBH, in cm) of a tree is known, the *basal area* (measured in cm^2) is calculated as

$$ba = \frac{\pi}{4} \cdot dbh^2 \tag{3.1}$$

The basal area for n trees is thus

$$BA = \frac{\pi}{4} \cdot \sum_{i=1}^{n} dbh_i^2 \text{ in cm}^2, \text{or } BA = \frac{\pi}{40000} \cdot \sum_{i=1}^{n} dbh_i^2 \text{ in m}^2 \tag{3.2}$$

To be able to compare the densities of different communities, the basal area must refer to the same forest area. The common standard is one hectare (ha). The basal area per hectare is calculated by dividing the total basal area of the forest by the forest area.

Assignment
Each of 1000 trees in a forest of 2 hectares has a DBH of 20 cm. Calculate the basal area per ha.

Answer: $1000\frac{\pi}{40000}\cdot 20^2 = 31.4\text{m}^2$; the basal area per ha is thus 31.4/2 = 15.7 m^2

The basal area of a forest is constantly changing as a result of tree growth, mortality, ingrowth (representing the trees that have reached breast height), and removals during a harvesting operation.

3.1.1.2 Leaf Area Index

The leaf area index (LAI) is defined as the sum of the leaf areas per unit of forest area. Total leaf area is intuitively appealing as a measure of density, because it reflects the degree of shading and the utilization of an available three-dimensional space. Accordingly, Vose and Allen (1988) proposed that the maximum potential leaf area should be used as a measure of density. Unfortunately, the leaf area is difficult to assess in practice. The leaf area is subject to fluctuations during the vegetation period, and even the most precise measurements are problematic. While the basal area can be expected to remain constant for extended periods of time, the leaf area is subject to continuous changes during the vegetation period. Fluctuations of the LAI are caused by changing levels of precipitation and the resulting fluctuations of soil conditions, that is, by several natural physiological processes during the vegetation period. Leaf area is unstable and therefore unsuitable as a measure of forest density. Of particular interest are density measures that remain stable for longer periods of time.

3.1.1.3 Stand Density Index

A popular measure of forest density is the *Stand Density Index* (SDI), which is based on the relation between the number of trees per hectare and the quadratic mean DBH (Dq). The quadratic mean DBH can be calculated if the basal area per hectare (G) and the number of trees per hectare (N) are known:

$$Dq = \sqrt{\frac{40000}{\pi}\cdot\frac{G}{N}} \tag{3.3}$$

The SDI is calculated assuming a very specific relation between N and Dq. Reineke (1933) defined this relation as follows:

$$N = \beta_0 \cdot Dq^{\beta_1}, \text{with } \beta_1 = 1.605 \tag{3.4}$$

Equation 3.4, which is also known as the Reineke model, is used to estimate the maximum number of trees per hectare, provided the parameters β_0 and β_1 are known.

The theoretical SDI is an abstract quantity that defines the number of trees per hectare for a *Dq* of 25 cm:

$$N = \beta_0 \cdot 25^{\beta_1} \tag{3.5}$$

By combining (3.4) and (3.5), the empirical SDI, which is based on the observed N and the observed Dq, is obtained:

$$N = \beta_0 \cdot \left(\frac{25}{Dq} \right)^{\beta_1} \tag{3.6}$$

Assume that the relation between *Dq* and the maximum number of trees per hectare *N* is defined by the parameters $\beta_0 = 100{,}000$ and $\beta_1 = -1.5$. The theoretical SDI is 100,000 $(25)^{-1.5} = 800$, which implies that the maximum number of trees in a forest with $Dq = 25$ cm is 800.

Assignment
What is the SDI for a forest with $Dq = 12$ cm? Calculate the maximum number of trees for a forest with parameters β_0 = 100 000, β_1 = −1.5, and $Dq = 12$ cm.

When we use observations from long-term spacing studies, we can show that the exponent β_1 can assume a rather wide range of values, depending on tree species and site conditions (Gadow 1986, 1987). These findings were confirmed in several other empirical studies (Zeide 1991, 1995; Dickel et al. 2010; Gadow and Kotzé 2015). The exponent β_1 is a highly variable quantity; the SDI is therefore not very suitable as a general standard for measuring forest density.

3.1.1.4 Nilson's Sparsity

Nilson (2006) argued that the most simple and logical relation should be that between variables of the same dimension, which is not the case in the Reineke model (Eq. 3.4). Assuming a regular spatial distribution of the trees within a forest, the average distance between the trees (*L*) can be estimated by the square root of the number of m² in a hectare divided by the number of trees per hectare (*N*):

$$L = \frac{10000}{N} \tag{3.7}$$

Nilson estimated the minimum distance among the survivors by using the following relationship:

$$L\min = a + b \cdot Dq \tag{3.8}$$

where *Lmin* is *Nilson's stand sparsity* (m); *Dq* is the quadratic mean diameter (cm) of the trees; and a and b are empirical parameters. The two variables have the same dimension, and the relationship is assumed to be linear. An example of this approach is given in the section dealing with maximum density.

3.1.1.5 Relative Spacing

A simple measure of density that has been applied in even-aged plantations is the *relative spacing index* (RS) that is calculated with the observed number of trees per hectare *(N)* and the dominant stand height *(Hdom):*

$$RS = \frac{\sqrt{\frac{10000}{N}}}{Hdom} \tag{3.9}$$

The numerator in Eq. 3.9 estimates the average distance between neighboring trees. Forest density increases with an increasing number of trees per hectare when the dominant height remains the same. The result is a decrease in the value of *RS* when there is an increase in density. An example of the development of *RS* over age can be seen in Fig. 3.1, where the relative spacing indices of the three communities follow a consistent trend.

The minimum *RS* of 0.095 was observed at age 18 in the plot with the highest planting density of 2965 trees per hectare (lowest point in the black line in Fig. 3.1; see Gadow and Kotzé 2015).

Assignment

The average inter-tree distance in a forest was 3 m ($N = 1111$) before a thinning. The average distance increased to 4 m ($N = 625$) after thinning, but the dominant height (20 m) did not change. Calculate the decrease in forest density.

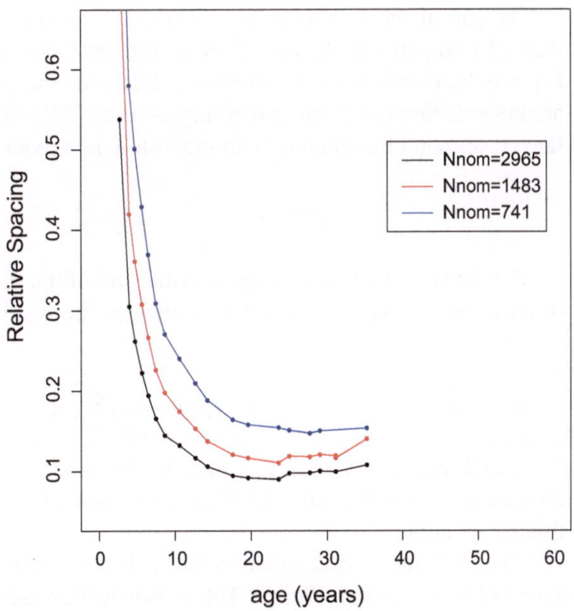

Fig. 3.1 Development of relative spacing for *Pinus patula* and three different planting densities (2965, 1483 and 741 trees per hectare)

Fig. 3.2 Hypothetical example of two estimates of CCF in a multispecies forest, one before and one after a selective harvest

An example of a recent application of the relative spacing index is a height-to-diameter ratio study involving several tree species and using data obtained from the Czech National Forest Inventory (Sharma et al. 2019).

3.1.1.6 Crown Competition Factor

A measure of density that assumes that forest density increases with increasing crown cover is the *crown competition factor* (Krajicek et al. 1961). The crown competition factor (CCF) can be used to estimate the density of a mixed forest with multiple species and is defined by the ratio of the maximum crown surface area over the area of a forest. It is necessary to calculate the crown surface areas of all the trees in a given forest if we want to estimate the total crown cover. An appropriate model is a linear relation involving the DBH (cm) and the crown diameter (CD, m) of a solitary tree of a particular species:

$$CD = \alpha_0 + \alpha_1 \cdot dbh \tag{3.10}$$

The parameters α_0 und α_1 are estimated by using solitary trees that have grown free of competition. Lässig (1991), for example, estimated the parameters of Eq. 3.10 by means of measurements of free-growing street trees. If these parameters are known, the theoretical crown surface area (CS, m^2) of a solitary tree of a particular species can be estimated with the following model:

$$CS = \pi \cdot \left(\frac{CD}{2}\right)^2 = \frac{\pi}{4} \cdot (\alpha_0 + \alpha_1 \cdot dbh)^2 \tag{3.11}$$

The density of a forest is equal to the sum of the theoretical crown surface areas of all its trees divided by the forest area (A, in m^2). Equation 3.12 is used for the calculation:

$$CCF = \frac{1}{A} \cdot \sum_{i=1}^{n} CS_i \tag{3.12}$$

A CCF equal to 1 indicates that the forest area is completely covered by tree crowns, assuming that all individuals are solitary trees. Figure 3.2 presents a hypothetical example.

Table 3.1 shows a calculation for a dense forest with two species (A and B) assessed in an area of 100 m^2. The crown surface areas of species A and B can be estimated with the following relations: $CS_A = 1.39 + 0.18(DBH)$ and

Table 3.1 Example of the change of the CCF caused by a harvest event in a forest with two species A and B. Before the harvest event CCF = 150.4/100 = 1.5. After the harvest event CCF = 111.2/100 = 1.1. The removed trees are marked with a + symbol

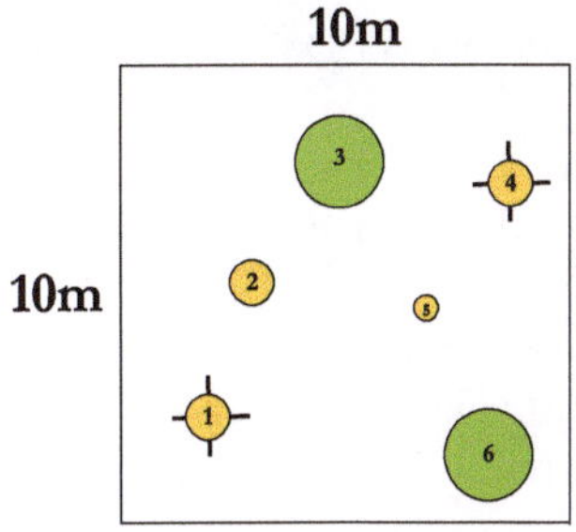

Tree No.	dbh (cm)	Crown width (m)	Crown surface area (m^2)
1 (A)+	20	5.0	19,6
2 (A)	20	5.0	19,6
3 (B)	40	7.3	41,8
4 (A)+	20	5.0	19,6
5 (A)	10	3.2	8,0
6 (B)	40	7.3	41,8
	CCF before the harvest event:		150.4
	CCF after the harvest event:		111.2

$CS_B = 1.64 + 0.14(DBH)$. No competition is assumed if CCF <1, because the sum of the estimated maximum crown surface areas is less than the area of the forest.

Any interpretation of the CCF value needs to consider the fact that the model is based on estimates of the *theoretical* (*maximum*) crown surface areas, not the *real* crown surface areas. It is therefore possible that a forest may have a CCF > 1. Even if the tree crowns do not touch, the real expansion of the crowns may be less than the potential expansion as the model used the measurements of solitary trees.

Assignment
Calculate the change in basal area for the example in Table 3.1.

3.1.1.7 Patterns of Density

The distribution density reveals patterns of patches with low or high density. The cell.f function can be used to calculate the cell-wise distributions of crowding (N), density (BA), richness (S), and quadratic mean DBH (Dq). The data are provided in the file OGF_1ha.csv. The function cell.f calculates basal areas (m^2/ha), crowdings (trees per hectare), number of species (per hectare), and mean squared DBHs (Dq, cm) for small cells within a large plot.

```
# z = number of columns(10),zz = number of rows (10), s=sidelength of a square cell
cell.f=function (dat,z,zz,s)                                          # dat=data.frame
{hh=0; BA=c(); Dq=c(); N=c(); S=c()
 for(g in 0:(z-1)){for(h in 0:(zz-1)){hh=hh+1
   r=dat[dat$x>=s*g&dat$x<s*(g+1)&dat$y>=s*h&dat$y<s*(h+1),]$dbh/2
   S[hh]=length(unique(dat[dat$x>=s*g&dat$x<s*(g+1)&dat$y>=s*h&dat$y<s*(h+1),]$Species))
   BA[hh]=sum(pi*(r^2))/10000*(10000/(s^2)); N[hh]=length(r)*10000/s^2
   S[hh]=S[hh]; Dq[hh]=sqrt(BA[hh]/N[hh]*40000/pi)
}}
return(data.frame(BA,S,N,Dq))}                                       # end function cell.f
```

Assignment
Use the function cell.f to calculate the maximum density, based on crowding and Dq, for the multispecies natural forest OGF and plot the results of the density pattern, as well as the fitted 99% percentile maximum density line.

Solution

```
cell=cell.f(dat,10,10,10)                # calling the cell-wise function
BAc=cell[,1]; Sc=cell[,2]; Nc=cell[,3]; Dqc=cell[,4]
# percentile regression: estimating Maximum density (N over Dq)
library(quantreg)
par(mar=c(5,6,1,1) + 1)                  # plot Margins
plot(Dqc,Nc,pch=19, ylab="N/ha", xlab="Dq(cm)",xlim=c(8,40),cex.lab=2)
start.values<-c(a=200000, b=-1.8)
dat=NULL; dat$x= Dqc; dat$y= Nc; Nformula<-y ~ a* x ^b
nlrq0.99 <- nlrq(Nformula, data= dat, tau=0.99, start.values)
tt =8:40; lines(tt, predict(nlrq0.99, list(x=tt)), col="red")
```

The total density calculated for a large forest area is not always a very meaningful quantity. A mixture of gaps and very dense patches may give the same total density as an even distribution of local densities.

3.1.2 Estimating Maximum Forest Density

Maximum forest density can be an indicator of shifts in the ecosystem's response to climate changes (Grace et al. 2002; Greenberg et al. 2009). In addition, observations about maximum forest density can provide a basis for an environmental gradient analysis, for example, to answer the question: What are the site conditions with the greatest effect on the potential density of a forest ecosystem? Knowledge of maximum forest density permits comparisons of potential biomass and carbon estimates with observed values. The ability to predict maximum forest density will also enable foresters to quantify the reduction of biomass and the loss in carbon sequestration by timber harvesting. This will allow more accurate and comprehensive comparisons of different residual structures, or alternative development "paths," for forest management (Gadow 2013). Hence, there are good reasons for improving estimates of the potential density of managed and unmanaged forest ecosystems.

3.1.2.1 Planted Forests

Estimates of the maximum density of a planted forest can be obtained from observations in unthinned and sufficiently large field plots maintained over long periods of time. The plots should include planting densities ranging from very low to very high stocking. The study sites should be observed throughout the life of the trees to eliminate age effects, and they should cover an area that is large enough to carry a sufficient number of surviving large trees at maturity. Such a design will provide information about the maximum density of a planted forest independent of tree age or average tree size.

Direct assessment of the potential density and development toward that elusive state requires densely stocked, unmanaged, and sufficiently large field plots that are enumerated regularly during long observation periods. A rare example of such an elaborate experiment is the *correlated curve trend* (CCT) spacing study established by O'Connor (1935) in South Africa (see Chap. 2). The eight nominal planting densities for the unthinned series (plots 1–8) and a map of the CCT spacing study Nelshoogte are presented in Fig. 3.3.

The original objective of the CCT study was to predict yields from plantations of various species of pine and eucalyptus for a wide range of planting densities. Most of the experiments were established between 1936 and 1938. Some relevant details of eight of these unique spacing trials for *Pinus patula* and *P. elliotti* are presented in Table 3.2.

The altitudes of the CCT studies range from 53 (Kwambonambi) to 1400 m.a.s.l. (Nelshoogte); mean annual temperatures range from 15.9 (Weza) to 21.8 °C

Nominal Densities	
Plot	**trees/ha**
1	2965
2	1483
3	988
4	741
5	494
6	371
7	247
8	124

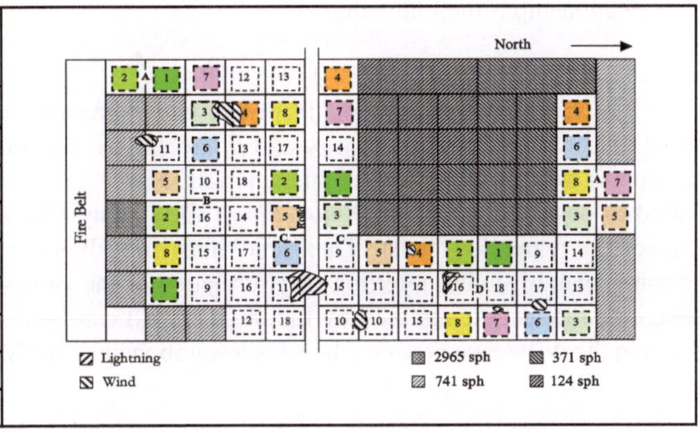

Fig. 3.3 Left: Nominal planting densities for the CCT unthinned series. Right: Map of the CCT spacing study Nelshoogte. The colored plots represent the unthinned series. Each of the four replications of one planting espacement covers an area of 0.081 ha. (See Gadow and Bredenkamp 1992)

Table 3.2 List of some CCT experiments for *Pinus elliottii* and *P. patula* and their geographical location

No	Location	Species	Latitude	Longitude	Altitude	MAT	MAP	soil texture	Sitecode	Soil depth class	Lithology	SI20
					m.a.s.l					cm		m
5	Weza	P.elliottii	-30,53333	29,73333	1133	15,9	830	ClLm	CT8	120	shale	19,1
6	Dukuduku	P.elliottii	-28,35000	32,26667	61	21,8	892	LmSa	ST7	250	sediment	17,0
7	Kwambonambi	P.elliottii	-28,66667	32,18333	53	21,5	1260	Sa	ST9	250	sediment	20,3
8	MacMac	P.elliottii	-24,98864	30,82073	1250	17,0	1463	Cl	WT3	>150	quartzite	23,9
11	Weza	P.patula	-30,60000	29,70000	1181	15,9	830	ClLm	CT8	120	shale	20,6
12	Entabeni	P.patula	-23,01262	30,00180	850	20,0	785	SaClLm	WT7	>150	Basalt	23,7
13	MacMac	P.patula	-24,98933	30,82041	1300	17,0	1328	Cl	WT6	>150	Oaktree	24,3
14	Nelshoogte	P.patula	-25,79687	30,80328	1400	17,0	937	SaClLm	CT9	120-150	Granite	23,9

MAT **mean annual temperature (°C),** ***MAP*** **mean annual precipitation (mm); SI20 = mean height of dominant trees at age 20**

(Dukuduku) and mean annual precipitation ranges from 830 (Weza) to 1463 mm (MacMac). Soil depths and geology were also assessed in each of the eight experiments. The soil was deep in all the experiments—measurements ranged from 120 cm to 150 cm. Humic soils on granite were encountered in the MacMac and Nelshoogte experiments, sandy soils in Dukuduku and Kwambonambi and red apedal dystrophic soils on basalt in Entabeni. The mean height of dominant trees at age 20 (SI20) was lowest for *P. elliottii* at Dukuduku (17.0 m) and highest for *P. patula* at MacMac (24.3 m).

Populations of trees growing at high densities are subject to density-dependent mortality or *self-thinning*. Data from the CCT experiments are therefore suitable for analyzing maximum densities of planted forests in response to age and site (Dickel et al. 2010). For any given average tree size, there is a limit to the number of trees that can coexist. The relationship between the average tree size (increasing over time) and the number of live trees per unit area (declining over time) can be described as a "limiting relationship" or "limiting line." A convenient model for estimating this relationship is the following:

$$N\max = \beta_0 \cdot Dq^{\beta_1} \tag{3.13}$$

where *Nmax* is the maximum number of surviving trees per hectare, *Dq* is the quadratic mean diameter (cm), and β_0 and β_1 are empirical parameters, which can be estimated from fully stocked, unthinned trials. Figure 3.4 presents the limiting lines fitted to observations in the *Pinus elliottii* CCT study MacMac and with a 0.975 percentile regression (left). The same lines are shown for the MacMac and Kwambonambi studies after log transformation of the variables. The difference in maximum density (expressed by the intercept β_0) can be considerable, as can be seen in the difference between the sites Kwambonambi and MacMac (Fig. 3.4).

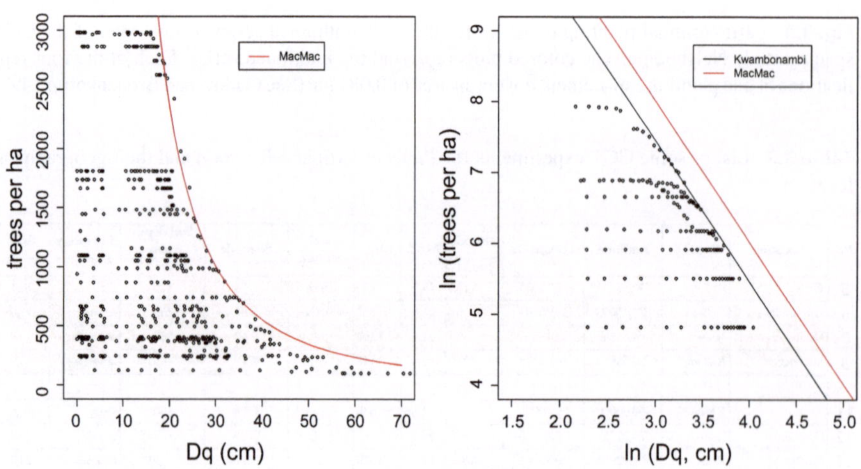

Fig. 3.4 Limiting line fitted to *Pinus elliottii* MacMac, using the 0.975 percentile data (left), and the same lines for MacMac and Kwambonambi after log transformation of the variables (right)

Nilson's sparsity is an alternative way of evaluating the limiting density of trees and therefore also an alternative to the limiting line. When the parameters a and b are known, Eq. 3.14 can be used to estimate maximum density:

$$N\max = \frac{10000}{\left(a + b \cdot Dq\right)^2} \tag{3.14}$$

An exhaustive evaluation on the assumption of the linearity of Nilson's sparsity is not possible, because the CCT material covers neither the seedling stage nor stands that are older than about 60 years. The tables published by Shvidenko et al. (2006) for Russian *Pinus sylvestris* forests on site quality II show that Nilson's sparsity is linear up to the age of 180 years and "curving upward" after that age. Analyzing deviations from linearity would require observations at higher ages.

3.1.2.2 Natural Forests

A different approach needs to be followed in a natural forest where the local densities are not controlled by planting espacement but by natural processes. Knowledge regarding maximum densities in unevenly aged natural forests, where patches of high and low density are often found in proximity, is still very limited. One approach is to divide sufficiently large field plots with mapped trees into small cells and to estimate a limiting relationship for the entire study area. The example of Zhang et al. (2015) shows how maximum density estimates can be obtained for an unmanaged natural forest (Fig. 3.5).

These results provide further evidence that invalidates the assumption of the universal validity of the exponent −1.605 of maximum density studies (Reineke 1933). It would be interesting to study the relationship between cell density and structure and to find answers to questions such as: What is the relation between forest density and species richness, or between density and the diameter coefficient of variation?[1] What are the density patterns in the canopy and understory layers? Another interesting topic is a comparison between the shapes of the self-thinning lines in planted and natural ecosystems. Figure 3.6 presents a comparison between the self-thinning lines of a natural plot Jiaohe and a *Pinus patula* plantation.

Maintaining a series of large field plots over long periods of time is expensive. One attempt to reduce these high costs is a design proposed by Nelder (1962). The compact Nelder design consists of a number of spokes of an imaginary wheel with trees planted at different radial distances at the intersection of circular arcs and linear spokes (Barry 2013). Parrot et al. (2012) discuss various aspects of the Nelder design that allow the effects of different planting densities to be examined in a single experiment. The experimental unit in a Nelder design is the individual tree where research usually involves relating the available growing area of each tree to growth rates or tree architecture (Mabvurira and Miina 2002; Aphalo and Rikala 2006). According to Affleck (2001), the analysis of Nelder experiments is

[1] High-density cells with a basal area mean value of 45 m^2/ha had a significantly higher species richness and higher diameter coefficient of variation than low density cells with a basal area mean value of 18 m^2/ha.

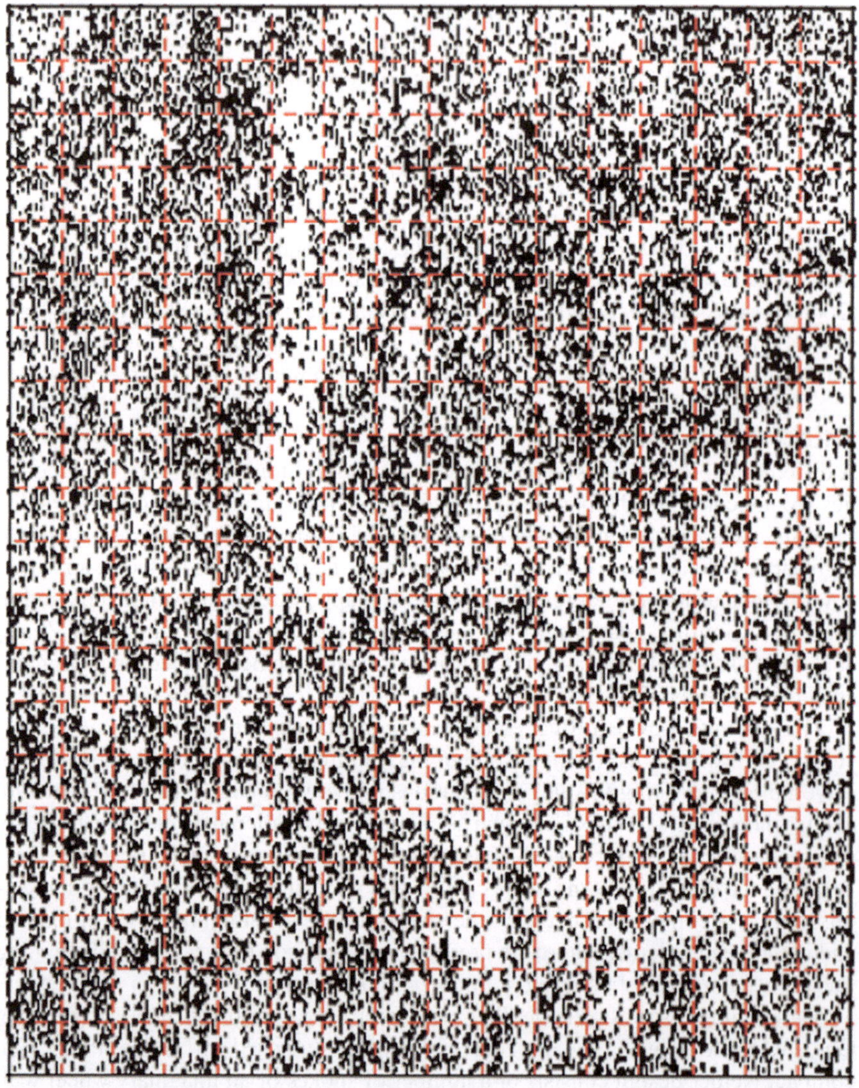

Fig. 3.5 Plot Jiaohe, with mapped trees covering 28.8 hectare, is subdivided into 320 cells to establish a self-thinning line; each cell measures 20 cm × 20 m

problematic because the compact arrangement of trees can result in significant correlations among neighboring variables. Hall (1994) notes that seedling survival is a primary concern when Nelder plantings are established because a major drawback of the Nelder design is the sensitivity of the analysis to tree mortality (see also Stape and Binkley 2010).

Although the compact Nelder design has a cost advantage, it only allows observations about the relationship between available growing space, tree growth, and

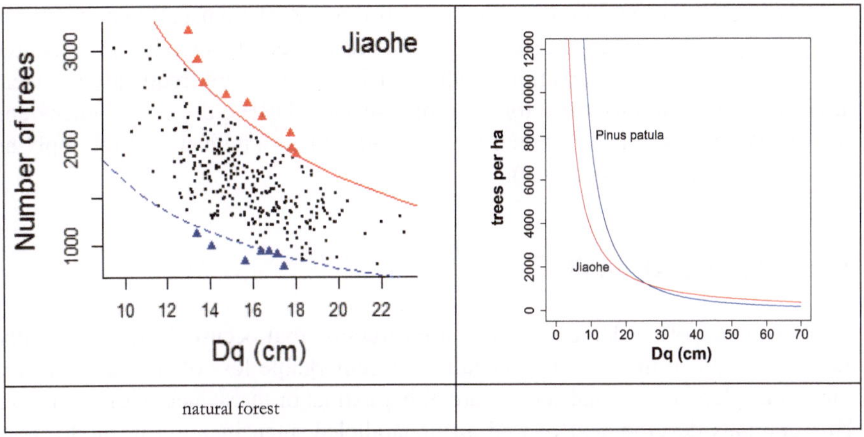

Fig. 3.6 Left: Maximum density estimates for the Jiaohe unmanaged temperate forests in north-eastern China ($N_{max} = 56163.5 \cdot Dg^{-1.17}$). Right: Comparison of the maximum number of surviving trees in a planted and a natural forest. For $Dq = 20$ cm, the estimated maximum number of surviving trees in the Pinus patula plantation is 2015 while it is 1644 in the Jiaohe natural forest. For $Dq = 40$ cm, the estimated maximum number of surviving trees is 504 in the Pinus patula plantation and 716 in the Jiaohe natural forest

tree architecture. The Nelder wheel does not provide information about maximum density. The main disadvantage of this design is the limited space, and the analysis is therefore limited to the juvenile stage (Gaul and Stüber 1996). As mentioned before, data about the maximum density of planted forests are very scarce because the cost of maintaining a series of unthinned, densely-stocked, and sufficiently large field plots over long periods can be prohibitive. However, the CCT experiments are a valuable source of information for the analysis of maximum density and tree survival. Surprisingly few studies have used these unique data sets.

3.2 Forest Heterogeneity and Structure

Structure is a fundamental notion and refers to patterns and relationships within a system. Structure may develop as a result of a planned design or through a process of self-organization. Some structures are fixed while others are open and dynamic. Open structures are characterized by hierarchy and networking as they evolve and adjust (Luhmann 1995). The term "structure" is open-ended, offering itself to a variety of interpretations within many disciplines of the sciences, arts, and humanities (Pullan and Bhadeshia 2000).

Heterogeneity is associated with a specific forest structure and refers to a range of phenomena, including richness (the number of species), diversity (the species frequency distribution), and the variation of tree dimensions.

Despite advances in remote sensing, mapped tree data are usually not available, except in specially designed research plots. Forest inventories tend to provide tree data samples in small observation windows. Tree locations are frequently not measured, and the tree attributes, for example, species, DBHs, diversity indices and other marks, are established directly in the field. This is a typical case for applying NNS (nearest neighbor statistics).

3.2.1 Species Diversity

A consequence of the large-scale forest destruction that occurred especially since the second half of the twentieth century is a remarkable loss of species diversity. Numerous plant and animal species are either extinct or threatened with extinction. This ominous development caused an unparalleled intensification in biodiversity research. Scientists began to use measures of species diversity during the early years of the twentieth century and among the first to propose indices of diversity were Fisher (1943) and Simpson (1949). The Shannon index[2] was first applied in studies on community species diversity by Margalef (1957). Whittaker (1972) divided the diversity indices into four spatial scales, namely, α, β, γ, and δ diversity. The literature on biodiversity seems endless and new ideas continue to emerge.

Species richness (S) refers to the number of different species in a species list. Richness does not reflect species abundance and is not the same as diversity, which considers the relative frequency of the different species in a community. The variety of biodiversity indices is bewildering, and many are viewed with suspicion. However, most diversity indices are entropies, not diversities, and their mathematical behavior does not always correspond to our intuitive concept of diversity. For example, if rare species count as much as common species, then community A is more diverse than community B in Fig. 3.7. If the common species are the important ones that shape the community, then B is more diverse than A.

When different diversity indices are applied to a system with S equally common species, some refer to S, some to log S, and some to 1–1/S. Obviously, these indices are not interchangeable in formulas that use "diversity" as a variable. A method that gives good results with one index may give meaningless results with another. Different interpretations of "diversity" are therefore appropriate in different situations.

There seems to be general agreement among statisticians that "numbers equivalent" (Hill 1973) and not entropy should be the diversity measure of choice (Elliot 2011). The first three Hill numbers depend on the weight that is assigned to rare species: $q = 0$ refers to species richness where rare species carry the same weight as common ones; $q = 1$ refers to the exponential of Shannon's entropy index; and $q = 2$ to the inverse of Simpson's concentration index. For example, if $q = 1$, the Shannon index $H' = -\sum p_i \cdot (\log p_i)$ is used, and the corresponding Hill effective number is

[2]Confusingly referred to in different ways, for example, as the Shannon-Weaver, the Shannon-Weiner, or the Shannon-Wiener index.

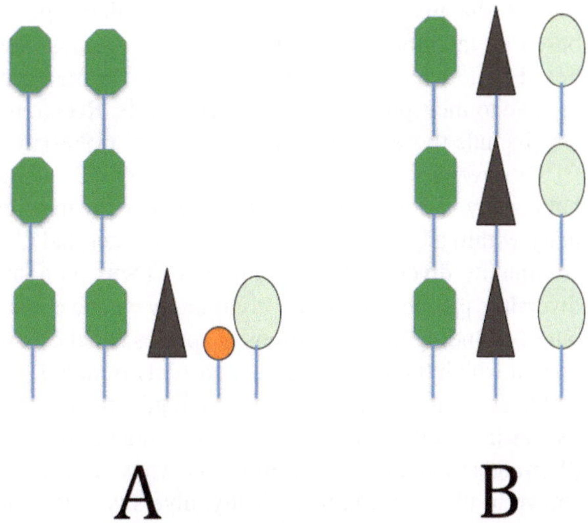

Fig. 3.7 Two examples to illustrate different interpretations of diversity: If rare species count as much as common species, then community A is more diverse than B; if the common species are the important ones that shape the community, then B is more diverse than A

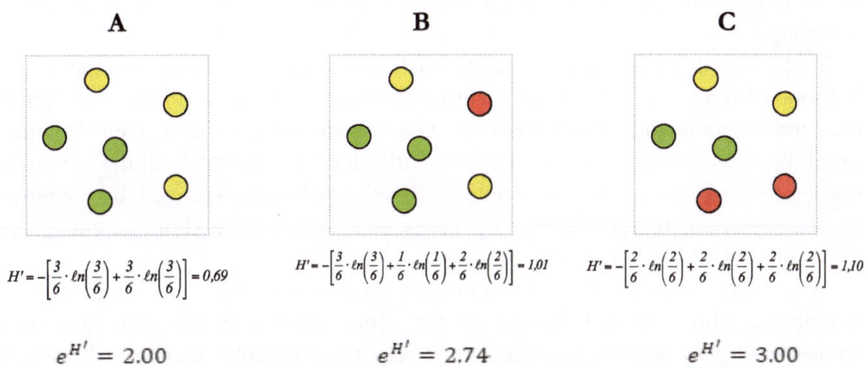

Fig. 3.8 Shannon indices for three hypothetical diversity patterns with the corresponding effective Hill numbers. Hill numbers are equal to richness when species occur evenly (examples A and C), but are less than richness when species do not occur evenly

$N = e^{H'}$. Hill numbers are equal to richness when species occur evenly, but less than richness when species do not occur evenly (Fig. 3.8).

Effective Hill numbers are used increasingly instead of entropies to characterize the taxonomic, phylogenetic, or functional diversity of a community. Various processes can create the same spatial pattern. Obviously, estimates of Hill numbers are a function of the sampling effort. Characterizing a spatial pattern is therefore insufficient to elucidate the mechanisms that generated it (McIntire and Fajardo 2009). It is also necessary to identify the possible role of exogenous or endogenous variables. Biological factors such as species dispersal abilities, aggregation behavior, and temporal variation of species population because of mortality and fecundity can all affect population patterns.

Ganeshaiah and Shaanker (2000, 2003) developed the avalanche index, which is based on taxonomic differences between species. Scholes and Biggs (2005) proposed a biodiversity intactness index, which measures the decline of populations relative to their presumed premodern levels. Recent reviews of biodiversity indices also include those by Ferris and Humphrey (1999) and Spanos and Feest (2007).

The diversity of a forest ecosystem is not only characterized by species richness, but also by various other attributes that portray the significant features of heterogeneity within a community. Following Leinster and Cobbold (2012), we can extend community diversity beyond the naïve focus on a simple index. Accordingly, the diversity "profile" of a forest community can be characterized by a number of attributes, including the proportions of species and the distribution tree sizes, and by the way in which these attributes are spatially mingled (Hui et al. 2011a, b).

Diversity measures that are meaningful for fish at certain river sections or butterflies in specific habitats may not be suitable for forest communities. Tree species do mingle spatially, and in different ways. Two patterns of species mingling are shown in Fig. 3.9. Mathematically, diversity is the same in both communities, but the two patterns are very different. The species in C are spatially segregated, resulting in large monospecific groups, while those in D show a high degree of spatial mingling.

Tree-based diversity patterns are established by evaluating the immediate neighborhood of trees that have specific attributes in common. Such a cohort of "reference trees" can include all the trees of a given species or trees of a given size class or all the trees that share a combination of attributes, for example, all the dominant canopy trees of several species. A neighborhood may be characterized, for example, by the variation in tree sizes or by the degree of *mingling* (Pommerening and Grabarnik 2019: 27; Fig. 3.10).

For a given reference tree, we can define species mingling as the proportion of neighbors, which do not belong to the same species as the reference tree. Implementing the mingling concept in R requires a function that identifies the n nearest neighbors of a reference tree. Once these have been identified, calculating the proportions and the "mingling" value is straightforward.

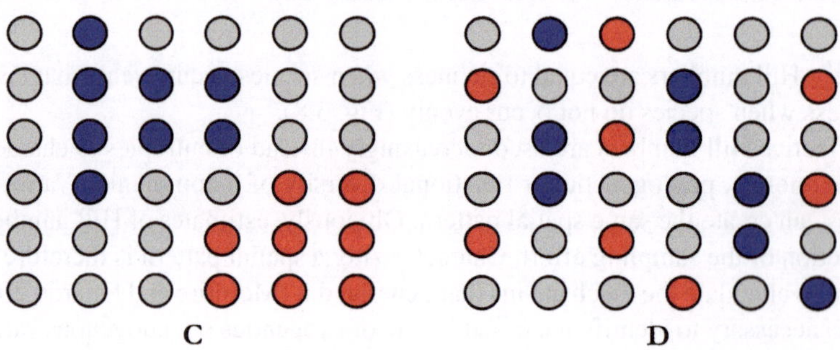

Fig. 3.9 Two communities, each with 36 trees and three species ($p1 = 0.55$; $p2 = 0.22$; $p3 = 0.22$); both communities have the same mathematical diversity, but differ in spatial mingling

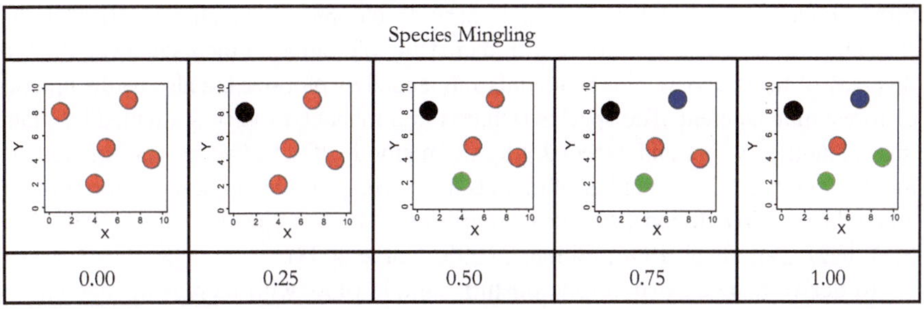

Fig. 3.10 Five possible constellations in the neighborhood of a given reference tree involving its four nearest neighbors. The numbers refer to the proportion of neighbors of the reference tree (which is shown in red in the center of each of the five graphs)

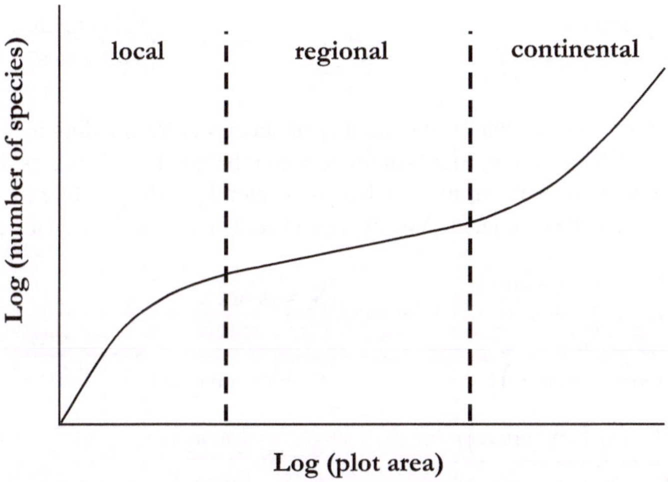

Fig. 3.11 General relationship between the sample plot area and the number of species. (Hubbell 2001)

3.2.2 The Species–Area Relationship

The relationship between the contiguous area of an ecosystem and the number of species, known as the species–area relationship (SAR), is an important outcome of earlier studies. Lawton (1999) referred to this relationship as one of the few fundamental laws in ecology. The differences in micro-site conditions generally increase with area, and accordingly, the variety of species that can be supported generally increases with increasing contiguous area (Williams 1964; Barkman 1989). However, the particular form of the SAR may change when the scale changes. An example is Hubbell's three-phase model of biodiversity (2001; see Fig. 3.11).

At the local level, the number of species increases quickly with increasing plot area. At the regional level, the cumulative relation is not so much determined by the relative species frequency as by the balance between species formation, species extinction, and spatial distribution. The continental and intercontinental biogeographic dimension, in turn, produces new and distinct evolutionary developments, which can cause another sharp rise of the species diversity with increasing plot area. At the local level, the biodiversity of forest ecosystems is not only influenced by plot size but also by the practice of forest management. The species richness in a managed forest is affected by plantings, thinning cycles, and especially by the methods of selective harvesting (Jenssen and Hofmann 2002). A number of models have been proposed to describe the species–area relationship (Monod 1950; de Caprariis et al. 1976; Lepě and Stursa 1989; Gitay et al. 1991; Buys et al. 1994; Williams 1995; Tjørve 2003).

To derive a species–area relationship, sample plots with increasing size can be assigned to random positions within the field plot. The number of species is determined within each sample. The sample areas and associated number of species are used to derive a species–area relationship for the whole study area. We can use the following models:

Asymptotic function	Power function
$S = A * (1 - \exp(-b * \text{area}))^c$	$S = A * \text{area}^b$

where S is richness, that is, the number of species corresponding to "area" measured in m^2. When the empirical parameters and the number of species S_1 observed in a given area$_1$ are known, the number of species S_2 corresponding to area$_2$ (e.g., one hectare) can be estimated. Assume two specific cases with known richness S_1:

$S_1 = A * (1 - \exp(-b * \text{area}_1))^c$	$S_1 = A \cdot \text{area}_1^b$
$A = \dfrac{S_1}{(1-\exp(-b*\text{area}_1))^c}$	$S_2 = A \cdot \text{area}_2^b$ and $S_2 = S_1 \dfrac{area_2^b}{area_1^b}$
$S_2 = S_1 \dfrac{(1-\exp(-b*\text{area}_2))^c}{(1-\exp(-b*\text{area}_1))^c}$	$S_2 = S_1 \dfrac{(\text{area}_2)^c}{(\text{area}_1)^c}$

where S_1 is the number of species observed in a field plot with area$_1$ m^2, and area$_2$ is 10,000 (one hectare). For example, if we assume that 14 species were observed on a 6000 m^2 field plot, we can estimate the number of species per hectare (with known parameter $c = 0.3750$ of the power function):

$$S_{ha} = 14*(10000 \wedge 0.3750)/(6000 \wedge 0.3750) = 17 \text{species per ha.}$$

The following example can be used to estimate the species–area relationship using *Lensahn.csv*:

```
# Sampling: Use the data.frame in Lensahn.RData     Plot area is 60 x 100m
maxX<-60; maxY<-100; maxcircle_r<-28; S<-array(0); area <-array(0);
sample<-100000; set.seed(123)
for(r in 3:maxcircle_r){                                          # sample radius
   x<-runif(sample,r,maxX-r); y<-runif(sample,r,maxY-r)           # n centres
   area[r]<-pi*r^2;
S[r]<-length(unique(dat$Art[(dat$x-x)^2 + (dat$y-y)^2 <= r^2]))}  # species in circle
S.f<-function(x,p1,p2,p3) {p1*(1- exp(p2*x))^p3}                  # Chapman Richards
Sreg <- nls(S ~ S.f(area,p1,p2,p3), start=list(p1=15,p2=-0.0005,p3=0.5),trace=TRUE)
plot(area,S, xlim=c(0,3000), ylim=c(0,15))                        # plot the relation
lines(1:3000, predict(Sreg, newdata=list(area=1:3000)), col=2)
```

The parameter estimate of the Chapman Richards function, estimated for six forest plots of different areas, are presented in Table 3.3.

Graphs of the fitted power functions are shown in Fig. 3.12.

In this way, the species area relation can be applied to derive a common standard that can be used to compare different ecosystems: the *species richness per hectare*. For example, if we use the Chapman Richards parameters of Lensahn (Table 3.3), 12 species on 400 m^2 would be equivalent to 26.2 species per ha. Alternatively, 14 species per 400 m^2 would be equivalent to 30.6 species per hectare and 8 species per 400 m^2 would be equivalent to 17.5 species per ha (Fig. 3.12).

These models allow us to determine the *minimum species area*, that is, the smallest area required to capture all the species present within a given contiguous region. Using relatively large contiguous sample areas with known tree positions, assessed in different climatic regions with varying tree species abundances, Gadow and Hui (2007) found a specific relationship between the maximum number of tree species

Table 3.3 Parameters of the Chapman Richards function for six forests and estimates of richness per ha. S_{obs} is the total observed richness in the plot

Study	plotarea (ha)	p_1	p_2	S_{obs}	$S_{per\ ha}$
Jiaohe	1.00	re934***	0.2343***	22	22
St Barbara	4.00	3.123***	0.1732***	18	
Lensahn	0.60	0.5478***	0.4139***	14	18.5
Diepwalle	4.56	2.8941***	0.2418***	26	
Jiaohe	21.00	13.99***	0.1087***	46	
PeruRA4	1.00	0.7946***	0.5437***	113	113

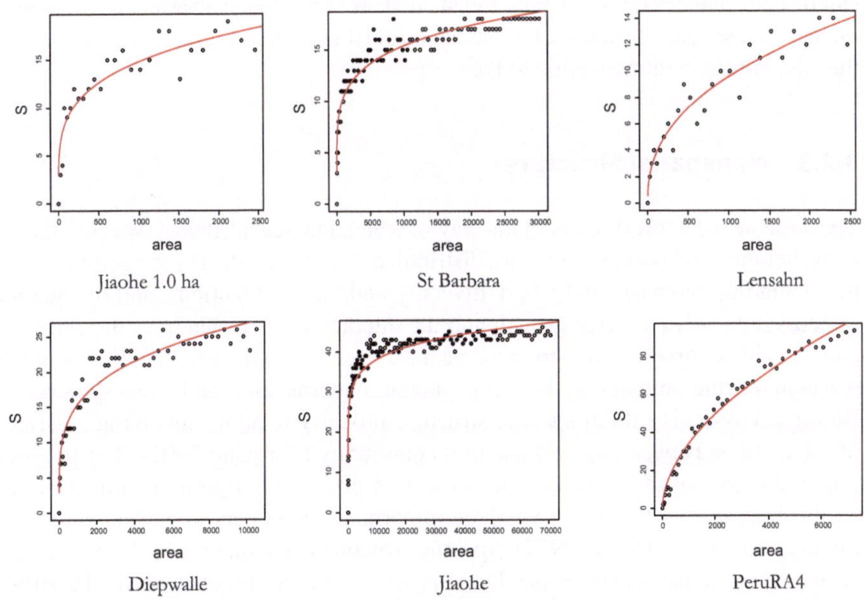

Fig. 3.12 Fitted power functions for the six plots in different areas

within a forest (S_{max}) and the minimum area required to capture all of them. The minimum species area (A_{min}) was estimated by using the power function $A_{\min} = 487.8 \cdot S_{\max}{}^{0.524}$. This particular analysis showed that, for contiguous forest areas, the form of the species–area relationship is directly defined by the species abundance and the maximum number of species. This result confirms assumptions made by Preston (1962), May (1975), and Hubbell (2001).

Undisturbed reference areas, which are large enough to include the essential features of an unmanaged forest, are useful for evaluating specific near-natural forest management systems based on selective harvesting. One of the key features of an unmanaged forest is the tree species distribution. Therefore A_{min}, the smallest area, which represents the species richness of the entire population, has practical relevance. Minimum species areas are related to balanced structure areas (BSA), which were defined by Koop (1981) as the minimum contiguous area that includes all tree developmental stages. Wehenkel et al. (2011) have shown that a Mexican multisized selection forest represents a balanced structure if a specific relationship between harvest and growth can be maintained, if there is a defined target diameter distribution, and if major natural disturbances are disregarded. When the BSA is used as an indicator of demographic sustainability, a range of goods and services can therefore be consistently produced over time, thus realizing a vision proposed by Nyland (2002).

Jenssen and Hofmann (2002) have shown relationships between different successional stages of a beech ecosystem and plant species richness. They used plots of the same size and found that the average number of species in that particular study grew from a dense sapling stage to a stage dominated by mature trees. The highest number of plant species was observed in the senescent stage of the trees where the small-scale changes between gaps and shaded areas produced a variety of particular niches. The species diversity of a managed forest is therefore greatly influenced by the silvicultural treatment applied there.

3.2.3 Nonspatial Structure

The structure of a forest refers to the way in which the tree attributes (species, diameters, heights, and root systems) are distributed. Structure affects a range of properties, including biomass production, diversity and habitat functions, and the quality of ecosystem services. Tree growth and the interactions between trees depend on a considerable degree on the structure of the forest. Analytical tools allow detailed research of the interaction between spatial patterns and biological processes. Biological systems exhibit specific structure-property relations involving a variety of interactions between individuals in a community (Torquato 2002). The production and dispersal of seeds and the associated processes of germination, seedling establishment, growth, and survival are important factors that influence plant community structuring (Harper 1977). Specific structures generate particular processes, such as growth and regeneration. These processes in turn produce particular structural arrangements.

The structure of a forest is in the first place the result of natural processes, including tree growth, mortality, and natural disturbances such as fire, wind, or snow damage. In addition, human disturbance such as felling or planting has a major structuring effect. Furthermore, the invasion of natural ecosystems by non-native species is another structuring element of increasing importance. As more of the earth's land surface becomes transformed by deliberate or inadvertent human activity, such *novel* ecosystems are increasing in importance (Hobbs et al. 2006; Richardson et al. 2011a, b).

Forest structure is not only the result of past events, also defines the starting condition for future developments. This is naturally of interest to scientists studying the spatial and temporal evolution of complex forest communities (McComb et al. 1993; Jaehne and Dohrenbusch 1997; Kint et al. 2003). Important ecosystem functions, as well as the potential and limitations of human use, are defined by current forest structures. Spatial patterns affect the competition status, seedling growth, and survival, as well as the crown formation of forest trees (Moeur 1993; Pretzsch 1995). The vertical and horizontal distributions of tree sizes determine the distribution of microclimatic conditions, the availability of resources, and the formation of habitat niches, and thus, directly or indirectly, the biological diversity within a forest community. Information about forest structure therefore contributes to improved understanding of the history, functions, and future development potential of a particular forest ecosystem (Harmon et al. 1986; Ruggiero et al. 1991; Spies 1997; Franklin et al. 2002).

Forest structure is of interest to students of ecology, but it also has economic implications. Simple bioeconomic models have been criticized for a lack of realism as they tend to focus on even-aged monocultures and to disregard natural hazards and risk. Knoke and Seifert (2008) used published studies to evaluate the influence of the tree species mixture on forest stand resistance against natural hazards, productivity and timber quality. They used Monte Carlo simulations in mixed forests of Norway spruce and European beech, assumed site conditions and risks, and found superior financial returns on mixed stand variants, mainly because of significantly reduced risks.

A first impression of forest structure is provided by the frequency distributions of tree sizes and tree species. Accordingly, this section introduces methods that can be used for describing structure based on tree diameters and diameter–height relationships.

3.2.3.1 Unimodal Diameter Distributions

The diameter at breast height of a forest tree is easy to measure and therefore a frequently used variable for growth modeling, economic decision-making, and silvicultural planning. Frequency distributions of tree diameters measured at breast height (DBH) are often available and provide a useful basis for an initial analysis of forest structure.

Unimodal diameter distributions are commonly used to describe forest structure. An example of a DBH frequency distribution of a 116-year-old spruce research plot located in the Solling Mountains in Germany is shown in Fig. 3.13. The plot has 41 trees on an area of 0.16 ha. The two-parameter cumulative Weibull function is defined by the following equation:

$$F(x) = P(X \le x) = 1 - e^{-\left(\frac{x}{scale}\right)^{shape}} \tag{3.15}$$

where $F(x)$ is the probability that a randomly selected DBH X is smaller or equal to some defined DBH **x**. To obtain maximum likelihood estimates of the Weibull pdf, we can use the function *fitdistr* in package *MASS* of the *Comprehensive R Archive Network (http://cran.r-project.org/).*

Using the listed DBH observations (Fig. 3.13, center), the ***R*** code for the two-parameter Weibull function would be:

```
n.trees<-c(1,4,2,6,6,7,6,4,2,2,0,1); dbh.class<-c(32,34,36,38,40,42,44,46,48,50,52,54)
tree.d<-NULL;
 for(i in 1:length(dbh.class))
 {tree.d<-c(tree.d,rep(dbh.class[i],times=n.trees[i]))}
 hist(tree.d,freq=FALSE,ylim=c(0,0.09),col="green", main=NULL)
 library(MASS)                                        # loading package MASS
w<-fitdistr(tree.d,"weibull")                        # fitting weibull pdf parameters
lines(32:54,dweibull(32:54,w$e[1],scale=w$e[2]),col=2,lwd=2)
```

Assignment

If the Weibull parameters scale = 13.4 and shape = 2.6, calculate the relative frequency in the DBH-class 42–44 cm.

Solution

P(42<BHD≤44) = P(BHD≤44) - P(BHD≤42) = F(44) - F(42) = 0.68 – 0.52 = 0.17

List of DBHs (cm)	*DBH class*	Number of trees
41, 41, 38, 53, 44, 42, 50, 43, 40, 44, 40, 33, 39, 32, 49, 47, 38, 40, 37, 34, 47, 37, 41, 38, 38, 43, 40, 42, 34, 39, 41, 44, 41, 45, 43, 36, 36, 46, 46, 34, 50	32	1
	34	4
	36	2
	38	6
	40	6
	42	7
	44	6
	46	4
	48	2
	50	2
	52	0
	54	1

Histogram of density

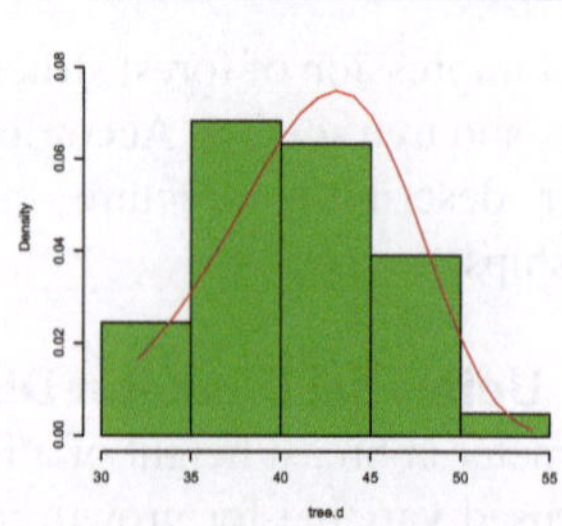

Fig. 3.13 List of spruce DBHs, table of DBH class frequencies, and histogram with fitted Weibull density function with parameters scale = 43.32; shape = 8.76

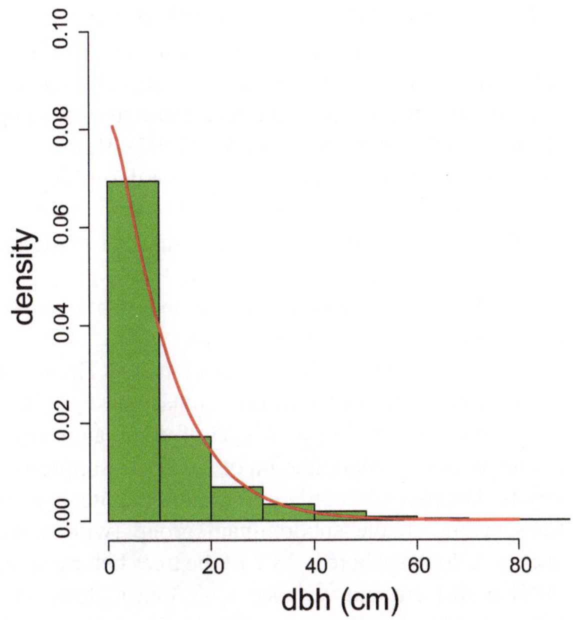

Fig. 3.14 Diameter distribution in plot OGF with fitted Weibull function (parameters: scale = 10.35; shape = 1.073)

A second example is a Weibull distribution fitted to an empirical diameter distribution in plot OGF by using the following code (it generates Fig. 3.14):

```
library(MASS);
dat<-read.csv("/Users/kg/Desktop/OGF_1ha.csv", header=TRUE)
my.wb<-fitdistr(dat$dbh,"weibull")
c=my.wb$e[1]                    # shape
b=my.wb$e[2]                    # scale
yl="density";xl="dbh (cm)" # labels
par(mar=c(5,6,1,1) + 0.1)   # plot Margins
hist(dat$dbh,freq=FALSE,ylim=c(0,0.1), xlab=xl,ylab=yl, cex.lab=2, col="green",main=NULL)
lines(1:80,dweibull(1:80,c,b),col=2,lwd=3)                  # c= 1.073; b=10.35
```

The Weibull function can also be inverted, which is useful for simulating a variety of forest structures characterized by the Weibull parameters:

$$F(x) = 1 - e^{-\left(\frac{x-a}{b}\right)^c} \text{ inverting gives} x = a + b \cdot \left[-\ell n\left(1 - F(x)\right)\right]^{\frac{1}{c}} \qquad (3.16)$$

where x is tree DBH and $F(x)$ is the probability that a randomly selected DBH is smaller than x. Assuming parameter values $a = 30$, $b = 13.7$, and $c = 2.6$, we can now answer the following question when we use the inverted three-parameter Weibull function: What is the DBH of a tree if 50% of the trees have a bigger DBH? When we use our data set, the answer is: $x = 30 + 13.4 \cdot \left[-\ell n(0.5)\right]^{\frac{1}{2.6}} = 41.6$ cm. Equation 3.16 is therefore useful for simulating diameter distributions with random numbers in the interval 0 and 1.

3.2.3.2 Bimodal Diameter Distributions

A great variety of empirical forest structures can be described by means of a theoretical diameter distribution. DBH frequencies can occur as bimodal distributions, which represent more irregular forest structures. Examples are given by Puumalainen (1996), Wenk (1996), and Condés (1997). Hessenmöller and Gadow (2001) found the bimodal DBH distribution useful for describing the diameter structures of beech forests which are often characterized by two distinct subpopulations, namely, fully developed canopy trees and suppressed, but shade-tolerant understory trees. They used the general form $f(x) = g \cdot f_u(x) + (1 - g) \cdot f_o(x)$, where $f_u(x)$ and $f_o(x)$ refer to the functions of the suppressed and dominant trees, respectively; g is an additional parameter, which links the two parts.

Mixtures of shade tolerant trees (such as *Fagus sylvatica* in Germany and *Acer mandshuricum* in north-eastern China) and light-demanding trees typically show two subpopulations: a population of dominant canopy trees with a wide range of diameters and a subpopulation of small and suppressed trees with a narrow range of DBHs. The two subpopulations are often quite distinct. In our example, 72% of the trees belong to the subdominant group (with mean DBH = 9.5 cm and variance = 20.6 cm), whereas 28% of the trees belong to the dominant group (with mean DBH = 38.1 cm and variance = 93.3 cm). However, the proportions of trees that belong to either the suppressed or the dominant group can differ, depending mainly on the silvicultural treatment history. Figure 3.15 shows the graph of a bimodal diameter distribution fitted to a hypothetical list of tree diameters.

The R library mclust may be used to estimate the bimodal distribution of the DBH measurements:

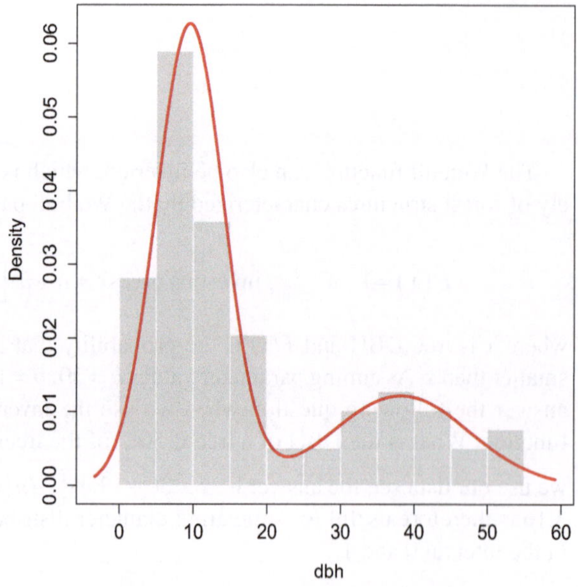

Fig. 3.15 Example of a bimodal diameter distribution of a shade-tolerant tree species involving a dominant subpopulation and suppressed subpopulation

```
library(mclust)
dbh<-c(28,34,9,6,11,7,8,5,6,9,38,5,6,5,12,35,3,4,2,3,12,9,
14,13,6,12,12,14,16,18,18,16,20,35,36,4,3,3,34,14,42,37,38,44,41,43,18,24,9,50,49,7,51,11,9
,54, 7,7, 6,5,9, 10,16,7,11,36,51,7,15,26,8,6,16, 10,11,10,12,21)
dens = densityMclust(dbh)
plot(dens, dbh, col="red",lwd=3,main=NULL)
```

Diameter structures become increasingly difficult to interpret as the number of tree species increases. This is one of the reasons why much work has focused on the DBH structures of monocultures—they are easy to describe and the methods are simple. Five common DBH and height distributions are shown in Fig. 3.16.

Structural analysis of the tree sizes can be rather challenging in multispecies forests where each individual species is characterized by its own particular diameter distribution.

3.2.3.3 Specific Diameter–Height Relationships

The size-class distribution of trees is one of the most characteristic attributes of an unmanaged natural forest. An important element of forest structure is the relationship between tree diameters and heights. The size-class distribution and the specific relationship between tree diameters and heights influence the growth potential and hence the current and future economic value of a forest (Knoebel and Burkhart 1991). Height distributions can be quantified by means of a discrete frequency distribution or a density function, in the same way as in a diameter distribution. However, height measurements in the field are considerably more time-consuming than DBH measurements, despite greatly improved measurement technology. For

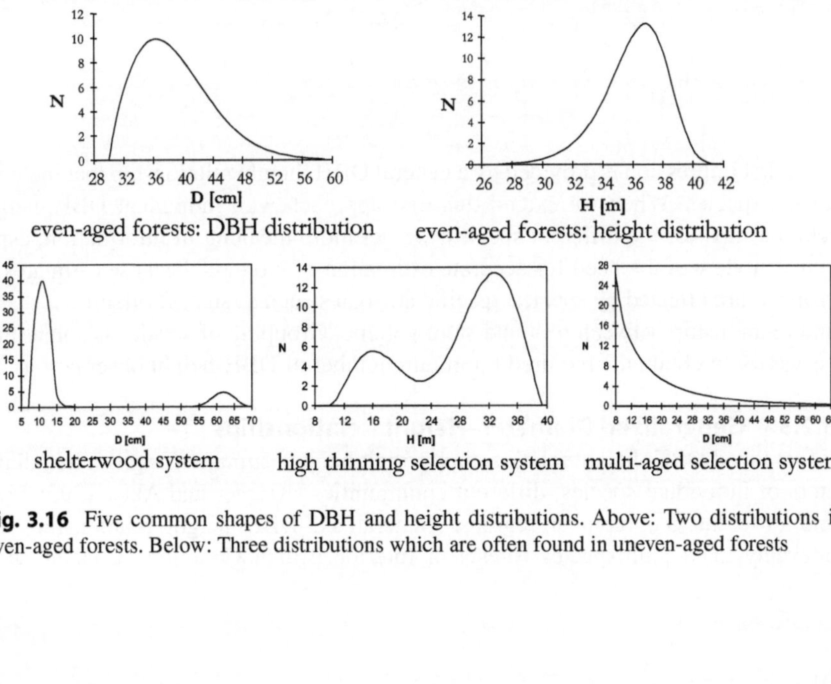

Fig. 3.16 Five common shapes of DBH and height distributions. Above: Two distributions in even-aged forests. Below: Three distributions which are often found in uneven-aged forests

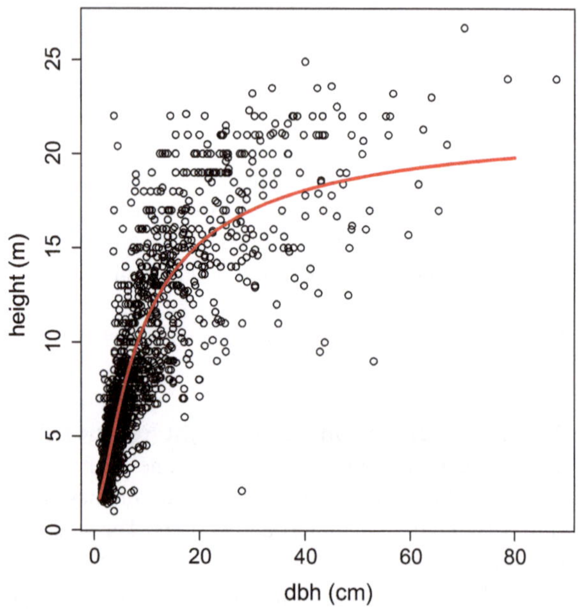

Fig. 3.17 Graph of the DBH–height relationship generated by using the R code above

this reason, simple DBH–height relationships are being developed, which permit height estimates for given tree diameters. The example in Fig. 3.17 was developed by using the following R code:

```
d=dat$dbh; h=dat$height
petterson.f<-function(x,p0,p1) {1.3+(x/(p0+p1*x))^3}
reg1 <- nls(h ~ petterson.f(d,p0,p1), start=list(p0=1, p1=0.27), trace=TRUE)
par(mar=c(5,6,1,1) + 0.1)   # Margins
plot(d, h, xlab="dbh (cm)", ylab="height (m)",cex.lab=2)
lines(1:80, predict(reg1, newdata=list(d=1:80)), col=2,lwd=3)
# p0=1.001; p1=0.365
```

A first impression is provided by a general DBH–height relationship that includes all the species. When we extend this first approach with individual DBH–height relationships for the different species, we get more meaningful information, especially in view of the need for accurate estimation of biomass, because estimates of biomass are affected by species-specific attributes such as stem form, crown shape, and relationship between root and stem volume. Grouping of species is sometimes necessary to obtain the required minimum number of DBH-height observations.

3.2.3.4 Generalized Diameter–Height Relationships

Generalized DBH–height relationships permit height estimates for given tree diameters of individual species, different communities (Kramer and Akça 1995: 138), and different geographical regions. Generalized DBH–height relationships for unevenly aged multispecies forests in Interior British Columbia, Canada, were developed, for example, by Temesgen and Gadow (2003). The analysis was based on permanent research plots. Eight tree species were included in the study: aspen (*Populus tremuloides* Michx.); western red cedar (*Thuja plicata* Donn.); paper birch (*Betula papyrifera* March.); Douglas fir (*Pseudotsuga menziesii* Mirb. Franco); larch (*Larix occidentalis* Nutt.); lodgepole pine (*Pinus contorta* Dougl.); ponderosa pine (*Pinus ponderosa* Laws.); and spruce (*Picea engelmanii* Parry * *Picea glauca (Monench) Voss*). Five models for estimating tree height as a function of tree diameter and several plot attributes were evaluated. The following model was found to be the best:

$$\hat{H} = 1.3 + a \cdot \left(1 - e^{b \cdot DBH^{c}}\right) \tag{3.17}$$

with $a = a_1 + a_2 \cdot BAL + a_3 \cdot N + a_4 \cdot G$ and $c = a_5 + a_6 \cdot BAL$.

BAL is the basal area of the larger trees (m^2/ha); *G* is the plot basal area (m^2/ha); *N* is the number of trees per ha; a_1 ... a_7; *b* and *c* are species-specific coefficients listed in Table 3.4.

A generalized DBH–height relationship such as the one described above can considerably improve the accuracy of height estimates in multispecies natural forests. In a managed forest, however, the parameter estimates should be independent of *G* and *BAL* because density changes at each harvest event, and this affects the model parameters.

3.2.3.5 Spatially Explicit Diameter–Height Relationships

Schmidt et al. (2011) modeled tree height–diameter relationships for Scots pine (*Pinus sylvestris*) in multisize and mixed-species stands in Estonia. The data set included 22,347 trees from the Estonian Permanent Forest Research Plot Network (Kiviste et al. 2007). The authors used the following model called the "Petterson function" in central Europe (Kramer and Akça 1995) and the "Näslund function" in Scandinavia (Kangas and Maltamo 2002):

Table 3.4 Parameter estimates for Eq. 3.17, based on multispecies forests in Interior British Columbia, according to Temesgen and Gadow (2003)

	A				B	c	
Tree species	a_1	a_2	a_3	a_4	B	a_5	a_6
Aspen	20.655			0.08724	0.01509	1.399	
Cedar	17.947		−0.0009	0.14087	0.03497	1.304	
Paper birch	20.446		−0.0007	0.13355	0.03576	1.262	
Douglas fir	32.037	−0.3504	−0.0007	0.18308	0.01797	1.093	0.00802
Larch	41.792				0.01709	1.118	0.00404
Lodgepole pine	20.852	0.3168	−0.0004	0.23962	0.03184	1.087	−0.0014
Ponderosa pine	32.208				0.01738	1.107	
Spruce	17.080	0.0932		0.34276	0.01073	1.462	

$$h_{ijk} = \left(\frac{d_{ijk}}{\alpha + \beta \cdot d_{ijk}} \right)^{\gamma} + 1.3 \tag{3.18}$$

where h_{ijk} and d_{ijk} are the total height (m) and breast-height diameter (cm) of the *k*th tree on the *i*th plot at the *j*th measurement occasion, respectively; and α, β, and γ are empirical parameters. The following R code shows an implementation using hypothetical diameter and height data for estimating the parameters of the Petterson model:

```
d<-c(15,18,45,27,31,17,25,64,15,55,48,28,22,37,36)
h<-c(12,17,29,19,20,14,20,33,12,34,33,24,22,30,31)
petterson.f<-function(x,p0,p1) {1.3+(x/(p0+p1*x))^3}
reg <- nls(h ~ petterson.f(d,p0,p1), start=list(p0=2.5, p1=0.25), trace=TRUE)
par(mar=c(5,6,1,1) + 1) # Margins
plot(d, h, xlab="dbh (cm)", ylab="height (m)",xlim=c(0,80), ylim=c(0,45),cex.lab=2)
lines(1:80, predict(reg, newdata=list(d=1:80)), col=2,lwd=3)
# p0=2.57; p1=0.268
```

The resulting graph is shown in Fig. 3.18.

Diameter–height relationships are often linearized in order to apply a mixed model approach (Lappi 1997; Mehtätalo 2004; Kinnunen et al. 2007). In addition, the use of generalized additive models (GAM) requires a specification of a linear combination of (nonlinear) predictor effects. The Petterson (Näslund) function can be linearized by using a constant exponent, for example, $\gamma = 3$ (Kramer and Akça 1995):

$$y_{ijk} = \frac{d_{ijk}}{\left(h_{ijk} - 1.3\right)^{1/3}} = \alpha + \beta \cdot d_{ijk} \tag{3.19}$$

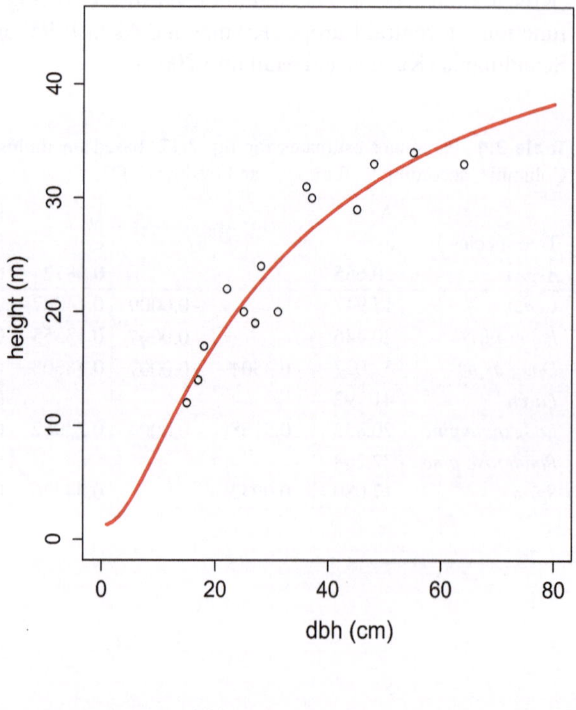

Fig. 3.18 Graph of the Petterson model based on the 15 hypothetical pairs of DBH-height values

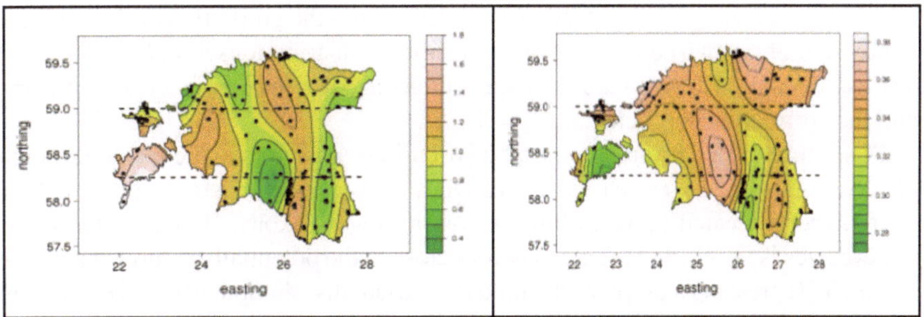

parameter alpha for northing/easting	parameter beta for northing/easting

Fig. 3.19 Two maps of Estonia showing plot locations and geographically explicit estimates of the parameters alpha and beta of the Petterson model (Eq. 3.6) for *Pinus sylvestris* trees in multi-size and mixedspecies stands in Estonia

Schmidt et al. (2011) fitted a two-level mixed model with random effects on stand and measurement occasion level. In this way, they could quantify the between-plot variability, as well as the measurement occasion variability. Figure 3.19 shows a map of Estonia and the geographical distribution of the two parameters of the DBH–height relationship.

Predictions of the coefficients of the diameter–height relationship were based on the quadratic mean diameter, as well as on the geographic coordinates x and y of the Estonian plots. The problem of spatially correlated random effects was solved by applying the specific methodology of Wood (2006) for two-dimensional surface fitting. The aim of this particular study was to use a spatially explicit approach, which would allow for a high accuracy prediction from a minimum set of predictor variables. Model bias was small, despite the somewhat irregular distribution of plot locations.

3.2.3.6 Bivariate Diameter–Height Distributions

Unmanaged forests are used as a standard for comparison with different types of managed stands. There are numerous examples showing that virgin beech forests exhibit structures that include more than one layer of tree heights (Korpel 1992). The vertical structure in a managed beech forest depends on the type of thinning that is applied. In a high thinning, which is generally practiced in Germany, only bigger trees are removed, while the smaller ones are allowed to survive for a very long time, resulting in a typical pattern of two subpopulations with different diameter–height relationships. Thus the population of trees is composed of a mixture of two subpopulations with different diameter–height distributions. Zucchini et al. (2001) presented a model for the diameter–height distribution that is specifically designed to describe such populations. The parameters all have familiar interpretations. If $f(d, h)$ denotes the bivariate probability density function of diameter and height, the proposed model is:

$$f(d,h) = \alpha \cdot n_1(d,h) + (1-\alpha) \cdot n_1(d,h) \qquad (3.20)$$

where α, a parameter in the interval (0, 1), determines the proportion of trees belonging to each of the two component bivariate normal distributions $n_1(d, h)$ and $n_2(d, h)$. The parameters of $n_j(d, h)$ are the expectations u_{dj}, u_{hj}; the variances σ^2_{dj} *and* σ^2_{hj}, and the correlation coefficients, ρ_j, $j = 1, 2$.

A mixture of two bivariate normal distributions was fitted to the diameter-height observations of 1242 beech trees with a DBH greater than 7 cm in the protected forest Dreyberg, located in the Solling region in Lower Saxony, Germany. The dominant species is beech, and the stand is very close to the potentially natural vegetation. Figure 3.20 presents a graph of the mixed bivariate distribution. Mixed DBH-height distributions can be fitted with the "flexmix" package (Leisch 2004) of the *Comprehensive R Archive Network.*

The larger subpopulation (approximately 80% of the entire population of *Fagus sylvatica*) comprises dominant trees in which the height–diameter relationship is quite different from that of the smaller subpopulation (approximately 20% of the population). Interestingly, the graphs of both distributions are very similar, although the species are different and although the forest communities occur in very different climates and represent different forest types. The occurrence of two distinct subpopulations is a common feature in both communities: a cohort of dominant trees with a high DBH–height ratio and a cohort of "slender" understory trees with a low DBH–height ratio.

3.2.4 Spatial Structure

Scientific literature abounds with studies of diameter distributions and diameter–height relationships of even-aged forests. More recently, the two-dimensional pattern of forest tree positions has generated much interest among statisticians.

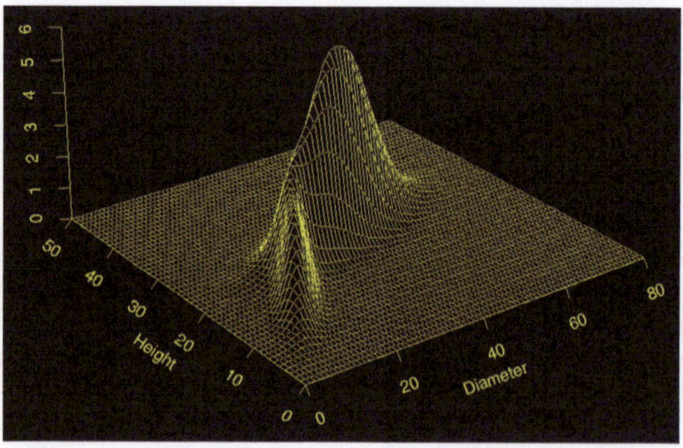

Fig. 3.20 Graph of a mixed bivariate normal distribution involving diameters and heights of a community of *Fagus sylvatica* trees in the Dreyberg forest. (See Zucchini et al. 2001)

Second-order characteristics (SOCs) were developed within the theoretical framework of mathematical statistics and then applied in various fields of the natural sciences, including the field of forestry (Illian et al. 2008; Møller and Waagepetersen 2007). The authors describe variability and correlations in marked and nonmarked point processes. SOCs depend on a distance variable r and quantify correlations between all pairs of points with a distance of approximately r between them. This allows SGO's to be related to various ecological scales and also, to a certain degree, to account for long-range point interactions (Pommerening 2002).

In statistical analysis, it is often assumed that the underlying point process is *homogeneous* (or *stationary*) and *isotropic*, that is, the corresponding probability distributions are invariant to translations and rotations (Diggle 2003; Illian et al. 2008). Although methods have also been developed for inhomogeneous point processes (see Møller and Waagepetersen 2007; Law et al. 2009 among others), those patterns are preferred in the analysis for which the assumptions of stationarity and isotropy holds. The latter simplifies the approach and allow a focused analysis of interactions between trees by ruling out additional factors such as, for example, varying site conditions. In this context, the choice of size and location of an observation window is crucial.

A popular second-order characteristic is Ripley's K-function (Ripley 1977). $\lambda K(r)$ denotes the mean number of points in a disc of radius r centered at the typical point i (which is not counted) of the point pattern where λ is the *intensity*, that is, the mean density in the observation window. This function is a cumulative function. Besag (1977) suggested transforming the K-function by dividing it by π and by taking the square root of the quotient, which yields the L-function that has both statistical and graphical advantages over the K-function. The L-function is often used for testing the CSR (complete spatial randomness) hypothesis, for example, in Ripley's L-test (Stoyan and Stoyan 1994; Dixon 2002).

Cumulative functions are not always easy to interpret. That is why functions of the nature of derivations are preferred for detailed structural analysis. One of these is the pair correlation function $g(r)$, which is related to the first derivative of the K-function according to the interpoint distance r (Eq. 3.1; see Illian et al. 2008):

$$g(r) = \frac{K'(r)}{2\pi r}. \tag{3.21}$$

For a heuristic definition, consider $P(r)$ as the probability of one point of the point process being at r and another one at the origin o. Let dF denotes the area of infinitesimally small circles around o and r, and $\lambda\, dF$ the probability that there is a point of the point process in a circle of area dF. Then

$$P(r) = \lambda dF \lambda dF \;\; g(r) \tag{3.22}$$

In this approach, the pair correlation function $g(r)$ acts as a correction factor for Poisson processes and for large distances where the point distributions are stochastically independent $g(r) = 1$. In the case of attraction of points $g(r) > 1$ (leading to a cluster process) and inhibition between points $g(r) < 1$ (resulting in regular patterns).

The partial pair correlation function $g_{ij}(r)$, which by analogy to $K_{ij}(r)$ (Lotwick and Silverman 1982) can also be referred to as the intertype pair correlation function, is a tool for investigating multivariate point patterns. Its interpretation is similar to that of $g(r)$. We will not present examples here, mainly because the focus of this section is on unmarked processes (refer to Penttinen et al. 1992; Stoyan and Penttinen 2000; Illian et al. 2008; Kint et al. 2003; Suzuki et al. 2008).

Complex spatial structures of uneven-aged multispecies forests are a challenging topic. A range of tools are available for detailed structural analysis; one of them is known as *nearest neighbor statistics* (NNS). NNS methods assume that the spatial structure of a forest is largely determined by relationships within neighborhood groups of trees. These methods have important advantages over classical spatial statistics, including low-cost field assessment and cohort-specific structural analysis. A cohort refers to a group of reference trees that share one or more common attributes. We distinguish species cohorts, gender cohorts, and size cohorts. When, for example, the immediate neighborhoods of the tall males of *Fraxinus mandshurica* in a deciduous forest in north-eastern China are compared to the neighborhoods of the juvenile females of the same species, significantly different neighborhood constellations are revealed.

Staupendahl and Zucchini (2006) presented an insightful review of a variety of forest structure indices that were specially developed or adapted for forestry use (Upton and Fingleton 1989; Upton and Fingleton 1990; Biber 1997; Gleichmar and Gerold 1998; Smaltschinski 1998; Gadow 1999). For practical purposes, the distance methods are particularly useful where complete mappings of populations are normally not feasible. These methods, which are also used for density estimation, are based on nearest-neighbor distances. The measurements are of two basic types: distances from a sample point to a tree or distances from a tree to a tree. Some of these methods are particularly useful, such as the *T*-square sampling (Besag and Glaves 1973) and its modifications (Hines and Hines 1979), especially regarding the test of complete spatial randomness (Diggle et al. 1976; Byth and Ripley 1980).

3.2.4.1 The Uniform Angle Index

Assunção (1994) described the spatial characteristics of a population of forest trees by calculating the angle between the vectors joining a particular sample point to its two nearest neighboring trees. Unaware of Assunção's work, but with the same intention, Gadow et al. (1998) also calculated the angles between neighboring trees. They did not use a reference point but a reference tree, thus allowing cohort-specific structural analysis. Cohort-specific structure refers to the specific structure in the vicinity of a cohort of reference trees, for example, the structure in the vicinity of a given species or in the vicinity of very large trees. They also considered more than two neighbors, thus extending the neighborhood range. Finally, they proposed that the angles should not be measured exactly but roughly classified, as exact measurements would be time-consuming during field assessments. They defined a standard angle to be equal to 360/(n + 1). For example, the standard angle in the case of four neighbors is 72° (Hui and Gadow 2002). The number of observed angles between two neighboring vectors, which is smaller than the standard angle is added and divided by the total number of angles, as follows:

$$W_i = \frac{1}{n}\sum_{j=1}^{n} v_{jk} \text{ with } v_{jk} = \begin{cases} 1, & \alpha_{jk} < \alpha_0 \\ 0, & \text{otherwise} \end{cases} \text{ and } 0 \le W_i \le 1 \tag{3.23}$$

When there are four neighbors W_i can assume five values: 0.0; 0.25; 0.5; 0.75 and 1.0. The following ***R***-function calculates the uniform angle index for a reference tree and its n nearest neighbors:

```
# input: x0;y0 (ref. tree coordinates), st_angle (standard angle)
# n=number of neighbors; x, y (vectors of neighbor coordinates)
uai<-function (x0,y0,x,y, n, st_angle)
{
   x1<-x-x0; y1<-y-y0                          # array of vectors for all neighbors
   c1<-sqrt(y1*y1 + x1*x1)                     # the lengths of the bevel edges
   phi<-acos(y1/c1)* sign(x1) * 180/pi + 360   # calc the angles
   for (i in 1:n) if (phi[i]>=360) phi[i]<- phi[i]-360
   sa<- order(phi,x1,y1,x,y); sa<-rbind(phi,x1,y1,x,y)[,sa]      # sorting the neighbors
   phi<-sa[1,]; x1<-sa[2,]; y1<- sa[3,]; x<-sa[4,]; y<-sa[5,]
   count<-0; if ((phi[4]- phi[1])< st_angle) count<-count+1
   for (i in n:2) if ((phi[i]- phi[i-1])< st_angle) count<-count+1
   w<-count/length(x); return(w)
} # end of function uai
```

The constellations for two reference trees located in the center of the square are shown in Fig. 3.21. The following ***R*** code generates the plots for the two groups and calculates the uniform angle indices, using the function *uai*:

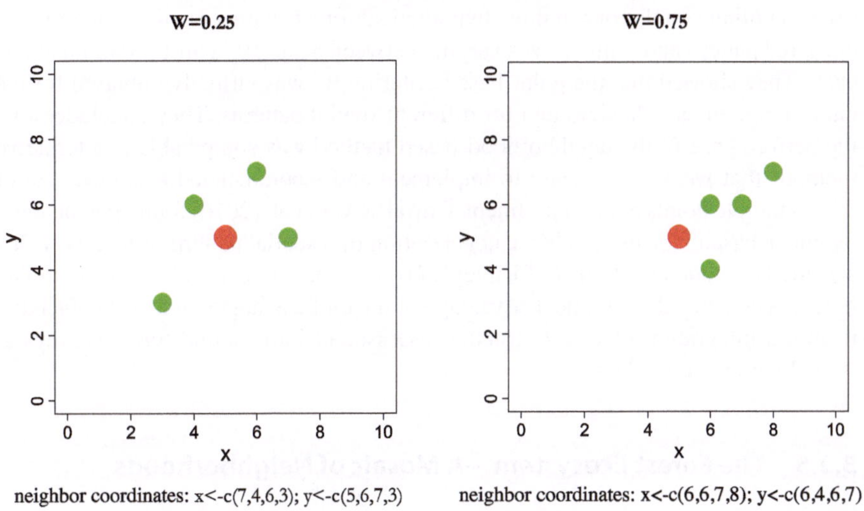

Fig. 3.21 Hypothetical structural patterns in the vicinity of two reference trees located in the center of a 10 m × 10 m square plot

```
x0=5; y0=5; x<-c(6,6,7,8); y<-c(6,4,6,7); stangle<-72    # stangle = standard angle
x0=5; y0=5;x<-c(7,4,6,3); y<-c(5,6,7,3)
par(mar=c(5,5,1,1) + 0.1)
plot(x,y, xlim=c(0,10),ylim=c(0,10), pch=19, cex=4,cex.lab=2,col="green") # plot them
points(x0, y0, pch=19, cex=5, col = "red")                  # pch=19: solid circle!!
w<-uai(x0,y0,x,y,4,stangle); w                              # call the function
```

Figure 3.21 shows the hypothetical structural patterns in the vicinity of two reference trees located in the center of a 10 m × 10 m square plot. The treatment of edge trees near the plot border can affect the estimation of structural indices since they can involve off-plot neighbors. Pommerening and Stoyan (2006) investigated the circumstances in which edge-correction methods were necessary and evaluated different approaches. They proposed a variable buffer zone around the edge of the plot. Only those trees that were located further away from the plot edge than the distance to their *nth* nearest neighbor were selected as reference trees.

Staupendahl and Zucchini (2006) presented an index considering the three trees nearest to a reference point. A structure unit can consist of a reference tree or a reference point and the n nearest neighbors in the vicinity of the tree or point. A reference-tree-based structure unit will reveal spatial patterns in the vicinity of a particular type of tree cohort (species and/or size).

The expected value of the uniform angle index is the average of an infinite number of realizations of *W* obtained by repeatedly randomizing orientations of the sample grid:

$$\mathrm{E}\left(\hat{W}\right) = \frac{1}{m} \cdot \sum_{r=1}^{m} \hat{W}_r \qquad for \;\; m \to \infty \tag{3.24}$$

where $\hat{W}$ is the estimate of *W* in the *r*th simulation. Staupendahl and Zucchini (2006) used sampling simulations in three hypothetical forests with a regular, random, and clustered pattern and sample sizes varying between 5 and 50, with 1000 replications each. They showed that the point-based criterion $\hat{W}_p$ was virtually unbiased for the entire range of sample sizes and for different spatial patterns. They concluded that the performance of the neighborhood-based method was comparable to alternative methods that were more costly to implement and recommended a sample size of 20–30 sample points per compartment. Corral-Rivas et al. (2010) found that nearest-neighbor-based indices enable a categorization of a spatial pattern of forests with a sensitivity comparable to that of Ripley's *L(t)* test, and at finer scales. As mentioned before, a considerable practical advantage of the uniform angle index is the fact that it allows integration of forest structure assessment into routine forest surveys at virtually no additional cost.

3.2.5 The Forest Ecosystem—A Mosaic of Neighborhoods

Traditionally, a forest ecosystem is characterized by area-based attributes (basal area, biomass, number of trees per hectare), mean values and distributions (diameters), and relationships (DBH–height regression). These classical attributes of a forest ecosystem are usually assessed in the field by sampling in field plots of specified shape and size (Kleinn et al. 2010). The most common method is the angle count method in fixed area plots of circular, square, or rectangular shape and sample points. Refer to standard forest mensuration textbooks for details (e.g., Van Laar and Akca 2007; Burkhart and Tomé 2012). In traditional forest sampling schemes, variables that characterize the spatial structure and diversity of an ecosystem, are either included in a rather rudimentary way or not considered at all.

Traditional forest sampling concentrates on the assessment of forest density, volume, and timber products. Information about the distributions of aggregation, species mingling, and size differentiation is required to describe and evaluate spatial structure and diversity in uneven-aged multispecies forests. *Aggregation* refers to the regularity of tree positions. High aggregation is often described as "clumped." *Species mingling* defines the degree of spatial segregation of the tree species in a forest. Low species mingling means high segregation, no matter how many species occur in the forest. *Size differentiation* measures the degree in which trees of different sizes are spatially mingled. High size differentiation implies that trees of varying size occur in close vicinity of one another. A convenient sampling scheme for assessing these variables is distance sampling, a popular method among ecologists despite the potential for bias (Krebs 1999; Nothdurft et al. 2010).

Assessing complex forest structures at affordable cost in practical applications is always an added bonus. This section therefore includes a brief introduction to methods that can be used to incorporate structural assessment in routine forest inventories. We also give examples of methods for measuring tree size diversity. Finally, we deal with species diversity, again in a spatial context. Forest spatial structure can be characterized by nearest neighbor statistics (NNS) that include the variables *aggregation, species mingling,* and *size differentiation* (Fig. 3.22). The interested reader is referred to an excellent and comprehensive treatment of neighborhood structures by Pommerening and Grabarnik (2019).

A convenient sampling unit is the *n-tree structure unit* (NSU)[3], which consists of either a sample point or a reference tree that is located closest to the sample point, and the n neighbors nearest to the point or tree (Fig. 3.22). In spatial statistics, the terms "point related" and "test location related" for summary characteristics are sometimes used (Pommerening 2008; Illian et al. 2008).

When the NSU is used, it will therefore be possible to assess forest structure and diversity plus all the other traditional variables without having to measure tree-to-tree distances or tree coordinates. Examples of *point*-based structure variables are shown below:

Aggregation (W_P)	Species mingling (M_P)	Size differentiation (T_P)
Proportion of angles $\alpha_{i1}, \ldots, \alpha_{in}$ between the vectors from reference point i towards the n nearest neighbors (clockwise), which are smaller than some standard angle α_0	Number of tree species of the n nearest neighbors as a proportion of the maximum possible number of species (n).	DBH coefficient of variation of the n nearest neighbors around a reference point i.

[3] The *n-tree structure unit* is not to be confused with the *n-tree sample,* where the plot radius is equal to the distance to the nth tree, as proposed by Prodan (1966).

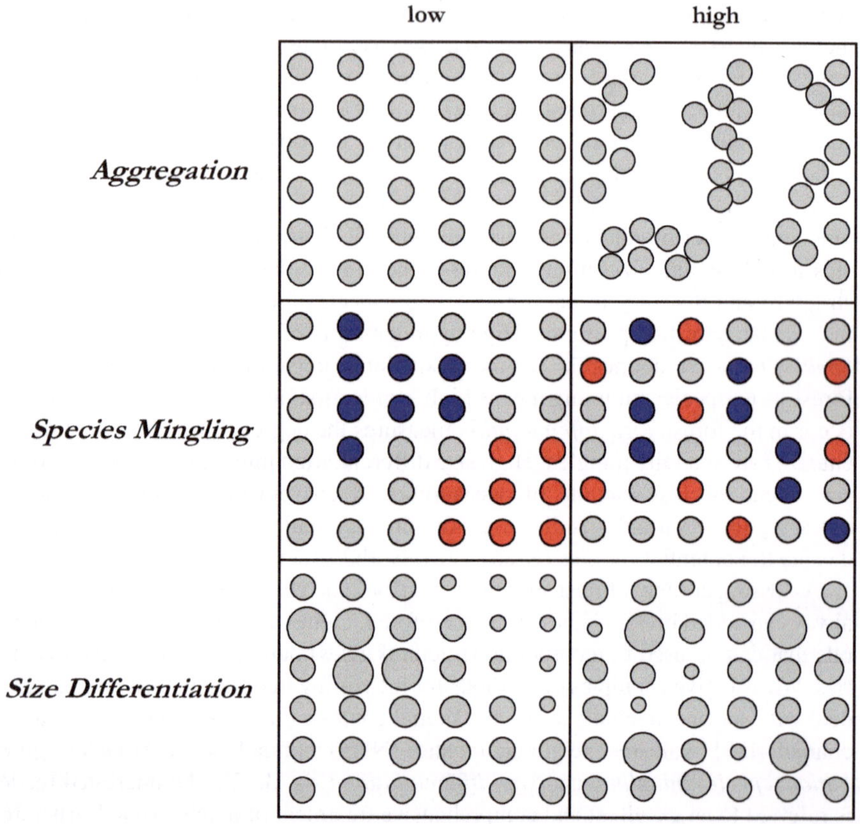

Fig. 3.22 Schematic representation of the three variables—aggregation, species mingling and size differentiation—that are used to describe forest structure and diversity (as proposed by Vanclay)

Examples of *tree*-based structure variables are shown below:

Symmetry (W_T)	Sameness (M_T)	Dominance (U_T)
Proportion of angles $\alpha_{i1}, \ldots, \alpha_{in}$between the vectors from reference tree *i* towards the *n* nearest neighbors (clockwise), which are smaller than some standard angle α_0	Proportion of species of the *n* nearest neighbors which do not belong to the species of the reference tree.	Proportion of *n* nearest neighbors with a DBH or height smaller than the DBH or height of the reference tree.

In a sampling design where the azimuth and distance to the plot center (or a reference corner) are known for each tree, the attributes of an *n-tree structure unit* can be calculated by using the x/y coordinates. In sampling designs where tree coordinates are not known, the n nearest neighbors need to be identified in the field, and the NSU can simply be integrated into the normal plot assessment. Measuring distances between trees is costly and unnecessary. Figure 3.23 shows two

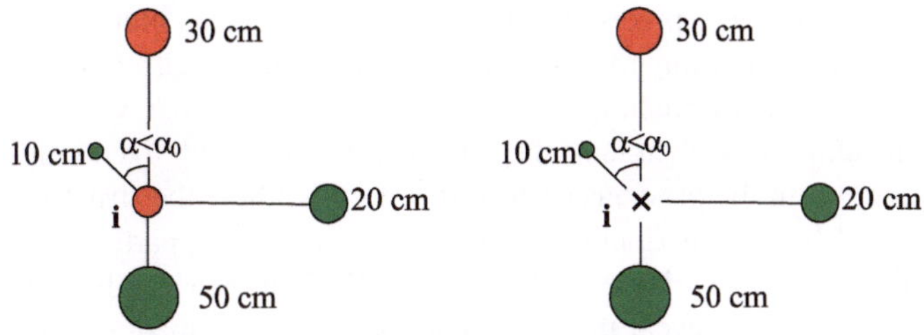

Fig. 3.23 Two neighborhood units used to assess the distributions of aggregation, species mingling, and size differentiation in a forest: a tree-based neighborhood unit (left) and a point-based unit (right). The numbers represent the tree DBHs

neighborhood units, one tree-based and the other point-based, and the four nearest neighbors around the reference point or reference tree. Both units feature the same two species *A* (one or two trees) and *B* (three trees). The numbers represent the tree DBHs.

Only one angle α_{ik} is smaller than the standard angle α_0 (which was chosen to be 72°). The numbers in the diagrams represent each tree's DBH. The following structure variables can be calculated for a tree-based and a point-based unit, respectively:

Tree-based structure unit	Point-based structure unit
$W_T = \frac{0+0+0+1}{4} = 0.25$	$W_P = \frac{0+0+0+1}{4} = 0.25$
$M_T = \frac{0+1+1+1}{4} = 0.75$	$M_P = \frac{2}{4} = 0.5$
$U_T = \frac{0+0+0+1}{4} = 0.25$	$T_P = \frac{\sqrt{\left\{(50-7.5)^2 + (10-27.5)^2 + (30-27.5)^2 + (20-27.5)^2\right\}/3}}{(50+10+30+20)/4}$ $= 0.65$

Hui and Albert (2004) proposed a slightly modified sampling design. Their sampling unit consists of the four trees closest to a sample point. Each of these four trees is a reference tree. This gives four *n-tree structure units* instead of just one. Thus, the structure of a forest is characterized by the distributions of the variables assessed in all the field sample points. An advantage of the tree-based approach is the fact that the analysis of the forest structure can be cohort-specific in this case. "Cohort-specific" means that the neighborhood constellations are evaluated for all the trees of a given cohort: all the trees of a particular species or all large individuals of a given species, for example.

Motz et al. (2010) carried out a study to investigate how tree diversity measures can be estimated as extensions of existing forest resource inventories. They compared the precision-of-angle count and fixed-radius plot sampling with respect to nine representative diversity indices in three different forest types at stand, enterprise, and national forest level. Their results indicated that most of the spatially explicit indices are more precisely estimated by fixed radius plots. The superiority of fixed radius plot sampling to angle count sampling increased significantly with an increasing diameter differentiation of the forests. Basal areas were estimated by angle count sampling with at least the same precision as with fixed radius plots.

The examples in the previous sections have shown how the structure of a forest can be defined by the distribution of tree sizes and by the particular relationship between different size variables (DBH-height; DBH-crown width). Such distributions, however, do not reveal the particular spatial arrangement of these variables (Schütz 2002). It requires little imagination to realize that a variety of spatial patterns of tree diameters is possible for any particular diameter frequency distribution. We can refer to a high degree of size mingling when large and small trees occur in close vicinity to each other, or to low size mingling when large and small trees are spatially segregated. Such spatial arrangements can be assessed in the field by evaluating sample plots or neighborhood groups of trees.

There are two main categories of edge-correction: *plus-sampling* and *minus-sampling* (Illian et al. 2008). Plus-sampling makes full use of all the data within the observation window, either by recording objects outside the observation window or by simulating them. Minus-sampling (also referred to as the border method) only makes use of an inner subset of the observation window. Pommerening (2008) summarized edge-correction methods according to these two categories.

We define the trees in such a group as a "structure unit." Fixed-area sample plots are preferred for an unbiased assessment of area-based variables, for example, number of trees or basal area per hectare. Neighborhood groups are useful if the target information must be independent of density, as in the case of the spatial mingling of tree diameters. Each structure unit has certain attributes. The distribution of the DBH mingling values of the sampled structure units will reveal the particular spatial arrangement of tree sizes in a larger forest area. Within a given diameter distribution, a greater homogeneity of structure units reveals spatial segregation: Trees of the same size are found in proximity to each other.

Examples of the segregation of tree sizes can be found in a forest with dense groups of saplings within gaps adjoining areas of mature trees. On the other hand, a high degree of spatial mingling can be found in a forest of shade tolerant species where all tree sizes occur in proximity to one another. Both forests can have the same diameter distribution but completely different diameter mingling patterns. DBH mingling within a structure unit can be described by means of a variety of indices (Kint et al. 2003). Six examples of indices calculated for neighborhood groups are listed in Table 3.5. The distribution of each index can facilitate the analysis of the structural composition of the entire forest; see reviews by Staudhammer and LeMay (2001) and McElhinny et al. (2005) for more information.

Sterba and Zingg (2006) found close correlations and specific relationships between the four heterogeneity indices (H', CV_{DBH}, *Gini*, *Skew*) and the two neighborhood indices (T_i, U_i). Their findings were based on data from a great variety of forest types, including coppice forests of different ages, and even-aged and uneven-aged forests. The four heterogeneity indices can be used to evaluate the evenness of tree sizes within a particular structure unit (a sample plot or a neighborhood group).

Table 3.5 Four indices of heterogeneity (H, CV_{DBH}, Gini, Skew) and two neighborhood indices (T_i, U_i), which can be used for describing DBH mingling within a structure unit (as proposed by Sterba and Zingg 2006)

Formula/symbol	Explanation	References
$H' = -\sum_{i=1}^{S} p_i \ell n(p_i)$	S = number of DBH classes; $p_i = n_i/N$ with n_i = number of trees in DBH class i; N = all trees	Shannon and Weaver (1949)
$CV_{DBH} = \frac{SDD}{D_m}$	CV_{DBH} = DBH coefficient of variation; SDD = standard deviation of DBHs (cm); D_m = mean DBH (cm)	Any textbook on statistics
$Gini = \frac{0.5 - L_A}{0.5}$	L_A = area under the Lorenz curve (obtained by plotting the cumulated basal areas over the cumulated tree numbers)	Stöcker (2002)
$Skew = \frac{\Sigma(D - D_m)^3 / (n-1)}{SDD^3}$	$Skew$ = skewness of DBH distribution; SDD = standard deviation of DBHs (cm); D_m = mean DBH (cm)	Sterba and Zingg (2006)
$T_i = 1 - \frac{1}{n} \cdot \sum_{i=1}^{n} \frac{\min(D_i, D_j)}{\max(D_i, D_j)}$	T_i = DBH differentiation; n = number of neighbors of reference tree i; D_i, Dj = DBH (cm) of ref. tree i and neighbor j	Füldner (1995)
U_i	U_i = dominance, the proportion of neighboring trees which are smaller than the reference tree in a structure unit	Hui et al. (1998)

The two neighborhood indices can be used to evaluate the structural attributes of a cohort, that is, a specific subpopulation of trees that are similar in some way.

Hyytiäinen and Haight (2012) employ a specific form of the Shannon index (S) to evaluate habitat quality. They describe the simultaneous species richness and tree size variation of a forest as follows:

$$S = -w_{sp} \sum_{u=1}^{y} \frac{B_u}{B} \ln\left(\frac{B_u}{B}\right) - w_{size} \sum_{v=1}^{z} \frac{B_v}{B} \ln\left(\frac{B_v}{B}\right) \tag{3.25}$$

The weights of species and size diversity are denoted by w_{sp} and w_{size}, respectively. B is the total basal area, B_u is the basal area of trees of species u, and B_v is the basal area of trees in diameter class v. The number of species is y, and the number of diameter classes is z. A negative characteristic of the Shannon-Weaver index is the fact that the value increases with increasing evenness of the relative frequencies, while rare species are often considered to contribute more to diversity than common species (see Hui et al. 2011a, b).

Nearest-neighbor statistics are particularly useful in the study of cohort-specific structural attributes. We use the data of two experimental field plots to illustrate the analysis of the diversity of tree sizes in the neighborhood of all the trees in a specific cohort of reference trees:

1. Plot Jiaohe1 is located in the *Jiaohe* forest in Jilin Province of north-eastern China. The forest is managed by selective harvesting. The plot area is one hect-

are. The tree diameters at breast height (DBH) range from 0.4 to 79.2 cm. The following tree species occur in the plot: *Betula costata, Carpinus cordata, Fraxinus mandshurica, Acer mandshurica, Tilia amurensis, Acer mono, Syringa reticulata var. mandshurica, Juglans mandshurica, Abies holophylla, Pinus koraiensis, Ulmus laciniata, Juglans mandshurica, Phellodendron amurense,* and *Ramus davurica.*

2. Plot 1080 is located in the southern central part of Estonia. The plot shape is circular and the plot radius is 30 m. Three tree species occur in the plot: *Pinus sylvestris, Betula pendula,* and *Picea abies.* Plot 1080 has no fixed buffer zone. Instead, only those trees that are located further away from the plot perimeter than the distance to their third nearest neighbor are considered in the analysis (see Pommerening and Stoyan 2006).

The following ***R*** code was used to identify the neighbor indices of a specific cohort of trees (*tree0*) inside the buffer zone (*inside*) with the *nnwhich()* function of the *spatstat* library. The dataframe *dat* refers to all the trees, including those located in the buffer zone. Their coordinates are *dat$X* and dat$Y.

```
tree0<-subset(inside, inside$d_cm>1)      # specify all reference trees with a dbh>1cm
nb<-4                                         # number of nearest neighbors
width<-100; leng<-100                         # size of plot (m)
library(spatstat)                             # specifics may change
allxy<-ppp(dat$X,dat$Y, c(0,width), c(0,leng))  # create a planar point pattern dataset
m4<-nnwhich(allxy, k=1:nb)                    # all nb nearest neighbors
ref<-m4[tree0$Tree,]                          # neighbor indices of cohort
```

It is important to note that this particular analysis does not require mapped data sets. The important advantage of using neighborhood groups is the fact that data can be assessed in routine forest surveys at practically no additional cost. More results of the analysis of the two plots are presented at the end of this section.

3.2.6 Species Segregation and Mingling

A major structural feature of a forest with only one species is the spatial distribution of tree sizes. We may be interested to know whether big and small trees are segregated or spatially mingled. An important structural attribute of a forest with more than one tree species is the way in which the species are spatially mingled or segregated. An index of segregation involving two species was proposed by Pielou (1977). The information required for calculating Pielou's index of segregation for two species is shown in Table 3.6.

When we use the quantities specified in Table 3.9, we find that the expected number of mixed pairs is equal to the sum of the joint probabilities $r_A s_B + r_B s_A$:

$$S = 1 - \frac{p_{AB} - p_{BA}}{r_A \cdot s_B + r_B \cdot s_A} \tag{3.26}$$

Table 3.6 Information required for calculating Pielou's index of segregation for two species A and B. The p_{AA}., p_{AB}, p_{BA}, and p_{BB} are the observed relative frequencies of one species having the other as nearest neighbor; r_A, and r_B are the proportions of A and B's frequencies; s_A, and s_B are the proportions of any tree having A or B as its nearest neighbor

	Nearest neighbor		
	Species A	Species B	Total
Species A	p_{AA}	p_{AB}	r_A
Species B	p_{BA}	p_{BB}	r_B
Total	s_A	s_B	1

Table 3.7 Index of segregation S for two species groups (beech and other species) in a mixed deciduous beech forest in northern Germany, before and after a selected thinning

	Plot 1	Plot 2	Plot 3	Plot 4	Plot 5	Plot 6	Mean
Before	−0.105	0	0	−0.233	−0.154	+0.095	−0.066
After	−0.111	0	0	−0.438	+0.065	0	−0.080

S takes on a negative value when the observed number of mixed pairs exceeds the expected one, that is when there is mutual attraction among the two species or groups of species. A value of zero indicates that the distributions of the two species are independent of each other. $S > 0$ indicates spatial segregation. The segregation index was calculated by Füldner and Gadow (1994) for six sample plots in a mixed deciduous beech forest in northern Germany. For their calculations, they used two groups of species (beech and others) before and after a thinning. The results are shown in Table 3.7.

The index assumes zero values in several plots. Some of the index values were not at all affected by the thinning, which means that neither segregation nor repulsion between the beech trees and the other species (mainly maple and ash) nor any change in this condition was observed. Mutual attraction was enhanced by the thinning in Plot 4 while the opposite was found in Plot 5.

Pielou's index can be used to test whether an observed value is significantly different from zero. However, its use is limited to first-neighbor constellations. Theoretically, a tree of species A can have up to five nearest neighbors of species B. Each B in turn can have an A as its nearest neighbor. This kind of constellation, which can give an inflated impression of the degree of attraction, is shown in Fig. 3.24.

The index of segregation can be inflated by either a clustered or a regular distribution of tree positions (Upton and Fingleton 1990: 244). Another problem is the fact that Pielou's index in this form does not lend itself to evaluating species-specific mingling, which may be important information in certain forest types. We will attempt to solve this problem by considering not one, but n nearest neighbors of a given tree. *Mingling* of the *i*th tree (M_i) is then defined as the relative proportion of neighboring species that are different.

An overall test of random labeling of a 2 × 2 nearest-neighbor contingency table was used by Zhang et al. (2010a, b, c) to evaluate the spatial segregation of genders

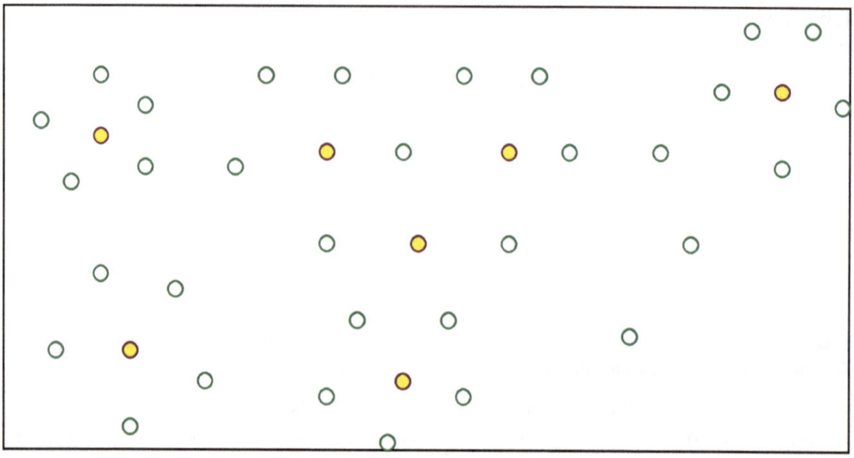

Fig. 3.24 Hypothetical forest showing inflated attraction between two species

in the tree species *Fraxinus mandshurica*. The objective was to test the null hypothesis that all counts in the nearest-neighbor contingency table were equal to their expected counts. The quadratic form proposed by Dixon (2002) is:

$$C = (N - EN) \cdot Sigma^{\wedge} \cdot (N - EN) \tag{3.27}$$

where *N* is the vector of all cell counts and *EN* is the expected count in the contingency table. *EN* has an asymptotic normal distribution with mean 0 and variance 1. *Sigma* is the variance-covariance matrix of those counts, and *Sigma*^ is a generalized inverse matrix of *Sigma*. Under the null hypothesis of random labelling of points, *C* has an asymptotic Chi-square distribution with k(k-1) degrees of freedom. p-Values are computed from the probability of observing equal or larger values of *C*. The overall statistic *C* can be partitioned into k specific test statistics *C*[*i*], where $i = 1 \ldots k$. Each *C*[*i*] tests whether the frequencies of the neighbors of species *i* are similar to the expected frequencies, which is the case if the points are randomly labeled. The analysis is performed with the "dixon" package in the R software.

A useful measure of tree species diversity reflects not only the species richness of a community, but also its particular spatial structure. It would be an important advantage for any index of forest diversity if the assessment effort can be integrated easily into routine management surveys aimed at improved land-use planning and ecological monitoring. Developing more effective variables which not only reflect species richness and evenness but also include a spatially relevant structural component is a worthwhile challenge (Pretzsch 2003; Xia 2007).

The literature on biodiversity is extensive and new ideas emerge continually.[4] Statistical methods for analyzing beta diversity include principle component

[4] For an exhaustive review of biodiversity indices, see Ferris and Humphrey (1999) and Spanos and Feest (2007).

analysis and redundancy analysis (Legendre and Gallagher 2001). Some studies presented sophisticated and advanced evaluations of beta diversity even in mega-diverse tropical forests (e.g., Condit et al. 2002). Many of these approaches require tree coordinates that are normally not available, a major limitation when their practical use is considered. If a costly assessment of tree coordinates is not required for diversity assessment, spatially relevant biodiversity analyses based on data from routine forest surveys are a distinct possibility.

The main structural feature of a forest of a single species is the distribution and spatial mingling of tree sizes. A multispecies forest is also characterized by tree species richness and the spatial mingling of tree species. A direct consequence of large-scale forest destruction, especially since the second half of the twentieth century, is a serious depletion of tree species diversity. Many species have already become extinct or are threatened with extinction, and because of these threats scientists have been increasing their activities in the area of biodiversity research.

Hui and Albert (2004) defined the spatial relationship between a particular reference tree and its n nearest neighboring trees as a "forest spatial structure unit." Theoretically, n could be any reasonable number. However, a series of field studies led them to the conclusion that the optimum group size of such a spatial structure unit consisted of a reference tree and its four nearest neighboring trees. Pommerening (2006) found that the optimum number of nearest neighbors is the factor that allows the best spatial reconstruction of a given forest from sampled NNS (see details below).

The species spatial mingling within a structure unit is equal to the proportion of neighbors, which are not of the same species as the reference tree. Prior work in this regard only considered the spatial mingling of the reference tree by calculating the proportion of neighbors that are not of the same species as the reference tree. The spatial mingling index proposed by Gadow (1993) and used by Füldner (1995), Pommerening (2002) and several other authors is defined as follows:

$$M_i = \frac{1}{n}\sum_{j=1}^{n} v_{ij} \tag{3.28}$$

where n is the number of nearest neighbors considered, $v_{ij} = 1$ if the jth neighboring tree is not of the same species as the ith reference tree and $v_{ij} = 0$ otherwise. The distribution of the M_i values allows a detailed study of the spatial diversity within a forest in conjunction with the species proportions within a given tree population.

However, the number of different tree species in the structure unit was not taken into account, and this was a shortcoming of the original mingling index.

Expected values of spatial indices may be of interest to students involved in the comparative analysis of forest structures. According to Lewandowski and Pommerening (1997) expected mingling (***E***M), can be calculated as

$$\mathrm{E}M = \sum_{i=1}^{s} \frac{p_i\left(p - p_i\right)}{p\left(p-1\right)} \tag{3.29}$$

where s is the number of species, p is the number of trees in the observation window, and p_i refers to the number of trees of species i.

Expected mark diameter differentiation is not as straightforward as expected mark mingling. Using tree diameters (DBH) as example marks Pommerening (1997: 18) proposed that DBH should be sorted in ascending order, that is, $i < j \Rightarrow DBH_i \leq DBH_j$. As a result, the index set J of a given forest is obtained. Then the auxiliary measure R is defined as

$$R_i = \begin{cases} 0 & \text{for } i = 1 \\ \sum_{j=1}^{i-1} DBH_j & \text{for } i = 2,\ldots,p \end{cases} \tag{3.30}$$

The expected mark differentiation for tree DBH, **E***T*, can now be calculated as

$$\mathrm{E}T = 1 - \frac{2}{p\left(p-1\right)} \sum_{j=1}^{p} \frac{R_j}{DBH_j}. \tag{3.31}$$

Details of the derivation of Eqs. (3.30) and (3.31) can be found in Pommerening (1997).

3.2.7 A System of Nearest Neighbor Statistics

The nearest-neighbor statistics (NNS) include the variables *symmetry, mingling, differentiation,* and *dominance.* These variables had been described in several previous studies, albeit not always with identical designations. Synonyms for *symmetry*, for example, are *aggregation* and *uniform angle index* (Gadow and Hui 2002; Gadow et al. 2011). Figure 3.25 presents a visual impression of the possible neighborhood constellations for a given reference tree located in the center of each of the 20 imaginary 10 m × 10 m plots.

Symmetry refers to the regularity of tree positions in the neighborhood of a reference tree and is expressed by the proportion of angles $\alpha_{i1}, \ldots, \alpha_{in}$between the vectors pointing from reference tree i toward the n nearest neighbors (clockwise). These angles must be smaller than some standard angle α_0. *Mingling*, which describes the species diversity in the neighborhood, is equal to the proportion of species of the n nearest neighbors that are not of the same species as the reference tree.

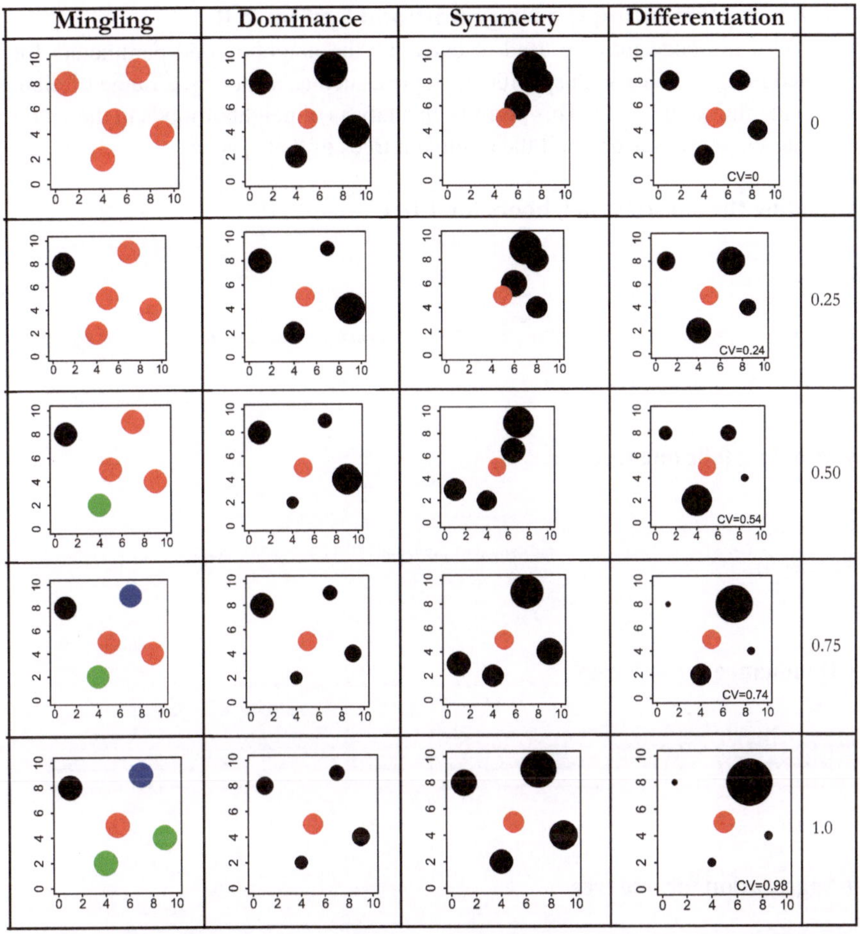

Fig. 3.25 Circles represent tree diameters; colors represent different tree species (the reference tree in each graph is located at the center). Each graph represents a specific value based on neighborhood spatial arrangements and tree diameters; 625 neighborhood constellations are possible with four variables, each of which can assume five discrete values

The variable *Differentiation* (T_i) measures the variation of tree dimensions in the neighborhood. T_i is equal to the diameter coefficient of variation (CV) of all the trees in the neighborhood group, constrained to a maximum value of 1 and expressed in five classes (0–0.25; 0.25–0.50; 0.50–0.75; 0.75–1.0).

Finally, *Dominance* measures the dominance of the reference tree in relation to its neighbors and is equal to the proportion of n nearest neighbors that are smaller than the reference tree (in height or in diameter if heights are not known). In a neighborhood group of five trees (a reference tree and its four nearest neighbors), each of these four variables can therefore assume five values resulting in $5^4 = 625$ possible neighborhood constellations (on condition that the diameter coefficient of variation can only assume five discrete values).

3.2.7.1 Implementing the Neighborhood System in R

The following code may be used to calculate the neighborhood parameters for a selected reference tree with specific attributes, such as a species, a range of diameters, and minimum height. This permits an analysis of neighborhoods in the vicinity of a particular species or of all the dominant trees of that species.

 # Finding the 4 nearest neighbors for 1 tree

```
Near.f=function(x0,y0,data1){
    x1=abs(data1$x-x0); y1=abs(data1$y-y0)
    dis=sqrt(x1*x1+y1*y1); nord=order(dis); nord[1:5]
data2=rbind(data1[nord[1],],data1[nord[2],],data1[nord[3],],data1[nord[4],],data1[nord[5],]
)}
```

 # Mingling fo0r one tree

```
Ming.f<-function (sp0,spNB){
count<-0 ; for (i in 1:4) { if (! identical(sp0,spNB[i])) count<-count+1}
m<-count/4}
```

 # Dominance for one tree

```
Dom.f<-function (D0,hNB){
count<-0; for (i in 1:4){if (hNB[i]<=D0) count<-count+1}
u<-count/4}
```

 # Aggregation for one tree

```
# Calculate the angle between two trees
Angle.f=function(x0,y0,xi,yi){
  deltax=xi-x0; deltay=yi-y0
  if (deltay>0 &deltax>=0) {Angle=atan(deltay/deltax)} else
  if (deltay==0 &deltax>=0) {Angle=pi*0.5} else
  if (deltay<=0 &deltax>=0) {Angle=pi+atan(deltay/deltax)} else
  if (deltay<0 &deltax<0) {Angle=pi+atan(deltay/deltax)} else
  if (deltay==0 &deltax<0) {Angle=pi*1.5} else
  {Angle=pi*2+atan(deltay/deltax)}
  Angle}
# Calculate the difference of two angles & compare to 72°
AngleDiff.f=function(Angle1, Angle2){
  AngleDiff=Angle1-Angle2
  if (AngleDiff>pi) {AngleDiff=2*pi-AngleDiff}
  if (AngleDiff>=72*pi/180){AngleDiff=0} else
  {AngleDiff=1}}
# calculate the Angle Index for one tree
W.f=function(x0,y0,x,y){
  Angle4=c()
  for(i in 1:4){
    xi=x[i];yi=y[i]
    Angle4[i]=Angle.f(x0,y0,xi,yi)
  }
  Angle4o=order(Angle4)
  count12=AngleDiff.f(Angle4[Angle4o[2]],Angle4[Angle4o[1]])
  count23=AngleDiff.f(Angle4[Angle4o[3]],Angle4[Angle4o[2]])
  count34=AngleDiff.f(Angle4[Angle4o[4]],Angle4[Angle4o[3]])
  count41=AngleDiff.f(Angle4[Angle4o[4]],Angle4[Angle4o[1]])
  count=(count12+count23+count34+count41)/4
  w=count}
```

The structure functions are implemented by selecting a particular type of reference tree defined by species, size, or a combination of attributes of interest. The neighborhoods of these reference trees of interest can then be analyzed in detail and compared with the neighborhoods of other trees of interest, for example, dominant versus subdominant individuals.

```
# Implementing the Structure Functions
# Define reference trees: species and plot
dat=rect; dat$x=x; dat$y=y; specie="Bel"; marg=4                    # define margin to avoid
bias
dat$d1=dat$dbh; dat$sp=rect$Species; dat$cr=0.56+0.06*dat$dbh
#dat$y1<-max(dat$y)-marg; dat$y2<-min(dat$y)+marg; dat$x1<-max(dat$x)-marg; dat$x2<-
min(dat$x)+marg
dat$y1<-100-marg; dat$y2<- marg; dat$x1<-100-marg; dat$x2<- marg
reftree<-subset(dat, specie==dat$sp&dat$x>dat$x2& dat$x<dat$x1& dat$y>dat$y2& y<y1,
select=c(x,y,sp,d1,cr))
# Calculations
M <-array(NA); D <-array(NA); W <-array(NA); C<-array(NA); T<-array(NA)
for (i in 1:length(reftree$x)){
  x0=reftree[i,]$x;y0= reftree[i,]$y
  Near4=Near.f(x0,y0, dat)[2:5,]          # dataframe NB's for 1 reftree
  x=Near4$x; y=Near4$y                    # coordinates ref tree and neighbors
  W[i]=W.f(x0,y0,x,y)                             #function Aggregation
  sp0=reftree[i,]$sp; spNB=Near4$sp
  M[i]=Ming.f(sp0,spNB)                   #function Mingling
  d0=reftree[i,]$d1; d=Near4$d1
  D[i]=Dom.f(d0,d)                        #function Dominance
  cr0=reftree[i,]$cr; crNB=Near4$cr
  C[i]<-Cr.f(x0,y0,x,y,cr0,crNB)                  #function Crowding
  t0=c(reftree[i,]$d1, tNB=Near4$d1)
  T[i]=sqrt(var(t0))/mean(t0); if(T[i]>1) T[i]<-1         # dbh differentiation
}
 reftree$M=M; reftree$D=D; reftree$W=W; reftree$C=C; reftree$T=T; reftree
print(paste0(specie," M= ", round(mean(M),digits=2)," D= ", round(mean(D),digits=2)," W= ",
round(mean(W),digits=2)," T= ", round(mean(T),digits=2)))
# "Achar: M= 0.64 D= 0.54 W= 0.51 T= 0.61"
# "Bel M= 0.47 D= 0.55 W= 0.5 T= 0.55"
# "Mundi M= 0.86 D= 0.62 W= 0.59 T= 0.6"
```

Table 3.8 shows parameters based on an analysis of a one-hectare teak plot in India (file central.csv).[5] The columns show the four neighborhood variables (M: Mingling, D: Dominance, W: Aggregation and T: Size differentiation) for all the species that are represented by at least 10 individuals. *Mingling* (M) values are generally high to very high, indicating a high species mixture at close range within the immediate neighborhood of a given species. The most *dominant* (D) species within their neighborhoods are *Terminalia tomentosa* and *Tectona grandis*. The *spatial distributions* (W) of trees within the neighborhoods are usually random (i.e., near 0.5) or tending toward regular with slightly lower values (*Lagerstroemia parviflora* = 0.37; *Ougeinia dalbergioides* = 0.41). Diameter differentiation (T) values are generally greater than 0.5, indicating a high variability of DBHs at close range.

Once established, a tree has a fixed location throughout its life. Each population of trees therefore exhibits a very specific structure and diversity at a given point of time. This condition is not static, however. Tree growth, regeneration, and mortality are important structuring processes. Specific structures generate particular processes of growth and regeneration. These processes in turn produce particular patterns and constellations. Thus, structure and processes are mutually dependent.

[5] Data set provided with the kind permission of the Forest Survey of India.

Table 3.8 Parameters listed in the file central.csv. Columns 7–10 show the four neighborhood variables (M: Mingling, D: Dominance, W: Aggregation and T: Size differentiation) for all the species that are represented by at least ten individuals

Code	Species	MeanH (m)	N/ha	BA (m^2/ha)	IV	M	D	W	T
Achar	*Buchanania latifolia*	10.5	14	0.4	0.018	0.64	0.54	0.51	0.61
Bel	*Aegle marmelos*	12.2	187	3.7	0.204	0.47	0.55	0.50	0.55
Bhira	*Chloroxylon swietenia*	10.9	42	0.7	0.043	0.59	0.48	0.50	0.52
Kari	*Miliusa tomentosa*	8.6	37	0.4	0.034	0.90	**0.30**	0.49	0.68
Lendia	*Lagerstroemia parviflora*	7.2	10	0.1	0.009	0.82	**0.29**	0.39	0.63
Mode	*Lannea coromandelica*	14.8	15	0.6	0.023	0.89	0.50	0.47	0.59
Mundi	*Mitragyna parvifolia*	13.6	21	0.5	0.025	0.86	0.62	0.59	0.60
Saja	*Terminalia tomentosa*	21.4	18	1.4	0.041	0.95	**0.84**	0.62	0.62
Sehara	*Bauhinia retusa*	8.8	17	0.3	0.018	0.81	0.59	0.64	0.49
Teak	*Tectona grandis*	20.1	171	14.7	0.416	0.71	**0.78**	0.48	0.69
Tendu	*Diospyros melanoxylon*	6.4	145	0.9	0.119	0.52	0.33	0.53	0.60
Tinsa	*Ougeinia dalbergioides*	8.4	10	0.1	0.009	0.88	0.34	0.41	0.67

Some degree of heterogeneity or richness is always associated with a specific forest structure. However, in a forest ecosystem such diversity refers not only to species richness but also to a range of phenomena that determine the heterogeneity within a community of trees. Most important on a macro scale is the diversity of tree dimensions.

Despite advances in remote sensing, mapped tree data are usually not available in the practice of CCF management. Because forest inventories tend to provide tree data samples in small observation windows, nearest-neighbor statistics (NNS) are a realistic choice in an overwhelming number of cases. For this reason, a variety of NNS methods are described in this chapter, including unmarked, as well as market patterns and their use in structural analysis and description. NNS methods assume that the spatial structure of a forest is largely determined by the relationships within neighborhood groups of trees. These methods have important advantages over classical spatial statistics, such as low cost field assessment and cohort-specific structural analysis.

Trees are objects that can be conveniently described by their locations and attributes. When mapped data are available, methods of point process statistics are therefore useful. Especially important are the *second-order statistics* (SOCs) that were developed within the theoretical framework of mathematical statistics and then applied in various fields of research, including that of forestry. Examples of SOCs in forestry applications are Ripley's K- and Besag's L-function, pair correlation functions, and mark variograms. SOCs describe the variability and correlations in marked and nonmarked point processes. Functional second-order characteristics depend on a distance variable r and quantify correlations between all pairs of points with a distance of approximately r between them. This allows them to be related to various ecological scales and also, to a certain degree, to account for long-range point interactions.

An analysis of forest structure reveals the distribution of tree attributes, including the spatial distribution of tree species and their dimensions, crown lengths, and leaf areas. The assessment of these attributes facilitates a comparison between a managed and an unmanaged forest ecosystem. In CCF, it is especially important to evaluate structural modifications caused by harvest events. For this reason, analytical tools that describe structure and diversity are needed. The structure of a forest is the result of natural processes and human disturbance. Human disturbance in the form of selective tree removal in CCF has a major effect and as a result forest structure is determined to a considerable extent by silviculture.

Important ecosystem functions, as well as the potential and limitations of human use, are defined by existing forest structures. The vertical and horizontal distributions of tree sizes determine the distribution of microclimatic conditions, the availability of resources, and the formation of habitat niches, and thus, directly or indirectly, the biological diversity within a forest community.

Forest structure is of interest to students of ecology, but it also has economic implications. Uneven-aged multi-species forests were found to be more resistant to natural hazards, and thus may generate superior financial returns. To ensure long-term productivity, however, the management of uneven-aged forests requires a good understanding not only of species-specific responses to shading and competition on different growing sites but also of more sophisticated methods of sustainable harvest planning. A selective harvest event in an uneven-aged forest involving the removal of a variety of tree sizes within each of several species is much more difficult to quantify and prescribe than by standard descriptors of harvest events which are used in rotation management systems. The methods described in this chapter may therefore be helpful, not only to analyze complex structures but also to predict the dynamics of an uneven-aged multispecies forest.

A better understanding of *forest structure* is a key to an improved definition of harvest events and to the modeling of tree growth and recruitment after that event. A logical improvement of the mingling index would be to consider, in Eq. (3.28), spatial mingling, as well as the number of tree species. This can be achieved by multiplying M_i with S_i/n_{max}, where S_i is the number of tree species in the neighborhood of reference tree i, including tree i, and n_{max} is the maximum number of species in this structure unit i. In our special case, $n_{max} = 5$, and this was found to be a good compromise between scientific assertion and practicality. The reasons for $n_{max} = 5$ were given in several studies that evaluated different numbers of neighbors under field conditions (Gadow and Hui 2002; Hui and Albert 2004; Hui et al. 2007). A greater number of neighbors allow a more detailed and differentiated analysis. The smaller the number of neighbors, the lower the cost of assessment. The most convenient number of neighbors from a practical point of view is 1, simply because one nearest neighbor is very easy to identify in the field. The assessment effort increases with an increasing number of neighbors. The cost of identifying a fifth neigbor increases sharply (because of the need to evaluate many tree-to-tree distances). The cited field studies have shown that the use of four neighbors ($n_{max} = 5$) represents a practical compromise between explanatory value and assessment effort.

3.2.7.2 Extensions of the Mingling Concept

The spatial diversity status (MS_i) of a particular tree species is determined by the relative species richness within the structure unit i and the degree of mingling with the reference tree, which can be expressed as follows:

$$MS_i = \frac{S_i}{n+1} \cdot M_i \tag{3.32}$$

where M_i are the species mingling values as defined above and $n + 1$ is the maximum number of species in the structure unit i. S_i is the number of species actually occurring in unit i. Equation (3.32) therefore measures the tree species richness as well as an important spatial characteristic of a structure unit. A reference tree of a common species is more likely to have neighbors of the same species, which is reflected by low MS_i values (Fig. 3.26, graphic on the left). On the other hand, a rare species is likely to produce a high proportion of high MS_i values (Fig. 3.26, graphic on the right). Hence, MS_i is especially sensitive to rare tree species. When we consider the assessment effort, another important practical advantage is the fact that it is not necessary to measure tree coordinates in the field. A useful statistic reflecting

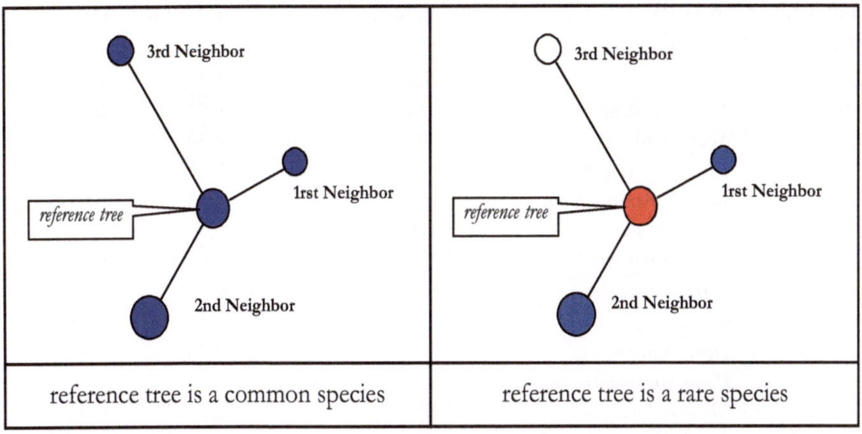

Fig. 3.26 Reference tree of a common species (left) and a rare species (right). A rare species tends to produce high MS_i values

the status of each individual tree species in the community is the species average spatial status (MS_{sp}), which is the average value of the MS_i for each tree species in the community:

$$MS_{sp} = \frac{1}{N_{sp}} \sum_{i=1}^{N_{sp}} (MS_i) \tag{3.33}$$

where N_{sp} is the number of trees of species *sp* in the community.

The commonly used diversity indices are represented by a single statistic, which combines the species richness and evenness (the ratio of observed to maximum richness). Assuming additivity, the spatial diversity of tree species in a tree population can be expressed as the sum of the average spatial diversity status of the different tree species. This sum is conveniently expressed by the TSS criterion proposed by Hui et al. (2011a, b):

$$TSS = MS_{sp1} + MS_{sp2} + \cdots + MS_{spn} = \sum_{spk=1}^{s} MS_{spk} \tag{3.34}$$

where *s* is the number of tree species. *TSS* assumes a maximum value of 1 when each tree is of a different species, in which case all MS_{sp} are equal to 1. When there is only one species of *N* trees in the community, the species richness is at a minimum, and the *TSS* value is zero. The *TSS* variable therefore represents the sum of the average spatial mingling values of all the tree species in the community. It is a measure not only of tree species richness but also of tree species spatial diversity within a given ecosystem. Table 3.9 shows the MS_{sp} values for the 21 tree species in Jiaohe1 (China) and the three species in plot 1008 (Estonia). The *TSS* value for the 17 tree species in Jiaohe1 is 11.22/17 = 0.66. The *TSS* value for the three tree species in 1080 is (0.02 + 0.46 + 0.44)/3 = 0.31. Because of its absolute dominance,

Table 3.9 Average MS_{sp} values and diameter coefficients of variation for the different tree species in plots Jiaohe1 and 1080

Plot	Species	Number of trees	MS_{sp}	av.cvd
Jiaohe1	*Acer mandshurica*	214	0.54	0.84
	Acer mono	202	0.50	0.81
	Ulmus propinqua	123	0.55	0.76
	Juglans mandshurica	112	0.63	0.74
	Fraxinus mandshurica	110	0.60	0.76
	Ulmus laciniata	65	0.66	0.81
	Syringa reticulata	60	0.63	0.70
	Carpinus cordata	55	0.67	0.81
	Tilia amurensis	38	0.70	0.89
	Pinus koraiensis	35	0.63	0.90
	Abies holophylla	16	0.68	0.90
	Betula costata	15	0.71	0.66
	Tilia mandschurica	8	0.71	0.78
	Phellodendron amurense	6	0.70	0.60
	Acer tegmentosum	5	0.84	0.74
	Prunus padus	3	0.67	0.68
	Qercus mongolica	3	0.80	0.88
1080	*Pinus sylvestris*	244	0.02	0.26
	Betula pendula	7	0.46	0.28
	Picea abies	7	0.44	0.45

Pinus sylvestris in plot 1080 has a very low MS_{sp} value. The average diameter coefficient of variation is 0.75 in Jiaohe1 and 0.33 in 1080. Both the TSS value and the average diameter coefficient of variation indicate a high spatial diversity in Jiaohe1. The spatial diversity of tree species and sizes is much lower in 1080.

TSS is sensitive to rare species and to variations in community structure, including species' spatial isolation and spatial mingling. For this reason, the TSS criterion is more effective in measuring tree species diversity than the commonly used indices. It allows detailed interpretation of forest spatial diversity and of forest structural modifications after selective thinnings in CCF systems.

A particular advantage of the TSS index is the fact that its assessment, which is based on neighborhood relations, can be integrated easily into routine forest management surveys at practically no additional cost.

3.3 Assessing Community Dissimilarity

Evaluating compositional differences among species assemblages is central to answering some of the most challenging questions concerning the mechanisms that generate and maintain biodiversity in forest ecosystems (Swenson et al. 2012; Legendre and Cáceres 2013; Hao et al. 2019a, b). The earliest measure of community dissimilarity was proposed more than 100 years ago by Jaccard (1900). Since that time, numerous indices have been published to estimate the dissimilarity among

Table 3.10 Prominent examples of indices for estimating the dissimilarity among species assemblages

Index Name	Equation	Description
Jaccard	$\frac{B+C}{A+B+C}$	A represents the number of species shared by two communities; B and C are the number of species unique to each of the two communities
Sørensen	$\frac{B+C}{2A+B+C}$	
Bray-Curtis	$\frac{\sum_{i=1}^{n}\left\|p_i^a - p_i^b\right\|}{\sum_{i=1}^{n}\left(p_i^a - p_i^b\right)}$ Or $\frac{1}{2}\sum_{i=1}^{n}\Delta_i^{a,b}$	n is the number of species; p_i^a and p_i^b are the relative frequencies of species i in plots a and b n is the number of species; $\Delta_i^{a,b}$ is the absolute difference on species frequencies; $\Delta_i^{a,b} = \|p_i^a - p_i^b\|$, p_i^a and p_i^b are the relative frequencies of species i in plots a and b

species assemblages; some prominent examples are given in Table 3.10. The variety of indices is bewildering, and many are viewed with suspicion. Chao et al. (2005) suggested that accurate estimators for the Jaccard and Sørensen indices are unattainable, as these two indices are based only on species-incidence (i.e., presence or absence) data. The Jaccard and Sørensen indices are also sensitive to sample size, especially for communities with numerous rare species.

The Bray-Curtis index is a modified version of the Sørensen index (Bray and Curtis 1957). Species abundance or frequency data are used for calculations, and the index conveys no information regarding the biological heterogeneity among species. A number of other species-based indices exhibit the same problem. The need for accurate statistical estimators to detect community dissimilarity has intensified, and new indices continue to appear (e.g., Legendre and Legendre 1998; Lennon et al. 2001; Magurran 2004; Luzopone and Knight 2005; Devictor et al. 2010; Swenson 2011).

Biological heterogeneity that reflects evolutionary or functional differences between species within communities constitutes an important component of community dissimilarity (Ganeshaiah et al. 1997; Ganeshaiah and Shaanker 2000; Webb 2000). One of the most popular approaches to quantifying evolutionary or functional dissimilarity in communities is the phylogenetic or functional beta diversity, which can be measured in different ways, including the UniFrac index (Devictor et al. 2010; Luzopone and Knight 2005), the PhyloSor index (Bryant et al. 2008), and Rao's dissimilarity index (Rao 1982; Ricotta and Burrascano 2009). Swenson (2011) has shown that most metrics of phylogenetic beta diversity are highly correlated, indicating that there is considerable redundancy among these indices. It is important to note that there is no generalized approach that unifies taxonomic, phylogenetic, and functional information within a coherent framework to measure community dissimilarity. In view of this need, Hao et al. (2019a, b) presented the *discriminating avalanche* to quantify the biological dissimilarity between species assemblages. The discriminating avalanche incorporates species abundance and any quantifiable biological difference between species, such as taxonomic, phylogenetic, and functional dissimilarities.

To demonstrate the approach, Hao et al. (2019a, b) subdivided a 30-hectare plot with mapped trees into 30 subplots, each with an area of one hectare. The taxonomic distances between species were estimated with the help of a table of classification (Clarke and Warwick 1998). The function "taxa2dist" from the R package "vegan" was used to calculate the taxonomic distances between species (Oksanen et al. 2013). Taxonomic distances are hierarchically structured, while phylogenetic distances are based on the branch lengths of a phylogenetic super-tree, as it provides a more accurate estimate of species evolutionary history. The phylogenetic distances between species were calculated by means of the mega-phylogeny of Qian and Jin (2016), which is based on the work of Zanne et al. (2014) and includes all families of extant seed plants in the world. The authors used the R package "S.PhyloMaker." Phylogenetic distances between species were calculated as the sum of the branch lengths separating each pair of species. In addition, 11 plant traits that are known to have functional significance for plant growth, survival, and regeneration were measured for the 38 species in the plot, including six leaf traits, four stem traits, and one regenerative trait (seed mass). The taxonomic, phylogenetic, or trait distances were calculated to discriminate between species i and j within a given community by means of the Euclidean distance:

$$d_{ij} = \sqrt{\sum_{k=1}^{c} \left(X_{ik} - X_{jk} \right)^2} \tag{3.35}$$

where X_{ik} and X_{jk} are the values of the ith and jth species for the kth character (trait), and c is the number of traits. The discriminating avalanche (dA) can be used as a measure of the biological "distance" between two communities (Gadow et al. 2018):

$$dA = \frac{1}{2} \sum_{i=1}^{s} \sum_{j=1}^{s} \Delta_i^{a,b} * d_{ij} * \Delta_j^{a,b} \tag{3.36}$$

where $\Delta_i^{a,b}$ refers to the absolute difference between the frequencies of species i in community a and b ($\Delta_i^{a,b} = |a_i - b_i|$; a_i and b_i represent the frequencies of species i in community a and b; $\Delta_j^{a,b}$ is the equivalent for species j); and d_{ij} represents the biological distance between species i and j. Hao et al. (2019a, b) used the normalized d_{ij} by dividing all pairwise distances by the maximum distance (Devictor et al. 2010). Principal component analyses (PCAs) of the functional traits showed that the first six principal components (PCs) jointly explained 85% of the total variation of the plant traits, and they were accordingly used in the analyses.

3.3.1 An Example Using Taxonomic Distances

To demonstrate the general approach, we use the information of a Linnean taxonomy to calculate the biological dissimilarity between species. Measuring the taxonomic distance is relatively straightforward compared to using more complex plant traits or phylogeny. The taxonomic distances between three one-hectare plots (La Victoria, Mil Diez, San Esteban) located in Durango in Mexico are estimated by means of a classification table that includes the five taxonomic levels (Table 3.11).

The standard botanical nomenclature follows *The Plant List* (TPL, www.theplantlist.org), an internationally accepted standard database for plant nomenclature. The taxonomic information can be extracted from the APG IV classification system (angiosperm phylogeny group 2016) and Christenhusz et al. (2011) for angiosperms and gymnosperms, respectively.

```
# use TPL to ensure that the naming is correct
library(plantlist)
a=unique(LaVic$Species); b=unique(MilDiez$Species); c=unique(SanEst $Species)
sp=unique(c(a,b,c))          # all species which occur in the 3 plots
sp2 <- TPL(sp)               # standardized species names and classification based on APG III
sp2                          # the first column should be species, delete fam. number)
# need to add new headers
x <- data.frame(sp2);names(x)=c("species","genus","family","order","family_number","class")
sp= subset(x,select=c("species","genus","family","order","class"))
write.csv(sp, ("/Users/kg/Desktop/tax_struc.csv"))
```

The distances between species are scaled in such a way that the longest path length between taxa is 1. For example, the distance is 1 if two individuals belong to different taxonomic groups (*angiosperms* and *gymnosperms*). The distance is 0.8 if two individuals belong to different orders, but share the same group (e.g., *Pinales* and *Fagales*). By this analogy, the distance will be 0.2 for two individuals that belong to different species but share the same genus. The normalized distances between two individuals that differ at different levels within the taxonomic hierarchy are shown in Table 3.12.

Table 3.11 Classification for the 11 species that occur in the three observational study areas

Species (SP)	Genus	Family	Order	Group
Arbutus arizonica	*Arbutus*	Ericaceae	Ericales	Angiosperms
Arbutus bicolor	*Arbutus*	Ericaceae	Ericales	Angiosperms
Arbutus tessellata	*Arbutus*	Ericaceae	Ericales	Angiosperms
Arbutus xalapensis	*Arbutus*	Ericaceae	Ericales	Angiosperms
Juniperus deppeana	*Juniperus*	Cupressaceae	Cupressales	Gymnosperms
Pinus strobiformis	*Pinus*	Pinaceae	Pinales	Gymnosperms
Pinus cooperi	*Pinus*	Pinaceae	Pinales	Gymnosperms
Pinus durangensis	*Pinus*	Pinaceae	Pinales	Gymnosperms
Pinus leiophylla	*Pinus*	Pinaceae	Pinales	Gymnosperms
Pinus teocote	*Pinus*	Pinaceae	Pinales	Gymnosperms
Quercus sideroxyla	*Quercus*	Fagaceae	Fagales	Angiosperms

Table 3.12 Normalized distances between two individuals that differ at different levels within the taxonomic hierarchy

Individuals differ by	Normalized distance
Species	0.2
Genus	0.4
Family	0.6
Order	0.8
Group	1.0

The absolute differences between the species frequencies in the three plots (the $\Delta_i^{a,b}$ and $\Delta_j^{a,b}$) and the normalized taxonomic distances between the species (the d_{ij}) provide the basic information for calculating the discriminating avalanche. The results are shown in Fig. 3.27 (Hao et al. 2019a, b).

The Jaccard, Sørensen, and Bray-Curtis community dissimilarity indices (Table 3.10) are also calculated. The different measures of biological dissimilarity exhibit different patterns. Compared to traditional measures, discriminating avalanche values show relatively moderate distances, that is, a greater similarity between plots (caused by the two dominant genera *Pinus* and *Arbutus* that include many closely-related species). The Jaccard and Sørensen dissimilarities exhibit similar tendencies: maximum distance between plots La Victoria and Mil Diez, minimum distance between plots San Esteban and Mil Diez (Fig. 3.28).

As expected, the values of the plain avalanche (pA) distances that are based only on species frequencies exceed those of the discriminating avalanche (dA) distances that incorporate species taxonomies. The pA distances exhibit a tendency similar to those of the Bray-Curtis indices: maximum distance between San Esteban and Mil Diez, minimum distance between La Victoria and San Esteban; whereas the dA distances exhibit a tendency different from all the others: maximum distance between La Victoria and Mil Diez, and minimum distance between La Victoria and San Esteban (Fig. 3.28).

Dissimilarity means "not the same." When we state that two forest communities are the same, we are in fact choosing to ignore some of the differences that do exist. Although numerous measures have been published, the development of new measures for quantifying community dissimilarity more accurately, more comprehensively and more objectively therefore remains a priority in community ecology (Chao et al. 2005). The discriminating avalanche is a new index that qualifies as an important measure of forest community dissimilarity.

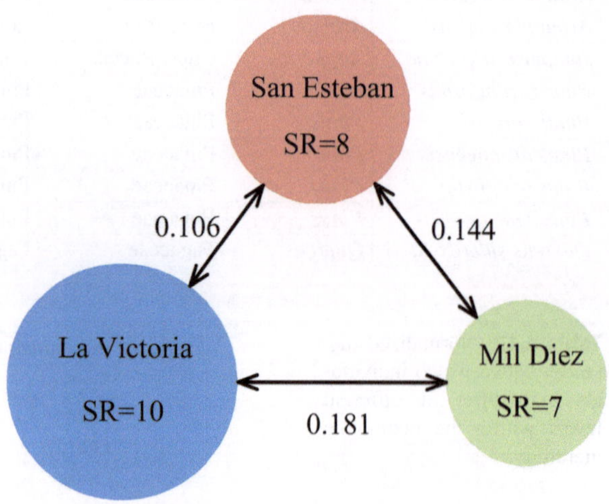

Fig. 3.27 Graphic representation of the results; circle size is proportional to species richness (SR) within each plot. The taxonomic distances between the plots are shown by the lengths of the three arrows

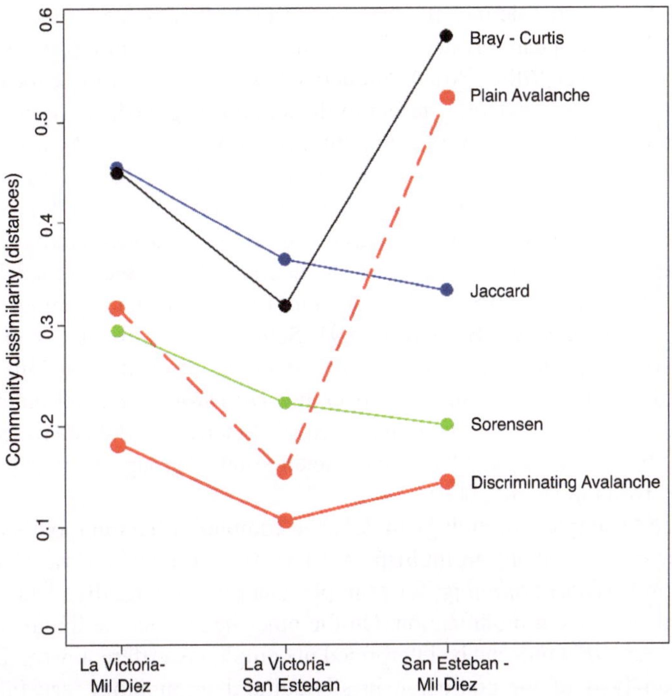

Fig. 3.28 Pair distances between the three forest plots in the case of five different measures of biological dissimilarity between communities

3.4 Analyzing Harvest Events

Many forest ecosystems are managed, and in those forests a harvest event is usually the most dramatic among all possible ecological disturbances. Harvesting involves regular reductions of density and the modification of forest structure. Harvest events modify the tree diameter distribution, the species composition, and the spatial patterns, and thus the distribution of radiation and temperature. A major objective of many CCF systems is to mimic natural ecosystem dynamics. Therefore, it is important to be able to describe the effects of a particular harvest event and to compare the after-harvest state of a particular ecosystem to some assumed ideal state. It is necessary to describe the structural changes caused by tree removals if we want to be able to assess the full impact of ecosystem modification caused by a particular harvest event (Puumalainen 1998).

3.4.1 Linguistic Variables

A harvest event is a silvicultural activity that modifies the structure of the ecosystem. A selective thinning, typical in CCF systems, is usually described by foresters with an adjective that defines the weight and type of ecosystem modification that occurred. The European forestry literature abounds with definitions of specific harvest events (Kramer 1988: 180). A "moderate low thinning" removes the understory trees; the adjective "moderate" refers to the amount removed. A "heavy high thinning" involves the removal of all competitors of some specially identified trees that are expected to produce future value (Schober 1991). The expected change of forest structure when a combination of two adjectives is used is shown in Fig. 3.29.

Harvest events are described by a great variety of linguistic variables, including "selective thinning" (German: *Auslesedurchforstung;* see Schädelin 1942; Leibundgut 1978: 116; Abetz 1976; Johann 1982), "plenter thinning" (German: *Plenterdurchforstung;* see Borggreve 1891; Schütz 1989), "qualitative group selection" (German: *qualitative Gruppendurchforstung*; see Kato and Mülder 1983), "structural thinning" (German: *strukturierende Durchforstung*; see Reininger 1987: 142). The "variable retention systems" in North America are defined by expressions such as "strip shelterwood," "irregular shelterwood," "group retention," or "group selection" (Maguire et al. 2006).

The use of simple terminology to describe complex structural modifications creates confusion, especially in multispecies forests. The linguistic variables *plenter thinning* and *selective thinning*, for example, can refer to virtually identical harvest events and ecosystem modification. On the other hand, a single linguistic variable such as *plenter thinning* can be interpreted in completely different ways. For a more detailed analysis of the confusion in silvicultural terminology, see Füldner and Gadow (1994). It is particularly challenging to describe the structural changes caused by different harvest events in uneven-aged multispecies forests that are selectively managed. Figure 3.30 presents an example of two harvest events and the corresponding structural changes. The harvest event resulting in the residual forest on the right was more moderate and a more clustered distribution of trees was retained than was the case with the other event shown below the before-harvest section of the forest. The understory was retained in both events.

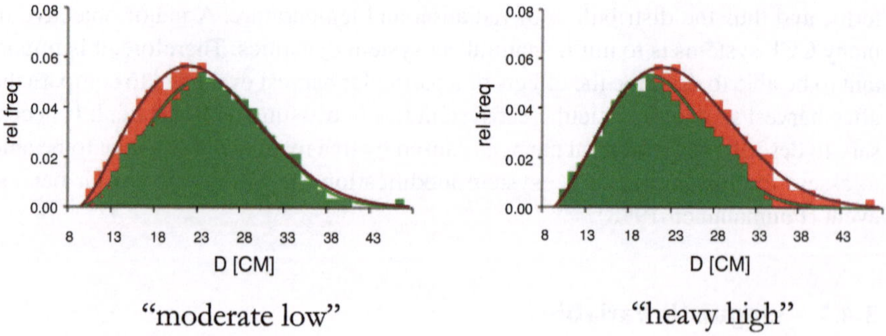

Fig. 3.29 Expected change of forest structure when two adjectives are combined to describe a specific harvest event. The remaining as well as the removed parts can be described by Weibull parameters

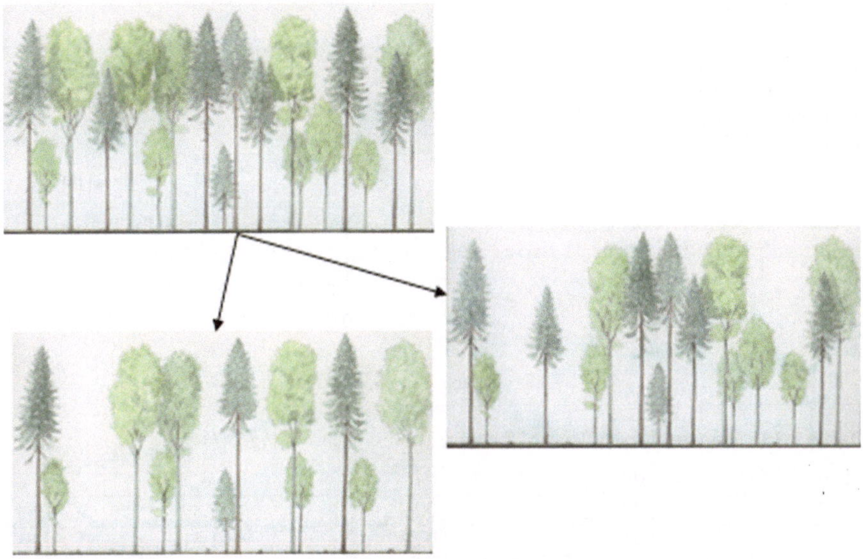

Fig. 3.30 Two hypothetical harvest events causing different structural changes in an uneven-aged forest

The example demonstrates that more specific adjectives are needed to differentiate between the two harvest events.

3.4.2 Simple Numerical Variables

A practical variable for defining the weight of a thinning in CCF is the portion of the basal area (m²/ha) removed during a harvest event. We denote this quantity as *rG* (Murray and Gadow 1993):

$$rG = \frac{basal\,area\,removed}{total\,asal\,area} \tag{3.37}$$

Correspondingly, the removed portion of the number of trees per hectare can be denoted as *rN*. The ratio of these two quantities, the *NG-ratio*, is a simple indicator of the type of thinning that occurred:

$$NG\_ratio = \frac{rN}{rG} \tag{3.38}$$

Both *rG* and *NG* can be related to the change of the parameters of the Weibull distribution (Staupendahl and Puumalainen 2000). In an uneven-aged multispecies forest the thinning weight (*rG*) and type (*NG*) can be calculated separately for each species or species group. Figure 3.31 shows a graphic describing the thinning weights in terms of basal area and stem number in three near natural forests in

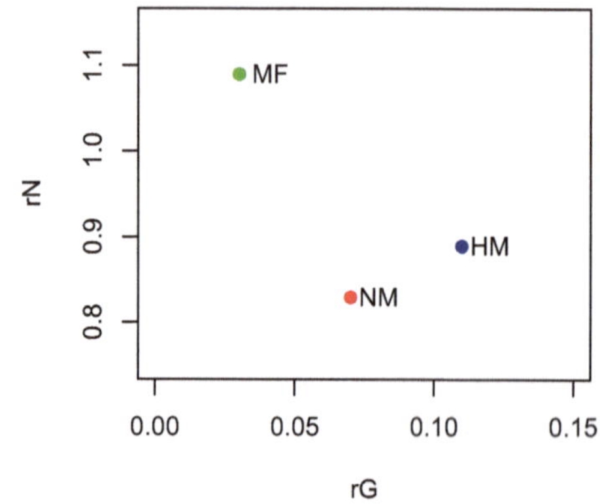

Fig. 3.31 Thinning weights in terms of basal area and stem number in three natural forests in north-eastern China, designated HM, NM, and MF

Table 3.13 Numerical equivalents of the three harvest events in Fig. 3.31

Forest	rG	rN	NG_ratio
HM	0.11	0.89	7.89
NM	0.07	0.83	11.51
MF	0.03	1.09	34.92

north-eastern China that had been subject to previous harvesting. The forests were designated as half-mature (HM), near mature (NM), and mature forest (MF), based on the length of time since the previous felling operations.

The numerical equivalents of the three harvest events is given in Table 3.13.

The weight of selective harvesting in the MF forest was very low. Although more than 1% of the trees were harvested, only 3% of the basal area was removed. An appropriate description of the selective harvest event would be "a weak, low thinning." The harvest event in the HM forest can be called a "moderately high thinning." These descriptions can be extended to include weights and selection preferences for individual species.

3.4.3 Removal Preferences in a Spatial Context

Sometimes we wish to know the spatial context of the trees that were removed during a harvest event: did the harvest event target the dominant trees or the suppressed trees in the vicinity of their immediate neighbors? Were the trees selected for removal mostly located in groups composed of one species or in mixed groups? To do this, we describe the neighborhood constellations of all the removed trees and compare the result with the neighborhoods of all the trees before the harvest. An example of a variable that can be used is species *mingling* (the proportion of the *n* nearest neighbors of a particular reference tree that are not of the same species as the reference tree). Another example is *dominance* (the proportion of the n nearest neighbors of a particular reference tree that are smaller than the reference tree). When four neighbors are considered, each of these two variables can assume five values. The relative proportion of the removed reference trees divided by the proportion of all the reference trees (before the harvest) within structural classes *i* and *j* is a measure of the removal preference (Pr_{ij}) within a given combination of spatial structures:

$$PR_{ij} = \frac{\text{removed}\, p_{ij}}{\text{before}\, p_{ij}} \tag{3.39}$$

with

PR_{ij} = selection preference in structure class *ij* ($i = 1\ldots m, j = 1\ldots n$)
removed p_{ij} = proportion of trees removed in structure class *ij*
before p_{ij} = proportion of trees in structure class *ij* before the thinning

Table 3.14 shows the removal preferences in the Lensahn experiment during the 2004 harvest event. The criteria *mingling* and *dominance* were used for beech and other deciduous species.

The beech trees selected for removal had occurred within a broad array of spatial constellations. These trees were either suppressed or dominant individuals and were located within pure or mixed species groups, or within regular or irregularly distributed neighborhoods. In Table 3.14, the highest removal preference (Pr_{ij} = 6,37) refers to neighborhood groups in which the removed beech tree was a) the smallest tree among its four nearest neighbors (U = 0.00) and b) surrounded by three neighbors that were not beech trees (M = 0.75). The other deciduous species were removed with high preference if they were codominant (U = 0.75) and surrounded by three out of four neighbors of a different species (M = 0.75).

Table 3.14 Removal preferences Pr_{ij} for five classes of spatial mingling (M) and dominance (U) for beech and other deciduous species for a thinning in 2004

<table>
<tr><th colspan="2" rowspan="3">Pr_{ij}</th><th colspan="10">Mingling</th></tr>
<tr><th colspan="5">Beech</th><th colspan="5">other Deciduous</th></tr>
<tr><th>0,00</th><th>0,25</th><th>0,50</th><th>0,75</th><th>1,00</th><th>0,00</th><th>0,25</th><th>0,50</th><th>0,75</th><th>1,00</th></tr>
<tr><td rowspan="5">Dominance</td><td>0,00</td><td>0.34</td><td>0.00</td><td>0.00</td><td>6.37</td><td>0.00</td><td>0.00</td><td>0.00</td><td>0.00</td><td>0.00</td><td>0.00</td></tr>
<tr><td>0,25</td><td>0.40</td><td>0.00</td><td>0.37</td><td>0.00</td><td>0.00</td><td>0.00</td><td>0.00</td><td>0.00</td><td>1.61</td><td>0.00</td></tr>
<tr><td>0,50</td><td>1.06</td><td>1.33</td><td>0.58</td><td>2.12</td><td>0.00</td><td>0.00</td><td>0.00</td><td>0.00</td><td>1.07</td><td>0.00</td></tr>
<tr><td>0,75</td><td>1.18</td><td>1.89</td><td>2.83</td><td>0.00</td><td>0.00</td><td>0.00</td><td>0.00</td><td>0.00</td><td>2.15</td><td>1.61</td></tr>
<tr><td>1,00</td><td>2.50</td><td>1.06</td><td>1.27</td><td>0.00</td><td>0.00</td><td>0.00</td><td>0.00</td><td>0.00</td><td>0.64</td><td>1.03</td></tr>
</table>

3.4.4 Simulating a Harvest Event

Numerous tools have been developed to facilitate the prediction of tree growth in uneven-aged multispecies forests. Fairly advanced growth models are available in many regions where forest ecosystems are selectively harvested. Harvest event models are indispensible for evaluating the specific dynamics of a selectively managed ecosystem. In our experience, developing a model that predicts the structural modification caused by a harvest event is a rather challenging task, one that requires effective algorithms that translate silvicultural prescriptions into tree selection algorithms. These algorithms result in complex structural modifications of the forest (Albert 1999; Hessenmöller 2002; Fig. 3.32).

It is not always possible to describe a harvest event by means of an index or a combination of indices that reflect the weight and the type of tree removals. Continuous cover forestry is characterized by a wide range of specific harvesting types, such as gap removals and *Z-tree* release thinnings. A "*Z-tree*" (*Zukunftsbaum*) in German is known as a "*Frame tree*" in the United Kingdom because of their role in providing the basic "frame" of a community of trees. Frame trees are selected for their outstanding vitality, stem quality, stability, and crown morphology. They are released from competition during a thinning by removing their immediate competitors. Z-tree release thinning is widely practiced in Germany and simulating this particular method allows a more detailed and meaningful quantitative analysis of a harvest event. An alternative to the schematic Z-tree release thinning method is a simulation approach to evaluate the effects of particular harvest events before its implementation. Albert (2001) simulated silvicultural alternatives by applying different thinning concepts, thinning intervals, and various silvicultural objectives. The simulations demonstrated a realistic approach to harvest event prediction. Hessenmöller (2002) simulated the structural modifications caused by different harvest events in his PhD thesis. Each harvest event was described by a set of rules published in silvicultural textbooks. The selective thinnings were also characterized by different removal intensities to see if the removal probabilities would be affected by the thinning weight. Hessenmöller (2002) simulated selective and gap thinnings for both an aggregated and a random forest. A harvest event modifies the tree size distribution, the species distribution, and the spatial pattern. Foresters who need to

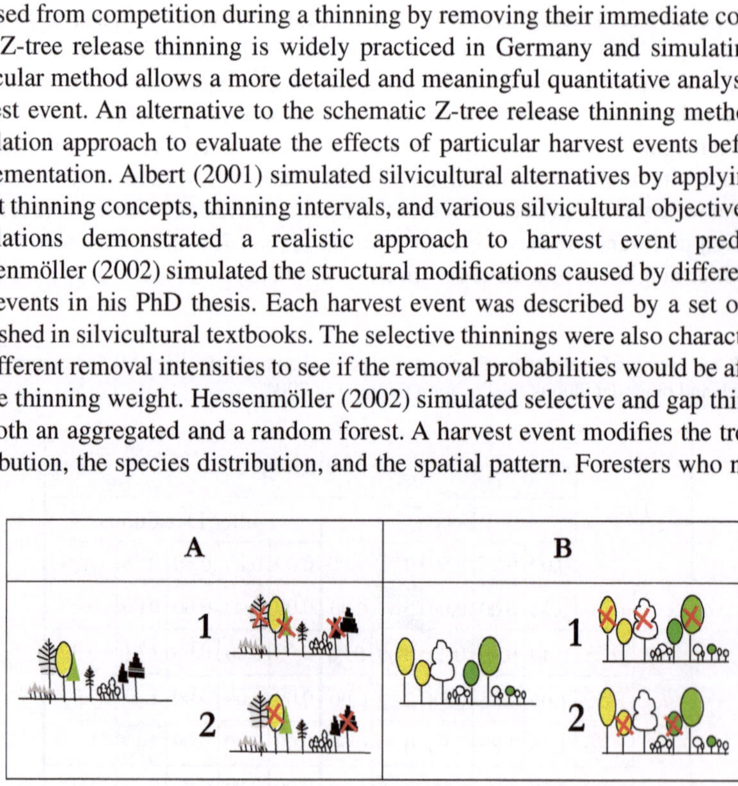

Fig. 3.32 Evaluating effects of particular harvest events before implementation in two hypothetical forests A and B

assess the impact of ecosystem modification caused by a particular harvest event have to be able to describe the structural changes caused by the tree removals (Mason et al. 2005).

3.4.5 The Gini Coefficient

A major objective of continuous cover silviculture is to mimic natural ecosystem dynamics. It is therefore often desirable to compare the current state of a particular ecosystem with some ideal state, such as an unmanaged virgin forest. Accordingly, we briefly introduce a method based on the Gini coefficient (*G*) which permits a comparison of forest structures. The Gini measures the area between the Lorenz curve (red line in Fig. 3.33) and a hypothetical line of absolute homogeneity, expressed as a percentage of the maximum area under the line. The *Gini* coefficient is equal to the ratio of the area between the *Lorenz* Curve and the line of absolute equality (numerator), and the whole area under the line of absolute equality (denominator). According to Fig. 3.33, the *Gini* Coefficient is therefore equal to the area *C* divided by the area of the triangle *0AB*.

It is possible to use the *Gini* coefficient to measure the degree of structural regularity, that is, the homogeneity of tree frequencies in m × m cells. In a situation of perfect homogeneity, the Lorenz curve would overlap the line of perfect equality and the Gini coefficient would be equal to zero. In the theoretical situation of one cell containing all the trees, the Lorenz curve would coincide with the axes and the *Gini* coefficient would equal unity.

Two 50 m × 50 m experimental plots are used to demonstrate the method. Uneven-aged and seminatural pine-oak stands are the predominant forest types.

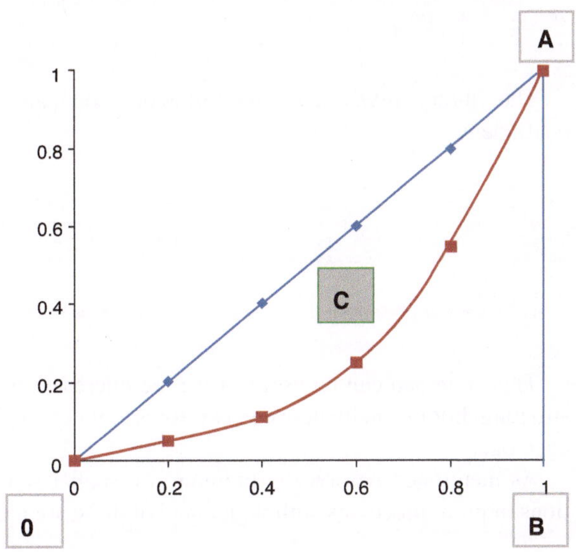

Fig. 3.33 The Gini Coefficient is equal to the area C divided by the area of the triangle 0AB

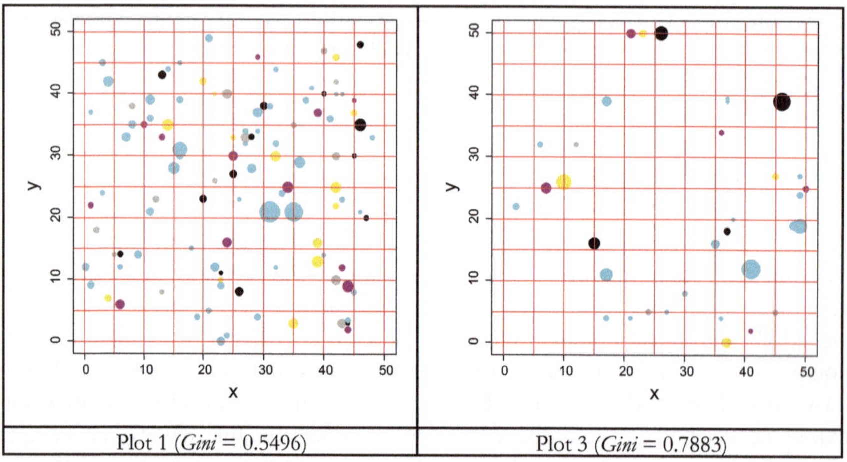

Fig. 3.34 The two Mexican plots used in the calculation of the Gini index, showing tree positions and superimposed cells

According to the *Método Mexicano de Ordenación de Bosques Irregulares* (Torres 2000) the forests are managed by selective removals. The following ***R*** code subdivides the area into one hundred 5 m × 5 m subplots and counts the number of trees in each cell (Fig. 3.34).

```
rm(list = ls())                 # remove everything
dat<-read.csv("/Users/kg/Desktop/Dur3.csv", header=TRUE); attach(dat)
M<-50                           # Length of side of square plot
n<-length(x)                    # Number of trees in plot
#set.seed(1); x<-runif(n,0,M); y<-runif(n,0,M) # Generate random trees if needed
m<-10                           # Number of row/column partitions
counts<-matrix(0,m,m)           # Set up matrix for counts
for (i in 1:n){                 # find row cell
    ir<-ceiling(m*y[i]/M); ic<-ceiling(m*x[i]/M); counts[ir,ic]<-counts[ir,ic]+1}
result<-counts[m:1,] # result; sum(result)
```

The library *reldist* and the following code are used to calculate the *Gini* coefficient:

```
library(reldist)
vec<-c(result)                              # generate the vector of cell frequencies
G<-gini(vec)                                # compute coefficient
The following code generates a plot of the trees in the cell grids
sp.tab<-read.csv("/Users/kg/Desktop/Dur_sp.Tablecsv", header=TRUE)
par(mar=c(5,5,1,1) + 0.1)
plot(x,y,xlim=c(0,M), ylim=c(0,M),pch=19,cex= dat$d*0.1,cex.lab=2, col=sp.tab[,3])
abline(v=seq(0,M,length=m+1), col="red"); abline(h=seq(0,M,length=m+1), col="red")
```

This approach can be useful if we are interested in measuring the difference in structural homogeneity between two forests, for example, a virgin forest and a managed one.

As mentioned before, the literature on spatial statistics abounds with descriptions of point processes, although many of these are merely of academic interest. In the real world, mapped tree coordinates are not always available, despite technological advances (e.g., laser scanning). It is therefore reasonable to suggest methods for analyzing spatial structures in situations where mapped tree data are not available. Fortunately, meaningful structural analysis can be performed at affordable cost when the neighborhood methods described in several sections of this chapter are used.

3.5 Reconstruction and Simulation

Several authors have proposed methods basedV on the structural information of a sample to reconstruct an entire forest (Pretzsch 1997; Lewandowski and Gadow 1997; Hui et al. 2003; Pommerening and Stoyan 2008). *Reconstructing* a forest in this way can improve the effectiveness of monitoring silvicultural operations. Reconstruction can create realistic ecosystems as a basis for spatially explicit analysis and visualization or facilitate the use advanced growth models based on known neighborhood constellations and competition status. A reconstruction is considered successful if there is a close resemblance between a real forest and its virtual counterpart. As it is usually impossible to measure all the tree positions in practice, the reproduction must be achieved with limited information. The information typically available to foresters involves distributions of neighborhood relations obtained during a routine forest inventory. A precondition for obtaining this information is the application of assessment techniques such as those described above. The simulation that is based on a sample of the structural variables described in the previous section generates a reproduction of a real forest.

Pretzsch (1997) proposed a method for producing a spatially explicit forest that was based on information about the manner in which two different tree species mingle (e.g., in clusters, rows, or groups). He used empirical functions to estimate the distance to the nearest neighbor and a set of probability functions combining an inhomogeneous Poisson process and a Hardcore process. Tree size mingling was not considered, and the species were limited to two.

According to Lewandowski and Gadow (1997), a reproduction of a forest is considered to be perfect if each tree in the real stand has a counterpart in the reproduced one, with exactly the same distances to its three nearest neighbors, and if all the species-mingling values and all size-differentiation values in the reproduction occur with the same frequencies as those in the real forest. Their simulation consists of four separate phases during which tree positions are generated and shifted and position attributes are exchanged until the distributions of the structural variables of the reproduction are close enough to those of the real stand. The initial positioning of the trees is done in phase 0. The positions can be allocated either at random or in some predefined manner. The denser the stand, the less important is the manner in which initial coordinates are generated. Phase 1 consists of three cyclic subphases: optimizing the distances to the first, the second, and the third neighbor.

The subphases are repeated until all the distances agree with those of the sample from the real forest. The purpose of phase 2 is to optimize *mingling*. This is achieved by successively exchanging two trees of different species (swapping trees of the same species will not change the mingling value).[6] The swap is retained if it reduces the difference between the real and the simulated distributions of the species-mingling values, otherwise it is repealed. Phase 2 is complete when the difference between the two distributions of the mingling values cannot be further reduced. The *diameter differentiation* is optimized in phase 3. A pairwise swapping of trees of the *same species* is done with the aim of attaining a distribution of the size-differentiation values that closely resembles that of the original forest. The algorithm was later refined by Pommerening (2006) who reversed the analysis, and so enabled the synthesis of forest structure from the indices derived. He investigated this idea with a simulation model that used the concept of cellular automata. The rules according to which the spatial pattern of tree positions developed in the stand matrix were deduced directly from the distributions of the structural indices of the input data. Pommerening used different combinations of indices to assess and simulate the structure of four sample stands. Simulations using species-specific distributions of indices and limiting the number of neighbors to three or four were most successful at reconstructing the original stand structure.

Hui et al. (2003) used the *uniform angle index* (UAI) to propose a method of reconstruction. Using the sampled UAI, they generated tree coordinates for random, clumped, and regular distributions. Pommerening and Stoyan (2008) proposed a method for synthesizing spatial point patterns from nearest-neighbor summary statistics sampled in small circular subwindows. They used a stochastic optimization technique based on simulated annealing and conditional simulation. The success of their reconstruction was tested by comparing tree point patterns reconstructed from sample data with the known patterns in three structurally different forests. Their validation has shown that it is possible to reconstruct complex forest structures successfully from nearest-neighbor sample data.

3.5.1 Applications in Spatstat

The package spatstat is a comprehensive open-source toolbox for analyzing spatial point patterns (Baddeley et al. 2019). The main focus is on two-dimensional point patterns. The library contains over 2000 functions for plotting spatial data, exploratory data analysis, model fitting, and simulation. Spatstat also has a comprehensive set of tools for analyzing forest structures. We will use a simple example involving the functions rpoispp, rNeymanScott, and runifpoint to generate a random, cluster, and regular pattern of tree positions without marks. Three simulated patterns are presented in Fig. 3.35, each with a 99% confidence envelope.

[6] A total enumeration would be a fairly hopeless endeavor. For example, in a forest containing 47 trees (with 20 trees of species 1, 15 of species 2, and 12 of species 3), there are $\binom{47}{20} * \binom{27}{15} * \binom{12}{12} = 1.697 * 10^{20}$ different ways in which the species can be assigned to available positions.

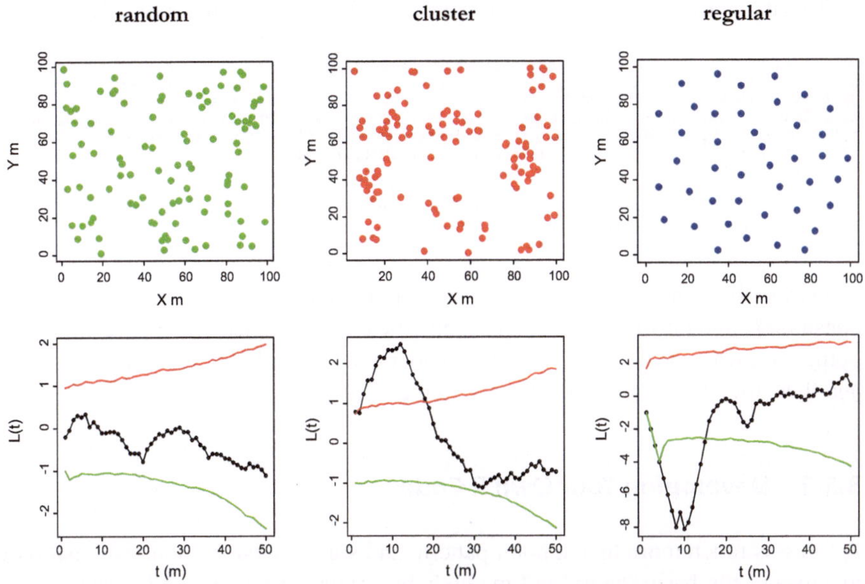

Fig. 3.35 Three spatial patterns simulated with different spatstat functions, each with a 99% confidence envelope

```
library(spatstat)
# Random
lamda=100/(100*100); random=rpoispp(lamda , win=owin(c(0,100),c(0,100)))
write.csv(random, file ="/Users/kg/Desktop/random.csv")                  # for re-use
rnd<-read.csv("/Users/kg/Desktop/random.csv", header=TRUE) # import data: 1-ha plot
plot(rnd$x,rnd$y,xlab="X m", ylab="Y m", pch=19,cex=2,col="green")
# Cluster
nclust <- function(x0, y0, radius, n) {return(runifdisc(n, radius, centre=c(x0, y0)))}
cluster<-(rNeymanScott(20, 0.1, nclust, radius=0.1, n=5)); plot(cluster)
write.csv(cluster, file ="/Users/kg/Desktop/cluster.csv")                # for re-use
clust<-read.csv("/Users/kg/Desktop/cluster.csv", header=TRUE) # import data: 1-ha plot
plot(clust$x*100,clust$y*100,xlab="X m", ylab="Y m", pch=19,cex=2,col="red")
# Regular
data(cells); x=cells$x*100;y= cells$y*100;regular<-data.frame(x,y)
write.csv(regular, file ="/Users/kg/Desktop/regular.csv")                # for re-use
reg<-read.csv("/Users/kg/Desktop/regular.csv", header=TRUE) # import data: 1-ha plot
plot(reg$x,reg$y,xlab="X m",ylab="Y m", pch=19,cex=2,col="blue")
```

The three distributions can be stored for re-use in three .csv files (random.csv, cluster.csv, and regular.csv). The *Comprehensive R Archive Network* libraries *spatstat* and *spatial* are used and the following code is applied to generate the 99% confidence limits for each pattern. The function *Lfn* calculates the *L*-function:

```
library (spatstat); library (spatial)
# Lfn calculates the L-function
Lfn=function(data,xu,yu,intervallength){  # interval between distances, scales.
ppregion(xl=0, xu=xu, yl=0, yu=yu)         # Sets the rectangular domain (xl= min, xu= max)
a=Kfn(data,min(xu,yu)/2,min(xu,yu)/2/intervallength) # K-Function in spatial package Ripley
Lfn=a$y-a$x; return(data.frame(t=a$x,Lfn=Lfn)) }       # L-Function;  t is the scale
#Critical values can be computed using Monte Carlo simulation.
# The following function Lenv calculates the envelope for L(t),
Lenv=function (data,xu,yu,intervallength){
x1=Sys.time(); ppregion(xl=0, xu=xu, yl=0, yu=yu); b=matrix(0, min(xu,yu)/2,1)
for (i in 1:10000) {a=Kfn(Psim(length(data$x)),min(xu,yu)/2,min(xu,yu)/2/intervallength)
   Lfn=a$y-a$x; Lfn=matrix(Lfn,,1); b=cbind(b,Lfn)}; low0.01=numeric(); up0.01=numeric()
   for (j in 1:dim(b)[1]){low0.01[j]=sort(b[j,])[51]; up0.01[j]=sort(b[j,])[9951] }
   x1=Sys.time()-x1; cat("run time:",x1,"\n")
return(data.frame(t=a$x,low0.01=low0.01,up0.01=up0.01))}
```

The above functions can be used to evaluate the spatial patterns of a forest plot.

```
dat<-read.csv("/Users/kg/Desktop/regular.csv", header=TRUE) # import data: 1-ha plot
pts<-list(x=dat$x, y=dat$y)                              # for regular and random only
#pts<-list(x=dat$x*100, y=dat$y*100)                     # for cluster
Lfn1=Lfn(pts,100,100,1); Lenv1=Lenv(pts,100,100,1)       # Lenv1 takes about 1-3 minutes
# Plot the function and 99% envelopes for distances up to 50 m (maximum circle radius)
plot(Lfn1$t,Lfn1$Lfn,ylim=c(min(range(Lenv1$ low0.01),range(Lenv1$
up0.01),range(Lfn1$Lfn)),max( range(Lenv1$ low0.01),range(Lenv1$
up0.01),range(Lfn1$Lfn))),xlab="t (m)",ylab="L(t)")
lines(Lfn1$t,Lfn1$Lfn); lines(Lenv1$low0.01, lwd=3, col="green"); lines(Lenv1$up0.01,lwd=3, col="red")
cbind(Lfn1,low0.01=Lenv1$low0.01,up0.01=Lenv1$up0.01)
```

Developing one's own code for simulating a forest is laborious and time-consuming, and may not be as elegant and efficient as a copied one. However, developing our own code contributes to deep understanding and increased independence, and thus often turns out to be the preferred choice.

3.5.2 Developing Your Own R Code

A forest is much more than a point pattern, and students can demonstrate that they understand the basic theory and methods by writing their own code. The dots in space have attributes, such as species and height. Each species has a specific diameter and height distribution. The following function Random_forest simulates a one-hectare random forest based on specific assumptions regarding the number of tree species and their respective diameter distributions:

```
# define a function that generates a random forest, based on certain assumptions.
# Generate positions - uniformly distributed - Poisson forest
# size = length of square plot (m); trees = number of trees in plot
# Nspec = number of species; Sprop = vector of species proportions
Random_forest<-function(size,trees, Nspec,Sprop,mean,sd) {
 x<-runif(trees,0,size); y<-runif(trees,0,size)          # random positions
 sp<-sample(1:Nspec,100,replace=TRUE,prob=Sprop); s <-matrix(NA,Nspec, trees)
 for (i in 1:Nspec) s[i,] <- sp==i          # for each species a vector of TRUE or FALSE
 D <-rep(NA, trees)
 for (i in 1:Nspec) {D[s[i,]] <- rnorm(sum(s[i,]), mean[i],sd[i]) }      # generate DBH's
return(cbind(x,y,D,sp))}
# Define the key parameters, apply the function and extract x,y,D,sp
size = 100; trees = 100; Nspec = 3; Sprop= c(50,20,30); mean=c(20,40,60); sd=c(3,6,4)
ww<-Random_forest(size,trees, Nspec,Sprop,mean,sd)
x<-ww[,1]; y<-ww[,2];D<- ww[,3];sp<- ww[,4]; forest<-data.frame(x,y,sp,DBH)
par(mar=c(5,5,1,1) + 0.1)                          # plotting parameters
plot(x,y,cex=D/20,pch=19,col=sp);
forest<-data.frame(sp,DBH,x,y)                     #
```

The data.frame "forest" can be used for further analysis. The number of trees can be specified as a realization of a Poisson-distributed random variable with the parameters mu = lambda * a * b, where lambda is the point density and a and b are the sides of a rectangular observation window (x < – random * a; y < – random * b). The results of three separate runs of the above code are shown in Fig. 3.36.

If the tree positions are regarded as constant, tree attributes (marks) can be assigned to them at random. To generate a random forest, the tree attributes are *permuted*, that is, randomly reassigned to the observed tree locations (Hartung

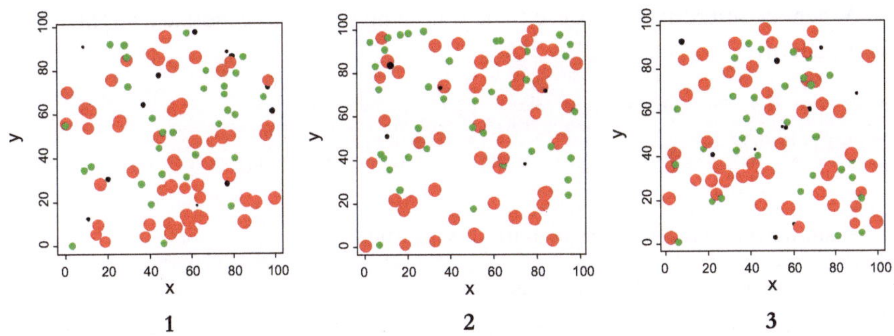

Fig. 3.36 Three one-hectare random forests using the same number of tree species with the same relative frequencies and associated species-specific diameter distributions

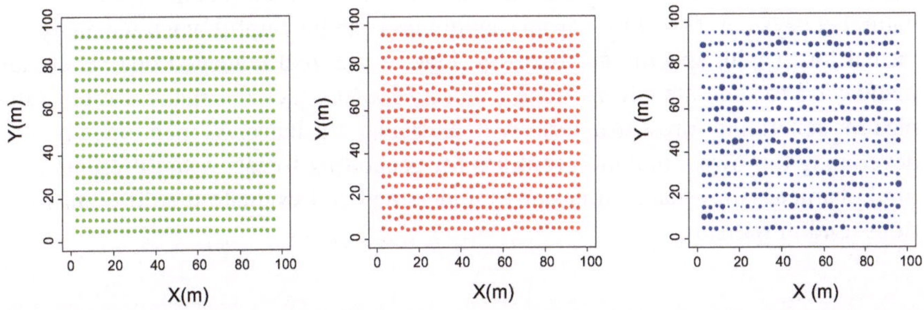

Fig. 3.37 Simulating a planted forest with a perfectly regular pattern (left); a slightly irregular pattern (center); and a slightly irregular pattern showing Weibull-distributed tree diameters (right)

1985: 96). There are *N!* ways in which the observed values of one attribute can be assigned to the *N* tree positions. It is assumed that each of these *N!* assignments has the same probability *1/N*!. Since the tree attributes are not modified, the distributions of tree species and diameters in the random forest are identical to that of the real forest.

3.5.2.1 Simulating a Planted Forest

Planted forests are usually characterized by a regular pattern of planting rows and a fixed distance between rows. The following lines will generate a perfectly regular pattern (Fig. 3.37):

```
#-generating a regular pattern of a planted forest
maxX=100; maxY=100; widthx=3; widthy=5     # set the scene
colsx<-seq(widthx, maxX-widthx, by = widthx) # vector of x positions
rowsy<-seq(widthy, maxY-widthy, by = widthy) # vector of y positions
dat <- expand.grid(x=colsx,y=rowsy)          # create a dataframe with all x/y combinations
# plot the perfectly regular pattern of the plantation
par(mar=c(5,5,1,1) + 0.1)                    # plotting parameters
plot(dat$x,dat$y,xlab="X(m)",ylab="Y(m)", xlim=c(0,100), ylim=c(0,100),pch=19,col="green")
```

When the jitter function is used, the following lines will generate a slightly irregular pattern:

```
# generating a slightly irregular pattern of the plantation
dat$xR<-jitter(dat$x,factor=0.3, amount=NULL); dat$yR<-jitter(dat$y,factor=0.5, amount=NULL)
par(mar=c(5,5,1,1) + 0.1)                    # plotting parameters
plot(dat$xR,dat$yR,xlab="X(m)",ylab="Y(m)",xlim=c(0,100),ylim=c(0,100),pch=19,col="red")
```

simulating a diameter distribution with jitter

```
dat$dbh<-rweibull(length(dat$x),2.5,12)
par(mar=c(5,5,1,1) + 0.1)                    # plotting parameters
plot(dat$xR,dat$yR,xlab="X (m)",ylab="Y (m)", xlim=c(0,100), ylim=c(0,100),pch=19,cex=
dat$dbh*0.08, col="blue")
```

The simulation technique enables us to generate random structures, which can be used as a basic reference for evaluating (or mimicking) developments in an observed natural ecosystem. Deviations from randomness can be helpful when we analyze a virtual forest with specific attributes in relation to a real managed one. There are many ways to simulate a forest. The R library *spatstat* includes many useful examples for generating forest structures with particular attributes. As mentioned before, developing our own simulation routines for generating forests with different structural characteristics makes it easier to understand and evaluate the different functions used in *spatstat*.

Literature

Abetz P (1976) Kann und soll die Standraumregulierung in Fichtenbeständen programmiert werden? FuH 31:117–119

Affleck DLR (2001) A comparative study of spatial analysis methods for forestry Nelder experiments. MSc thesis, The University of British Columbia, 70 pp

Albert M (1999) Analyse der eingriffsbedingten Strukturveränderung und Durchforstungsmodellierung in Mischbeständen. Dissertation Universität Göttingen. Hainholz Verlag, 201 p

Albert M (2001) Predicting the selection of elite trees in mixed- species stands – a rule-based algorithm for silvicultural decision support systems. Allg Forst- u J-Ztg 173(9):153–161

Angiosperm Phylogeny Group (2016) An update of the Angiosperm Phylogeny Group classification for the orders and families of flowering plants: APG III. Bot J Linn Soc 161:105–121

Aphalo PJ, Rikala R (2006) Spacing of silver birch seedlings grown in containers of equal size affects their morphology and its variability. Tree Physiol 26(9):1227–1237

Assmann E (1961) Waldertragskunde. Organische Produktion, Struktur, Zuwachs und Ertrag von Waldbeständen. BLV Verlagsgesellschaft, München, Bonn, Wien, p 490

Assunção R (1994) Testing spatial randomness by means of angles. Biometrics 50:513–537

Baddeley A, Turner R, Rubak E (2019) Spatial point pattern analysis, model-fitting, simulation, tests. https://cran.r-project.org/web/packages/spatstat/spatstat.pdf. Accessed on 4 Mar 2019

Barkman JJ (1989) A Critical evaluation of minimum area concept. Vegetatio 85:89–104

Barry RP (2013) Nelder plots. Encyclopedia of environmetrics 4

Begon M, Harper JL, Townsend CR (1996) Ecology: individuals, populations and communities. Blackwell Science, Oxford

Besag J (1977) Contribution to the discussion of Dr Ripley's paper. J R Statist Soc Ser B 39:193–195

Besag JE, Glaves JT (1973) On the detection of spatial pattern in plant communities. Bulletin of the International Statistical Institute 45:153–158

Biber P (1997) Analyse verschiedener Strukturaspekte von Waldbeständen mit dem Wachstumssimulator SILVA 2. Vortrag anläßlich der Jahrestagung 1997 der Sektion Ertragskunde im Deutschen Verband Forstlicher For- schungsanstalten. Tagungsbericht: 100–120.

Borggreve B (1891) Die Holzzucht, 2nd edn. Parey, Berlin

Bray JR, Curtis JT (1957) An ordination of the upland forest communities of southern Wisconsin. Ecol Monogr 27:325–349

Bryant JB, Lamanna C, Morlon H, Kerkhoff AJ, Enquist BJ, Green JL (2008) Microbes on mountainsides: contrasting elevational patterns of bacterial and plant diversity. Proc Natl Acad Sci U S A 105:7774–7778

Burkhart HE, Tomé M (2012) Modeling forest trees and stands. Springer, Dordrecht. 457 p

Buys MH, Maritz JS, Boucher C, Van Der Walt JJA (1994) A model for species-area relationships in plant communities. J Veg Sci 5:63–66

Byth K, Ripley BD (1980) On sampling spatial patterns by distance methods. Biometrics 36:279–284

Chao A, Chazdon RL, Colwell RK, Shen TJ (2005) A new statistical approach for assessing similarity of species composition with incidence and abundance data. Ecol Lett 8:148–159

Christenhusz MJ, Reveal JL, Farjon A, Gardner MF, Mill RR, Chase MW (2011) A new classification and linear sequence of extant gymnosperms. Phytotaxa 19(1):55–70

Clarke KR, Warwick RM (1998) A taxonomic distinctness index and its statistical properties. J Appl Ecol 35:523–531

Condés S (1997) Simulación de parcelas arboladas con datos del 2 Inventario Forestal Nacional. Tesis Doctoral. Tesis Doctoral Escuela Técnica Superior de Ingenieros de Montes de Madrid. 616 pp

Condit R, Pitman N, Leigh EG Jr, Chave J, Terborgh J, Foster RB (2002) Beta-diversity in tropical forest trees. Science 295(5555):666–669

Corral-Rivas JJ, Wehenkel C, Castellanos BH, Vargas LB, Diéguez-Aranda U (2010) A permutation test of spatial randomness: application to nearest neighbour indices in forest stands. J For Res 15:218–225

de Caprariis P, Lindemann RH, Collins CM (1976) A method for determining optimum sample size in species diversity studies. Math Geol 8(5):575–581

Devictor V, Mouillot D, Meynard C, Jiguet F, Thuiller W, Mouquet N (2010) Spatial mismatch and congruence between taxonomic, phylogenetic and functional diversity: the need for integrative conservation strategies in a changing world. Ecol Lett 13:1030–1040

Dickel M, Kotze H, Gadow K, Zucchini W (2010) Growth and survival of *Eucalyptus grandis* – a study based on modeling lifetime distributions. Int J Math Comput For Nat-Resour Sci 2(2):20–30. (available online at http://mcfns.com)

Diggle PJ (2003) Statistical analysis of spatial point patterns, 2nd edn. Arnold, London, 159p

Diggle PJ, Besag JE, Glaves JT (1976) Statistical analysis of spatial point patterns by means of distance methods. Biometrics 32:659–667

Dixon PM (2002) Encyclopedia of environmetrics, vol 3. Wiley, Chichester, pp 1796–1803

Elliott M (2011) Marine science and management means tackling exogenic unmanaged pressures and endogenic managed pressures—a numbered guide. Mar Pollut Bull 62(4):651–655

Ferris R, Humphrey JW (1999) A review of potential biodiversity indicators for application in British forests. Forestry 72(4):313–328

Fisher RA, Corbet SA, Williams CB (1943) The relation between the number of species and the number of individuals in a random sample of an animal population. J Anim Ecol 12:42–58

Franklin JF, Spies TA, Van Pelt R, Carey AB, Thornburgh DA, Berg DR, Lindenmayer DB, Harmon ME, Keeton WS, Shaw DC, Bible K, Chen JQ (2002) Disturbances and structural development of natural forest ecosystems with silvicultural implications, using Douglas-fir forests as an example. For Ecol Manage 155(1-3):399–423

Füldner K (1995) Strukturbeschreibung von Buchen-Edellaubholz-Mischwäldern. Cuvillier Verlag, Göttingen: 146 S

Füldner K, Gadow Kv (1994) How to define a thinning in a mixed deciduous beech forest. In: Proc IUFRO conference in Lousa, Portugal: mixed stands – research plots, measurements and results, models, pp 31–42
Gadow K (1986) Observations on self-thinning in pine plantations. S Afr J Sci 82(7):364–368
Gadow Kv (1987) Untersuchungen zur Konstruktion von Wuchsmodellen für schnellwüchsige Plantagenbaumarten. Forstl. Forschungsberichte, Universität München Nr. 77, 147 S
Gadow K (1993) Zur Bestandesbeschreibung in der Forsteinrichtung. Forst und Holz 48(21):602–606
Gadow K (1999) Waldstruktur und Diversität. Allg Forst- u Jagdzeitg 170(7):117–122
Gadow K (2013) Messung und Modellforschung – Grundlagen der Forsteinrichtung (Measurement and modeling research – basis for forerst design). Allgemeine Forst und Jagdzeitung 184(7/8):143–158
Gadow Kv, Bredenkamp BV (1992) Forest management. Academica Press, Pretoria, 151 p
Gadow Kv, Hui GY (2002) Characterizing forest spatial structure and diversity. In: Proceeding of the SUFOR conference, sustainable forestry in temperate regions, April: 7–9
Gadow K, Hui GY (2007) Can the tree species-area relationship be derived from prior knowledge of the tree species richness? For Stud/Metsanduslikud Uurimused 46:13–22
Gadow Kv, Kotzé H (2015) Tree survival and maximum density of planted forests – observations from South African spacing studies. For Ecosyst 1:21. https://doi.org/10.1186/s40663-014-0021-4
Gadow K, Hui GY, Albert M (1998) Das Winkelmass – ein Strukturparameter zur Beschreibung der Individualverteilung in Waldbeständen. Centralbl für das ges Forstw 115:1–10
Gadow Kv, Zhang CY, Wehenkel C, Pommerening A, Corral-Rivas J, Korol M, Myklush S, Hui GY, Kiviste A, Zhao XH (2011) Forest structure and diversity. In: Pukkala T, Gadow K (eds) Continuous cover forestry, Book series managing forest ecosystems, vol 24. Springer, Heidelberg, pp 29–84
Gadow Kv, Hao MH, Kahriman A, Göktürk A, Corral-Rivas J, Ganeshaiah KN (2018) Biodiversity and biodistance among forest communities – new methods based on observational plots from Turkey, Mexico and South Africa – new methods based on observational plots from Turkey, Mexico and South Africa. Presentation at the Beijing Forestry University during a Forest Ecosystems workshop on 18 September, 2018
Ganeshaiah KN, Chandrashekara K, Kumar ARV (1997) Avalanche index: a new measure of biodiversity based on biological heterogeneity of the communities. Curr Sci 73:128–133
Ganeshalah KN, Shaanker UR (2000) Measuring biological heterogeneity of forest vegetation types: avalanche index as an estimate of biological diversity. Biodivers Conserv 9:953–963
Ganeshalah KN, Shaanker UR (2003) A decade of diversity. University of Agricultural Sciences Bangalore/India. Internal report, 100 p
Gaul HT, Stüber V (1996) Der Eichen-Nelder-Verbandsversuch Göhrde. Forst und Holz 51:70–75
Gitay H, Rexburg SH, Wilson JB (1991) Species-area relation in a New Zealand tussock grassland, with implications for nature reserve design and for community structure. J Veg Sci 2:113–118
Gleichmar W, Gerold D (1998) Indizes zur Charakterisierung der horizontalen Baumverteilung. Forstwiss Centralbl 117:69–80
Grace J, Berninger F, Nagy L (2002) Impacts of climate change on the tree line. Ann Bot (London) 90(4):537–544
Greenberg JA, Dobrowski SZ, Vanderbilt VC (2009) Limitations on maximum tree density using hyperspatial remote sensing and environmental gradient analysis. Remote Sens Environ 113:94–101
Hall RB (1994) Use of the crown competition factor concept to select clones and spacings for short-rotation woody crops. Tree Physiol 14(7-8-9):899–909
Hao M, Ganeshaiah KN, Zhang C, Zhao X, von Gadow K (2019a) Discriminating among forest communities based on taxonomic, phylogenetic T and trait distances. For Ecol Manage 440(2019):40–47
Hao M, Corral-Rivas JJ, González-Elizondo MS, Ganeshaiah KN, Nava-Miranda MG, Zhang C, Zhao X, von Gadow K (2019b) Assessing biological dissimilarities between five forest communities. For Ecosyst 6:30

Harmon ME, Franklin JF, Swanson F, Sollins P, Gregory SV, Lattin JD, Anderson NH, Cline SP, Aumen NG, Sedell JR, Lienkaemper GW, Cromack K Jr, Cummins KW (1986) Ecology of coarse woody debris in temperate ecosystems. In: MacFadey A, Ford ED (eds) Advances in ecological research 15. Academic Press, Orlando, pp 133–302

Harper JL (1977) Population biology of plants. Academic Press, London

Hartung J (1985) Statistik. Lehr- und Handbuch der angewandten Statistik. R. Oldenbourg Verlag, München Wien, 806 p

Hessenmöller D (2002) Modelle zur Wachstums- und Durchforstungssimulation im Göttinger Kalkbuchenwald. Dissertation, University Göttingen, p 163

Hessenmöller D, Gadow K (2001) Beschreibung der Durchmesserverteilung von Buchenbeständen mit Hilfe der bimodalen Weibullfunktion. Allg Forst- u J-Ztg 172(3):46–50

Hill M (1973) Diversity and evenness: a unifying notation and its consequences. Ecology 54:427–431

Hines WGS, Hines RJO (1979) The Eberhardt index and the detection of nonrandomness of spatial point distributions. Biometrika 66:73–80

Hobbs RJ, Arico S, Aronson J, Baron JS, Bridgewater P, Cramer VA, Epstein PR, Ewel JJ, Klink CA, Lugo AE, Norton D, Ojima D, Richardson DM, Sanderson EW, Valladares F, Vilà M, Zamora R, Zobel M (2006) Novel ecosystems: theoretical and management aspects of the new ecological world order. Glob Ecol Biogeogr 15:1–7

Hubbell SP (2001) The unified neutral theory of biodiversity and biogeography. Princeton University Press, Princeton\Oxford, 375 p

Hui GY, Albert M (2004) Stichprobensimulationen zur Schätzung nachbarschaftsbezogener Strukturparameter in Waldbeständen. Allg Forst- u Jagdzeitg 175(10):199–209

Hui GY, Kv G (2002) Das Winkelmass: Herleitung des optimalen Standardwinkels. Allgemeine Forst-und Jagdzeitung 173(10):173–177

Hui GY, Albert M, Gadow KV (1998) Das Umgebungsmaß als Parameter zur Nachbildung von Bestandesstrukturen. Forstwiss Centralbl 117:258–266

Hui GY, Albert M, Chen BW (2003) Reproduktion der Baumverteilung im Bestand unter Verwendung des Strukturparameters Winkelmaß. Allg Forst und Jagdzeitung 174:109–116

Hui GY, Gadow Kv, Hu YB, Xu H (2007) Structure-based forest management. China Forestry Publishing House, Beijing, China, 200 pp. (in Chinese, with English abstract)

Hui C, Richardson DM, Robertson MP, Wilson JRU, Yates CJ (2011a) Macroecology meets invasion ecology: linking the native distributions of Australian acacias to invasiveness. Divers Distrib 17:872–883

Hui G, Zhao X, Zhao Z, Gadow K (2011b) Evaluating tree species spatial diversity based on neighborhood relationships. For Sci 57(4):292–300

Hyytiäinen K, Haight RG (2012) Optimizing continuous cover forest management. In: T Pukkala and Kv Gadow (eds), Continuous Cover Forestry. Springer, pp 195–227

Illian J, Penttinen A, Stoyan H, Stoyan D (2008) Statistical analysis and modeling of spatial point patterns. Wiley, Chichester, 534p

Jaccard P (1900) Contribution au problème de l'immigration post-glaciaire de la flore alpine. Bulletin de la Société Vaudoise des Sciences Naturelles 36:87–130

Jaehne S, Dohrenbusch A (1997) Ein Verfahren zur Beurteilung der Bestandesdiversität. Summary: a method to evaluate forest stand diversity. Forstwissenshaftliches Centralblatt 116:333–345

Jensen M, Hofmann G (2002) Entwicklungszyklen des baltischen Buchenwaldes. AFZ/Der Wald 19:1012–1018

Johann K (1982) Der „A-Wert" – ein objektiver Parameter zur Bestimmung der Freistellungsstärke von Zentralbäumen. [The "*A*-thinning index" – an objective parameter for the determination of release intensity of frame trees]. Tagungsbericht der Jahrestagung 1982 der Sektion Ertragskunde im Deutschen Verband Forstlicher Forschungsanstalten in Weibersbrunn, 146–158

Kangas A, Maltamo M (2002) Anticipating the variance of predicted stand volume and timber assortments with respect to stand characteristics and field measurements. Silva Fennica 36(4):799–811

Kato F, Mülder D (1983) Qualitative Gruppendurchforstung der Buche. Allg Forst- u J-Ztg 154:139–145

Kinnunen J, Maltamo M, Päivinen R (2007) Standing volume estimates of forests in Russia: how accurate is the published data. Forestry 80(1):53–64
Kint V, Meirvenne MV, Nachtergale L, Geuden G, Lust N (2003) Spatial methods for quantifying forest stand structure development: a comparison between nearest-neighbor indices and variogram analysis. For Sci 49(1):36–49
Kiviste A, Laarmann D, Hordo M, Sims A (2007) Factors influencing tree mortality on the basis of permanent forest sample plot data Paper presented at the International Workshop on disturbance regimes in a changing environment. 03. – 06. October 2007 in Tukums, Latvia
Kleinn C, Stahl G, Fehrmann L, Kangas A (2010) Issues in forest inventories as an input to planning and decision processes. Allg Forst- u JZtg 181(11/12):205–210
Knoebel BR, Burkhart HE (1991) A bivariate distribution approach to modeling forest diameter distributions at two points of time. Biometrics 47:241–253
Knoke T, Seifert T (2008) Integrating selected ecological effects of mixed European beech–Norway spruce stands in bioeconomic modelling. Ecol Model 210:487–498
Koop H (1981) Vegetatiesstructuur en dynamiek van twee natuurlijke bossen: het Neuenburger en Hasbrucher Urwald, Verslagen van landbouwkundige onderzoekingen 904 PUDOW Wageningen
Korpel S (1992) Ergebnisse der Urwaldforschung für die Waldwirtschaft im Buchen-Ökosystem. Allgemeine Forstzeitschrift 47(21):1148–1152
Krajicek JE, Brinkman KA, Gingrich SF (1961) Crown competition. A measure of density. For Sci 7:35–42
Kramer H (1988) Waldwachstumslehre. Verlag Paul Parey, Hamburg/Berlin, 374 p
Kramer H, Akça A (1995) Leitfaden zur Waldmesslehre. Sauerländer, J D, Frankfurt a. Main, 298p
Krebs CJ (1999) Ecological methodology, 2nd edn. Addison Wesley Longman, Inc, Menlo Park, 620p
Lappi J (1997) A longitudinal analysis of height/diameter curves. For Sci 43(4):555–570
Lässig R (1991) Zum Wachstum von Fichtensolitären [Picea abies (L.) Karst.] in Südwestdeutschland. Dissertation, Univ. Freiburg, 136 S. (In German)
Law R, Illian J, Burslem DFRP, Gratzer G, Gunatilleke CVS, Gunatilleke IAUN (2009) Ecological information from spatial patterns of plants: insights from point process theory. J Ecol 97:616–628
Lawton JH (1999) Are there general laws in ecology? Oikos 84:177–192
Legendre P, Cáceres MD (2013) Beta diversity as the variance of community data: dissimilarity coefficients and partitioning. Ecol Lett 16(8):951–963
Legendre P, Gallagher ED (2001) Ecologically meaningful transformations for ordination of species data. Oecologia 129:271–280
Legendre P, Legendre L (1998) Numerical ecology, 2nd edn. Elsevier, Amsterdam
Leibundgut H (1978) Die Waldpflege. Paul Haupt, Bern/Stuttgart
Leinster T, Cobbold CA (2012) Measuring diversity: the importance of species similarity. Ecology 93(3):477–489
Leisch F (2004) FlexMix: a general framework for finite mixture models and latent class regression in R. J Stat Softw 11(8):2004. (http://www.jstatsoft.org/v11/i08/)
Lennon JJ, Koleff P, Greenwood JJD, Gaston KJ (2001) The geographical structure of British bird distributions: diversity, spatial turnover and scale. J Anim Ecol 70:966–979
Lepě J, Stursa J (1989) Species-area curve, life history strategies, and succession: a field test of relationships. Vegetatio 83:249–257
Lewandowski A, Gadow Kv (1997) Ein heuristischer Ansatz zur Reproduktion von Waldbeständen [A heuristic method for reproducing forest stands]. Allg Forst- u J -Zeitung 168:170–174
Lotwick HW, Silverman BW (1982) Methods for analysing spatial processes of several types of points. J R Stat Soc Ser B 44:406–413
Luhmann N (1995) Social Systems. Stanford University Press, Stanford
Luzopone C, Knight R (2005) UniFrac: a new phylogenetic method for comparing microbial communities. Appl Environ Microbiol 71(12):8228–8235
Mabvurira D, Miina J (2002) Individual-tree growth and mortality models for Eucalyptus grandis (Hill) Maiden plantations in Zimbabwe. For Ecol Manage 161(1–3):231–245

Maguire DA, Mainwaring D, Halpern CB (2006) Stand dynamics after variable retention harvesting in mature Douglas-fir forests of western North America. Allgemeine Forst und_Jagdzeitung 177:120–131

Magurran AE (2004) Measuring biological diversity. Blackwell, Oxford

Margalef R (1957) Information theory in ecology. Genet Syst 3:36–71

Mason B, Kerr G, Pommerening A, Edwards C, Hale S, Ireland D, Moore R (2005) Continuous cover forestry in British conifer forests. In: Forest Research, 2005: Annual report and accounts 2003–2004:38–53

May RM (1975) Patterns of species abundance and diversity. In: Cody ML, Diamond JM (eds) Ecology and evolution of communities. Belknap Harvard University, Cambridge, pp 81–120

McComb WC, Spies TA, Emmingham WH (1993) Douglas-fir forests: managing for timber and mature forest habitat. J For 91(12):31–42

McElhinny C, Gibbons P, Brack C, Bauhus J (2005) Forest and woodland stand structural complexity: its definition and measurement. For Ecol Manage 218:1–24

McIntire EJB, Fajardo A (2009) Beyond description: the active and effective way to infer processes from spatial patterns. Ecology 90:46–56

Mehtätalo L (2004) A longitudinal height-diameter model for Norway spruce in Finland. Can J For Res 34:131–140

Moeur M (1993) Crown width and foliage weight of northern Rocky Mountain conifers, USDA, Forest Service, Intermountain Forest and Range Experiment Station, Ogden, Utah, Research Paper INT-283, p 14

Møller J, Waagepetersen RP (2007) Modern statistics for spatial point processes. Scand J Stat 34:643–684

Monod J (1950) La technique de culture continue, the´orie et applications. Ann l'Institut Pasteur 79:390–410

Motz K, Sterba H, Pommerening A (2010) Sampling measures of tree diversity. For Ecol Manage 260:1985–1996

Murray DM, Gadow K (1993) A flexible yield model for regional timber forecasting. Southern J Appl For 17(2):112–115

Nelder JA (1962) New kinds of systematic designs for spacing experiments. Biometrics 18:283–307

Nilson A (2006) Modeling dependence between the number of trees and mean tree diameter of stands, stand density and stand sparsity. In: Cieszewski CJ, Strub M (eds), Second international conference on forest measurement and quantitative methods and management & the 2004 Southern Mensurationists Meeting 15–18 June 2004 Hot Springs, Arkansas USA: 74–94. University of Georgia, Athens

Nothdurft A, Saborowski J, Nuske RS, Stoyan D (2010) Density estimation based on k-tree sampling and point pattern reconstruction. Can J For Res 40:953–967

Nyland RD (2002) Silviculture. Concepts and applications, 2nd edn. McGraw-Hill, New York

O'Connor AJ (1935) Forest research with special reference to planting distances and thinning. Govt. Printer, Pretoria

Oksanen J, Guillaume BF, Kindt R, Legendre P, Minchin PR, O'Hara RB, Simpson GL, Solymos P, Stevens MHH, Wagner H (2013) Vegan: community ecology package. R package version 2.0-9

Parrott D, Brinks J, Lhotka J (2012) Designing Nelder wheel plots for tree density experiments. New For 43(2):245

Penttinen A, Stoyan D, Henttonen HM (1992) Marked point process in forest statistics. For Sci 38:806–824

Pielou EC (1977) Mathematical ecology. Wiley, New York, 384 p

Pommerening A (1997) Eine Analyse neuer Ansätze zur Bestandesinventur in strukturreichen Wäldern [An analysis of new approaches towards stand inventory in structure-rich forests]. Ph.D. dissertation, Faculty of Forestry and Forest Ecology, University of Göttingen, Cuvillier Verlag Göttingen: 187 pp

Pommerening A (2002) Approaches to quantifying forest structures. Forestry 75(3):305–324

Pommerening A (2006) Evaluating structural indices by reversing forest structural analysis. For Ecol Manage 224:266–277

Pommerening A (2008) Analysing and modeling spatial woodland structure. Habilitationsschrift (D.Sc. dissertation), University of Natural Resources and Applied Life Sciences, Vienna, Austria

Pommerening A, Grabarnik P (2019) Individual-based methods in forest ecology and management. Springer Nature, 411 p

Pommerening A, Stoyan D (2006) Edge-correction needs in estimating indices of spatial forest structure. Can J For Res 36:1723–1739

Pommerening A, Stoyan D (2008) Reconstructing spatial tree point patterns from nearest neighbor summary statistics measured in small subwindows. Can J For Res 38:1110–1122

Preston FW (1962) The canonical distribution of commonness and rarity: Part I. Ecology 43:185–215

Pretzsch H (1995) Perspektiven einer modellorientierten Waldwachstumsforschung. Forstwissenschaft-liches Centralblatt 114:188–209

Pretzsch H (1997) Analysis and modeling of spatial stand structures. Methodological considerations based on mixed beech-Larch stands in Lower Saxony. For Ecol Manage 97:237–253

Pretzsch H (2003) Diversität und Produktivität von Wäldern (Diversity and productivity of forests). Allg Forst u Jagdztg 174(5/6):88–97

Prodan M (1966) Holzmeßlehre. J.D. Sauerländer's Verlag, Frankfurt a. M, 644 p

Pullan W, Bhadeshia H (eds) (2000) Structure in science and art. Cambridge University Press, Cambridge

Puumalainen J (1996) Die Beta-Funktion und ihre analytische Parameterbestimmung für die Darstellung von Durchmesserverteilungen. Arbeitspapier 15–96, Inst. f. Forsteinrichtung, Univ. Göttingen, p 12

Puumalainen J (1998) Optimal cross-cutting and sensitivity analysis for various log dimension constraints by using dynamic programming approach. Scand J For Res 13:74–82

Qian H, Jin Y (2016) An updated megaphylogeny of plants, a tool for generating plant phylogenies and an analysis of phylogenetic community structure. J Plant Ecol 9:233–239

Rao CR (1982) Diversity and dissimilarity coefficients: a unified approach. Theor Popul Biol 21:24–43

Reineke LH (1933) Perfecting a stand density index for even-aged forests. J Agric Res 46(7):627–638

Reininger H (1987) Zielstärken-Nutzung oder die Plenterung des Altersklassenwaldes. Österreichischer Agrarverlag, Wien

Richardson DM, Pyšek P, Carlton JT (2011a) A compendium of essential concepts and terminology in invasion ecology. In: Fifty years of invasion ecology – The legacy of Charles Elton. Wiley-Blackwell, Oxford, pp 409–420

Richardson DM, Carruthers J, Hui C, Impson FAC, Miller JT, Robertson MP, Rouget M, Le Roux JJ, Wilson JRU (2011b) Human-mediated introductions of Australian acacias – a global experiment in biogeography. Divers Distrib 17:771–787

Ricotta C, Burrascano S (2009) Testing for differences in beta diversity with asymmetric dissimilarities. Ecol Indic 9:719–724

Ripley BD (1977) Modelling spatial patterns. J R Stat Soc. Series B (Methodological), 172–212.

Ruggiero LF, Jones LC, Aubry KB (1991) Plant and animal habitats associations in Douglas-fir forests of the Pacific Northwest: an overview. In: Ruggiero LF, Aubry KB, Carey AB, Huff HM (tecn cords.) Wildlife and Vegetation of Unmanage Douglas-fir forest. USDA Forest Service, GTR-PNW-285. pp 447–462

Schädelin W (1942) Die Auslesedurchforstung als Erziehungsbetrieb höchster Wertleistung. 3. Aufl., Bern

Schmidt M, Kiviste A, Gadow K (2011) A spatially explicit height-diameter model for scots pine in Estonia. Eur J For Res 130:303–315. https://doi.org/10.1007/s10342-010-0434-8)

Schober R (1991) Eclaircies par le haut et arbres d'avenir. Rev For Fr XLIII:385–401

Scholes RJ, Biggs R (2005) A biodiversity intactness index. Nature 434(3):45–49

Schütz J-Ph (1989) Der Plenterbetrieb. Unterlage zur Vorlesung Waldbau III, ETH Zürich, 54pp

Schütz J-P (2002) Silvicultural tools to develop irregular and diverse forest structures. Forestry 75:329–338

Shannon CE, Weaver W (1949) The mathematical theory of communication. Univ. of Illinois Press, Urbana

Sharma RP, Zdenek V, Vacek S, Kucera M (2019) A nonlinear mixed-effects height-to-diameter ratio model for several tree species based on czech national forest inventory data. Forests 10(1):70. https://doi.org/10.3390/f10010070

Shvidenko AZ, Schepaschenko DG, Nilson S, Buluy YI (2006) Tables and models of growth and productivity of forests of major forests forming species of Northern Eurasia. Ministry of Natural Resources, Federal Agency of Forest Management, International Institute for Applied Systems Analysis, Vienna, 803 p

Simpson EH (1949) Measurement of diversity. Nature 163:688

Smaltschinski T (1998) Charakterisierung von Baumverteilungen. Forstwiss Centralbl 117:355–361

Spanos KA, Feest A (2007) A review of the assessment of biodiversity in forest ecosystems. Manag Environ Qual 18(4):475–486

Spies TA (1997) Forest stand structure, composition, and function. In: Creating a forestry for the 21st century: the science of ecosystem management. Oxford University Press, Oxford

Stape JL, Binkley O (2010) Insights from full-rotation Nelder spacing trials with Eucalyptus in Sao Paulo, Brazil. South For 72(2):91–98

Staudhammer CL, LeMay VM (2001) Introduction and evaluation of possible indices of stand structural diversity. Can J For Res 31:1105–1115

Staupendahl K, Puumalainen J (2000) Modellierung des Einflusses von Durchforstungen auf die Durchmesserverteilung von gleichaltrigen Fichtenreinbeständen. Centralblatt für das gesamte Forstwesen, 116. Jahrgang, Heft 4:249–262

Staupendahl K, Zucchini W (2006) Estimating the spatial distribution in forest stands by counting small angles between nearest neighbors. Allg Forst- u J-Ztg 177(8/9):160–168

Sterba H, Zingg A (2006) Abstandsabhängige und abstandsunabhängige Bestandesstrukturbeschreibung. Allg Forst- u J-Ztg 177(8/9):169–176

Stöcker G (2002) Analyse und Vergleich von Bestandesstrukturen naturnaher Fichtenwälder mit Lorenz-Funktionen und Gini-Koeffizienten. Centralblatt für das gesamte Forstwesen 119:12–39

Stoyan D, Penttinen A (2000) Recent applications of point process methods in forestry statistics. Stat Sci 15:61–78

Stoyan D, Stoyan S (1994) Fractals, random shapes and point fields. Methods of geometrical statistics. Wiley, Chichester, 394p

Suzuki SN, Kachi N, Suzuki J-I (2008) Development of a local size-hierarchy causes regular spacing of trees in an even-aged *Abies* forest: analyses using spatial autocorrelation and the mark correlation function. Ann Bot 102:435–441

Swenson NG (2011) Phylogenetic beta diversity metrics, trait evolution and inferring the functional beta diversity of communities. PLoS One 6:e21264

Swenson NG, Stegen JC, Davies SJ, Erickson DL, Forero-Montana J, Hurlbert AH, Kress WJ, Thompson J, Uriarte M, Wright SJ, Zimmerman JK (2012) Temporal turnover in the composition of tropical tree communities: functional determinism and phylogenetic stochasticity. Ecology 93:490–499

Temesgen H, Gadow K (2003) Generalized height-diameter models – an application for major tree species in complex stands of interior British Columbia. Eur J For Res 123(1):45–51

Tjørve E (2003) Shapes and functions of species-area curves: a review of possible models. J Biogeogr 30:827–835

Torquato S (2002) Random heterogeneous materials. Microstructure and macroscopic properties. Interdisciplinary applied mathematics 16. Springer, New York

Torres Rojo JM (2000) Sostenibilidad del volumen de cosecha calculado con el método de ordenación de montes. Madera y Bosques *6*(2):57–72

Upton G, Fingleton B (1989) Spatial data analysis by example. Vol. 2: Categorical and directional data. Wiley, Chichester, 416 p

Upton G, Fingleton B (1990) Spatial data analysis by example. Vol. 1: Point pattern and quantitative data, 2nd edn. Wiley, Chichester, 410 p

Van Laar A, Akça A (2007) Forest mensuration. Vol 13 of the Springer Book Series managing forest ecosystems, 418 p

Vose JM, Allen HL (1988) Leaf area, stemwood growth and nutrition relationships in loblolly pine. For Sci 34(3):547–563

Webb CO (2000) Exploring the phylogenetic structure of ecological communities: an example for rain forest trees. Am Nat 156:145–155

Wehenkel C, Corral-Rivas JJ, Hernandez-Díaz JC, Gadow Kv (2011) Estimating balanced structure areas in multi-species forests on the Sierra Madre Occidental, Mexico. Ann For Sci 68. https://doi.org/10.1007/s13595-011-0027-9

Wenk G (1996) Durchmesserverteilungen im Buchenplenterwald (als Bestandteil des Forschungsprojektes „Sortenorientiertes Wachstumsmodell"). Tagungsband des Deutschen Verbandes Forstlicher Versuchs-anstalten. Sektion Ertragskunde, Mai 1996 in Neresheim

West PW (1982) Comparison of stand density measures in even-aged regrowth eucalypt forest of southern Tasmania. Can J For Res 13:22–31

Whittaker RH (1972) Evolution and measurement of species diversity. Int Ass Plant Tax 21:213–251

Williams CB (1964) Patterns in the balance of nature. Academic, London

Williams MR (1995) An extreme-value function model of the species incidence and species-area relationship. Ecology 76:2607–2616

Wood SN (2006) Generalized additive models: an introduction. R. Chapman & Hall/CRC, Boca Raton, p 391

Xia FC (2007) Study on the plant biodiversity and spatial patterns of broad-leaved Korean Pine forest in Changbai Mountain. Doctoral dissertation, Beijing Forestry University, China, 198 pp

Zanne AE, Tank DC, Cornwell WK, Eastman JM, Smith SA, FitzJohn RG, McGlinn DJ, O'Meara BC, Moles AT, Reich PB, Royer DL, Soltis DE, Stevens PF, Westoby M, Wright IJ, Aarssen L, Bertin RI, Calaminus A, Govaerts R, Hemmings F, Leishman MR, Oleksyn J, Soltis PS, Swenson NG, Warman L, Beaulieu JM (2014) Three keys to the radiation of angiosperms into freezing environments. Nature 506:89–92

Zeide B (1991) Self-thinning and stand density. Forest Science 37:517–523

Zeide B (1995) Tolerance and self-tolerance of trees. For Ecol Manage 13:149–166

Zeide B (2005) How to measure stand density. Trees 19:1–14

Zhang X, Lei Y, Cao QV (2010a) Compatibility of stand basal area predictions based on forecast combination. For Sci 56(6):552–557

Zhang CY, Petráš R, Zhao X, Gadow K (2010b) Estimating beech growth and survival. A study based on longterm experiments in Slovakia. Allgemeine Forst- und Jagdzeitung 181(3/4):44–53

Zhang CY, Zhao XH, Gadow K (2010c) Partitioning temperate plant community structure at different scales. Acta Oecologica 36(2010):306–313

Zhang CY, Zhao XH, Gadow K (2015) Maximum density patterns in two natural forests: an analysis based on large observational field studies in China. For Ecol Manage 346:98–105

Zucchini W, Schmidt M, Gadow K (2001) A model for the diameter-height distribution in an uneven-aged beech forest and a method to assess the fit of such models. Silva Fennica 35(2):168–183

The Shape and Growth of Forest Trees

4

Abstract

This section introduces methods for describing the shape of forest trees, using models of tree crowns, stems, root systems, and biomass. The effect of competition on individual tree growth is also included in this chapter.

A *Fagus sylvatica* Ecosystem in Europe

© Springer Nature Switzerland AG 2021

K. von Gadow et al., *Sustaining Forest Ecosystems*, Managing Forest Ecosystems 37,
https://doi.org/10.1007/978-3-030-58714-7_4

4.1 The Shape of Forest Trees

Flowering plants exhibit a great variety of shapes, sizes, and habits. They have well-developed root and shoot systems with very specific physical shapes and external structures. The morphology of forest trees incorporates biomechanical and physiological characteristics that help us to understand the formative forces of their development. This section contains methods that can be used to describe the morphology of tree crowns, tree stems, and root systems.

4.1.1 The Tree Crown

Radiation absorption, transpiration, and photosynthesis are basic processes that determine the growth of a forest tree. These processes are affected by microclimatic variables, mainly wind speed, air temperature, and water-vapor pressure. The crown architecture, which is defined by species-specific branching and contacts with neighboring trees, determines the spatial distribution of leaf area, carbon gain, and water loss within the canopy (Sinoquet et al. 2001). Leuschner (2002) describes the tree crown as a unique habitat, a "treasure-box of species diversity," especially in tropical forests where more insect species are being discovered than in any other habitat on earth. The shape of the tree crown, which is largely determined by species-specific characteristics, changes through time as a result of changing tree age and competition (Kramer and Kozlowski 1979). In addition, the crown shape often provides information about silvicultural treatment in the past.

The crown shape has been described with different geometric models. Mohren et al. (1984), for example, used simple conical or parabolic shapes for assessing the net primary production of young Douglas fir trees. A parabolic form was also used by Biging and Wensel (1990) for mixed conifer species in California. Pretzsch (1992a, b) used a composite model involving two geometric forms, a cone and a paraboloid. Mawson et al. (1976) used 15 different geometric models to describe the profile of tree crowns. Ito et al. (1997) estimated the crown volume as an elliptic cylinder. Figure 4.1 shows a simple model of a tree crown with several characteristic parameters which are often used for estimating tree growth, competition, and biomass (see Burger 1939). Burger's model, which distinguishes between the shaded part of the crown ("shade crown") and the part exposed to sunlight ("light crown"), has been used in numerous studies about the response of crown shape to varying silvicultural treatments.

Crown lengths are measured with hypsometers while measurements of crown radii are obtained by using special devices such as a crown mirror (Van Laar and Akça 1997: 76). Other methods based on digital photographs have also been used (Gaffrey 1996), but it was found that these methods produce inaccurate measurements of crown radii because of the problems with delineating the crown perimeter in dense forests. Biging and Gill (1997) used a Criterion 400 Survey Laser Instrument (CLSI) for crown radius assessment. They hoped to obtain accurate crown profile measurements by assessing the three-dimensional coordinates of the branch tips and

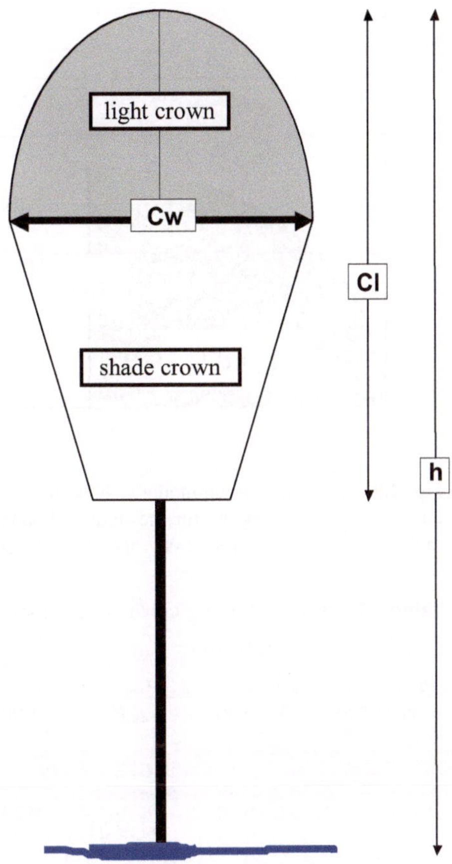

Fig. 4.1 Burger's model of a tree crown with Cw = crown width; Cl = crown length; and Cl/h = Cr relative crown length or crown ratio

other points along the crown perimeter. However, they found this method to be unreliable because the laser beam would often pass through the foliage. Furthermore, the flux gate compass of the device did not allow for a more than 15° tilt from the horizontal plane.

Figure 4.2 shows the model of a tree crown developed by Pretzsch (1992a, b: 112 *et seq.*). The profile of the upper part of the crown that is exposed to sunlight (the "light crown") is described by means of an exponential function. The shaded part (the "shade crown") is represented by a linear function

The advantage of this model is its simplicity. Parameter-parsimonious models allow generalizations, for example the modeling of species-specific crown shapes. Table 4.1 shows the parameters of the crown models for three tree species.

This simple approach allows a quantitative description of competition effects and radiation distribution in complex forests by means of species-specific (α, β, γ, and δ) and tree-specific (r_{max}, r_{base}, and L_o) parameters.

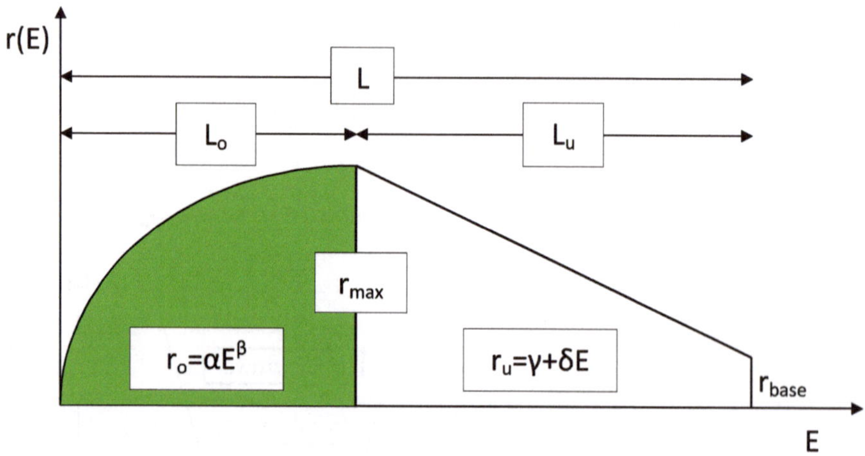

Fig. 4.2 Pretzsch's crown model (1992a, b). $r(E)$ = crown radius (m) at E m distance from tree top; r_o, r_u = crown radius for the light and shade crown, respectively; r_{max} = maximum crown radius; r_{base} = crown radius at crown base; and α,β,γ, and δ = model parameters

Table 4.1 Parameters of the crown models for three tree species (Pretzsch 1992a, b)

	Light crown			Shade crown		
Species	L_o	α	β	r_{base}	γ	δ
Picea abies	$L \cdot 0.66$	r_{max}/L_o	1.00	$r_{max} \cdot 0.50$	$r_{max}\text{-dbh}\cdot L_o$	$\frac{r_{base}-r_{max}}{L-L_o}$
Fagus sylvatica	$L \cdot 0.40$	$r_{max}/\sqrt[3]{L_o}$	0.33	$r_{max} \cdot 0.33$	$r_{max}\text{-dbh}\cdot L_o$	$\frac{r_{base}-r_{max}}{L-L_o}$
Abies alba	$L \cdot 0.50$	$r_{max}/\sqrt{L_o}$	0.50	$r_{max} \cdot 0.50$	$r_{max}\text{-dbh}\cdot L_o$	$\frac{r_{base}-r_{max}}{L-L_o}$

4.1.1.1 Crown Models of Higher Resolution

The possibilities for inventing more complex crown models are endless. Ottorini (1991), for example, describes the crowns of *Pseudotsuga menziesii* by means of two equations:

$$L = 11.6 \cdot \ln\left(\frac{E}{20}+1\right) \quad (4.1)$$

$$\text{HA} = 0{,}79 \cdot \text{L} \quad (4.2)$$

where L is the branch length (cm); HA is the horizontal distance from the base of a branch to its tip (cm); and E is the distance from the tree top. This method is effective for modeling the crowns of young conifer trees. See Fig. 4.3 for details.

More complex models of branching were developed by Früh (1992) and Kurth (1994) to improve their understanding of transpiration processes and obtain a realistic method for the visualization of tree architecture. Another method for modeling tree crowns at higher resolutions is based on *fractal structures*. This approach was used by Zeide and Gresham (1991) and Zeide and Pfeifer (1991). Hussein et al. (2001) used time series methods to model the crown profiles of beech trees. He selected the following quadratic trendline equation to describe the profile of a tree crown:

$$r_i = b_0 + b_1 \cdot h_i + b_2 \cdot h_i^2 \tag{4.3}$$

with r_i = crown radius at tree height h_i; h_i = tree height with crown radius (r_i); b_0; and b_1 and b_2 as estimated parameters. In addition, Hussein found a method based on moving averages rather effective. He investigated 10 different time series models, including a spatial autocorrelation approach developed by Biging and Gill (1997). He found Eq. (4.4) especially useful for modeling the residuals around the trend line:

$$X_{h_i} = \theta \cdot \varepsilon_{h_i - 1} + \varepsilon_{h_i} \tag{4.4}$$

where x_h is the quadratic difference between trend line and observation at height; θ is the coefficient of the function of moving averages of the first order; and ε_h is the random effect at height h. Examples of two beech crown profiles with (a) the fitted trend line and (b) the moving averages model are shown in Fig. 4.4.

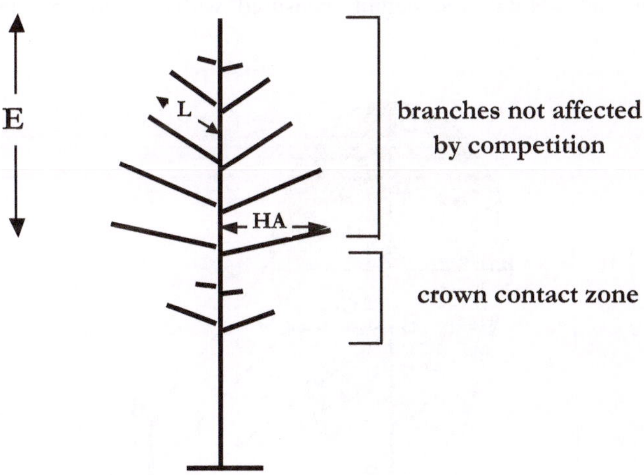

Fig. 4.3 Ottorini's crown model of *Pseudotsuga menziesii* (1991)

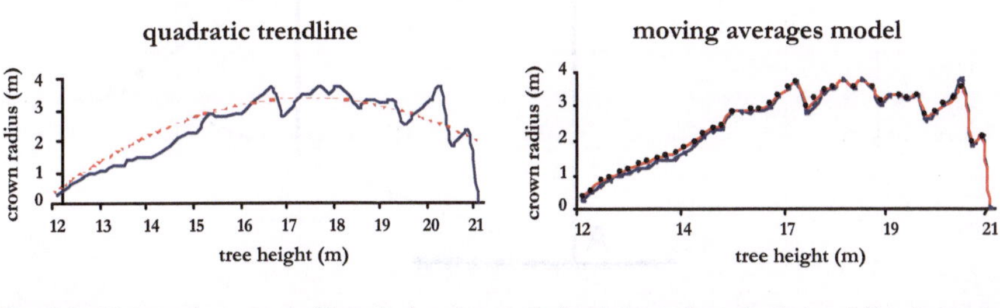

Fig. 4.4 Measured crown profiles of a beech tree. Left: Estimated quadratic trend line (parameters: $b_0 = -20.8670$; $b_1 = 2.7159$; and $b_2 = -0.0761$). Right: Moving averages model ($\theta = -0.890$)

As mentioned before, parameter-parsimonious approaches are especially effective, allowing a generalization in which parameters are estimated as a function of forest structure and density as well as certain tree variables such as tree height and diameter at breast height.

4.1.1.2 The Crown Window

Any device for assessing crown radii of varying heights must provide measurements of acceptable accuracy. One example is the crown window, a modification and improvement of a device which was first developed by Albert (1994), a transparent window divided into square sections of equal size. A perforated piece of aluminum (the head stabilizer), mounted to the end of a bar which is fixed to one side of the frame, is used to keep the distance between the operator's eye and the crown window constant. The frame is mounted on a tripod. Using a suitable horizontal distance, for example one tree length, the device is positioned parallel to the stem axis to ensure angular consistency. The geometric relations of the device are shown in Fig. 4.5.

The following variables are defined: O = position of operator's eye; OA = distance between the operator's eye and the base of the tree (m); Oa = distance between the operator's eye and the device (m); h = total tree height measured with the hypsometer (m); ad = total tree height measured with the crown window (cm);

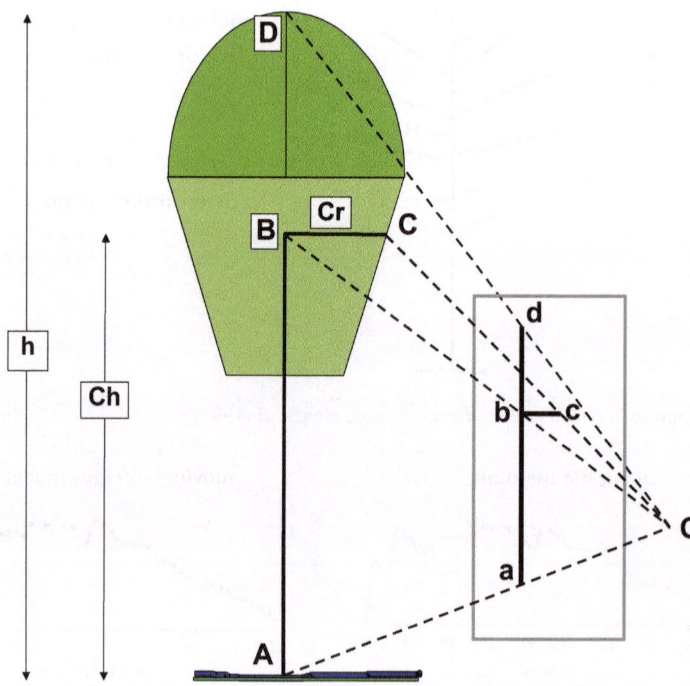

Fig. 4.5 Schematic presentation of the basic geometric relations of the crown window device

AB = Ch = tree height at which crown radius is measured (m); ab = tree height measured with the *crown window* (m); BC = Cr = crown radius at height Ch (m); and bc = crown radius measured with the crown window (cm). The total tree height h (m) is measured with a hypsometer and also in terms of the distance on the lattice (ad). The ratio ad/AD = ad/h gives the scale of the lattice to represent the reality. By definition, ad/h = ab/AB = bc/BC. Thus, AB = Ch = ab*h/ad and BC = Cr = bc*h/ad. The distances ad and bc are measured in the field. It is therefore possible to calculate the absolute values for various tree heights and the corresponding crown radii by measuring the length of different segments on the crown window and by multiplying these values with h/ad.

Measuring the lengths of the window height ad and the window radius bc in the field is time-consuming. It is therefore more practical to draw the outlines of the tree crown in the field on a transparency which is superimposed on the crown window grid, and to digitize the drawing by means of suitable software such as a specific ARC/INFO data model for representing maps in digital form (Environmental Systems Research Institute 1990).

Subsequently, the crown profile is represented as a polygon by recording a discrete number of data points. In the ARC/INFO system, such a polygon or digital map is known as a *coverage*. The real values of the crown radii are obtained by means of the h/ad ratio. The key of the method is the combination of the cost-effective field assessment and the automated calculation of the crown periphery.

4.1.2 The Stem

Estimates of assortment distributions in standing trees are primarily based on descriptions of stem form. A simple description of stem form is given by the *form quotient,* which represents the ratio of the tree diameters at two different heights (Prodan 1966; Akça et al. 1994). The logical extension of this concept is a vector of form quotients for describing an entire stem profile. Such form series have been in use for many years (Grundner and Schwappach 1942; Schober 1953; Gadow et al. 1998). An application of this concept was used by Sloboda (1984) and Gaffrey (1988) who developed a series of linear relations between the breast height diameter and the diameter at different relative tree heights. The parameters of the various relations were then predicted by means of a fifth degree polynomial.

Another popular method for describing stem profiles is based on applications of spline functions. The method requires diameter measurements taken at n heights $[(d_i, h_i)\ i = 1,\ldots, n]$ along the stem of a tree. The spline function describes a stem profile through a series of polynomials which are linked in a particular way. In order to obtain a complete description of the profile, all coordinates in the interval $[h_i, h_i + 1]$ must be determined by interpolation. The most common application is a cubic spline function representing n–1 third-degree polynomials

$$f_i(x) = \alpha_i + \beta_i(x - h_i) + \gamma_i(x - h_i)^2 + \delta_i(x - h_i)^3, x \in [h_i, h_i + 1] \quad (4.5)$$

that are smoothly linked at the nodes by means of the second derivatives. Given n nodes, this approach requires n–1 sets of coefficients, that is, a total of 3(n–1) coefficients (Späth 1973: 27 *et seq.*; Saborowski 1982: 7).

Series of form quotients, high-degree polynomials, and spline functions can describe a *particular* stem profile with high precision. However, a large number of parameters are needed for these methods. Consequently, what appears at first to be an advantage turns out to be a handicap. It is impossible, or at best unnecessarily difficult, to use taper series and spline functions for developing generalized stem profile models to describe stem form as a function of environment, forest density, and silviculture.

4.1.2.1 Stem Profile Functions

The stem represents an assortment of saleable timber products, and if we want to estimate the volume of those products we must be able to estimate the shape of the stem. A simple model of stem geometry is the *form factor*, that is, the stem volume expressed as a proportion of the volume of a theoretical cylindrical solid shape defined by the tree's height and diameter at 1.3 m. Form factors can be used to estimate the total tree volume, but they cannot describe a stem profile and are therefore not suited for product modeling. On the other hand, it is usually easy to establish a general relation between form factor, stand density, and tree age. Stem forms can be described in great detail when spline techniques or high-degree polynomials are used. Such very accurate profile descriptions are required in a sawmill where optimum cutting options are sought for individual logs. Too much precision, however, can be a serious disadvantage. The simpler model is often the more useful one, because it permits estimates of tree form in response to forest density and silviculture (Fig. 4.6).

When the model is too simple, such as the model of the *form factor*, it does not produce the desired minimum information for estimating timber product yields. The detailed models cannot be used to make general statements about the way in which silviculture affects the stem form, while the simple ones cannot produce profile descriptions that are sufficiently accurate. A suitable compromise is a parameter-parsimonious taper function.

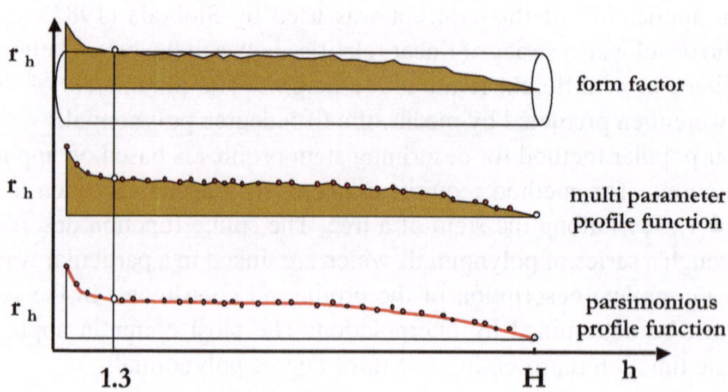

Fig. 4.6 Three approaches for describing a stem profile: breast height form factor (top); multiparameter stem profile function (center); and parameter-parsimonious stem profile function (bottom)

Kozak (1988) divided taper models into two major groups: single taper models that describe diameter changes from ground to tip with a single function (Behre 1923; Demaerschalk 1973; Brink and Gadow 1986; Riemer et al. 1995; Hussein et al. 2008; Arias Rodil et al. 2015) and segmented taper models that use different subfunctions for various parts of the stem (Ormerod 1973; Max and Burkhart 1976; Demaerschalk and Kozak 1977; Cao et al. 1980; Reed and Green 1984; LeMay et al. 1993; Lee et al. 2003). The following section describes examples of parameter-parsimonious stem profile functions. Demaerschalk (1973) published a model which is identical to the equation for general conoids that was first used by Pressler (1865) to describe the shape of a tree stem:

$$d_i^2 = \alpha \cdot dbh^2 \left(\frac{h - h_i}{h} \right)^{\beta} \tag{4.6}$$

with

d_i = stem diameter (normally under bark, cm) at stem height h_i (m)
DBH = breast height diameter (over bark, cm)
h = total tree height (m)
α, β = empirical model parameters to be estimated

The model can be fitted to measurements of tree diameters at different stem heights to obtain parameter estimates by using the following R-code:

```
dem.f<- function(dbhsqr,h,hi,alpha,beta) {alpha*dbhsqr*((h-hi)/h)^beta}
reg <- nls(dsqr ~ dem.f(dbhsqr,h,hi,alpha,beta), start=list(alpha=0.9, beta=1.5), trace=TRUE)
```

Equation 4.6, which is useful for demonstrating the general usefulness of a parameter-parsimonious taper function, can be rearranged to estimate the stem height corresponding to a given stem diameter:

$$h_i = h - h \left(\frac{d_i^2}{\alpha \cdot dbh^2} \right)^{1/\beta} \tag{4.7}$$

Example Assume the following parameter estimates: α = 0.9573; β = 1.2850. What is the stem diameter at a stem height of 10 m for a tree with a DBH of 20 cm and a total height of 22 m? Solution: 13.26 cm. What is the stem height where the stem diameter is 7 cm? Solution: 17.56 m.

If we divide the complete stem of a tree into a set of elemental disks with an infinitesimal height (dh_i), the volume of each disk (v_i) can be calculated as the product of the infinitesimal height and the section of the disk $[(\pi/4) \cdot d_i^2 \cdot dh_i]$. The total volume of the tree (V_{total}) is therefore the sum of the volumes of each elemental disk (v_i), that is, the integral of the elemental volumes. This integral can be obtained for the taper function defined in Eq. 4.6:

Table 4.2 Example demonstrating the calculation of individual log volumes

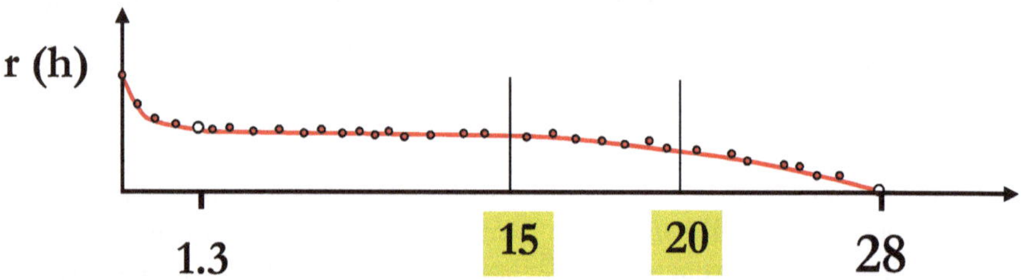

Height (m)	diameter squared	diameter (cm)	radius (cm)	disc area (cm²)	section vol. (cm³)	section vol. (m³)
15.0	205.7	14.3	7.2	161.6		
16.0	185.6	13.6	6.8	145.8	15368.1	0.01537
17.0	166.0	12.9	6.4	130.4	13807.3	0.01381
18.0	146.8	12.1	6.1	115.3	12284.9	0.01228
19.0	128.3	11.3	5.7	100.7	10803.3	0.01080
20.0	110.2	10.5	5.2	86.6	9365.7	0.00937
					Sum	**0.06163**

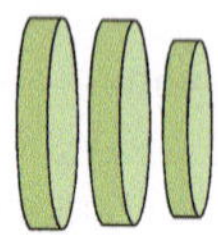

$$V_{\text{total}}\left(m^3\right)=\int_0^h v_i=\int_0^h \frac{\pi}{40000} d_i^2 \cdot dh_i=\int_0^h \frac{\pi}{40000} \cdot \alpha \cdot dbh^2\left(\frac{h-h_i}{h}\right)^\beta dh_i \\ =\frac{\pi}{40000} \cdot \frac{\alpha \cdot dbh^2 \cdot h}{(\beta+1)} \tag{4.8}$$

To obtain the volume in cubic meters, the heights and the diameters should be expressed in meters, and as the diameters are expressed in cm, we divided by 40,000 instead of by 4 in Eq. 4.8. The same method can be used to estimate the volume between two heights (h_i). In this case, we modify the limits of the integral:

$$V_{h_1-h_2}\left(m^3\right)=\int_{h_1}^{h_2} \frac{\pi}{40000} d_i^2 \cdot dh_i=\frac{\pi}{40000} \cdot \frac{\alpha \cdot dbh^2}{(\beta+1) \cdot h^\beta} \cdot\left[\left(h-h_1\right)^{\beta+1}-\left(h-h_2\right)^{\beta+1}\right] \tag{4.9}$$

Example Calculate the log volume between the heights **15 m** and **20 m** of a tree with a DBH of **24 cm** and a total height of **28 m** by using Eq. 4.9 with the parameter values defined in the previous example. Solution: 0.03659 m³. The calculations are demonstrated in Table 4.2.

Disk heights of 1 m were used in this particular example. Thinner disk slices will improve the accuracy of the estimate. (Note: There are five disks, but for demonstration purposes only three are shown.) When Eq. 4.9 is used, the volume estimate is:

$$V_{15-20}\left(m^3\right)=\frac{\pi}{40000} \cdot \frac{0.9573 \cdot 24^2}{2.2850 \cdot 28^{1.2850}} \cdot\left[13^{2.2850}-8^{2.2850}\right]=0.6161 \mathrm{m}^3$$

Most of the standard forest mensuration textbooks (Prodan 1966; Husch et al. 1982; Philip 1994; Van Laar and Akca 1997; Burkhart and Tomé 2012) agree that the form of a tree is convex neiloidal close to the ground, conical close to the top, and moderately concave paraboloidal or cylindal in between. Lee et al. (2003) developed two taper functions for *Pinus densiflora* that are based on these idealized shapes. Stem profile measurements were available for 2 439 trees growing in the central-eastern region of the Republic of Korea. The following taper function complies with the theory (Lee et al. 2003):

$$d_i = k \cdot \left(1 - \frac{h_i}{h}\right)^r \tag{4.10}$$

where d_i is the tree diameter at relative height h_i and k is a constant depending on the units used for d. k should be equal to the diameter at the base of the tree (d_{base}) to guarantee that d_i is equal to d_{base} when h_i is equal to zero. The value of r varies with the form of the solid: $r < 1$ means that the profile is parabolic, $r = 1$ corresponds to a cone-shaped profile, and $r > 1$ describes a neiloid shape (Fig. 4.7).

It may be assumed that r varies continuously with stem height. This relation can be described by the following polynomial:

$$r = f(h_i) = r_1 h_i^2 + r_2 h_i + r_3 \tag{4.11}$$

We can assume a relation between parameter k (equal to d_{base}) and the breast height diameter (DBH):

$$k = k_1 \cdot dbh^{k_2},\ (k_1, k_2 > 0) \tag{4.12}$$

Substituting Eqs. 4.10 and 4.11 in 4.9, one obtains the following taper function:

$$d_i = k_1 \cdot dbh^{k_2} \left(1 - \frac{h_i}{h}\right)^{r_1 h_i^2 + r_2 h_i + r_3},\ (k_1, k_2, r_1, r_3 > 0, r_2 < 0) \tag{4.13}$$

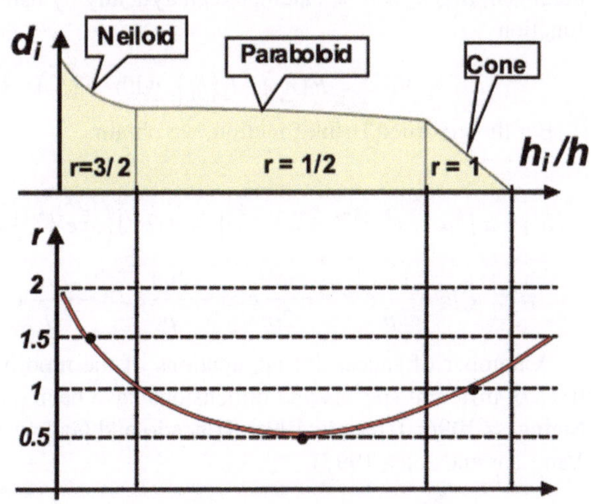

Fig. 4.7 Schematic representation of the taper model developed by Lee et al. (2003). The parameter r of Eq. 4.10 is assumed to vary continuously, as shown in the lower part of the diagram

Equation 4.12 is conditioned in such a way that $d_i = 0$ when $h_i = h$, but when h_i = *breast height (1.2 in Korea)*, $d_i \neq dbh$. Lee et al. (2003) found that this model gives good results when compared with the functions of Kozak (1988) and Max and Burkhart (1976).

4.1.2.2 The Brink Function

The following *modified Brink function* (named after Chris Brink, who contributed substantially in developing it; see Brink and Gadow 1986) is another example of a parameter-parsimonious stem taper model. The original function was conditioned by Riemer et al. (1995) to ensure that the function DBH is equal to the observed DBH, and that the DBH is zero when the height is equal to the total tree height, resulting in the following stem profile model:

$$r_i = u + v \cdot e^{-ph_i} - w \cdot e^{qh_i} \tag{4.14}$$

with

$$u = \frac{i}{1+e^{q(1.3-h)}} + (r_{1.3} - i)\left(1 - \frac{1}{1-e^{p(1.3-h)}}\right), v = \frac{(r_{1.3} - i)e^{p \cdot 1.3}}{1-e^{p(1.3-h)}}$$

and

r_i = tree radius (cm) at height h_i (m);
h = total tree height (m);
$r_{1.3}$ = tree radius at breast height (cm);
i = parameter (common asymptote);
p = parameter (describing lower part of stem);
q = parameter (describing upper part of stem).

As mentioned before, Eq. 4.14 complies with the constraints that $r_i = 0$ when $h_i = h$, and that $r_i = r_{1.3}$ when $h_i = 1.3$. The stem volume of a tree V_{h1-h2} between two heights h_1 and h_2 can be calculated analytically by using the integral of the squared function

$$V_{h_1-h_2} = F(h_2) - F(h_1) \text{ with } F(h_i) = \int \pi \cdot r_i^2 \cdot dh_i \tag{4.15}$$

For the modified Brink function, we obtain:

$$\begin{aligned} F(h_i) &= \pi \int \left(u^2 + v^2 e^{-2ph_i} + w^2 e^{2qh_i}\right) dh_i + 2\pi \int \left(uve^{-ph_i} - uwe^{qh_i} - vwe^{(q-p)h_i}\right) dh_i \\ &= \pi \left(u^2 h_i - \frac{v^2}{2p} e^{-2ph_i} + \frac{w^2}{2q} e^{2qh_i} - \frac{2uv}{p} e^{-ph_i} - \frac{2uw}{q} e^{qh_i} + \frac{2vw}{p-q} e^{(q-p)h_i}\right) \end{aligned} \tag{4.16}$$

A number of successful applications of the modified Brink function involving trees of different species and dimensions have been published (Riemer et al. 1995; Steingass 1996; Trincado 1996; Trincado and Gadow 1996; Hui and Gadow 1997; Van Laar and Akca 1997).

Steingass (1996) fitted the function to 2 882 *Pseudotsuga menziesii* trees.[1] Hence the DBH, the total height, and the three parameters *i*, *p*, and *q* of the modified Brink function were known for each tree. The sample trees representing the upper and lower 5% quantiles of the DBHs were selected to represent the biggest and smallest individuals. Tree no 3223316 represents the 5% quantile, tree no 3111077 the 50% quantile, and tree no 7120225 the 95% quantile. The fitted stem profiles of the 5, 50, and 95 percentile trees are shown in Fig. 4.8.

The parameters of the trees corresponding to the three stem profiles in Fig. 4.8 are listed in Table 4.3.

Assignment
Calculate the diameter at a height of 15 m for a tree with a breast height diameter of 41 cm and a height of 35 m. Use the modified Brink function with parameters $I = 17.5$, $p = 1.0$, and $q = 0.03$. Solution: 2*12.41 = 24.82 cm. For the same tree, calculate the height that corresponds to a diameter of 8 cm. Solution: 29.76 m.

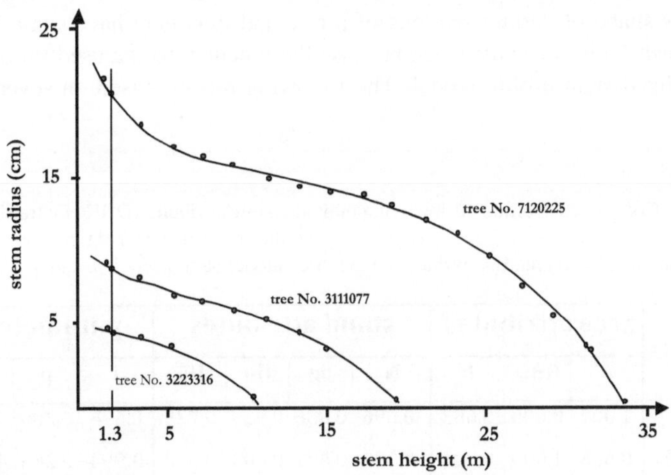

Fig. 4.8 Fitted stem profiles of the trees corresponding to the 5, 50, and 95 percentiles in a sample of 2 882 *Pseudotsuga menziesii* trees

Table 4.3 Parameters corresponding to the three stem profiles shown in Fig. 4.8

Tree	DBH (cm)	*h* (m)	*i*	*p*	*q*	Age
95 percentile No. 7120225	41.4	33.5	16.45	0.3259	0.1023	86
50 percentile No. 3111077	17.8	19.1	7.39	0.3096	0.1281	46
5 percentile No. 3223316	9.2	10.2	4.24	0.9999	0.1966	38

[1] The measurements were kindly provided by the Northwest German Forest Research Institute in Göttingen.

A practical advantage of the modified Brink function is the fact that volumes of logs can be calculated directly when analytical integration is used and certain specifications and top diameter limits (peeler, sawlog, pulpwood) are known. The model fits observed trees of different species and sizes very well, including the critical basal sweep.

Korol and Gadow (2003) fitted the modified Brink function to 2 682 diameter–height measurements of 250 *Picea abies* trees from 40 stands of different ages and densities. Table 4.4 shows the correlations between the individual tree parameters *i*, *p*, and *q* and various tree and stand attributes.

The high correlation between parameter *i* and the tree DBH (0.976) confirms the findings of Steingass (1996) for *Pseudotsuga menziesii*, and of Hui and Gadow (1997) for *Cunninghamia lanceolata*. Parameter *p*, which describes the shape of the lower part of the stem, is highly variable. Korol and Gadow (2003) could not establish a relation between *p* and any of the tree and stand attributes. However, the parameter *q*, which describes the upper part of the stem, was related to the total height of the sample trees and to the stand basal area. These results have shown that at least two of the three parameters can be predicted with reasonable accuracy from parameters that are usually known in pure stands.

The parameters of the modified Brink function were found to be related to the particular shape of certain sections of a tree and thus may have some biological significance. This is an advantage because the function can be used for developing a generalized stem profile model. The following model, based on several studies

Table 4.4 Correlation coefficients involving individual tree attributes (DBH cm; tree height *h* m), stand attributes (stand age A; trees per ha N, quadratic mean diameter dg cm; dominant stand height Ho m; and basal area BA m^2/ha), and the three model parameters *i*, *p*, and *q*

	tree attributes			stand attributes				parameters		
	A	BHD	h	N	dg	Ho	BA	i	p	q
A	1,000	0,605	0,683	-0,696	0,856	0,919	0,729	0,598	0,209	-0,338
DBH	0,605	1,000	0,905	-0,501	0,591	0,603	0,522	0,994	0,245	-0,552
h	0,683	0,905	1,000	-0,587	0,672	0,696	0,628	0,920	0,371	-0,592
N	-0,696	-0,501	-0,587	1,000	-0,873	-0,888	-0,897	-0,517	-0,283	0,443
dg	0,856	0,591	0,672	-0,873	1,000	0,972	0,917	0,595	0,278	-0,362
Ho	0,919	0,603	0,696	-0,888	0,972	1,000	0,919	0,607	0,269	-0,385
BA	0,729	0,522	0,628	-0,897	0,917	0,919	1,000	0,537	0,301	-0,375
i	0,598	0,994	0,920	-0,517	0,595	0,607	0,537	1,000	0,308	-0,598
p	0,209	0,245	0,371	-0,283	0,278	0,269	0,301	0,308	1,000	-0,475
q	-0,338	-0,552	-0,592	0,443	-0,362	-0,385	-0,375	-0,598	-0,475	1,000

described in the previous sections, represents a further development of the original Brink function:

$$d_i = 2\left[\begin{array}{l} \dfrac{p_0 dbh}{1-e^{p_2(1.3-h)}} + dbh\left(\dfrac{1}{2}-p_0\right)\left(1-\dfrac{1}{1-e^{p_1(1.3-h)}}\right)+ \\ +e^{p_1(1.3-h_i)}\left(\dfrac{dbh\left(\dfrac{1}{2}-p_0\right)}{1-e^{p_1(1.3-h)}}\right) - e^{p_2(h_i-h)}\left(\dfrac{p_0 dbh}{1-e^{p_2(1.3-h)}}\right) \end{array}\right] \tag{4.17}$$

with

d_i = tree diameter (cm) at tree height h_i (m);
h = total tree height (m);
DBH = tree diameter at breast height (cm);
p_0, p_1, and p_2 = empirical model parameters to be estimated.

Arias-Rodil et al. (2015) selected the equation for developing a taper function for *Pinus radiata* trees in Spain after evaluating eight candidate models. They did not recommend the use of their mixed-effects approach, unless an additional stem diameter measurement was available. However, if the DBH and the total tree height were known, the population average stem profile could be calculated for *Pinus radiata* trees by using their parameter estimates of a fixed-effects model.

To evaluate the function, a set of 232 *Eucalyptus grandis* trees with 8 815 measurements of under-bark diameters (d_{ubi}) for different tree heights (h_i) was kindly provided by Mr Heyns Kotze of Mondi Forests in South Africa. The structure of the object *dat* in the file *TaperGrandis.RData* is shown below:

```
'data.frame':        8815 observations of  8 variables:
 $ Tree: int  1 1 1 1 1 1 1           tree number
 $ h   : num  0.16 0.6 1.3            height corresponding to dubi
 $ d   : num  49.9 46 44.6            diameter under bark at height hi
 $ H   : num  45.1 45.1               total tree height (m)
 $ D   : num  49.6 49.6               the dbh (cm)
 $ HD  : num  0.909 0.909             ratio H(m)/D(cm)
```

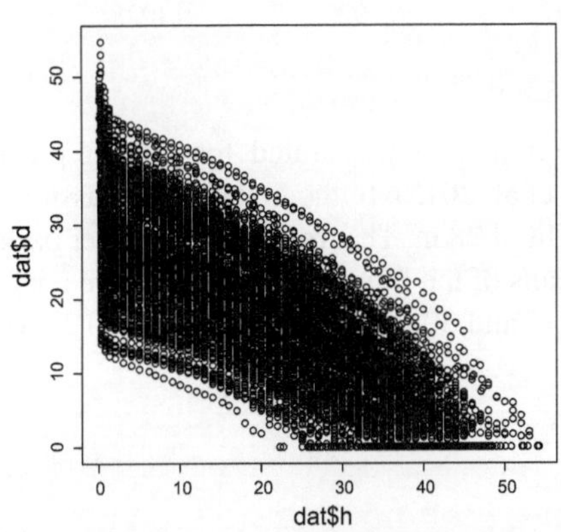

plot(dath, datd)

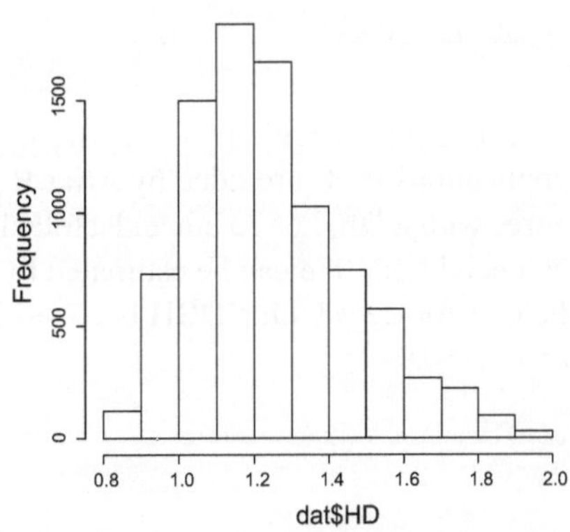

hist(dat$HD)

Table 4.5 Summary of a dataset involving 232 *Eucalyptus grandis* trees with 8 815 measurements, including under-bark diameters (d_{ubi}) at different tree heights (h_i). DBH is the diameter at breast height (cm), h is the total tree height (m), and HD is the ratio (H/DBH)

	DBH (cm)	h (m)	HD	h_i (m)	d_{ubi} (cm)
Min.	14.10	22.30	0.8051	0.12	0.00
1st Qu	27.40	35.80	1.0970	8.00	11.30
Median	33.50	40.40	1.2100	18.00	18.80
Mean	33.02	39.81	1.2506	18.99	18.93
3rd Qu	39.80	44.50	1.3740	28.80	26.20
Max.	49.70	54.20	1.9366	54.20	54.70

A summary of the dataset is given in Table 4.5.

The function parameters can be estimated by means of the following R-code, where D is the tree diameter at breast height, H is the total height of the sample trees, and HD is the ratio of tree height (m) over tree diameter (cm), which is assumed to affect stem form and therefore the model parameters:

```
brink.d <- function(D, H, h, b0,b1,b2){
d <- 2 * ((b0 * D) / (1 - exp(b2 * (1.3 - H))) + (D / 2 - b0 * D) * (1 - (1 / (1 - exp(b1 * (1.3 - H))))) + (exp(-b1 * h)) * (((D / 2 - b0 * D) *
exp(1.3 * b1))/(1 - exp(b1 * (1.3 - H)))) - exp(b2 * h)*((b0 * D * exp(-b2 * H)) / (1 - exp(b2 * (1.3 - H)))))
return(d)}
reg <- nls(d ~ brink.d(D,H,h,p0,p1,p2), start=list(p0=5, p1=0.3, p2=0.03),data=dat, trace=TRUE)
summary(reg)
reg.dbh.30<- nls(d ~ brink.d(D,H,h,p0,p1,p2), start=list(p0=5,p1=0.3,p2=0.03), data=dat[dat$D>30,], trace=TRUE)
summary(reg.dbh.30)
```

The object reg uses the function brink.d and a summary of this object returns the details of the nonlinear regression, including the parameter estimates, which in this particular case are all highly significant. The object reg.dbh.30 again uses the function brink.d to estimate the parameters of the Brink function for trees with a DBH higher than 30 cm. The estimates for the *Eucalyptus grandis* dataset are shown in the following table for DBHs greater than 30 cm, smaller than 20 cm and DBHs between 20 cm and 30 cm.

DBH	n	p_0	p_1	p_2
>30 cm	144	0.4092	0.2968	0.0430
>20 cm & <=30 cm	67	0.4079	0.2410	0.0495
<=20 cm	21	0.3112	0.0920	0.1228
Total sample trees	232			

The implementation of the Brink function is demonstrated for three typical applications in R (provided by Arias Rodil et al. 2015). In the first case, we assume a tree with a DBH of 25 cm and a total height of 28 m. The tree diameter under bark at a height of 10 m can be estimated by means of the function brink.d and the coefficients for trees with a DBH between 20 cm and 30 cm ($b_0 = 0.4079$; $b_1 = 0.2410$; and $b_2 = 0.0495$):

```
dib<- brink.d (25,28,10,0.4079,0.2410,0.0495)
```

The estimate is 16.96 cm. In the second application, we wish to estimate the tree height at which the diameter is 20 cm. When we use the function brink.h, we obtain an estimate of 5.9 m.

```
brink.h <- function(D, H, d, b0,b1,b2){
  h <- optimize(function(x) (d -brink.d(D, H, x, b0,b1,b2))^2, interval = c(0, H))$minimum
  return(h)}
ht<- brink.h (25,28,20,0.4079,0.2410,0.0495)                    # 5.9m
```

The third case involves a volume estimate for a specific section, between a given lower height hst and the height of a given thin-end diameter *d*. The function brink.v can be used to implement the estimate:

```
brink.v<- function(D, H, d, hst, b0,b1,b2){
  u.h <- optimize(function(x) (d -brink.d(D = D, H = H, h = x, b0=b0,b1=b1,b2=b2))^2, interval = c(0, H))$minimum
  v <- integrate(function(x) pi/40000 * (-brink.d (D = D, H = H, h = x, b0=b0,b1=b1,b2=b2))^2, lower = hst, upper = u.h)$value
  return(v)}
```

For example, if the thin-end diameter would be zero ($d = 0$) and the total tree volume would be estimated starting from the stump height at 10 cm (hst = 0.1), the total volume estimate for the tree would be 0.514 m^3. Similarly, the volume of a log between a height of 5 m and a thin-end diameter of 10 cm would be estimated at 0.274 m^3. To improve stem profile estimates, the parameters can be related to forest density, tree age, and the H–D ratio for a given species.

4.1.2.3 Modeling Stem Quality

Schmidt (2001) developed the following model for estimating branch thickness in *Picea abies* trees:

$$BD_{\max} = \left(b_0 d + a_1 C_{\text{depth}}\right) + \left(b_2 d + \frac{b_3 d}{h} + a_3 C_{\text{depth}}\right)\left(1 - \left(1 - C_{\text{depth\_rel}}\right)^{a_4 C_{\text{depth\_rel}}}\right) \tag{4.18}$$

where $BD_{\max}$ is the diameter of the biggest branch in a whorl (mm); C_{depth} is crown depth (distance from tree top, dm); $C_{\text{depth\_rel}}$ is C_{depth} in relation to crown length from top to lowest green branch; DBH is the diameter at breast height (mm) and h is the tree height (dm). Schmidt subdivided the branched part of a tree into different zones for a comprehensive description of its branchiness (Fig. 4.9).

Quality assessment of standing trees is a serious challenge. A major problem is the uncertainty associated with the attributes of trees regarding their invisible inner quality. Another predicament is the deficiency of inventory and forecasting methods. Stem quality attributes are hard to predict and have to be assessed in the field. Usually, this has to be done at minimum cost. A number of methods are available to assess stem quality. Assessment techniques can be divided into three groups (Fig. 4.10).

The *fixed lengths* method requires subdivision of the utilizable stem section into one or more lengths, each of which is then assigned to a quality type. The corresponding tree heights have to be measured or estimated, which may be costly. Assortment distributions are then calculated by means of the data collected and predefined quality classes. Mikulka (1955) used three five-meter sections in mature beech stands, while the Swiss National Forest Inventory (Schweizerisches

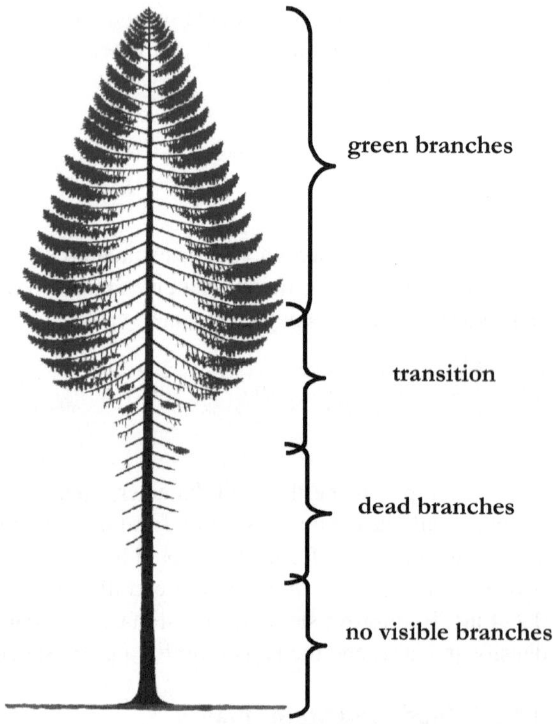

Fig. 4.9 Schmidt's model (2001) of branchiness. He subdivided the branched part of a tree into different zones for a comprehensive description of its branchiness

Landesforstinventar 1988) assigns one of the three possible quality classes to the first two four-meter sections of standing trees with a minimum breast height diameter of 20 cm. Düser (1978) estimated utilizable logs in beech stands by sampling the lengths up to the base of the crown.

The *relative lengths* method divides the whole tree into sections of equal length and assigns each section to one of a fixed number of predefined quality classes. It is fairly easy to establish relative tree lengths in the field, this being the main advantage of the technique which was used by Speidel (1955), Bernstorff et al. (1988), and Bachmann (1990).

The *butt log method* is based on the premise that the lower part of the stem contains a major share of the value of a tree, and that it is sufficient to limit the assessment effort to the lower four-meter or six-meter section. One of the first applications of the butt log method was Arnswaldt's large-scale inventory of valuable beech and oak stands in northern Germany (Arnswaldt 1950). The butt log method is one of the most effective techniques of stem quality assessment, but it can be improved further in two ways. Firstly, the volume calculations will be more precise if appropriate taper functions are used (Trincado and Gadow 1996). Secondly, the assignment of quality classes can be based on quality criteria measured in the field and the use of a key that shows buyer-specific requirements (Wiegard et al. 1997).

The essence of the butt log method is an appraisal of the first four-meter or six-meter log by applying species-specific and visible log quality criteria. For oak trees, the following five quality criteria are often used: *branches*, *lumps*, *wounds*,

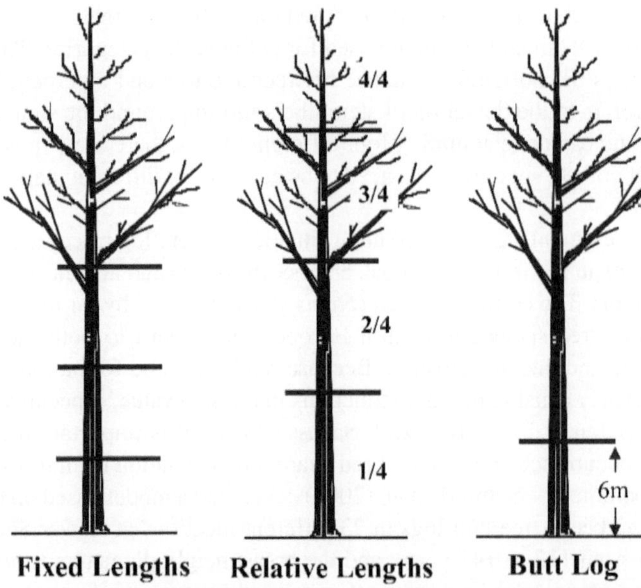

Fig. 4.10 Schematic representation of three methods of stem quality assessment

sinuosity, and *twist*. Such visible criteria can reveal 70–80% of the tree's inner defects (Schweizerisches Landesforstinventar 1988). The first defective *branches* are characterized by the branch diameters and by the number of branches occurring in the butt log.

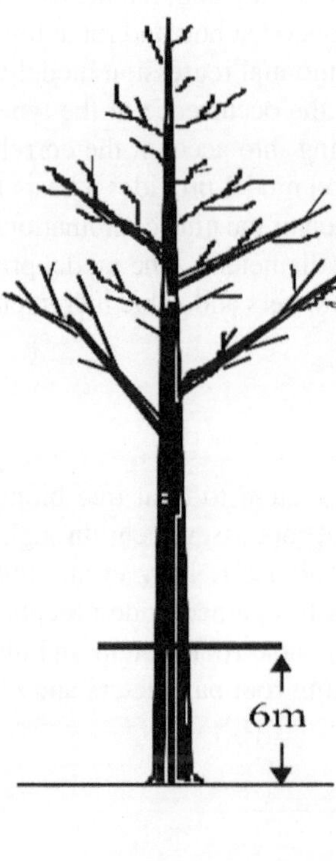

Lumps and open *wounds* are differentiated on the basis of their size, and threshold diameters of 8 cm and 10 cm are used for defining the categories. The criterion *sweep* measures the horizontal distance between the tree and a perpendicular rod, while *twist* refers to the deviation of stem fibers turning around the stem axis. With five criteria and each one assuming four different values, there are 4^5 possible combinations. Of course, specifications need to be adapted for different markets and tree species.

Quality assessments are a key issue in situations where log prices are influenced by value, for example in European oak and beech forests and also in some valuable pine ecosystems. The common beech (*Fagus sylvatica* L.) is by far the most important deciduous tree species in central Europe, with regard to both the harvested timber volume and the area covered. Because red heartwood formation is a serious quality deficiency causing distinct reductions in the sale value, especially in mature beech trees (Kramer 1988: 103, 220; Sachsse 1991), it is important to be able to estimate the occurrence and extent of red heartwood formation from forest and individual tree parameters. Schmidt et al. (2005) developed a model based on the assessment of 1 252 beech tree butt logs in 23 different locations in Lower Saxony. Tree ages ranged from 113 to 141 years, and the breast height diameters from 32 cm to 93 cm. A two-stage modeling approach was used because the lower bound of the dependent variable "heartwood diameter" can (often) assume the value zero Occurrence and extent therefore had to be dealt with separately. It was also necessary to distinguish between red heartwood and rot. Furthermore, the top and butt end of the logs had to be evaluated separately when there were differences in the heartwood diameter. Nine categories of red heartwood were identified: no heartwood, red heartwood only at the butt end, red heartwood only at the top end, rot only at the butt end, rot only at the top end, red heartwood at butt and top end, rot at the butt and top end, red heartwood at butt and rot at top end, and rot at butt and red heartwood at top end. A multinomial regression model was developed in this study to estimate the probability of the occurrence of the types red heartwood and rot in relation to the tree DBH, taking into account the correlation between the butt and top end of logs. The heartwood model provides a basis for estimating (a) the probability of the occurrence of each of the nine combinations and (b) the corresponding extent (red heartwood and rot diameters). The model provided a basis for decisions relating to optimum target diameters and value increment of individual beech trees.

4.1.3 The Root System

The contribution of the root system to total tree biomass and its carbon storage effect can be considerable, but root assessment through excavation is costly, especially in large trees (Grier et al. 1981; Kurz et al. 1996; Bolte et al. 2004). As a result, modeling root systems has gained wide acceptance. The relations between certain characteristics of the coarse root system and above-ground variables have been found useful for estimating root parameters and partitioning (Atkinson 2000;

Drexhage and Collin 2001; Hoffmann and Usoltsev 2001; Le Goff and Ottorini 2001; Kuiper and Coutts 1992).

Bolte et al. (2004) reported the results of a root study conducted in European beech and Norway spruce forests in the Solling Mountains in Lower Saxony in Germany. Forty-two spruce and 27 beech trees in six different forest stands were selected for coarse root sampling. The soil around each tree was soaked and then a skidder carefully pulled down each sample tree with a cable winch. The tree trunks were cut off and the extracted root systems were placed upside down on the ground and cleaned. The exposed root system was subdivided into four horizontal quadrants defined by the north-south and east-west axes in order to coordinate measurement and recording (Fig. 4.11).

Each quadrant was further partitioned into four horizontal layers and four vertical sections according to Nielsen's model (1995). Bolte et al. (2004) used this particular design to facilitate the assessment and description of the root volume at different depths and distances from the stem axis. In the hypothetical example presented in Fig. 4.11, there are 64 spatially defined segments. The position of each segment *ijk* is defined by its quadrant i $\{i = 1, \ldots, 4\}$, a horizontal section j $\{j = 1, \ldots, 4\}$, and a vertical section k $\{k = 1, \ldots, 4\}$. The length of every coarse root ($d \geq 2$ mm) within each section and the root diameter entering and leaving a section were recorded and their volumes calculated. The following equation, fitted to a sample of 42 spruce trees and 27 beech trees, describes the particular relation between the coarse root dry weight *cdw* (kg) and the tree diameter at breast height *dbh* (cm):

$$\ln(cdw) = a + b \cdot \ln(dbh) \tag{4.19}$$

with $a = 5.59$ and $b = 2.79$ for spruce and $a = 4.00$ and $b = 2.27$ for beech.

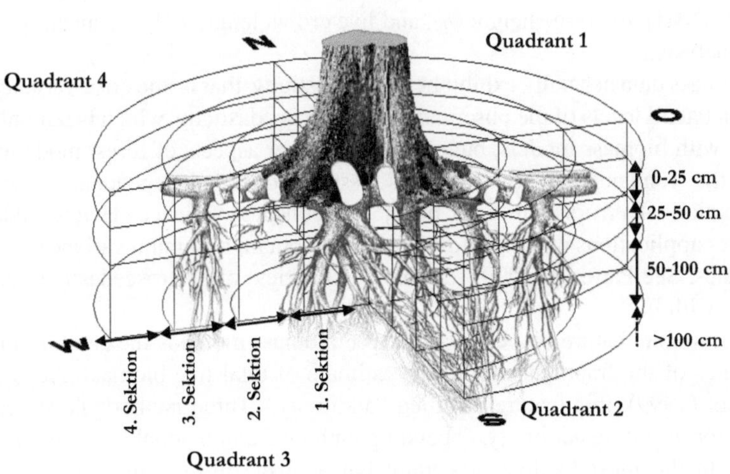

Fig. 4.11 Exposed root system, subdivided into four horizontal quadrants, which were then further subdivided into four vertical sections (1, …, 4) and four horizontal layers (0–25 cm; 25–50 cm; 50–100 cm; and >100 cm)

Interestingly, the differences between the two species turned out to be small. The root system volume and biomass of the spruce trees increase slightly with increasing breast height diameter when compared to the corresponding results for the beech trees.

Fehrmann et al. (2002) summarized the results for each section to describe the spatial distribution of coarse root volumes for several large sample trees. He found that in beech trees the vertical distribution reached a maximum close to the stem axis, while in the spruce trees the maximum occurred at a distance of about 0.7 m from the stem axis. Almost 85% of the total coarse root volumes were found in the first two sections.

4.1.4 Compatible Biomass Models

Biomass and carbon content are among the most important and intensively researched attributes of individual trees and forest communities. These variables are relevant for forest ecological studies and management decisions at varying geographical scales, while estimates of biomass stock and rates of biomass change are relevant in political discussions relating to climate change, the bio-economy, and the use of bio-energy.

Tree biomass is defined as the dry weight of the mass of a tree. Depending on the specific purpose of a study, specific fractions of total tree biomass, such as above-ground and below-ground biomass, or biomass of stems, branches, twigs, leaves, and roots may be assessed. Destructive sampling of selected trees provides essential observations of the dry weights of the different fractions (wood, bark, branches, leaves, or needles) and the total dry weight of each tree. These fractions can be related to easily measured variables of standing trees, such as the diameter at breast height (DBH), total tree height (h), and live crown length (Cl), by means of regression analysis.

Biomass data normally exhibit heteroscedasticity, that is, the error variance is not constant at all levels of the observations. Heteroscedasticity, which is not only associated with biomass models, but also affects other aspects of forest modeling, violates the fundamental least squares assumption of independence and equal distribution of errors with zero mean and constant variance. Heteroscedasticity requires applications of weighted regression to obtain minimum variance parameter estimates. See Parresol (1999) for a detailed review of heteroscedasticity and how to deal with it.

A desirable feature of a compatible tree biomass model is the condition that the estimates of the fractions sum to the estimate of total tree biomass. According to Parresol (1999), this feature is called "additivity." Three methods have been proposed for imposing additivity, depending on how the individual fractions are aggregated. In the most flexible procedure but also the most complicated one, joint generalized least squares, also called seemingly unrelated "regressions" (SUR), are used to fit a system of correlated biomass equations to cross-equation constraints. An example of this procedure is presented in the next section.

4.1.4.1 Developing a System of Compatible Biomass Equations

A database of 75 *Pinus pinaster* Ait. trees that were destructively sampled to estimate the biomass of different tree fractions will be used. A total of 35 permanent sample plots of even-aged *Pinus pinaster* stands were used in this study. These plots were selected throughout the inventory areas of Galicia in northwestern Spain to provide representative information of site quality, age, and stand density. Two or three trees were subjectively selected on each sample plot to represent the different combinations of the pooled distribution of diameters and heights. These observations will be used to develop a compatible system for simultaneously estimating three tree fractions (crown, wood, and bark biomass) and the total above-ground biomass in our example. To ensure additivity, the estimates of the total above-ground biomass model should be equal to the sum of the estimates of the three fractions (crown, wood, and bark).

The disaggregative approach, initially proposed by Tang et al. (2000) and adapted by Dong et al. (2015), will be used to fit the system. The approach is based on fitting a model for the total above-ground biomass and then disaggregating this biomass by means of multiplicative component-discrimination functions. In our example, an allometric model will be used to fit the total above-ground biomass (total_kg in the dataframe) using the diameter at breast height (d_cm in the dataframe) and the tree height (h_m in the dataframe) as independent variables:

$$\text{total}_{\text{kg}} = a_0 \cdot dbh^{a_1} \cdot h^{a_2} = \exp\left(a_0^{'} + a_1 \ln\left(dbh\right) + a_2 \ln\left(h\right)\right) \tag{4.20}$$

where ln is the natural logarithm.

The next step involves defining a function to discriminate between the crown (crown_kg in the dataframe) and stem wood and bark (wood_kg and bark_kg in the dataframe). In this example, an exponential model proposed by Affleck and Diéguez-Aranda (2016) will be used:

$$\frac{\text{crown\_kg}}{\left(\text{crown\_kg} + \text{stem\_kg}\right)} = \frac{\text{crown\_kg}}{\text{total\_kg}} = \frac{1}{1 + \exp\left(b_0 + b_1 \ln\left(dbh\right) + b_2 \ln\left(h\right)\right)} \tag{4.21}$$

Next, a function to discriminate between the wood and the bark of the stem biomass (stem_kg in the dataframe) is defined in the same way:

$$\frac{\text{wood\_kg}}{\left(\text{wood\_kg} + \text{bark\_kg}\right)} = \frac{\text{wood\_kg}}{\text{stem\_kg}} = \frac{1}{1 + \exp\left(c_0 + c_1 \ln\left(dbh\right) + c_2 \ln\left(h\right)\right)} \tag{4.22}$$

The system consists of these three equations that are fitted simultaneously by using the *seemingly unrelated regression* (SUR) method to account for error correlations. The R-code begins with a description of the data, a line to read the csv original file, an option to delete warning messages, and the command to load the library "systemfit" (Henningsen and Hamann 2007) to fit the models simultaneously:

```
# Import the database (75 destructively sampled trees of
# Pinus pinaster Ait. in Northwestern Spain)
biomass <- read.csv("biomass.csv", sep=";"); options(warn=-1)
library(systemfit)
```

Three new variables are then added to the dataframe: (1) the stem biomass (stem_kg) defined as the sum of wood and bark biomass; (2) the ratio between crown biomass and total biomass (r_crown_total) and (3) the ratio between wood biomass and stem biomass (r_wood_stem):

```
# New variables: Ratio definitions
biomass$stem_kg<-biomass$wood_kg+biomass$bark_kg
biomass$r_crown_total<-biomass$crown_kg/biomass$total_kg
biomass$r_wood_stem<-biomass$wood_kg/biomass$stem_kg
```

In the next step, the three models of the system are defined, and a list is created with the reference names of the three models ("total.eq," "ratio_wood.eq," and "ratio_wood.eq"):

```
# Model definitions
total.eq<-total_kg~exp(a0+a1*log(d_cm)+a2*log(h_m))
ratio_crown.eq<-r_crown_total~1/(1+exp(b0*+b1*log(d_cm)+b2*log(h_m)))
ratio_wood.eq<-r_wood_stem~1/(1+exp(c0+c1*log(d_cm)+c2*log(h_m)))
model<-list(total.eq,ratio_crown.eq,ratio_wood.eq)
```

After that, the models are fitted independently to get the initial values of the iterative fitting procedure. The three models are linearized to fit the system by means of linear ordinary least squares (OLS):

$$\text{total}_{\text{kg}} = a_0 \cdot d^{a_1} \cdot h^{a_2} = \exp\left(a_0' + a_1 \ln(dbh) + a_2 \ln(h)\right) \Rightarrow$$

$$\ln\left(\text{total}_{\text{kg}}\right) = a_0' + a_1 \ln(dbh) + a_2 \ln(h)$$

$$\frac{\text{crown}_{\text{kg}}}{\left(\text{crown}_{\text{kg}} + stem_{\text{kg}}\right)} = \frac{1}{1 + \exp\left(b_0 + b_1 \ln(dbh) + b_2 \ln(h)\right)} \Rightarrow$$

$$1 + \exp\left(b_0 + b_1 \ln(dbh) + b_2 \ln(h)\right) = \frac{\text{crown}_{\text{kg}} + stem_{\text{kg}}}{\text{crown}_{\text{kg}}} = 1 + \frac{stem_{\text{kg}}}{\text{crown}_{\text{kg}}} \Rightarrow$$

$$\exp\left(b_0 + b_1 \ln(dbh) + b_2 \ln(h)\right) = \frac{stem_{\text{kg}}}{\text{crown}_{\text{kg}}} \Rightarrow$$

$$\ln\left(\frac{stem_{\text{kg}}}{\text{crown}_{\text{kg}}}\right) = b_0 + b_1 \ln(dbh) + b_2 \ln(h)$$

$$\frac{\text{wood}_{\text{kg}}}{\left(\text{wood}_{\text{kg}}+\text{bark}_{\text{kg}}\right)}=\frac{1}{1+\exp\left(c_0+c_1\ln\left(dbh\right)+c_2\ln\left(h\right)\right)}\Rightarrow$$

$$1+\exp\left(c_0+c_1\ln\left(dbh\right)+c_2\ln\left(h\right)\right)=\frac{\text{wood}_{\text{kg}}+\text{bark}_{\text{kg}}}{\text{wood}_{\text{kg}}}=1+\frac{\text{bark}_{\text{kg}}}{\text{wood}_{\text{kg}}}\Rightarrow$$

$$\exp\left(c_0+c_1\ln\left(dbh\right)+c_2\ln\left(h\right)\right)=\frac{\text{bark}_{\text{kg}}}{\text{wood}_{\text{kg}}}\Rightarrow$$

$$\ln\left(\frac{\text{bark}_{\text{kg}}}{\text{wood}_{\text{kg}}}\right)=c_0+c_1\ln\left(dbh\right)+c_2\ln\left(h\right)$$

```
# Fit log-linear models independently for starting values
total.fit<-lm(log(total_kg)~log(d_cm)+log(h_m), data=biomass)
ratio_crown.fit<-lm(log(stem_kg/crown_kg)~log(d_cm)+log(h_m), data=biomass)
ratio_wood.fit<-lm(log(bark_kg/wood_kg)~log(d_cm)+log(h_m), data=biomass)

start.values<-c(coef(total.fit),coef(ratio_crown.fit),coef(ratio_wood.fit))
names(start.values)<-c("a0","a1","a2","b0","b1","b2","c0","c1","c2")
```

Finally, the system is fitted simultaneously to obtain the parameter estimates for each model:

```
# Parameter estimates fitting the systems simultaneously with additivity
fitsur<-nlsystemfit("SUR", model, start.values, data=biomass)
# Parameter estimates of total above ground biomass equation
fitsur$eq[[1]]$b
# Parameter estimates of proportion of crown versus total biomass
fitsur$eq[[2]]$b
# Parameter estimates of proportion of wood versus stem biomass (wood and bark biomass)
fitsur$eq[[3]]$b
```

The following results are obtained:

```
# Parameter estimates of total above ground biomass equation
> fitsur$eq[[1]]$b
      a0        a1        a2
-4.360091  2.000096  1.132805
> # Parameter estimates of proportion of crown versus total biomass
> fitsur$eq[[2]]$b
       b0         b1         b2
0.9284247 -0.8693954  1.4145428
> # Parameter estimates of proportion of wood versus stem biomass (wood and bark biomass)
> fitsur$eq[[3]]$b
        c0          c1          c2
1.32595402  0.05281007 -1.09560963
```

Now, a new dataframe is created which includes the observed biomass values and the estimates obtained with the models. The final expressions for estimating the total above-ground biomass, crown biomass, stem biomass, wood biomass, and bark biomass are as follows:

$$\text{Total}\,\widehat{\text{above}}\,\text{ground}=\exp\left(-4.3601+2.0001\cdot\ln\left(dbh\right)+1.1328\cdot\ln\left(h\right)\right)$$

$$\widehat{\text{Crown}} = \widehat{\text{Total above ground}} \cdot \frac{1}{1+\exp\left(0.9284 - 0.8694 \cdot \ln\left(dbh\right) + 1.4145 \cdot \ln\left(h\right)\right)}$$

$$\widehat{\text{Stem}} = \widehat{\text{Total above ground}} \cdot \left[1 - \frac{1}{1+\exp\left(0.9284 - 0.8694 \cdot \ln\left(dbh\right) + 1.4145 \cdot \ln\left(h\right)\right)}\right]$$

$$\widehat{\text{Wood}} = \widehat{\text{Stem}} \cdot \frac{1}{1+\exp\left(1.3260 + 0.0528 \cdot \ln\left(dbh\right) - 1.0956 \cdot \ln\left(h\right)\right)}$$

$$\widehat{\text{Bark}} = \widehat{\text{Stem}} \cdot \left[1 - \frac{1}{1+\exp\left(1.3260 + 0.0528 \cdot \ln\left(dbh\right) - 1.0956 \cdot \ln\left(h\right)\right)}\right]$$

```
# Database with observed and predicted values
predicted<-data.frame(biomass[,c("tree","total_kg","wood_kg","bark_kg","crown_kg")],
    total_predicted=fitsur$eq[[1]]$predicted,
    crown_predicted=fitsur$eq[[1]]$predicted*fitsur$eq[[2]]$predicted,
    wood_predicted=fitsur$eq[[1]]$predicted*(1-fitsur$eq[[2]]$predicted)
        *fitsur$eq[[3]]$predicted,
    bark_predicted=fitsur$eq[[1]]$predicted*(1-fitsur$eq[[2]]$predicted)
        *(1-fitsur$eq[[3]]$predicted))
```

The system is compatible, which means that the sum of the individual biomass fractions (crown, wood and bark) is exactly equal to the total tree biomass and the sum of wood and bark fractions is exactly equal to the stem biomass. For example, for a tree with a DBHdbh of 20 cm and a total height of 23 m, the biomass estimates are:

$$\widehat{\text{Total above ground}} = \exp\left(-4.3601 + 2.0001 \cdot \ln\left(20\right) + 1.1328 \cdot \ln\left(23\right)\right) = 178.32\text{kg}$$

$$\widehat{\text{Crown}} = 178.32 \cdot \frac{1}{1+\exp\left(0.9284 - 0.8694 \cdot \ln\left(20\right) + 1.4145 \cdot \ln\left(23\right)\right)} = 10.63\text{kg}$$

$$\widehat{\text{Stem}} = 178.32 \cdot \left[1 - \frac{1}{1+\exp\left(0.9284 - 0.8694 \cdot \ln\left(20\right) + 1.4145 \cdot \ln\left(23\right)\right)}\right] = 167.69\text{kg}$$

$$\widehat{\text{Wood}} = 167.69 \cdot \frac{1}{1+\exp\left(1.3260 + 0.0528 \cdot \ln\left(20\right) - 1.0956 \cdot \ln\left(23\right)\right)} = 146.82\text{kg}$$

$$\widehat{\text{Bark}} = 167.69 \cdot \left[1 - \frac{1}{1+\exp\left(1.3260 + 0.0528 \cdot \ln\left(20\right) - 1.0956 \cdot \ln\left(23\right)\right)}\right] = 20.87\text{kg}$$

20.87 + 146.82 + 10.63 = 178 kg (total above-ground biomass) and 146.82 + 20.87 = 167.69 kg (stem biomass). The observed versus estimated values of total above-ground biomass and the three fractions including the bisector and the value of the coefficient of determination are plotted (Fig. 4.12).

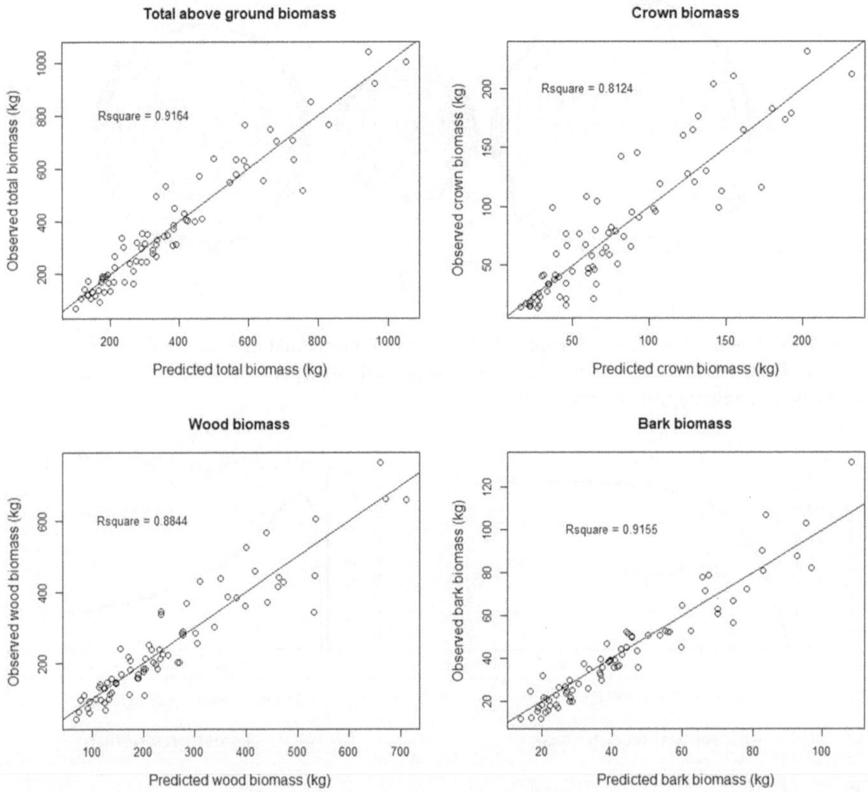

Fig. 4.12 Observed versus estimated values of total above-ground biomass and the three fractions including the bisector and the value of the coefficient of determination

The graph of the relation between observed and predicted biomass values for the four compartments includes the bisector and the value of the coefficient of determination (R^2) estimated as the square of the correlation coefficient between observed and predicted values.

4.2 Individual Tree Growth Models

The growth of individual trees is affected significantly by density and competition. An example of this effect is presented in Fig. 4.13, which shows disks taken at breast height of 39-year-old *Pinus patula* trees grown in the Weza CCT experiment in South Africa (Craib 1939).

Tree no 1 developed under conditions of free growth (plot 8 of the CCT experiment), while tree no 2 came from a high density plot (plot 1 of the CCT experiment). Tree no 3 was grown at medium density and was thinned at age 7. The response to the release from competition was an immediate increase in ring width.

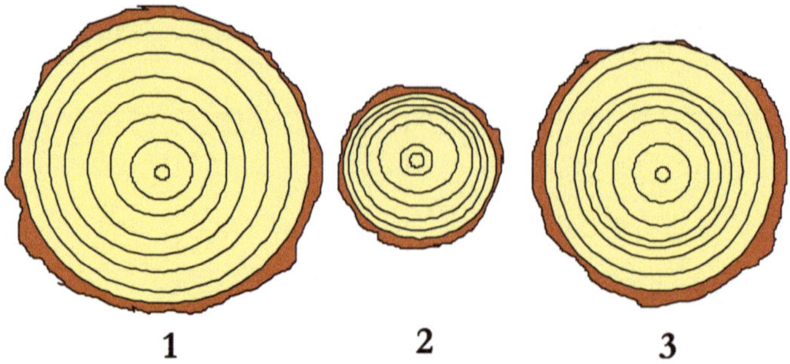

Fig. 4.13 Disks taken at breast height of 39-year-old Pinus patula trees in the Weza CCT experiment in South Africa (Craib 1939). They show the growth in response to different densities and the response to a release from competition

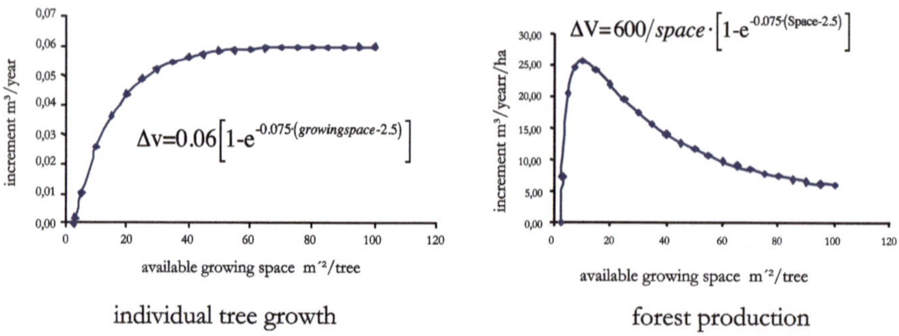

Fig. 4.14 Relation between available growing space per tree and volume increment of individual trees (Δv, m^3/year) and of the entire plot (ΔV, m^3/year/ha) in a 39-year-old *Picea abies* forest (see Thomasius and Thomasius 1976)

Figure 4.14 shows the relations between available growing space (m^2/tree) and (a) volume increment of individual trees (m^3/year) and (b) the production of the entire plot (m^3/year/ha) in 39-year-old *Picea abies* forest plots at different densities (see Thomasius and Thomasius 1976)

In this particular example, the initial sharp response of individual tree growth to increased growing space per tree levels off as soon as the growing space per tree exceeds about 30 m^2. However, 30 m^2 growing space per tree is not optimal in terms of forest production. Forest production reaches a maximum at about 10 m^2 growing space per tree.

4.2.1 Modeling Tree Survival

Survival analysis entails a group of statistical methods dealing with the occurrence and timing of events such as death (e.g., Cox and Oakes 1984). According to Klein and Moeschberger (1997, p. 21), the distribution of the nonnegative random variable T, which describes forest age, can be characterized by four functions (Staupendahl 2011; Staupendahl and Zucchini 2011). The probability density function $f(t)$ describes the frequency distribution of the points in time when trees die. In the case of continuously measured time, it is defined by

$$f(t) = \lim_{\Delta t \to 0} \frac{P(t \le T < t + \Delta t)}{\Delta t}, \quad \text{with } t \ge 0 \tag{4.23}$$

For small $\boldsymbol{\Delta} t$, $f(t) \cdot \boldsymbol{\Delta} t$ may be thought of as the approximate unconditional probability that tree death will occur at time t. The cumulative distribution function $F(t)$, the integral of the density function, gives the probability that a death has occurred by time t:

$$F(t) = P(T \le t) = \int_0^t f(x)\,dx \tag{4.24}$$

The survival function $S(t)$ is the complement of $F(t)$. Therefore, it gives the probability that a tree survives at least until time t:

$$S(t) = P(T > t) = 1 - F(t) \tag{4.25}$$

The probability that the event occurs at a certain time, on condition that the tree has survived until that time, is also important. The density of this conditional death probability is called the risk or hazard function (hazard rate) $h(t)$ and is defined by

$$h(t) = \lim_{\Delta t \to 0} \frac{P(t \le T \langle t + \Delta t | T \ge t)}{\Delta t} = \frac{f(t)}{S(t)} \tag{4.26}$$

For small $\boldsymbol{\Delta} t$, $h(t) \cdot \boldsymbol{\Delta} t$ gives an approximation of the conditional probability of tree death. In the case of discrete time, which is normally considered in forest research, the unconditional probability that a planted tree dies at age t_i is given by

$$f(t_i) = P(T = t_i) = S(t_{i-1}) - S(t_i) \tag{4.27}$$

where $S(t_i)$ gives the discrete survival function, with $i = 1, 2, \ldots, t_0 = 0$ and $S(t_0) = 1$. Accordingly, the discrete time hazard rate (Fahrmeir 2007: 26) gives the conditional probability that a tree which survived until age t_i dies at this age:

$$h(t_i) = P(T = t_i \mid T \ge t_i) = \frac{f(t_i)}{S(t_{i-1})} \tag{4.28}$$

In other words, $h(t_i)$ indicates the proportion of trees in age t_{i-1} that die on average at age t_i. According to Pienaar and Shiver (1981), Gadow (1987), Kouba (2002), and Dickel et al. (2010), the Weibull distribution (Weibull 1951) is useful to model individual tree mortality because it requires only two parameters while at the same time being quite flexible. If T is Weibull distributed with scale parameter β and shape parameter α, the density, distribution, survival, and hazard functions are given as follows (Staupendahl 2011):

$$f(t) = \frac{\alpha}{\beta} \cdot \left(\frac{t}{\beta}\right)^{\alpha-1} \cdot \exp\left[-\left(\frac{t}{\beta}\right)^{\alpha}\right] \tag{4.29}$$

$$F(t) = 1 - \exp\left[-\left(\frac{t}{\beta}\right)^{\alpha}\right] \tag{4.30}$$

$$S(t) = \exp\left[-\left(\frac{t}{\beta}\right)^{\alpha}\right] \tag{4.31}$$

$$h(t) = \frac{\alpha}{\beta} \cdot \left(\frac{t}{\beta}\right)^{\alpha-1} \tag{4.32}$$

with $t > 0$. These functions are directly derivable from one another. The survival function (Eq. 4.28) was fitted to observed surviving *Pinus patula* trees n years after planting in the Nelshoogte spacing experiment. The parameters of the survival function were evaluated for five initial planting densities. The results of the parameter estimates of the survival function are presented in Table 4.5. Each of the parameters α and β is a linear function of the number of planted trees per ha. The estimated linear function of the shape parameter is α = 5.38–0.001012 (N_0). For the scale parameter, it is β = 49.36–0.005206 (N_0), where N_0 refers to the initial number of planted trees per hectare (Table 4.6).

This relation is potentially useful, in that it allows us to estimate the lifetime distribution for a range of planting densities. A graph of the survival functions (Eq. 4.31), and the corresponding mortality (Eq. 4.29) and hazard functions (Eq. 4.32) are shown in Fig. 4.15 for five different initial espacements.

Figure 4.15 shows that the probability of tree survival is different for the different planting espacements. As expected, the probability of tree survival for the same age increases with a decreasing number of planted trees per unit area. The peaks of tree mortality occur earlier and are more lasting in the higher planting densities. The differences are notable for the entire range of ages. Examples of a survival function fitted to four planting densities at the CCT study area MacMac are shown in Fig. 4.16

Assignment

We assume the following ages and survival rates (defined as the ratio N_2/N_1):

```
age<-c(7.0,10.2,15.2,19.2,23.3,25.2,32.6,36.1,37.8,39.8)
surv<-c(0.97,0.96,0.89,0.82,0.63,0.56,0.44,0.34,0.26,0.26)
```

Fit a survival function to these observations using the two-parameter Weibull (Eq. 4.30)

Solution:

```
Surv.f<-function(x,beta,alpha) {exp(-(x/beta)^alpha)}
model<-nls(surv~Surv.f(age,beta,alpha), start=list(beta=30, alpha=2), trace=TRUE,control=nls.control(warnOnly=TRUE))
plot(age, surv, xlim=c(0,65), ylim=c(0,1),main="Plot1")
tt<-1:65; lines(tt, predict(model, list(age=tt)), col=2)
```

Table 4.6 Weibull parameter estimates of Eq. 4.28 for five planting densities in the Nelshoogte CCT experiment (MEF = model efficiency.)

Planted trees per ha (N_0)	α	β	MEF	Bias
2965	2.35	34.18	0.9646	0.0026
1483	3.71	41.04	0.9382	0.0033
988	5.04	44.01	0.8574	0.0044
741	4.36	45.79	0.7098	0.0008
494	4.71	47.06	0.9115	-0.0071

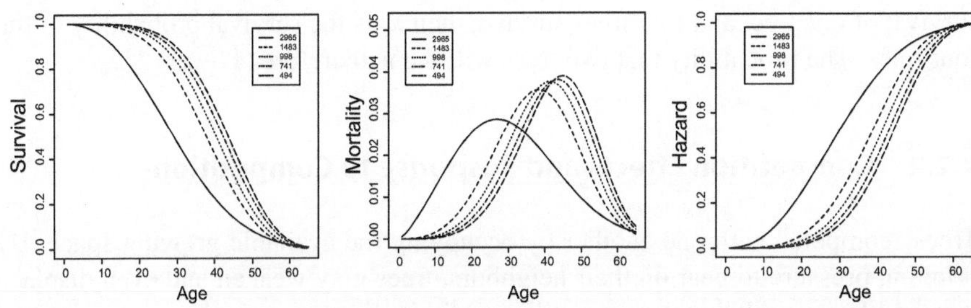

Fig. 4.15 Estimated survival (left) and corresponding mortality density (center) and hazard (right) functions for five different planting espacements N_0, expressed in terms of planted trees per hectare

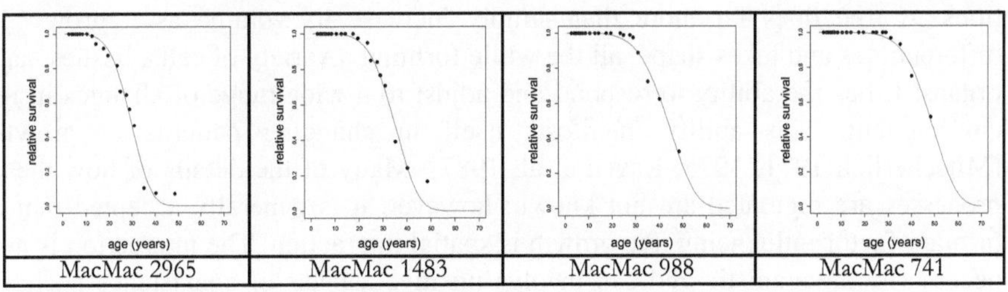

MacMac 2965	MacMac 1483	MacMac 988	MacMac 741

Fig. 4.16 Examples of a survival function fitted to four planting densities at MacMac

Dickel et al. (2010) showed that if there were N_1 trees at the initial time point, and assuming that the lifetime distribution of the trees is Weibull distributed, it follows that the number of surviving trees t time units later is binomially distributed with parameters N_1 and $\pi = 1 - F(t; \alpha, \beta)$ where α and β can be estimated from the initial stems per hectare planted.[2]

$$P(N_2 = n) = \binom{N_1}{n} \cdot \pi^n \cdot (1-\pi)^{N_1 - n} \tag{4.33}$$

This result may be explained as follows: Supposing we have an initial population of six trees (a,b,c,d,e,f). Four of the six initial trees survive at the end of a given period of time. There are 15 possible combinations of four of the six initial trees surviving: (a,b,c,d)(a,b,c,e)(a,b,c,f)(a,b,d,e)(a,b,d,f) (a,b,e,f)(a,c,d,e) (a,c,d,f) (a,c,e,f)(a,d,e,f)(b,c,d,e)(b,c,d,f) (b,c,e,f)(b,d,e,f)(c,d,e,f). This can be obtained by the following expression:

$$\binom{N_1}{n} = \frac{N_1!}{n!(N_1 - n)!} = \frac{6!}{4!2!} = \frac{6 \bullet 5}{2} = 15.$$

There are 15 possible outcomes, and the probability for each of these outcomes is equal to the product of the probabilities for each tree. If π_i is the probability of survival of one tree, and four trees survive, then π^4 is the survival probability of the four trees. The probability that two trees will die is therefore $(1 - \pi)^{(6-4)}$.

4.2.2 Competition Effects and Response to Competition

Trees "compete" with one another by occupying the available growing space. By bringing pressure to bear on their neighbors, trees may weaken and even displace them, with the result that they can utilize additional resources in the newly acquired space. This process is costly, as additional phytomass needs to be assembled and maintained, but the reward is access to new resources (Matyssek 2003). Trees need light, carbon dioxide, water, and minerals. These elementary substances are converted into complex organic molecules by means of very specific chemical reactions. A tree does far more than simply increase its volume as it grows. It differentiates and takes shape, all the while forming a variety of cells, tissues, and organs. It has the ability to respond and adjust to a wide range of changes in its environment. This ability manifests itself in changing patterns of growth (Mitscherlich 1971, 1975; Raven et al. 1987). Many of the details of how these processes are regulated are not known; however, it is generally accepted that a primary factor influencing tree growth is spatial interaction. The interaction is not necessarily antagonistic and can involve positive aspects of coexistence such as mutual adaptation, dependence, and protection. Forests are highly integrated communities where the relations between individual trees are competitive as well as facilitative.

[2]When N_1 is not the number of trees at the initial time point, the left truncated Weibull distribution has to be used.

Single-tree spatial models use information about the distances and sizes of neighboring trees, and this effect is designated as *competition* or *point density*.[3] Trees do not always compete for water, nutrients, or light. Competition is often simply a matter of physical obstruction or constriction by neighboring individuals that grow so close to one another that they impede the expansion of neighboring tree crowns.

Numerous measures to quantify tree competition have been designed, and there are endless possibilities for inventing new ones. Most of the existing competition indices incorporate the area potentially available (AVA) to a tree, or the sizes, distances, and directional distributions of neighboring trees, using either two-dimensional or three-dimensional spatial data (Clutter et al. 1983; Tomé and Burkhart 1989; Pukkala 1989; Holmes and Reed 1991; Biging and Dobbertin 1992, 1995; Vanclay 1994: 58 *et seq.*). Competition in forest communities therefore refers to the relation between trees of the same or different species in which individuals are adversely affected by neighboring trees that have similar resource requirements (Sahney et al. 2010; Seifert et al. 2014). The literature on competition is extensive and covers a wide range, from simple indices to complex search algorithms.

The literature differentiates between competition effects and response to competition. A competition effect involves the reduction in the performance of a specific reference tree by the aggregate competitive strength of neighboring trees. Both above- and below-ground competition effects can be caused by reduced resource availability (Cahill 1999; Cahill 2002; Song et al. 2012). A direct assessment of a change in resource availability is usually very labor-intensive, requiring an assessment of changes in radiation, moisture, or nutrients under field conditions. For this reason, the aggregate effect is generally estimated by means of a competition index which may include a measure of density, the relative size, or available growing space of a reference tree, the number and proximity of immediate neighbors, and estimates of shading and constriction.

The degree to which the performance of a reference tree is modified by the competition effect of the neighbors is known as *response to competition*. Some trees are more sensitive to competition than others, but the degree of indifference or sensitivity depends on various factors, including the species (light-demanding or shade-tolerant), resource limitation (dependent on or independent of high moisture supply), and tree age. Light-demanding trees are usually less sensitive to shading at a young age, but become gradually more sensitive with increasing age (Valladares and Niinemets 2008; Seifert et al. 2014). The response to above-ground competition includes changes in the biomass structure and allometry of above-ground components (Seifert 2003; Wang et al. 2011; Rötzer et al. 2012). However, these factors may be insufficient to explain the response to below-ground competition.

---

[3] A distinction must be made between *competition* and *density*. Competition refers to the spatial interaction between neighboring trees, while density is a population attribute, a measure of how completely a site is occupied by trees. In the literature, the term *point density* is sometimes used as a synonym for *competition*.

Because of the technical problems regarding sampling and measurement, below-ground competition has been given little attention compared to above-ground competition (Casper and Jackson 1997; Bauhus et al. 2000; Curt and Prevosto 2003; Bolte et al. 2004). Early successional, fast-growing, and light-demanding species are usually very responsive to the removal of root competition and are capable of rapid morphological adaptation (Messier and Nikinmaa 2000; Gunaratne et al. 2011). However, isolating the effects of above-ground competition and identifying specific responses to below-ground competition are major challenges.

Facilitation, which involves positive interactions between individuals, can complicate the interpretation of a particular response to competition. Certain mixtures of tree species can increase nutrient availability, nutrient uptake, and nutrient use efficiency (Rothe and Binkley 2001; Richards et al. 2010; Forrester 2014). However, identifying specific responses related to specific species mixtures and at varying levels of density presents a challenge.

When the spatial pattern is stationary across space, that is, when the density of individuals $\lambda(x)$ does not depend on the location x, the competition effects can be assessed by using distance-independent measures of competition. One of the earliest indices used to measure competition is the "Crown-DBH ratio" (CDR = crown diameter of a tree divided by its DBH) proposed by Seebach (1846). The CDR was evaluated by Spiecker (1994) for *Prunus avium* trees, resulting in the relation (DBH increment, cm/year) = 0.54·(CDR)–4.7. Interestingly, this relation was found to be independent of the type of forest.

4.2.2.1 Overtopping Basal Area: BAL and BALMOD

Normally, a distance-independent index is assumed to require both the stand density (a population attribute) and the relative social rank (a tree attribute). An example is the overtopping basal area or the basal area of larger trees (BAL). The BAL index is defined as the sum of the basal areas of all trees larger than the reference tree within a defined competition zone (Wykoff et al. 1982). The BAL thus mainly estimates the effects of overtopping and radiation exclusion. We need the individual DBHs (cm) of the n trees in a hectare of forest if we want to calculate the total basal area (BA, m^2 per ha):

$$BA = \frac{\pi}{40000} \cdot \sum_{i=1}^{n} dbh_i^2 \tag{4.34}$$

The basal area percentile p_i of tree i represents the proportion of the total basal area occupied by trees which have an individual tree basal area less than or equal to the basal area bai of tree i:

$$p_i = \frac{1}{BA} \cdot \sum_{j=1}^{n} ba_j, \forall j \quad ba_j \leq ba_i \tag{4.35}$$

The basal area percentile is an expression of the social rank of the reference tree i. To be suitable as a measure of competition, the density of the forest must be included. The BAL index for tree i (BAL_i) can be derived from the cumulative basal area distribution:

$$p_i = 1 - \left(\frac{BAL_i}{BA} \right), \text{ and thus } BAL_i = BA \cdot (1 - p_i) \quad (4.36)$$

where BAL_i is the sum of the basal areas of all trees with a DBH greater than the DBH of the reference tree i (m^2/ha) and BA is the total basal area of the forest (m^2/ha).

Assignment

The table below gives details of 10 trees on a sample plot of 0.025 ha.

The basal area of the plot is BA = 0.752 m2 or 0.752/0.025 = 30.08 m^2/ha. Calculate the basal area percentile and the BAL index of tree no 4.

i	1	2	3	4	5	6	7	8	9	10
dbh_i (cm)	20	21	23	27	31	32	33	37	37	41
ba_i (m^2)	0.031	0.033	0.042	0.057	0.075	0.080	0.086	0.108	0.108	0.132

Solution

p_4=(0.031+0.033+0.042+0.057)/0.752 = 0.217. 21.7 % of the total basal area are occupied by trees with a basal area less or equal to 0.057 m^2, and BAL_4 is thus 30.08(1-0.217)= 23.55 m^2/ha.

Numerous studies have shown that the BAL index is a highly significant explanatory variable requiring only standard information that is usually readily available. However, this advantage does not always apply in multispecies forests. The BAL index is "species-blind," which means that neither the species of the reference tree nor the species of the competitors are considered. Accordingly, Albert (2004) proposed a species-sensitive index that he called *BALMIX*. BALMIX improved the estimates of competition effects in a specific set of circumstances, and further studies in this regard may prove to be promising.

The BAL index is a linear function of basal area. There are alternative ways to combine the relative dominance (the basal area percentile) of the reference tree with the density of a forest. In the modified BAL (BALMOD) index, the expression $1-p_j$ is weighted with the *Relative Spacing* (*RS*) (Schröder and Gadow 1999):

$$BALMOD_i = \left(\frac{1-p_i}{RS} \right), \text{with } p_i = 1 - \left(\frac{BAL_i}{BA} \right) \text{and } RS = \frac{\sqrt{\frac{10000}{N}}}{\text{Hdom}} \quad (4.37)$$

The biggest diameter tree in the forest has a basal area percentile = 1 and its BALMOD is zero. For the same basal area percentile, the BALMOD index increases exponentially with decreasing RS (i.e., increasing forest density), and there is a sharp increase after RS = 0.15 (Fig. 4.17).

The BALMOD index combines the relative dominance of each individual tree (expressed by the basal area percentile) with the density-variable relative spacing that is based on dominant height. The BALMOD performed better than BAL when differences in the growth of *Pinus sylvestris* in Spain (Schröder and Gadow 1999) and *Quercus petraea* in northern Germany (Struck 1999; Table 4.7) had to be explained.

The basal area percentile for tree no 1 (DBH = 2 cm; height = 3.7 m) is calculated as follows:

$$p_1 = (3.1)/319.48 = 0.01.$$

The mean height of the three most dominant trees is Hdom = (10.5 + 9.1 + 8.6)/3 = 9.4 m. The number of trees per ha is 23 810 and RS is equal to 0.0689. The BALMOD of tree no 1 is therefore (1–0.01)/0.0689 = 14.37.

The most common forms of physical impediment within a group of interacting trees are constriction of the growing space and shading (Assmann 1953). Competition indices focusing on overtopping therefore proved to be fairly reliable predictors when tree growth was simulated in environments where light is a limiting resource. Canham et al. (1999), for example, developed a gap light index for mixed stands in North America. The index takes into account the open areas of the sky where the tree crown can receive photon flux. As an extension of Canham's idea, hemispherical projections of neighboring trees similar to fisheye-photos were simulated to describe competition for trees (Courbaud 1995; Biber 1996) and individual branches (Seifert 2003).

4.2.2.2 Local Crowding

The local crowding index has been defined as the basal area of all trees, excluding the reference tree, within a small circle of influence (Steneker and Jarvis 1963; Seifert et al. 2014). The reference tree is located at the center of the circle of influence and the circle radius is very small, therefore emphasizing root competition. Seifert et al. (2014) selected a radius equal to twice the mean distance between trees (calculated with a squared grid distribution of the trees in mind and, therefore, equal

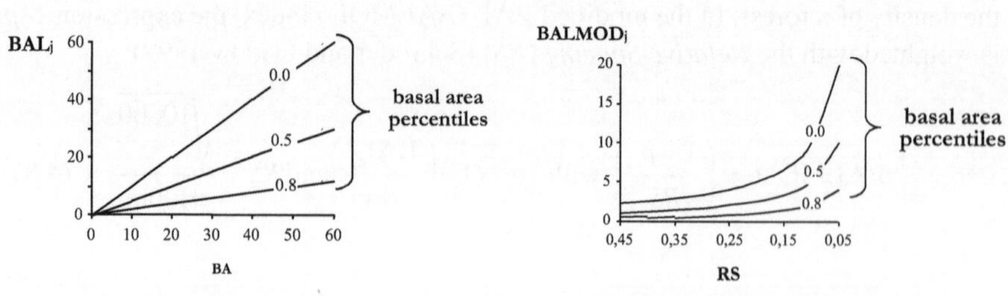

Fig. 4.17 Effects of basal area (BA) and relative spacing (RS) on BAL and BALMOD for different p_j

Table 4.7 Observations of 15 young oak trees (see Struck 1999), with calculations of the Balmod for each tree

Tree no	1	2	3	4	5	6	7	8	9	10	11	12	13	14	15
dbh_i (cm)	2.0	3.0	3.0	3.5	4.0	4.0	4.0	4.0	5.0	5.0	5.5	6.5	7.0	8.0	8.5
h_i (m)	3.7	3.4	3.7	6.2	4.6	5.7	6.1	6.3	6.6	7.2	7.1	8.4	9.1	8.6	10.5
ba_i (cm^2)	3.1	7.1	7.1	9.6	12.6	12.6	12.6	12.6	19.7	19.7	23.8	33.2	38.5	50.4	56.8
BAL_i	316	302	302	293	242	242	242	242	203	203	179	146	107	57	0
BA (cm^2)	319.48														
p_i	0.01	0.05	0.05	0.08	0.24	0.24	0.24	0.24	0.36	0.36	0.44	0.54	0.66	0.82	1
N	23 810														
Hdom (m)	9.4														
RS	0.0689														
$Balmod_i$	14.4	13.8	13.8	13.4	11.0	11.0	11.0	11.0	9.3	9.3	8.1	6.7	4.9	2.6	0

to the square root of 10 000 divided by the number of trees per hectare). Coomes and Allen (2007) and Seifert et al. (2011) used this index to estimate the effects of below-ground competition.

Numerous indices have been proposed for analyzing competition in forests with mapped trees, and the potential for developing new ones is boundless (Biging and Dobbertin 1992; Bachmann 1998). Distance-dependent competition indices include available growing space (Brown 1965; Hessenmöller 2002), overlapping zones of influence (Gerrard 1969), and distance-weighted size ratios (Hegyi 1974; Lee and Gadow 1997). Five examples of distance-dependent measurement of the competition effect are described here.

4.2.2.3 Area Potentially Available

Various methods have been put forward for estimating the area potentially available (APA) for a specific tree to expand and grow on a horizontal plane.[4] The APA, also known as the *topological space*, can be defined as a set of points as well as a set of neighborhoods for each point that satisfies particular axioms for relating points and neighborhoods. The APA borders are defined as a convex polygon (Fig. 4.18).

A Voronoi polygon includes the set of points that are located closer to a particular tree than to any other tree. A Voronoi diagram represents a partitioning of a plane into polygons that are based on the distance to points in a specific subset of the plane (Brown 1965; Jack 1967; Stöhr 1963; Klier 1969; Hessenmöller 2002). The associated methods, known as *Dirichlet* or *Voronoi Tesselation*, are named after the two scientists who developed area subdivisions based on specific geometric patterns.

To construct a *topological* APA, the polygons are formed by line segments that are equidistant from two neighboring trees positioned on either side of the segments. If a line is drawn from a tree to its immediate neighbor, the distance between the two trees will be perfectly bisected by the line segment that separates the two

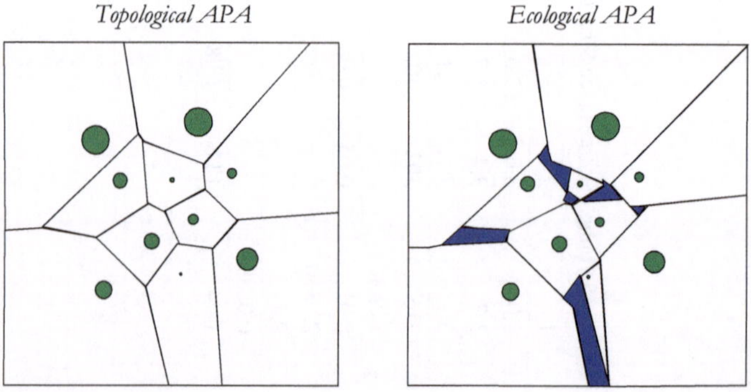

Fig. 4.18 Examples of topological and ecological APA patterns of 10 trees with different dimensions (see Hessenmöller 2002)

[4] See, for example, Brown 1965; Pielou 1977: 81; Lee 1980; Matèrn 1986: 39; Matsumura 1988; Nance et al. 1988; Staupendahl 2011; Hessenmöller 2002.

polygons. An *ecological* APA is obtained by weighting the distances between neighboring trees with their diameters or crown volumes (Stöhr 1963; Klier 1969). The ecological APAs are also convex polygons, but in this case the subdivision of the forest is incomplete.

Pukkala (1989) did a simulation study based on a mapped stand of *Pinus sylvestris* by applying the ecological field theory, which assumes that each tree reduces the availability of radiation, water, and nutrients by means of its roots, crown, and stem. Each tree generates a field of resource use that can be described mathematically by a distribution function for that specific tree. The fields are then added and projected to the ground where they form a map of resource-use fields.

4.2.2.4 Overlapping Zones of Influence

Many studies of competition assume that each tree is surrounded by a competition-influence zone. The assumption is that the area over which a tree competes for the resources of the site can be represented, in horizontal projection, by a circle. The radius of the circle is a function of tree size. The amount of competition to which a given tree is exposed is sometimes assumed to be a function of the extent to which its influence zone overlaps the zones of neighboring trees. Fig. 4.19 shows a hypothetical influence zone for each of the six trees and the shaded areas of overlap for the reference tree *i* and three close competitors.

Different functions of tree size for calculating an influence-zone radius and different procedures for weighting the overlap areas with the relative sizes of the competitors have led to a proliferation of indices. Gerrard (1969) defined the prototype as follows:

$$GI_i = \left(\frac{1}{\pi \cdot r_i^2}\right) \cdot \sum_{j=1}^{m} \mathrm{AO}_{ij} \qquad (4.38)$$

where AOij is the area of overlap between the reference tree i and competitor tree j. A reasonable approach for calculating the diameter of the influence zone is to use the potential crown width of a free-growing tree.

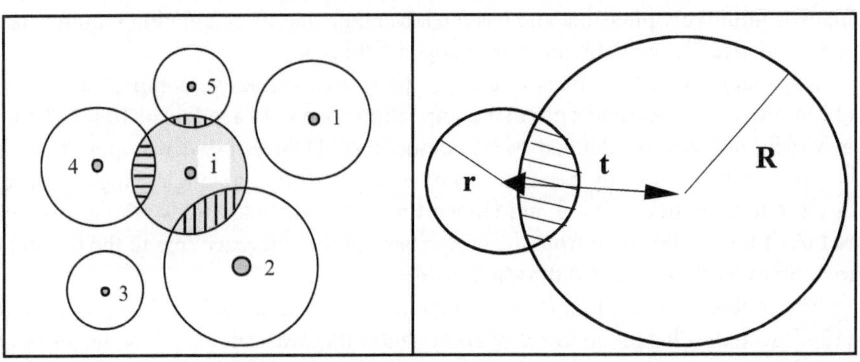

Fig. 4.19 Left: Reference tree i and five potential competitors with the influence zones of three of those; shading indicates the area of overlap with competitors 2, 4, and 5. Right: Variables required for calculating the area of overlapping zones of influence

4.2.2.5 The KKL Index

An important step in developing indices focused on the combination of light and crowding (Canham et al. 2004). Coates et al. (2009) applied a rigorous analysis to test these effects for different conifer species in British Columbia. Major outcomes of their study were that tree sizes and species react in different ways to competition. Whereas competition for nutrients affected trees of all sizes, competition for light mainly affected smaller trees in *Nothofagus* stands in New Zealand. Coates et al. (2009) provided further evidence that competitive regimes can change during the lifetime of a tree.

The KKL index (Pretzsch 1992a, b, 2009) requires information about tree heights and crown dimensions. The search for potential competitors is based on a cone with an angle of 60° that is inserted with its tip at 60% of the tree height. The crown tips of all neighboring trees that are presumed to intrude inside that cone are considered to be competitors. The angle from the tree tip to the virtual cone mantle of those neighbors is calculated. This is weighted by the cross-sectional crown area of the competitor in relation to the subject tree and then used as a measure of competition.

4.2.2.6 Distance-Weighted Size Ratio

Another group of competition indices are known as distance-weighted size ratios. These indices are based on the hypothesis that the competition effect exerted by a neighboring tree increases with an increasing size of the competitor and a decreasing distance between the competitor and reference tree. The index proposed by Hegyi (1974) can be used in situations where tree heights and crown dimensions are not known:

$$HG_i = \left(\frac{1}{D_{ij}}\right) \cdot \sum_{j=1}^{m} \frac{dbh_j}{dbh_i} \tag{4.39}$$

where *HGi* is the *Hegyi index* for reference tree *i*; *m* is the number of competitors; DBH_j and DBH_i are the DBHs of the competitor and the reference tree, respectively (cm); and D_{ij} is the distance between reference tree *i* and competitor *j*. The Hegyi index is intuitively plausible and has the advantage that its calculation requires data which are usually available in stem-mapped field plots.

Obviously, not all the trees in a large forest area are active competitors. Hegyi (1974) therefore specified a circular competition zone with a radius of 10 ft (3.05 m) in which all trees are defined to be competitors. Daniels (1976) proposed that a competitor should be any tree counted by a factor-10 ft^2/acre angle gauge centered at the reference tree. Biging and Dobbertin (1992) included trees within a search radius of which the angle from the crown base of the reference tree to the competitor's tip exceeds a specified threshold value.

The problem of defining *active* competitors was addressed by Lee and Gadow (1997) who developed the *Iterative Hegyi Index* that was calculated by means of an iterative procedure for utilizing sample plot data in a *Pinus densiflora* forest in Korea. Initially, all the trees within a specified competition zone were considered potential competitors of a given reference tree. The competition zone for a particular reference tree was a circle with radius (CZR):

$$\mathrm{CZR} = k \cdot \sqrt{\frac{10000}{N}} \tag{4.40}$$

with $k = a$ constant defining the radius of the competition zone and N = number of trees per hectare. The value that was selected was based on simulations to identify the maximum correlation between tree growth and competition. Initially, all the neighbors within the competition zone were defined to be active competitors. In an iterative search, some of those potential competitors were identified as "passive" if they were located "behind" another tree within a competition-elimination sector defined by a competition-elimination angle.[5] This simple principle is explained in Fig. 4.20 where the competition-elimination sector is indicated by the dark shading.

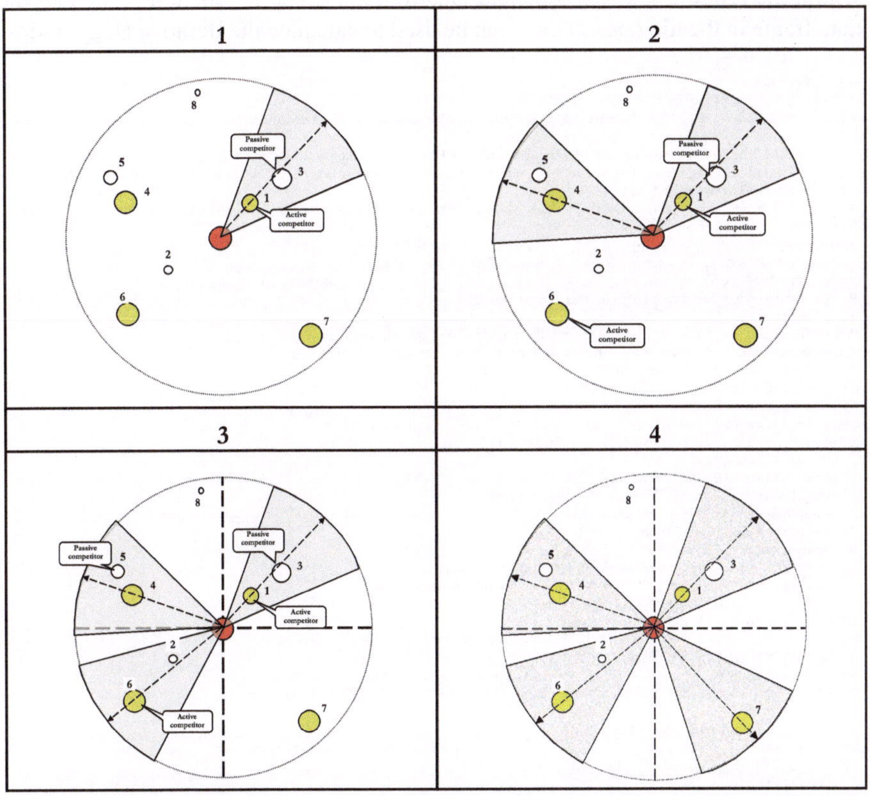

Fig. 4.20 Example with four iterations to identify active competitors (tree nos 1, 4, 6, and 7) in this hypothetical example. Tree nos 3 and 5 are passive competitors; the competition effect of tree no 2, a very small tree, is considered negligible

[5] Lee and Gadow (1997) used a competition-elimination angle of 30° and $k = 3$ in a study involving the diameter growth of *Pinus densiflora*.

Figure 4.20 shows a hypothetical example involving four iterations to identify the active competitors. The nearest neighbor of the reference tree is tree number 1, an active competitor. After the selection of tree number 1, an elimination sector is marked by using an eliminating angle from the reference tree to competitor number 1. Number 3 is located "behind" number 1 within the elimination sector; therefore, number 3 is a passive competitor. The next nearest neighbor is tree number 2, but that tree is very small and its competition effect is considered negligible. The next nearest neighbor is tree number 4, a new active competitor. As number 5 is behind number 4 within the competition sector designated for number 4, number 5 is also a passive competitor. Tree numbers 6 and 7 are also selected as active competitors by continuing with the same approach

The effectiveness of the Iterative Hegyi Index is affected by the parameter *k*. CZR is equal to the average distance between trees if $k = 1$, so k should be greater than 1. The optimum value, found by simulation, was identified to be somewhere within the range $2 < k < 4$ (Lee and Gadow 1997). The R function *it.heg* and the data.frame in the file *Lensahn.csv* can be used to calculate the Iterative Hegyi Index:

```
# Function it.heg
# These four references may be cited when using the function:
# Lee, W.K. and Gadow, K. v. 1997: Iterative Bestimmung der Konkurrenzbäume in Pinus densiflora Beständen. Allgemeine Forst und Jagdzeitung 168 (3/4): 41-44.
# Gadow, K. v. 2013: R-Function it.heg for calculating the Iterative Hegyi Index. Unpublised R-code.
# Lee, W.-K, Gadow, K. v., Chung, D-J. & Lee, J-L., 2004: Individual tree dbh-growth in Pinus densiflora and Quercus variabilis mixed stands in central Korea. Ecological modelling. 176: 187-200.
# Seifert, T., S. Seifert, A. Seydack, G. Durrheim and K. v. Gadow, 2014: Competition Effects in an Afrotemperate Forest. Forest Ecosystems 1:13.
rm(list = ls())                                             # clear the console
dat<-read.csv("/Users/kg/Desktop/Lensahn.csv", header=TRUE)        # import .csv
dat$id= (dat$d_04- dat$d_99)/5                                          # annual growth rate
# 2. Function to calculate iterative Hegyi (changed asin to atan!!)
it.heg<- function (x0,y0,d0, cz,CEA) {
# dat is the original dataframe; x0,y0,d0 are the coordinates and dbh of the reference tree
# cz is the dataframe with details of the competitors; CEA is the competition elimination angle
distlist=sort(sqrt((cz$x-x0)^2+( cz$y-y0)^2))
distlist=distlist[2:length(distlist)]          # first is zero (ref.tree)
# determine active competitors
z=0; for (i in 1:(length(distlist)-1))
  {object1=cz[ sqrt((x0-cz$x)^2+(y0-cz$y)^2)==distlist[i],]
  for (j in (i+1):length(distlist))
  {object2=cz[sqrt((x0-cz$x)^2+(y0-cz$y)^2)==distlist[j],]
  a=sqrt((object1$x-object2$x)^2+(object1$y-object2$y)^2)
  b=sqrt((x0-object1$x)^2+(y0-object1$y)^2)
  c=sqrt((x0-object2$x)^2+(y0-object2$y)^2)
   acosarg=(b^2+c^2-a^2)/(2*b*c)
   acosarg=min(1,acosarg) # there maybe problems through rounding
   acosarg = max(-1,acosarg)          # the argument must be between -1 and 1
  A=acos(acosarg); B=atan(object2$d_99/(100*c))
if ((abs(A)+abs(B))<((CEA/2)/180)*pi) z=c(z,j)}}
HgCI=0                                                # calculate the index for all active competitors
if (length(z)>1) {z=z[-1]; distlist=distlist[-z]}
for (i in 1:length(distlist))                         # Calculate the iterative HgCI
 {object=cz[ sqrt((x0-cz$x)^2+( y0-cz$y)^2)==distlist[i],]; HgCI=HgCI+(object$d_99/d0)/distlist[i]}
 return(HgCI) }
# 3. defining the Competition Zone Radius (CZR)
k<-2; plotarea<-0.6                                      # the area of plot is 0.6ha in Lensahn
CZR<-sqrt(10000/(length(dat$d_99)/plotarea))*k  # 8.36 m
# 4. defining the reference trees
marg<-CZR; specie="Bu"
y1<-max(dat$y)-marg; y2<-min(dat$y)+marg;
x1<-max(dat$x)-marg; x2<-min(dat$x)+marg
reftree<-subset(dat, dat$sp_code==specie&dat$x>x2&dat$x<x1&dat$y>y2&dat$y<y1)

# 5. calling the function
hg<-rep(NA)                                              # set up an empty array
for (i in 1:nrow(reftree))
 {cz=unique(dat[sqrt((dat$x-reftree$x[i])^2+( dat$y- reftree$y[i])^2)<=CZR,])
 hg[i]<-it.heg(reftree$x[i], reftree$y[i], reftree$d_99[i], cz, 30)}
# 6. plotting the results
par(mar=c(5,6,2,2) + 0.1)                                 # plot Margins!!!!!
plot(hg,reftree$id, xlim=c(0,18), ylim= c(0,1.7), pch=19,xlab="Iterative Hegyi",ylab="dbh growth (cm/year)", cex.lab=1.5,)
legend(10,1.4,paste("species=", specie),bty="n",cex=1.5)
```

4.2.2.7 Other Competition Indices

Numerous distance-dependent competition indices have been developed over the years, and the potential for developing new ones is boundless. To demonstrate the variety of possible approaches, four additional examples of modeling competition effects for a specific reference tree are shown in Fig. 4.21. Biging and Dobbertin (1992) estimated the sum of the crown volumes (calculated as the volume of a cone: $V = \pi \cdot Cr^2 \cdot h/3$) of competitors by using a competition-effect angle (between 50° and 60°) to define the parts of neighboring crowns that are included. Their index is calculated as the ratio (total inside crown volume of neighbors) to (crown volume of the reference tree).

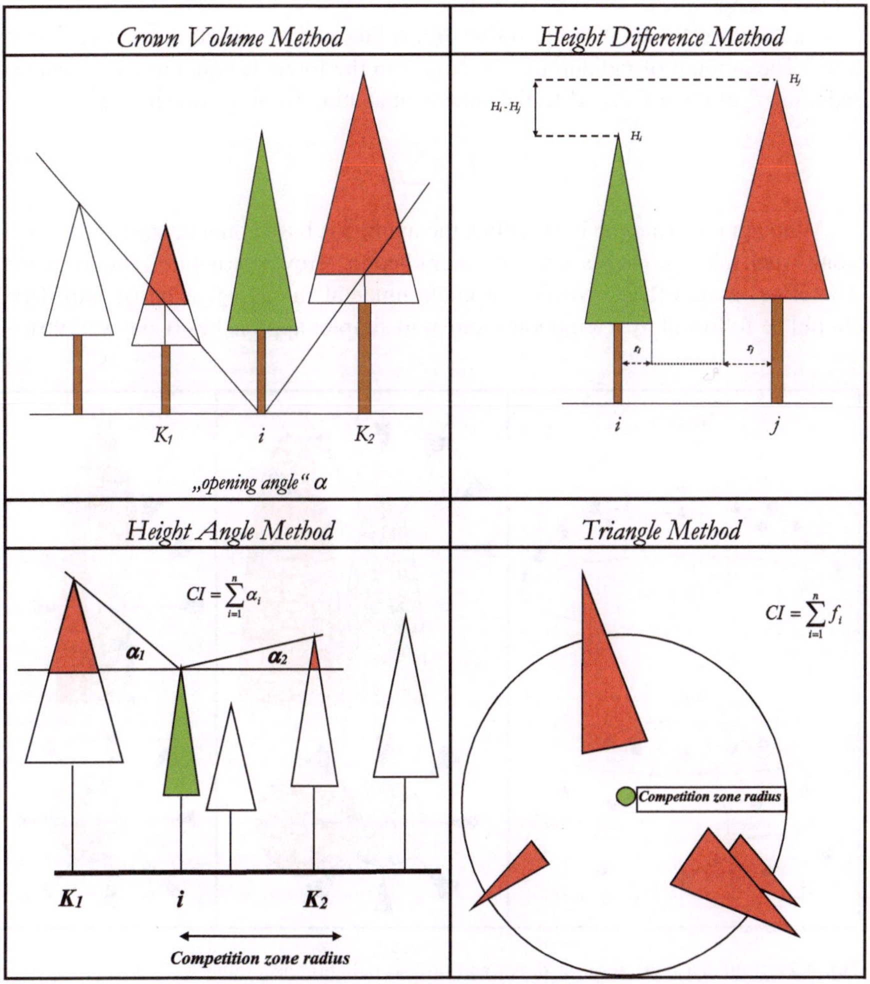

Fig. 4.21 Four methods for modeling competition effects for a specific reference tree: the Crown-Volume Method (Biging and Dobbertin 1992), the Height-Angle Method (Rautiainen 1999), the Height-Difference Method (Schütz 1989), and the Triangle Method (Ekö and Ågestam 1994)

Schütz (1989; see also Ung et al.; 1997) calculated the competition effect by means of tree height (h), crown radius (Cr), and distance between the reference tree and a neighbor. Competition effect is estimated by:

$$C_i = \sum_{j=1}^{m} \left(0.65 \cdot \frac{H_j - H_i}{Cr_i + Cr_j} + 0.5 \cdot \frac{D_{ij}}{Cr_i + Cr_j} \right) \tag{4.41}$$

A neighbor is considered to be a competitor if its distance to the reference tree is less than or equal to $0.5(Cr_i+Cr_j)+0.65(h_j–h_i)$. Competition models based on shading with a spatially explicit simulation of radiation have also been used to estimate competition effects. Courbaud (1995) and Biber (1996) estimated the amount of radiation reaching a particular point within a forest by discretizing direct and diffuse radiation (Fig. 4.22).

Competition for light is the result of shading and constriction by neighboring trees. The amount of radiation L_i for point i in the forest is equal to the amount of radiation l_{ij} of the discrete celestial sphere emanating from the direction j:

$$L_i = \sum_{j=1}^{n} \ell_{ij} \tag{4.42}$$

Although the principle is plausible, the method is based on numerous simplifications, such as continuous changes in radiation, crown structure, and foliation. Therefore, as is often advisable in environmental modeling, a thrust into detail should be followed by a temporary retreat to simpler approaches (Gadow 1996).

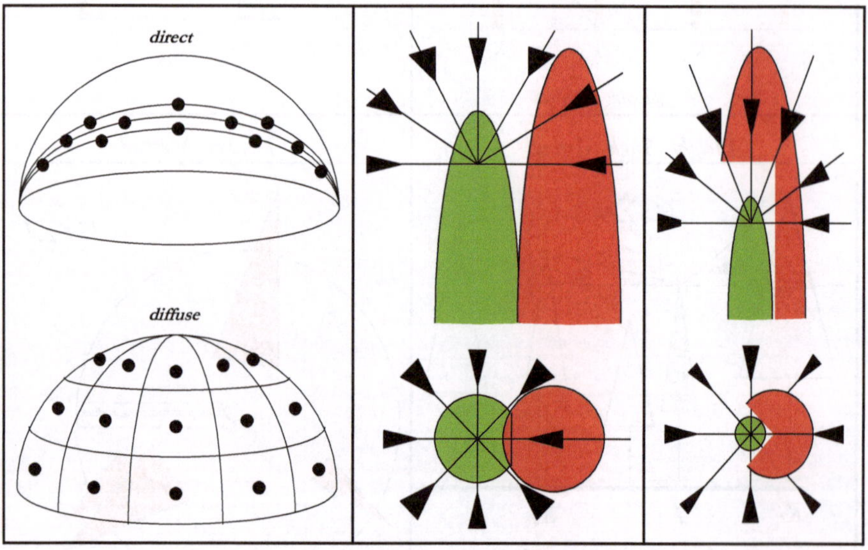

Fig. 4.22 Left: Discretization of direct and diffuse radiation according to Courbaud's method (1995). Right: Competition for light is the result of shading and constriction by neighboring trees

4.2.3 Examples of Individual Tree Growth Models

Numerous individual tree growth models have been developed, all of which are based on some measure of competition that affects potential tree growth. The following is an example of a model developed by Pukkala (1989) for predicting the basal area growth of individual *Pinus silvestris* trees in Finland:

$$\log(''g_i) = -3.093 - 0.1815 \cdot dbh_i + 2.052 \cdot \sqrt{dbh_i} + \frac{38.85}{(t_i + 10)}$$
$$-0.009561 \sum_{j=1}^{n} \frac{dbh_j}{D_{ij}} + 1.043 \frac{dbh_i}{dg} - 0.01173 \cdot BA \quad (4.43)$$

where Δg_i is the future 5-year basal area growth (cm^2) of a tree with breast diameter DBH$_i$ (cm) and breast height age t_i (years); DBH$_j$ (cm) is the diameter of a neighboring tree closer than 5 m; D_{ij}(m) is its distance to the reference tree i; and dg is the quadratic mean diameter (cm) of the trees within a radius of 5 m from tree i (including the reference tree i) and BA is the stand basal area (m^2/ha).

Noteworthy examples of individual tree growth models are the growth simulators *FOREST* and *MOSES,* which are based on similar assumptions and principles, and the three-dimensional model *SILVA*. In principle, the use of these models is not limited to a particular type of forest. They require empirical parameter estimates and (real or assumed) tree data. One of the possible applications is to provide some insight into growth processes that less detailed models cannot give.

FOREST, one of the first spatially explicit growth simulators, was developed by Ek and Monserud (1974), who used the so-called *potential-modifier* approach. The assumption is that maximum or potential growth derived from observations of solitary trees is reduced by competition. The effect of competition is modeled by using growth modifiers. One of the noteworthy contributions of the FOREST concept is the simultaneous use of modifiers that embody past and present competition:

$$\Delta W = \Delta W_{pot} \cdot \mathrm{CR}^b \cdot \mathrm{OVS} \quad (4.44)$$

with

ΔW = estimated change of size variable W
ΔW_{pot} = potential change of size variable W
CR = crown ratio (crown length/tree height)
b = species-specific parameter
OVS = overstocking multiplier

The model assumes that the current crown ratio is the result of *historical* stress or previous competition. A small crown ratio indicates high historical stress and less current performance. *Current* competition is described by the *overstocking multiplier* (OVS), which is a function of the area of overlapping influence zones:

$$\text{OVS} = \left(1 - e^{\frac{1}{\beta \cdot \text{CIA}}}\right)^{\alpha} \tag{4.45}$$

with

$$\text{CIA} = 1 + 0.1 \cdot \text{CI} \cdot \left(1 - \beta_1^{\,h+1}\right) \cdot \left(1 + \beta_2 \cdot \Delta CI\right)^{-\beta_3};$$

and CI = competition index according to Bella (1971):

$$\text{CI}_i = \sum_{j=1}^{n} \left(\frac{a_{ij}}{A_i} \cdot \frac{S_j}{S_i} \right);$$

a_{ij} = area of the influence-zone overlap between subject tree i and competitor j,
A_i = area of the influence zone of subject tree i,
S_j/S_i = size-weighting factor (S = tree height multiplied by the potential crown radius of a solitary tree),
ΔCI = change of CI caused by thinning,
H = tree height,
β, α, β_1, β_2, β_3 = species-specific parameters.

The equation for calculating CIA includes a term for predicting thinning effects. The conceptual design of FOREST is based on simple logic, which explains its intuitive appeal. Not surprisingly, a number of subsequent models are based on the same concept. One example is the growth simulator *PROGNAUS*, developed by Sterba (1990), for mixed forests of spruce (*Picea abies*) and Scots pine (*Pinus silvestris*) in Austria. Another growth simulator for individual trees that exhibits some resemblance to the FOREST structure is the SILVA model that utilizes three-dimensional spatial data for estimating tree growth in mixed spruce and beech forests in Germany (Pretzsch 1992a, b). Competition is measured by shading and constriction and by using a specific crown model developed by Pretzsch (1992a, b: 112 *et seq.*) for beech (*Fagus sylvatica*), spruce (*Picea abies*), and fir (*Abies alba*).

Individual tree models provide high-resolution estimates, but predictions on stand level are often problematic. Stand level models provide higher accuracy, but inadequate resolution. The *telescope principle* (Burkhart 1987) is based on the premise that all available information should be used, subject to an important condition: predictions on stand and tree levels must be compatible; therefore, the estimates at different levels of resolution must be exactly equal.

4.3 Selecting Trees for Harvest

Selecting trees for removal is one of the most important tasks in the small-scale *forest gardening* systems practiced in central Europe. This task is increasingly delegated to a harvesting machine operator, on the assumption that the resulting modifications of the stand structure are similar or even identical to those done by a professional forester. As a result, the question arises: Are there differences in the "marking behavior" of different foresters or between a forester and a harvesting machine operator who are given the same set of silvicultural objectives? If there are differences, it would not be sufficient to model average thinning effects. One would need to model the individual tree *marking behavior* of each forester or machine operator who is engaged in selecting trees for removal.

4.3.1 The Four-Forester Problem

The first problem that needs to be addressed is the method of measuring differences in tree-selection patterns. As an example, consider an exercise involving four teams of experienced foresters who were asked to mark a forest stand comprising 300 trees, principally *Abies alba* and *Fagus sylvatica*. Each team was given the same set of objectives outlining the silvicultural principles and specific aims of the thinning. They then proceeded to mark the stand by indicating the suggested removals on a numbered tree list. The results of the trial thinnings are available as a list of 300 trees, together with the number of foresters who marked each of the trees, that is, the number of marks, out of 4, assigned to each tree. The results are summarized in Table 4.8.

Thus, for example, 191 of the 300 trees were not marked by any of the four foresters and only one was marked by all four foresters. The problem of finding a way to quantitatively assess the degree of agreement or disagreement between the foresters was addressed by Zucchini and Gadow (1995). As the responses of the individual foresters are not available, it is not possible to make comparisons between the marking behavior of individual foresters. For example, it is not possible to determine whether differences are mostly attributable to one of the four foresters. Only indices of overall agreement can be used.

In general, an index is a measure designed to facilitate comparison between a given situation and one or more standard situations that provide interpretable reference points. For example, the correlation coefficient which measures the strength of the linear relation between two variables has three reference points, namely 1, 0, and −1, and they indicate, respectively, a perfect linear relation with positive slope; no linear relation; and a perfect linear relation with negative slope.

For the purpose of interpretation, the most important consideration in the construction of an index of agreement has to do with the choice of reference points, that is, how the index is standardized. It is possible to regard the frequencies in Table 4.7 from very different points of view. In order to reflect a particular point of view when an index is used, it is necessary to standardize the latter accordingly. It would seem

Table 4.8 Frequency distribution of the number of marks on 300 trees

Number of marks	0	1	2	3	4
Frequency	191	70	30	5	1

Table 4.9 All the possible values of absolute differences, given four foresters

x	$n–x$	$\|2x - n\|$
0	4	4
1	3	2
2	2	0
3	1	2
4	0	4

that the general type of index that is employed does not materially affect the assessment of overall agreement, but that the way in which the index is standardized does affect the assessment.

Suppose there are K trees and that each of n foresters either marks or does not mark each tree. Let x denote the number of foresters who mark a particular tree, and $(n–x)$ those who do not mark it. The absolute difference between x and $n–x$ provides a measure of agreement for a given tree. For $n = 4$ foresters, the possible values of the absolute difference $|x - (n - x)| = |2x - n|$ are listed in Table 4.9.

Suppose that the trees are numbered $k = 1, 2, \ldots, K$ and that xk denotes the number of foresters who marked tree k. Then

$$C = \sum_{k=1}^{K} |2x_k - n|$$

can be used to construct an index of agreement among the foresters for the stand of trees under consideration. The minimum and maximum possible values of C are, respectively, 0 and $C_{max} = nK$. A possible index is then:

$$I_1 = \frac{C}{C_{max}}$$

There is no difficulty in interpreting the upper bound of this index; if $I_1 = 1$, there is complete agreement. However, the interpretation of the lower bound (0) is less straightforward. Note that in the case of $n = 2$ foresters, the lower bound is achieved if there is *perfect disagreement*, that is, if the two foresters assess every single tree in the opposite way. In other words, $I_1 = 0$ implies a greater degree of disagreement than would occur if, for example, the foresters were to select the trees to be marked *at random*. We will, therefore, make a distinction between the terms "*disagreement*" and "*lack of agreement*." The former will be used for a situation in which the foresters are reaching *contrary conclusions* and the latter for a situation in which their *conclusions are unrelated.*

In view of this interpretation, it may be preferable to standardize the index in such a way that a situation in which the selection is made independently and at random receives the value 0. One such index is:

$$I_1(p) = \frac{C - C_{ind}(p)}{C_{max} - C_{ind}(p)} \tag{4.46}$$

where $C_{\text{ind}}(p)$ is the expected value of C with the condition that the foresters select the trees to be marked independently and with probability p. The expectation C_{ind} is computed by means of the binomial probability function:

$$C_{\text{ind}}(p) = K\sum_{x=0}^{n} |2x - n| \binom{n}{x} p^x (1-p)^{n-x} \tag{4.47}$$

No *assumptions* are made about the behavior of the foresters; but a different baseline is set for the index. If each forester *were* to mark trees at random with probability p, then the expected value of $I_1(p)$ would be zero. Values of $I_1(p)$ that are less than zero reflect *disagreement rather than lack of agreement.*

The value of p has a substantial influence on the index, because it determines its zero level and hence its interpretation. Ideally, p should be set equal to the proportion of the trees in the stand that the foresters were supposed to mark, if the proportion was (either explicitly or implicitly) preassigned. For example, the foresters may have understood that approximately 15% of the trees should be marked, in which case the appropriate index would have been $I_1(0.15)$.

Table 4.10 gives the observed frequencies with the corresponding expected frequencies obtained in the scenario where each forester uses a marking strategy, and where each tree is marked independently and at random with probability $p = 0.15$. (expected frequency $\dot{=}$ 300 · [(4 ! /(n° marks ! · (4 − n° marks)!) · $0.15^{n^{\circ}\text{ marks}}$ · $(1 - 0.15)^{4 - n^{\circ}\text{ marks}}$]

Note that 191 out of 300 trees were marked by none of the foresters. This may suggest a high degree of agreement. However, the corresponding expected frequency in the hypothetical scenario where the foresters mark about 15% of the trees at random is 156.6, indicating that the degree of agreement is not as high as it might appear to be at first glance. When a decision is made on how to assess the degree of agreement, it is therefore important to specify the reference point to be used. For the frequencies in Table 4.9, one obtains:

$$I_1 = 0.77$$
$$I_1(0.15) = 0.18$$

The gross discrepancy between the values of the two indices reflects the very different assumptions on which they are based. The second index is appropriate in situations where the proportion of trees to be marked is (perhaps implicitly) prescribed *a priori* to be approximately 0.15. From this perspective, there is almost a lack of agreement, but since $I_1(0.15)$ is greater than zero there is no indication of disagreement.

Table 4.10 Observed and expected frequencies for the number of marks ($p = 0.15$)

No of marks	0	1	2	3	4
Observed freq.	191	70	30	5	1
Expected freq.	156.6	110.5	29.3	3.4	0.2

The first index I_1 would be appropriate if each forester felt entirely free to mark as many or as few trees as they wished; that is, with absolutely no prior notion of the percentage to be marked. From that perspective, as indicated by the high value (0.77), the degree of agreement is high. In cases where p is not known, but in which it is known that such a value was either actually preassigned or implied, it would be reasonable to use the index $I_1(\hat{p})$ where $\hat{p} = (x_1 + x_1 + \cdots x_k) / nK$, which is the average proportion of trees marked by the n foresters. Here $\hat{p} = (191{\cdot}0 + 70{\cdot}1 + 30{\cdot}2 + 5{\cdot}3 + 1{\cdot}4)/(4{\cdot}300) = 0.124$. The associated index value is $I_1(0.124) = 0.02$ which, again, indicates that there is a lack of agreement, but no disagreement.

It is necessary to keep in mind that the assessment refers to the performance of the foresters regarding a particular stand. In some stands, the trees are easier to classify and therefore the above-mentioned four foresters can easily be in complete agreement. Conversely, if the trees are difficult to classify, there will probably be fewer agreements. Therefore, the variation of tree attributes, that is, the degree of homogeneity in the stands, needs to be considered.

Now, consider a similar exercise involving five foresters. In this instance, the attributes of the 300 numbered trees are available. It is therefore possible to use a generalized linear model for binary variables to evaluate the results (Füldner et al. 1996). Two options, *remove* or *remain*, are available for each tree. The decision is influenced by a number of factors (also known as *covariates*) such as the intrinsic attributes of the tree itself and the neighborhood constellation, including the attributes of the competitors in the immediate vicinity of the tree. This influence must be quantified. First, we define $Z_1, Z_2, \ldots Z_n$ to be the destinies of the trees in the stand with:

$$Z_i = \begin{cases} 1, & \text{if tree } i \text{ is removed} \\ 0, & \text{if tree } i \text{ remains, with } i = 1,2,\ldots,n. \end{cases}$$

In addition, the variable $\pi_i = P\{Z_i = 1\}$ is introduced, which represents the probability that tree i is removed. With two decision alternatives, *remove* or *remain,* the outcome is comparable to throwing a coin. Heads means remove, tails remain. However, the coins are not identical. A different coin is thrown for each tree, because the probability that *heads* is thrown is a particular one for the *i*th tree, namely π_i. If, on account of its attributes, it is undesirable that tree i remains, the value of π_i will be closer to 1 with a higher probability that the coin shows *heads*. However, if the tree should remain, on account of its attributes, the coin will be biased toward *tails.*

The random variables (Z_i) can only assume one of the two values (0 or 1). The commonly used linear models, such as regression analysis, are therefore not applicable for specifying a relation between i and a tree attribute. A suitable approach in this case is a so-called *generalized linear model* (McCullagh and Nelder 1989). First, a *LINK*-function is defined. The *logit link* function, for example, can be written as follows:

$$\pi_i = \frac{e^{\lambda_i}}{1 + e^{\lambda_i}}, i = 1,2,\cdots,n \tag{4.48}$$

with $\lambda_i = \beta_0 + \beta_1 \cdot dbh_i$, defining a linear relation between the tree diameter DBH_i and the transformed values of π_i. The parameters β_0 and β_1 can be estimated by using available statistical software such as *S-Plus, GLIM, SAS,* or *R.* See McCullagh and Nelder (1989) and Dobson (1990) for details about theory and algorithms.

Assignment

We assume the following probabilities of a tree to be removed and tree diameters

removed<-c(1,0,0,0,0,0,0,0,0,1,0,0,0,0,1,0,0,1,0,0)
dbh<-c(23.1,16.5,18.9,12.1,18.4,23.6,17.2,19.1,18.7,32.5,22.1,18.4,
19.5,22.2,29.4,20.1,22.5,24.1,19.9,17.5)

Fit the logit function to these observations and estimate, for each tree, the probability that it would be removed.

Solution

```
prob_removed<-glm(removed~dbh, family="binomial")
summary(prob_removed)
prob<- exp(-56.235+2.404*dbh)/(1+exp(-56.235+2.404*dbh))
```

The parameter values for the logit link function for the five-forester example are listed in Table 4.11 and the estimated probabilities are represented in Fig. 4.23.

The probability that a tree will be removed, depending on the dbh and associated with each of the five foresters (A … E), based on the parameters for the logit link function in Table 4.10, is shown in Fig. 4.23.

There are obvious differences among the foresters relating to their selection probabilities. The foresters D and B do not show any diameter preference. Their selection probability is indifferent to the tree diameter, amounting to about 10% in any tree size. The selection probability for foresters A, C, and E increases with increasing tree diameter. Forester A has a very strong preference for the big trees. Furthermore, it is noticeable that the foresters apply different thinning weights. C takes out considerably more trees than B, for example

Table 4.11 Parameter values for the logit link function

Forester	β_0	β_1
A	−3.213	0.03543
B	−2.178	0.00312
C	−2.023	0.02330
D	−2.313	0.00739
E	−2.013	0.01639

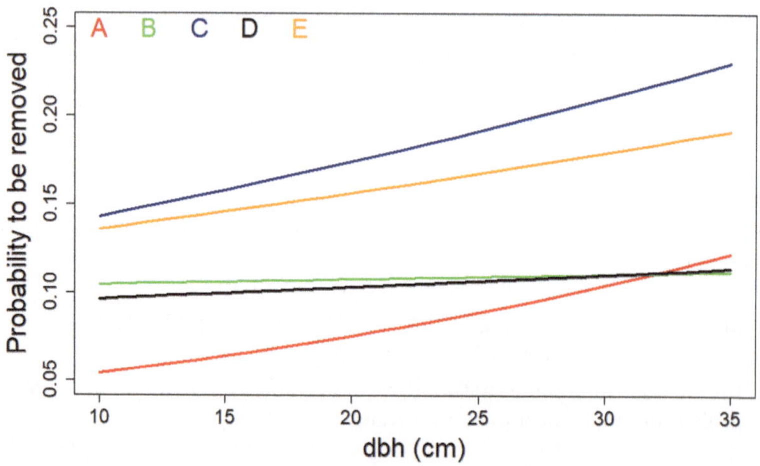

Fig. 4.23 Probability that a tree would be removed associated with each one of the five foresters according to the parameters for the logit link function as presented in Table 4.10

The example shows that, given the same silvicultural objectives, considerable person-specific variability in thinning procedures can occur.[6] This observation leads to several conclusions and considerations. Firstly, the commonly used silvicultural specifications for describing the objectives of a thinning are obviously not operationally viable (otherwise they would all have had the same selection probabilities). Secondly, it may be desirable from an ecological perspective to allow such variations in tree-selection probabilities, as they reflect the differences in the individual experiences of the foresters and lead to a greater diversity of forest structures. Thirdly, if such variability is permitted or even encouraged, it should be made transparent; otherwise, it will be impossible to forecast forest development, which is not only influenced by natural growth, but also by thinnings. Models describing individual forester-specific modifications of forest structures are therefore an essential part of any growth modeling effort.

4.3.2 Spatial Thinning Models

Forest development is the result of natural growth as well as modifications caused by silvicultural operations and other disturbances. In a managed forest, thinning operations, which are often carried out at regular intervals, are very important. A model of forest development would therefore be incomplete without a thinning component. Thinnings can be modeled either by imitating the way in which a forester normally selects trees for removal (*imitation*) or by prescribing thinning rules

[6] However, Kahle et al. (1996) found a great degree of consistency displayed by the same forester who was asked to simulate a thinning in the same stand several times at monthly intervals.

which must be obeyed, thereby defining the specific manner of selection (*prescription*).

Thinning decisions are normative, intrinsically fuzzy and influenced by a variety of criteria (Kahn 1994). Hence, the ability to mimic the decisions that led to the removal of individual trees will provide a better understanding of a thinning process (Daume 1995). The fuzzy character of a thinning is exemplified by vague descriptions (*high* thinning, *low* thinning), the personalized nature of the decisions, and the fact that exact measurements of tree attributes are usually not available. Whereas a single attribute, such as the species, may be sufficient reason for *not* removing a tree, multiple attributes are required for explaining a positive selection for removal. Various combinations of *thinning-relevant* attributes can be used to quantify removal probabilities, depending on the kind of forest and the specific objectives of the thinning operation. After these quantities have been assigned, trees can be sorted and those with the highest probabilities removed.

Typical thinning-relevant attributes in a nonspatial situation are tree species and variables defining the relative size or the economic value of a tree. A spatial model permits a much greater choice of criteria, such as the structural attributes of specific cells within a forest or the point density in the immediate vicinity of a given tree (Table 4.12).

Table 4.12 shows the relative frequencies of the common distribution of the structural variables *mingling* (*M*) and *diameter differentiation* (*T*), both based on the three nearest neighbors for a mixed stand of beech and spruce in southern Germany. The stand was thinned in 1987 and the before-thinning (*total*) and removed data are shown separately for beech and spruce. The shading in the cells increases with increasing relative frequency. Removal preferences can be calculated by means of the data in Table 4.12. Before the thinning, only 13% of the spruce trees were part

Table 4.12 Relative frequencies of the distribution of species mingling and DBH differentiation for the total population before thinning and for the removed trees

Spruce						**Beech**					
Total		*M*				Total		*M*			
		0	**0.33**	**0.67**	**1**			**0**	**0.33**	**0.67**	**1**
T	**0 - 0.3**	0.00	0.32	0.15	0.00	*T*	**0 - 0.3**	0.19	0.08	0.04	0.00
	0.3-0.5	0.04	0.13	0.00	0.00		**0.3-0.5**	0.07	0.07	0.04	0.01
	0.5-0.7	0.00	0.02	0.09	0.02		**0.5-0.7**	0.04	0.06	0.07	0.01
	0.7-1.0	0.00	0.06	0.11	0.06		**0.7-1.0**	0.13	0.07	0.08	0.03
Removed		*M*				Removed		*M*			
		0	**0.33**	**0.67**	**1**			**0**	**0.33**	**0.67**	**1**
T	**0 - 0.3**	0.00	0.25	0.00	0.00	*T*	**0 - 0.3**	0.20	0.20	0.00	0.00
	0.3-0.5	0.00	0.75	0.00	0.00		**0.3-0.5**	0.20	0.20	0.20	0.00
	0.5-0.7	0.00	0.00	0.00	0.00		**0.5-0.7**	0.00	0.00	0.00	0.00
	0.7-1.0	0.00	0.00	0.00	0.00		**0.7-1.0**	0.00	0.00	0.00	0.00

of the mingling class [0.33][7] and the differentiation class [0.3–0.5]. However, 75% of all the trees in this class that had one beech tree as one of its three nearest neighbors were removed during the thinning. The selection preference in this class, calculated as the ratio of removed to total cell values, is very high.

Similar interpretations can be developed for the beech trees. For example, 40% of all the removed beech trees were found in pure clusters containing only beech trees. In another 40%, two beech trees and one spruce tree were among the three nearest neighbors. The corresponding *total* cell values were not particularly high. Again, this points to a high selection preference in the two mingling classes. The preference is also abnormally high in the lower two differentiation classes, indicating preferred removal of trees surrounded by neighbors of equal or approximately equal size. For the purpose of *imitating* a particular thinning operation, we can therefore define a preference index as follows (see Daume 1995):

$$\mathrm{PR}_{ij} = \frac{\text{removed}\, p_{ij}}{\text{before}\, p_{ij}} \tag{4.49}$$

with

PR_{ij} = selection preference in structure class ij ($i = 1\ldots m, j = 1\ldots n$);
removed p_{ij} = proportion of trees removed in structure class ij;
before p_{ij} = proportion of trees in structure class ij, before the thinning.

The PR_{ij} values are obtained by dividing the cell values in the *removed* class by the corresponding cell values in the *total* class. For example, the high value of 5.77 in the *M* class 0.3 (precisely 0.33) and the *T* class 0.3 (actually 0.3–0.5) was obtained by calculating the ratio 0.75/0.13. In this particular forest, the structure variables *mingling* and *differentiation* obviously provide an excellent basis for modeling the selection of trees for removal. The suitability of these variables is explained by the fact that they implicitly describe the immediate neighborhood constellation of each tree in a manner which must be highly relevant to the selection process.

The individual decisions involving the selection of trees for removal are taken in *thinning cells*, that is, areas which are sufficiently small for judging the effect of a particular removal on the remaining survivors, and large enough so that at least one tree is a potential candidate for removal. In practice, the forester who selects trees for removal moves from cell to cell and makes individual and independent decisions in each one. Consequently, it is possible to use structure variables for calculating selection preferences. However, the variables should involve not only one, but several neighbors. The number of neighbors should be appropriate for the desired cell size, and the size of the cell should resemble the size that is used by the forester in the field.

Daume (1995) conducted an analysis of a thinning operation in a section of a 60-year-old mixed beech forest and found that knowledge of the structural attributes

[7] Note that with three neighbors, the variable *M* can assume four possible values; refer to Chapter 2 for details.

is not sufficient for modeling a thinning and that additional spatial information was required. The objective was to predict removal decisions for specific trees within the context of a *thinning cell*. This approach imitates the way in which a forester marks trees for removal, namely moving from cell to cell and making independent decisions that are influenced only by the point density and structure in each particular cell. In practice, it is impossible for a forester to base his removal decisions on a comparison of all the trees with similar attributes in the forest. The tree-selection decision only applies to a small assemblage of trees within a specific local cell. The trees within each cell carry cell-specific selection preferences, and a selective thinning represents the sum of individual decisions that can be spatially defined.

The simplest approach would involve a subdivision of the total forest area into a number of square cells containing a group of trees for which a collective decision can be made. This approach was described in detail by Daume et al. (1997). His example shows that the use of preferences based on two-structure variables can be a very good basis for modeling the thinning process, provided that at least three neighbors are used. The identification of candidates suitable for removal can be further improved by using additional spatial information, such as preference values within thinning cells. The removed trees are those which carry the highest structural preference within their particular cells. This approach appears to produce good results when it is based on the assumption that thinning cells are square, but further research, using data from a variety of forest types, is needed to identify optimum cell shapes and sizes.

Literature

Affleck DLR, Dieguez-Aranda U (2016) Additive nonlinear biomass equations: a likelihood-based approach. Forest Science 62(2):129–140

Akça A, Gadow Kv, Mench A, Mann P, Pahl A Setje-Eilers U (1994) Überprüfung des Formquotienten q7 (d7/BHD) in der Bundeswaldinventur für die Hauptbaumarten Buche, Eiche, Fichte und Kiefer im landeseigenen Wald Nordrhein-Westfalens. Project report, 23 p

Albert M (1994) Erhebung von Strukturinformationen in einem Buchen-Edellaubholz Mischbestand. Diplomarbeit am Institut für Forsteinrichtung und Ertragskunde, Universität Göttingen. 109 p

Albert M (2004) Modifikationen des BAL-Konkurrenzindex zur Prognose des Durchmesserzuwachses in Mischbeständen. Jahrestagung der Sektion Ertragskunde des DVFFA in Schlägl vom 24–26 Mai 2004

Arias-Rodil M, Diéguez-Aranda U, Rodríguez Puerta F, López-Sánchez CA, Canga Líbano EC, Cámara Obregón AC, Castedo-Dorado F (2015) Modelling and localizing a stem taper function for *Pinus radiata* in Spain. Can J For Res 45(6):647–658

Arnswaldt HJ v (1950) Wertkontrolle in Laubholzrevieren. Forstarchiv:130–135

Assmann E (1953) Zur Bonitierung süddeutscher Fichtenbestände. AFZ 10:61–64

Atkinson D (2000) Root characteristics: why and what to measure. In: Smit AL, Bengough AG, Engels C, Van Noordwijk M, Pellerin S, Van de Geijn SC (eds) Root methods: a hand book. Springer, Berlin/Heidelberg, pp 2–32

Bachmann P (1990) Produktionssteigerung im Wald durch vermehrte Berücksichtigung des Wertzuwachses. Eidgenössische Forschungsanstalt für Wald, Schnee und Landschaft, Birmensdorf

Bachmann M (1998) Indizes zur Erfassung der Konkurrenz von Einzelbäumen – Methodische Untersuchung in Bergmischwäldern. LMU/LWF, München
Bauhus J, Khanna PK, Menden N (2000) Aboveground and belowground interactions in mixed plantations of Eucalyptus globulus and Acacia mearnsii. Can J For Res 30(12):1886–1894
Behre CE (1923) Preliminary notes on studies of tree form. J For 21:507–511
Bella IE (1971) A new competition model for individual trees. For Sci 17:364–372
Bernstorff A v, Kurth A, Grilc J, Günter R (1988) Betriebswerk Gartow. Unpublished management plan: 163p
Biber P (1996) Konstruktion eines einzelbaumorientierten Wachstumssimulators für Fichten-Buchen-Mischbestände im Solling. Diss. Forstw. Fak. d. Ludwig-Maximilians-Univ. München: 239 S
Biging GS, Dobbertin M (1992) A comparison of distance-dependent competition measures for height and basal area growth of individual conifer trees. For Sci 38(3):695–720
Biging GS, Dobbertin M (1995) Evaluation of competition indices in individual tree growth models. For Sci 41(2):360–337
Biging GS, Gill SJ (1997) Stochastic models for conifer tree crown profiles. For Sci 43(1):25–33
Biging GS, Wensel LC (1990) Estimation of crown form for six conifer species of northern California. Can J For Res 20(8):1137–1142
Bolte A, Rahmann T, Kuhr M, Pogoda P, Murach D, Gadow K (2004) Relationships between tree dimensions and coarse root biomass in mixed stands of European beech (Fagus sylvatica L.) and Norway Spruce (Picea abies [L.] Karst.). Plant and Soil 0:1–11
Brink C, Gadow Kv (1986) On the use of growth and decay functions for modelling stem profiles. EDV in Medizin und Biologie 17(1/2):20–27
Brown GS (1965) Point density in stems per acre. Forestry Research Notes No. 38, New Zealand, p 11
Burger H (1939) Kronenaufbau gleichaltriger Nadelholzbestände. Mitt Schweiz Anst f d forstl Vers Wesen 21:5–58
Burkhart HE (1987) Data collection and modelling approaches for forest growth and yield prediction. In: Predicting forest growth and yield – current issues, future prospects. Inst. of Forest Resources. Univ. of Washington. Contribution Nr. 58: 3-16
Burkhart HE, Tomé M (2012) Modeling forest trees and stands. Springer, Dordrecht. 457 p
Cahill JF (1999) Fertilization effects on interactions between above- and below- ground competition in an old field. Ecology 80:466–480
Cahill JF (2002) Interactions between root and shoot competition vary among species. Oikos 99:101–112
Canham CD, Coates KD, Bartemucci P, Quaglia S (1999) Measurement and mdoelling of spatially explicit variation in light transmission through interior cedar-hemlock forests of Bristish Columbia. Can J For Res 29:1775–1783
Canham CD, lePage PT, Coates KD (2004) A neighborhood analysis of canopy tree competition: effects of shading versus crowding. Can J For Res 34:778–787
Cao QV, Burkhart HE, Max TA (1980) Evaluation of two methods of cubic volume prediction of loblolly pine to any merchantable limit. For Sci 26:71–80
Casper BB, Jackson RB (1997) Plant competition underground. Annu Rev Ecol Syst 28:545–570
Clutter JL, Fortson JC, Pienaar LV, Brister GH, Bailey RL (1983) Timber management – a quantitative approach. Wiley, New York, 333 p
Coates KD, Canham CD, LePage PT (2009) Above- versus below-ground competitive effects and responses of a guild of temperate tree species. J Ecol 97:118–130
Coomes DA, Allen RB (2007) Effects of size, competition and altitude on tree growth. J Ecol 95:1084–1097
Courbaud B (1995) Modélisation de la croissance en forêt irrégulière, – perspectives pour lespessières irrégulières de montagne. Rev For Fr XLVII:173
Cox DR, Oakes D (1984) Analysis of survival data. Chapman & Hall, London
Craib IJ (1939) Thinning, pruning and management studies on the main exotic conifers grown in South Africa. Govt. Printer, Pretoria

Curt T, Prévosto B (2003) Root biomass and rooting profile of naturally regenerated beech in mid-elevation Scots pine woodlands. Plant Ecol 167(2):269–282

Daniels RF (1976) Simple competition indices and their correlation with annual loblolly pine tree growth. For Sci 22:454–456

Daume S (1995) Durchforstungssimulation in einem Buchen-Edellaubholz Mischbestand. Diplomarbeit, Fachbereich Forstwissenschaft, Universität Göttingen

Daume S, Füldner K, Gadow K (1997) Zur Modellierung personenspezifischer Durchforstungenin ungleichaltrigen Mischbeständen. AFJZ 169(2):21–26

Demaerschalk JP (1973) Integrated systems for the estimation of tree taper and volume. Can J For Res 3(1):90–94

Demaerschalk JP, Kozak A (1977) The whole bole system: a conditioned dual-equation system for precise prediction of tree profiles. Can J For Res 7:488–497

Dobson AJ (1990) An introduction to generalized linear models. Chapman and Hall, Statistic Text Series, London

Dickel M, Kotze H, Gadow K, Zucchini W (2010) Growth and survival of *Eucalyptus grandis* – a study based on modeling lifetime distributions. Int J Math Comput For Nat-Resour Sci 2(2):20–30. (available online at http://mcfns.com)

Drexhage M, Colin F (2001) Estimation root system biomass from breast-height diameters. Forestry 74:491–497

Düser R (1978) Programmierte Berechnung stehenden Holzes mit dem Taschenrechner nach der Düser-Flori-Formel. Beilage der AFZ, Nr. 36

Ek AR, Monserud RA (1974) Dimensional relationships of forest and open-grown stems in Wisconsin. Univ. Wis. Forestry Res. Note 181, 7

Ekö PM, Ågestam E (1994) A comparison of naturally regenerated and planted Scots pine (*Pinus sylvestris* L.) on fertile sites in southern Sweden. For u Landsc Res 1:III–126

Environmental Systems Research Institute Inc (1990) Understanding GIS, The ARC/INFO Method

Fahrmeir L (2007) Lebensdauer- und Ereignisanalyse [Survival and Event History Analysis]. In: Vorlesungsskript, Institut für Statistik. Ludwig-Maximilians-Universität, München. 141 pp. (in German)

Fehrmann L, Kuhr M, Gadow Kv (2002) Zur Analyse der Grobwurzelsysteme großer Waldbäume an Fichte [*Picea abies* (L.) Karst.] und Buche [*Fagus sylvatica*]. Forstachiv 74(3):96–102

Forrester DI (2014) The spatial and temporal dynamics of species interactions in mixed-species forests: from pattern to process. For Ecol Manage 312:282–292

Früh T (1992) C-Bilanz auf der Basis der Pipe-Modell-Theorie, Produktionsmodell. Reports of the For-schungszentrum Waldökosysteme (Göttingen), Band 31:355–367

Füldner K, Sattler S, Zucchini W, Gadow K v (1996) Modellierung personenabhängiger Auswahlwahrscheinlichkeiten bei der Durchforstung. AFJZ 167(8):159–162

Gadow Kv (1987) Untersuchungen zur Konstruktion von Wuchsmodellen für schnellwüchsige Plantagenbaumarten. Forstl. Forschungsberichte, Universität München Nr. 77, 147 S

Gadow K (1996) Modelling growth in managed forests-realism and limits of lumping. Sci Total Environ 183:167–177

Gadow K, Hui GY, Albert M (1998) Das Winkelmass – ein Strukturparameter zur Beschreibung der Individualverteilung in Waldbeständen. Centralbl für das ges Forstw 115:1–10

Gaffrey D (1988) Forstamts- und bestandesindividuelles Sortimentierungsprogramm als Mittel zur Planung, Aushaltung und Simulation. Diplomarbeit, Fachbereich Forstwissenschaft, Universität Göttingen

Gaffrey D (1996) Sortenorientiertes Bestandeswachstums-Simulationsmodell auf der Basis intraspezifischen, konkurrenzbedingungten Einzelbaumwachstums-insbesondere hinsichtlich des Durchmessers- am Beispiel der Douglasie. Diss. Forschungszentrum Waldökosysteme d. Uni. Göttingen, 413 p

Gerrard DJ (1969) Competition quotient – a new measure of the competition affecting individual forest trees. Mich State Univ Agr Exp Stn Res Bull No 20

Grier CC, Vogt KA, Keyes ML, Edmonds RL (1981) Biomass distribution and above- and below-ground production in young and mature Abies amabilis zone ecosystems of the Washington Cascades. Can J For Res 11:155–167

Grundner F, Schwappach A (1942) Massentafeln zur Bestimmung des Holzgehaltes stehender Waldbäume und Waldbestände. 9th edn, Berlin

Gunaratne AMTA, Gunatilleke CVS, Gunatilleke IAUN, Madawala Weerasinghe HMSP, Burslem DFRP (2011) Release from root competition promotes tree seedling survival and growth following transplantation into human induced grasslands in Sri Lanka. For Ecol Manag 262(2):229–236

Hegyi F (1974) A simulation model for managing jackpine stands. In: Fries J (ed) Growth models for tree and stand simulation. Royal College of Forestry, Stockholm, pp 74–90

Henningsen A, Hamann JD (2007) Systemfit: A package for estimating systems of simultaneous equations in R. J Stat Softw 23(4):1–40. http://www.jstatsoft.org/v23/i04/

Hessenmöller D (2002) Modelle zur Wachstums- und Durchforstungssimulation im Göttinger Kalkbuchenwald. Dissertation, University Göttingen, p 163

Hoffmann CW, Usoltsev VA (2001) Modelling root biomass distribution in Pinus sylvestris forests of the Turgai Depression of Kazakhstan. For Ecol Manage 149:103–114

Holmes MJ, Reed DD (1991) Competition indices for mixed species northern hardwoods. For Sci 37(5):1338–1349

Hui GY, Gadow Kv (1997) Entwicklung und Erprobung eines Einheitsschaftmodells für die Baumart Cunninghamia lanceolata. Forstw Cbl 116(5):315–321

Husch B, Miller CI, Beers TW (1982) Forest mensuration. Wiley, New York

Hussein KA, Albert M, Gadow K (2001) The crown window- a simple device for measuring tree crowns. Forstw Cbl 119:43–50

Hussein KA, Schmidt M, Kotzé H, Gadow K (2008) Parameter-parsimonious taper functions for describing stem profiles. Scientia Silvae Sinicae 44(6):1–8

Itô H, Sumida A, Isagi Y, Kamo K (1997) The crown shape of an evergreen Oak, Quercus glauca, in a hardwood community. J For Res 2:85–88

Jack WH (1967) Single tree sampling in even-aged plantations for survey and experimentation. IUFRO-Proceedings, Section 25, Munich, pp 379–403

Kahle M, Stüber V, Pommerening A (1996) Strukturbeschreibung in Buchen-Edellaubholzbeispiel-beständen. Institut für Forsteinrichtung und Ertragskunde, Georg-August-Universität Göttingen, 61 p. (In German)

Kahn M (1994) Modellierung der Höhenentwicklung ausgewählter Baumarten in Abhängigkeit vom Standort. Dissertation, Forstwissenschaftliche Fakultät, Universität München

Klein JP, Moeschberger ML (1997) Survival analysis: techniques for censored and truncated data. Springer, New York, 502 pp

Klier G (1969) Beitrag zu den Baumabstandsmethoden in der Forstwirtschaft. Arch. Forstwes., Berlin 18 Jg. Heft 12, S. 1257–1264

Korol M, Gadow K (2003) Ein Einheitsschaftmodell für die Baumart Fichte (A generalized taper model for Spruce). Forstw Cbl 122:175–182

Kouba J (2002) Das Leben des Waldes und seine Lebensunsicherheit [Forest life and its temporal uncertainty]. Forstwiss Centralbl 121:211–228 (in German, with English summary)

Kozak A (1988) A variable exponent taper equation. Can J For Res 18:1363–1368

Kramer H (1988) Waldwachstumslehre. Verlag Paul Parey, Hamburg/Berlin, 374 p

Kramer PJ, Kozlowski TT (1979) Physiology of woody plants. Academic press, New York, 811 p

Kuiper LC, Coutts MP (1992) Spatial disposition and extension of the structural root system of Douglas-fir. For Ecol Manage 47:111–125

Kurth W (1994) Morphological models of plant growth: possibilities and ecological relevance. Ecol Modell Int J Ecol Modell Syst Ecol 75/76:299–308. Verlag Elsevier

Kurz WA, Beukema SJ, Apps MJ (1996) Estimation root biomass and dynamics for the carbon budget of the Canadian forest sector. Can J For Res 26:1973–1979

Le Goff N, Ottorini J-M (2001) Root biomass and biomass increment in a beech (Fagus sylvatica L.) stand in North-East France. Ann For Sci 58:1–13

Lee DT (1980) Two-dimensional Voronoi diagram in the LP-metric. J ACM 27(4):604–618

Lee WK, Gadow K (1997) Iterative Bestimmung der Konkurrenzbäume in Pinus densiflora Beständen. AFJZ 168(3/4):41–44

Lee WK, Gadow K, Chung D-J, Lee J-L (2003) Individual tree dbh-growth in Pinus densiflora and Quercus variabilis mixed stands in central Korea. Ecol Model 176:187–200

LeMay V, Kozak A, Muhairwe K, Kozak RA (1993) Factors affecting the performance of Kozak's (1988) variable-exponent taper function. Wood u Wiant 1993:34–53

Leuschner C (2002) Lebensraum Baumkrone – Schatzkiste der Biodiversität. Georgia Augusta, Wissenschaftsmagazin der Georg-August-Universität Göttingen; No. 1: 18–25

Matèrn B (1986) Spatial variation, Lecture notes in statistics, 2nd edn. Springer, Berlin, p 151 S

Matsumura N (1988) Entwicklung und Anwendungsmöglichkeiten eines einzelbaumorientierten Simulationsmodells für Douglasie. Docvtoral Diss. Georg-August-Universität Göttingen

Matyssek R (2003) Kosten und Nutzen von Raumbesetzung und –ausbeute. AFZ/Der Wald 17:862–863

Mawson JC, Thomas JW, Degraff RM (1976) Program HTVOL. The determination of tree crown volume by layers. USDA For Serv Res Pap NE-354

Max TA, Burkhart HE (1976) Segmented polynomial regression applied to taper equations. For Sci 22(33):283–289

McCullagh P, Nelder JA (1989) Generalized Linear Models, 2nd edition. Chapman and Hall, London

Messier C, Nikinmaa E (2000) Effects of light availability and sapling size on the growth, biomass allocation, and crown morphology of understory sugar maple, yellow birch, and beech. Ecosci 7:345–356

Mikulka B (1955) Versuche zur zahlenmäßigen Erfassung der Qualität von Waldbeständen. Mitt. der Schw. Anst. f. d. forstl. Vers. Wes. 23. Bd. 1

Mitscherlich G (1971) Wald, Wachstum und Umwelt – Waldklima und Wasserhaushalt. J.D. Sauerländer's Verlag

Mischerlich G (1975) Waldwachstum und Umwelt. Volume 3: Boden, Luft und Produktion. J.D. Sauerländer's Verlag, Frankfurt am Main

Mohren GMJ, Van Gerwen CP, Spitters CJT (1984) Simulation of primary production in even-aged stands of Douglas-fir. For Ecol Manage 9:27–49

Nance W, Grissom JE, Smith WR (1988) A new competition index based on weighted and constrained area potentially available. In: Ek AR, Shifley SR, Burke TE (eds) Forest growth modelling and prediction, vol NC-120, USDA For Serv Gen Techn. Rep., pp 134–142

Nielsen CNN (1995) Detailed instructions for root architecture assessments with the Rootarch-method. Arboretum, Royal Vet. and Agriculture University, Int. Rep. No. 7, Hoersholm (DK), 12 p

Ormerod DW (1973) A simple bole model. For Chronicle 49:136–138

Ottorini JM (1991) Growth and development of individual Douglas Fir stands for applications to simulation in silviculture. Annales des Sciences Forestieres 48

Parresol BR (1999) Assessing tree and stand biomass: a review with examples and critical comparisons. Forest Science 45:573–593

Philip MS (1994) Measuring trees and forests. Wallingford, Cab International. 310 S

Pielou EC (1977) Mathematical ecology. Wiley, New York, 384 p

Pienaar V, Shiver B (1981) Survival functions for site-prepared slash pine plantations in the flatwoods of Georgia and Northern Florida. South J App For 5:59–62

Pressler R (1865) Zur Verständigung über den Reinertragswaldbau und dessen Betriebsideal. AFJZ 36, pp 41–55 and pp 173–191 and pp 261–271

Pretzsch H (1992a) Konzeption und Konstruktion von Wuchsmodellen für Rein- und Mischbestände, vol 115. Schriftenreihe d. Forstw. Fak. University München, p 332

Pretzsch H (1992b) Modellierung der Kronenkonkurrenz von Fichte und Buche in Rein-und Mischbeständen. Allg Forst-u Jagd Zeitung 163(11-12):203–213. (In German)

Pretzsch H (2009) Forest dynamics, growth and yield. Springer, Berlin, p 664

Prodan M (1966) Holzmeßlehre. J.D. Sauerländer's Verlag, Frankfurt a. M, 644 p

Pukkala T (1989) Prediction of tree diameter and height in a Scots pine stand as a function of the spatial pattern of trees. Silva Fenn 23:83–99

Rautiainen O (1999) Growth dynamics and management of Shorea robusta forests in Southern Nepal. Diss. Faculty of Fortestry, University of Joensuu
Raven PH, Evert RF, Eichhorn SE (1987) Biology of plants. Worth Publishers Inc., New York, 685 p
Reed DD, Green EJ (1984) Compatible stem taper and volume ratio equations. For Sci 30(4):977–990
Richards AE, Forrester DI, Bauhus J, Scherer-Lorenzen M (2010) The influence of mixed tree plantations on the nutrition of individual species: a review. Tree Physiol 30(9):1192–1208
Riemer T, Gadow K, Sloboda B (1995) Ein Modell zur Beschreibung von Baumschäften. Allg Forst- und J Ztg 166(7):144–147
Rothe A, Binkley D (2001) Nutritional interactions in mixed species forests: a synthesis. Can J For Res 31:1855–1870
Rötzer T, Seifert T, Gayler S, Priesack E, Pretzsch H (2012) Effects of stress and defence allocation on tree growth: Simulation results at the tree and stand level. In: Matyssek R, Schnyder H, Ernst D, Munch JC, Oßwald W, Pretzsch H (eds) Growth and defence in plants: resource allocation at multiple scales, vol 220, Ecol Stud., pp 401–432
Saborowski J (1982) Entwicklung biometrischer Modelle zur Sortimentenprognose. Dissertation, Universität, Göttingen
Sachsse H (1991) Kerntypen der Rotbuche. Forstarchiv 62:238–242
Sahney S, Benton MJ, Ferry PA (2010) Links between global taxonomic diversity, ecological diversity and the expansion of vertebrates on land. Biol Lett 6(4):544–547
Schmidt M (2001) Prognosemodelle für ausgewählte Holzqualitätsmerkmale wichtiger Baumarten. Dissertation, Institut für Waldinventur und Waldwachstum, Universität Göttingen: 302 p
Schmidt M, Gadow K, Höfle H (2005) Vorkommen und Ausprägung von fakultativen Kerntypen bei Rotbuche auf niedersächsischen Kalk- und Rötstandorten. Allg Forst und Jagdzeitung 176(11/12):200–211
Schober R (1953) Ertragstafeln und Forsteinrichtung. Forstwiss Cbl 72:1–13
Schröder J, Gadow K (1999) Testing a new competition index for maritime pine in North-Western Spain. Can J For Res 29:280–283
Schütz J-Ph (1989) Der Plenterbetrieb. Unterlage zur Vorlesung Waldbau III, ETH Zürich, 54pp
Schweizerisches Landesforstinventar (1988) Berichte Nr. 1986: Ergebnisse der Aufnahme 1982–1986. Eidgenössische Anstalt für das forstliche Versuchswesen. Birmensdorf
Seebach C (1846) Ertragsuntersuchungen im Buchenhochwalde. Krit Blätter f Forst- u Jagdwiss 23:74–88
Seifert T (2003) Integration von Holzqualität und Holzsortierung in behandlungssensitive Waldwachstumsmodelle. Technical University Munich, Dissertation, p 314
Seifert T, du Toit B, Jooste GH, Seibold S, Durrheim G, Seydack A (2011) Towards an individual tree forest growth model for natural forests in South Africa: testing competition indices to model individual tree diameter growth in afro-temperate forests. In: Proceedings of the fifth national national forest and woodlands symposium, Richards Bay South Africa, Richards
Seifert T, Seifert S, Seydack A, Durrheim G, Gadow Kv (2014) Competition effects in an afrotemperate forest. For Ecosyst 1:13
Sinoquet H, Le Roux X, Adam B, Ameglio T, Daudet FA (2001) RATP: a model for simulating the spatial distribution of radiation absorption, transpiration and photosynthesis within canopies: application to an isolated tree crown. Plant Cell Environ 24:395–406
Sloboda B (1984) Bestandesindividuelles biometrisches Schaftformmodell zur Darstellung und zum Vergleich von Formigkeit und Sortimentausbeute sowie Inventur. Tagungsbericht d. Sektion Ertragskunde, Neustadt
Song M, Hu Q, Tian Y, Ouyang H (2012) Seasonal patterns of root and shoot interactions in an alpine meadow on the Tibetan Plateau. J Plant Ecol 5:182–190
Späth H (1973) Spline-Algorithmen zur Konstruktion glatter Kurven und Flächen. Oldenbourg, München
Speidel G (1955) Die Wertklasse als Gütemaßstab in der Forsteinrichtung. Forstarchiv:217–224
Spiecker M (1994) Wachstum und Erziehung wertvoller Waldkirschen. Mitt. d. Forstl. Vers. u. Forschungsanstalt Baden-Württemberg, No 181

Staupendahl K (2011) Modellierung der Überlebenswahrscheinlichkeit von Waldbeständen mithilfe der neu parametrisierten Weibull-Funktion [Modelling the survival probability of forest stands using the reparametrised Weibull function]. Forstarchiv 82:10–19. (in German, with English summary)

Staupendahl K, Zucchini W (2011) Schätzung von Überlebensfunktionen der Hauptbaumarten auf der Basis von Zeitreihendaten der Rheinland-Pfälzischen Waldzustandserhebung [Estimating survival functions for the main tree species based on time series data from the forest condition survey in Rheinland-Pfalz, Germany]. Allg. Forst- Jagdztg. 182(5/6). (in German, with summaries in English and French)

Steingass F (1996) Beschreibung der Schaftprofile von Douglasien. Diplomarbeit, Fachbereich Forstwissenschaft, Universität Göttingen

Steneker GA, Jarvis JM (1963) A preliminary study to assess competition in a white spruce – trembling aspen stand. For Chron 39:334–336

Sterba H (1990) Validitätsprüfung des Einzelbaumsimulators für Fichten-Kiefern-Mischbestände. Tagungs-bericht. Dt. Verb. Forstl. Forschungsanstalten – Sektion Ertragskunde, pp 101–117

Stöhr FK (1963) Erweiterungsmöglichkeiten der Winkelzählprobe. Dissertation, Albert-Ludwig-University, Freiburg/Br. 83 S

Struck G (1999) Effekte von Pflanzverband, Kulturdüngung und Schlagräumung am Beispiel junger Traubeneichenbestände. Dissertation, Fak. f. Forstwiss. u. Waldökologie, University of Göttingen, 196 p

Tang S, Zhang H, Xu H (2000) Study on establishing and estimating a compatible biomass model. Scientia Silvae Sinicae 36(Suppl. 1):19–27. (In Chinese with English abstract)

Thomasius HO, Thomasius H (1976) Anwendungsbeispiel zu einem Verfahren der Berechnung ertragskundlich optimaler Bestandesdichtewerte und Diskussion dieses Verfahrens. Informationen TU Dresden, Sektion Forstwirtschaft, Bereich Waldbau und Forstschutz, Tharandt, pp 21–76

Tomé M, Burkhart HE (1989) Distance-dependent competition measures for predicting growth of individual trees. For Sci 35(3):816–831

Trincado GF (1996) Modellierung der Schaftform von Fichten (Picea abies Karst.) und Buchen (Fagus Sylvatica L.). MSc thesis. University of Göttingen, 50 p

Trincado G, Gadow K (1996) Zur Sortimentschätzung stehender Laubbäume. Clbl ges Forstw 1:27–38

Ung C-H, Raulier F, Oullet D, Dhôte J-F (1997) L'indice de compétition interindividuelle de Schütz. Can J For Res 27:521–526

Valladares F, Niinemets Ü (2008) Shade tolerance, a key plant feature of complex nature and consequences. Annu Rev Ecol Evol Syst 39:237–257

Van Laar A, Akça A (1997) Forest mensuration. Cuvillier Verlag Göttingen, 418 pp. ISBN 3-89588-874-5

Vanclay JK (1994) Modelling forest growth – applications to mixed tropical forests. CAB International, Wallingford, 312 pp. ISBN 0 85198913 6

Wang J, Zhang C, Xia F, Zhao X, Wu L, Kv G (2011) Biomass structure and allometry of Abies nephrolepis (Maxim) in Northeast China. Silva Fenn 45(2):211–226

Weibull W (1951) A statistical distribution of wide applicability. J Appl Mech 18:293–297

Wiegard C, Netzker D, Gadow K (1997) Die Erdstückmethode der Wertinventur. Forstarchiv 68:144–148

Wykoff WR, Crookston NL, Stage AR (1982) User's guide to the stand prognosis model. USDA Forersty Service General Technical Report No. INT–133, p 112

Zeide B, Gresham CA (1991) Fractal dimensions of tree crowns in three loblolly pine plantations of coastal South Carolina. Can J For Res 21(8):1208–1212

Zeide B, Pfeifer P (1991) A method for estimation of fractal dimension of tree crowns. For Sci 37(5):1253–1265

Zucchini W, Gadow K (1995) Two indices of agreement among foresters selecting trees for thinning. Forest Landsc Res 1:199–206

Forest Production

5

Abstract

Quantitative models are important for understanding and predicting the structure and dynamics of complex systems. This chapter introduces site models and models of forest production. Harvest events that cause abrupt modifications in the distribution of temperature and radiation with associated changes in competition, growth, soil processes, and habitat availability are among the most important ecological disturbances. Modeling harvest events is therefore also included in this chapter.

A *Eucalyptus grandis* Ecosystem in South Africa

© Springer Nature Switzerland AG 2021

K. von Gadow et al., *Sustaining Forest Ecosystems*, Managing Forest Ecosystems 37,
https://doi.org/10.1007/978-3-030-58714-7_5

5.1 Site Models

Forest production and individual tree growth, mortality, and successful recruitment are affected by the environment. Most of the environmental variables become more important when their availability becomes a limiting factor. Humidity is the limiting factor on dry sites, whereas soil aeration is limiting on wet sites, especially when nutrient availability is good. Light status and nutrient status are key environmental factors on moist and nutrient-rich sites. The relation between the availability of some site factors and the productive potential (V) is sketched in Fig. 5.1.

The light reaction curve is asymptotic while the relations with temperature, water, and nutrients are defined by a maximum. Several authors have attempted to quantify the relations depicted in Fig. 5.1 usually by applying a multiple regression model (e.g., Moosmayer and Schöpfer 1972; Schönau 1988). What complicates the application of available experience is the fact that site characteristics are often only available in the form of categorical variables, for example, "very wet" and "nutrient-rich." Kahn (1994) used nine independent semantic variables to form three complex ecological factors (Table 5.1).

Kahn used nine variables to determine the state of three complex ecological factors: *nutrient status, temperature,* and *humidity*. Specific values of the nine basic variables are combined to determine the effect of three "complex ecological factors" NUTRIENT, TEMP, and MOIST. These factors are assembled by means of specific transformation functions (Fig. 5.2).

The linguistic variables, which are available on ordinal scales, are transformed into a continuous scale by means of specific transformation functions and numerical approximation, as proposed by Chen and Hwang (1992). The effect on the dominant height of each of the nine individual site variables is estimated by means of a response function $\mu(X)$. X represents the category of a linguistic variable and $\mu(X)$ the effect on height growth. Different kinds of function are used, the most common being the unimodal trapezoidal function (Fig. 5.3).

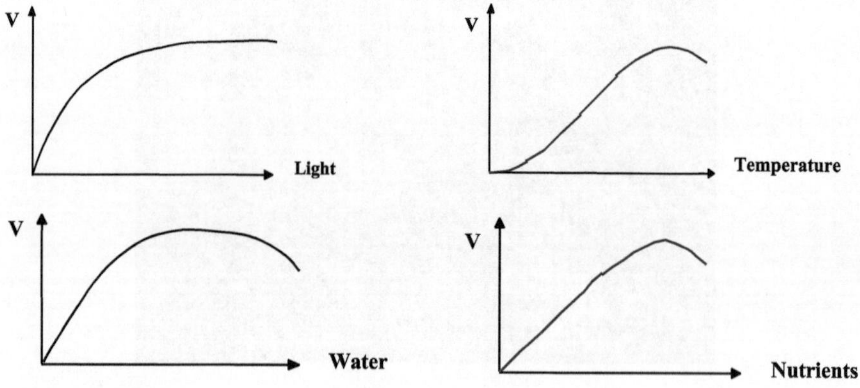

Fig. 5.1 Relations between the availability of some site factors and the production potential (V) (see Kimmins 1990)

Table 5.1 Nine independent variables combined to form three complex ecological factors

Complex Ecological Factor	Variable	Description
nutrient status	NOx	nitrogen oxide concentration
	CO_2	CO_2-concentration
	NST	nutrient status (linguistic variable)
temperature	DT10	days per year with temperature > 10 C
	TVAR	difference between temperature extremes [°C]
	TV	mean annual temperature [°C]
humidity	MV	index of aridity
	Frisch	soil humidity (linguistic variable)
	NV	total rainfall in vegetation period (mm)

Kahn (1994)

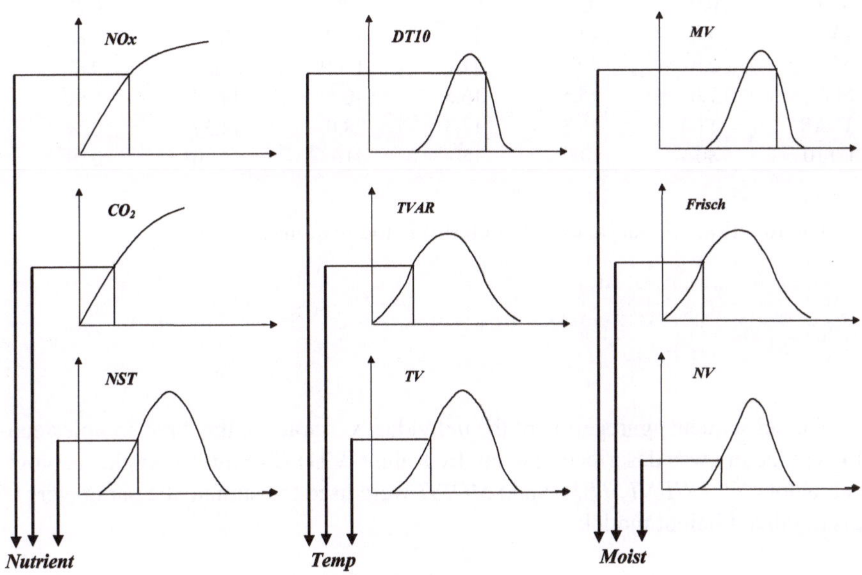

Fig. 5.2 Graphs of nine transformation functions for estimating the three complex ecological factors NUTRIENT, TEMP, and MOIST (Kahn 1994:115)

The unimodal function increases steadily toward a maximum and decreases for X values beyond the maximum. The segmented linear trapezoidal function shown in Fig. 5.3 is defined by four parameters c1 … c4, threshold values that need to be specified. The parameters for *Picea abies* as well as an example with observed X values and the corresponding μ(X) are shown below (Table 5.2).

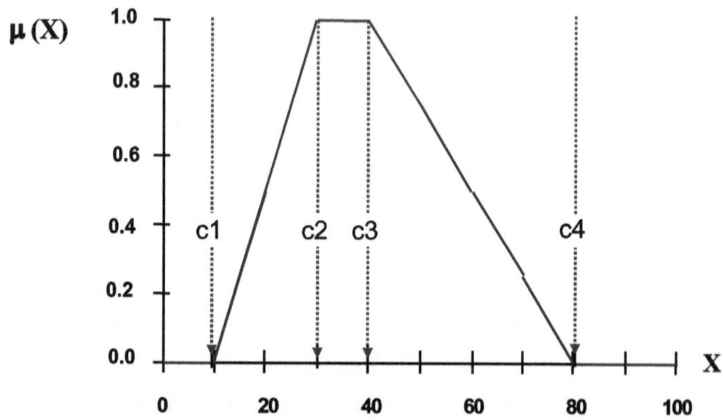

Fig. 5.3 Graph of a unimodal segmented linear trapezoidal response function with four segments defined by the parameters c1 … c4

Table 5.2 Example of the segmented linear trapezoidal function defined by four parameters

Variable	c1	c2	c3	c4	X	μ(X)
Frisch	0.0	0.6	0.8	1.0	0.37	0.62
NST	0.0	0.4	0.8	1.0	0.115	0.29
TV	4.0	14	14.5	20.0	12.5	0.85
NV	230	390	410	1500	330	0.63
MV	7.0	15.5	16.5	40	14.67	0.90
TVAR	11.5	16.8	17.3	28.0	14.35	0.54
DT10	80	145	150	210	141.01	0.94

The function e.f can be used to determine the response value.

```
e.f<-function(X,c1,c2,c3,c4) {
  if ((X>=c1)&(X<c2)) e=(X-c1)/(c2-c1) else if ((X>=c2)&(X<c3)) e=1
  else if ((X>=c3)&(X<c4)) e=(c1-X)/(c1-c3) else e=0;
return(e)}
NV=e.f(330,230,390,410,1500)
```

The subsequent aggregation of the individual variables of the three complex ecological factors was described in detail by Kahn (1994). The three complex ecological factors *NUTRIENT*, *TEMP*, and *MOIST* were used to estimate the parameters of his dominant height model:

$$H_o = A \cdot \left(1 - e^{-k \cdot t}\right)^2 \tag{5.1}$$

The effect of site attributes on the asymptote A as well as the parameter *k* of this growth function was found to be significant (Kahn 1994). Another useful review of modeling site quality was published by Vanclay (1992).

5.1.1 Site Index Systems Based on Dominant Height

Height growth is relatively independent of forest density. For this reason, the dominant stand height at a given reference age, known as *site index,* is often used to estimate the productive potential of a growing site. The development of the dominant height can be described by means of a three-parameter asymptotic function such as the Chapman-Richards generalization of Bertalanffy's growth model (Eq. 5.2; Pienaar and Turnbull 1973; Clutter et al. 1983). The site index system in Fig. 5.4, which was based on observations of *Pinus densiflora* in Korea, is an example of a site index system. The reference age is 50 years. Site index 50 can be calculated if the dominant height Ho, the age (t), and the parameter k are known:

$$SI_{50} = H_o \cdot \left(\frac{1 - e^{-k \cdot 50}}{1 - e^{-k \cdot t}} \right)^m \tag{5.2}$$

The parameter k scales the time axis, while m offers additional flexibility concerning the shape of the growth curve. The value of m is usually close to 2 for height and approximately 3 for volume.

The parameters of the Chapman-Richards model are estimated by nonlinear regression. It is often assumed that the shape of the growth curve is independent of site index, that is, the model parameters are constant for any site (Kiviste et al. 2002). A polymorph site index system, in contrast to such an anamorph system, has parameters that are affected by site.

Desirable attributes of a site index model are: (1) a logical behavior (height should be zero at age zero and equal to site index at the specified reference age); (2) a sound theoretical basis, which includes polymorphism, that is, the curves should not be proportional; (3) the model asymptote should increase with increasing site index; (4) existence of an inflection point; and (5) base-age invariance to ensure that

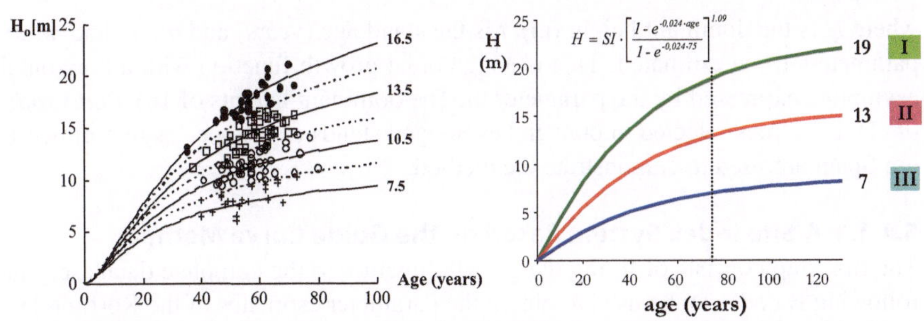

Fig. 5.4 Left: Site index system for *Pinus densiflora* in Korea; the reference age is 50 years (Chung 1996). Right: Site index system for *Pinus pinea* in Spain; the reference age is 75 years (Montero et al. 2004)

the height at any age can be estimated from the height at any other age. Whether or not these requirements can be met depends on both the construction method and the growth function used to develop the curves.

According to Clutter et al. (1983), most of the approaches used to fit site index curves can be viewed as special cases of three general development techniques: (1) the guide curve method; (2) the parameter-prediction method; and (3) the difference-equation method. The guide curve method assumes proportionality (anamorphism) among curves for different site qualities and is used to generate a set of anamorphic site index curves. The parameter prediction method is based on fitting a growth function tree-by-tree or plot-by-plot and relating the parameters of the fitted curves to site index. A height-over-age series can be obtained from stem analysis or long-term growth trials. The difference equation method (3) is based on the assumption that observations of the same plot or dominant tree belong to the same site index curves. An algebraic difference form of the growth function is developed where height H_2 at age t_2 is expressed as a function of t_2, and height H_1 at age t_1 and t_1. The algebraic difference form is obtained by substituting one parameter in the growth function. The choice of this parameter determines the behavior of the model: substitution of the asymptote parameter generally produces anamorphic or proportional curves, while substitution of any other parameter produces polymorphic or nonproportional curves with either common or different asymptotes, depending on the growth function used.

The advantages of the difference equation method in comparison to the parameter prediction methods are: (1) short observation periods of temporary plots or stem analysis data from trees whose total age is less than the reference age can be used; (2) the curve estimates are identical to site index at the reference age; and (3) the equations are base-age invariant. Many growth functions can be employed to develop site index curves. In the following example, the *Korf* function is used:

$$H = a_1 \cdot \exp\left(-a_2 \cdot t^{-a_3}\right), \tag{5.3}$$

where H is the dominant height (m), t is the stand age (years) and a_1, a_2, and a_3 are parameters to be estimated. This is a sigmoidal growth function with a horizontal asymptote expressed by the parameter a_1. The dominant heights of 161 *Pinus radiata* D. Don trees selected in pure and even-aged stands of the species in northwestern Spain are used to demonstrate the method.

5.1.1.1 A Site Index System Based on the Guide Curve Method

The first step consists of fitting the growth function to the complete database. The following R code can be used to obtain the parameter estimates of the Korf model:

```
f00 <- read.table("/Users/kg/Desktop/site_index_data.txt", header=TRUE)
korf<-function(age,a1,a2,a3){
  yhat<-a1*exp(-a2*age^(-a3))}
nlmod.korf<-nls(height~korf(age,a1,a2,a3), data=f00, start=list(a1=80, a2=7, a3=0.7))
summary(nlmod.korf)
```

The following guide curve is obtained:

$$H = 65.8462 \cdot \exp\left(-7.6115 \cdot t^{-0.6036}\right) \tag{5.4}$$

After the guide curve is fitted, the next step is to represent a family of site index curves, with the guide curve used as reference. Each individual curve of the system is defined by its site index (SI), that is, by the value of the dominant height at a specific age called "reference age" (t_{ref}). In this example, the reference age is fixed at 20 years, which represents the middle of the rotation age of the *Pinus radiata* ecosystems in northwestern Spain. A family with four different curves defined by site index 26 m, 21 m, 16 m, and 11 m at 20 years will be represented by means of the growth function. The SI 26 curve is defined by

$$26 = a_1 \cdot \exp\left(-a_2 \cdot 20^{-a_3}\right) \tag{5.5}$$

To solve this equation, the estimates of two parameters of the guide curve have to be substituted. There are three different options and therefore three different families can be represented. The first family is obtained when the estimates of the parameters a_2 and a_3 are fixed and the value of the parameter a_1 (the asymptote) is estimated for each site index:

$$\begin{aligned}
26 &= a_1 \cdot \exp\left(-7.6115 \cdot 20^{-0.6036}\right) \Rightarrow a_1 = \frac{26}{\exp\left(-7.6115 \cdot 20^{-0.6036}\right)} = 90.544 \\
21 &= a_1 \cdot \exp\left(-7.6115 \cdot 20^{-0.6036}\right) \Rightarrow a_1 = \frac{21}{\exp\left(-7.6115 \cdot 20^{-0.6036}\right)} = 73.1320 \\
16 &= a_1 \cdot \exp\left(-7.6115 \cdot 20^{-0.6036}\right) \Rightarrow a_1 = \frac{16}{\exp\left(-7.6115 \cdot 20^{-0.6036}\right)} = 55.7196 \\
11 &= a_1 \cdot \exp\left(-7.6115 \cdot 20^{-0.6036}\right) \Rightarrow a_1 = \frac{11}{\exp\left(-7.6115 \cdot 20^{-0.6036}\right)} = 38.3072
\end{aligned} \tag{5.6}$$

In our first example, the family consists of four anamorphic (proportional) curves with different asymptotes (a_1). A graph of the system (Fig. 5.5) can be developed by means of the following R code:

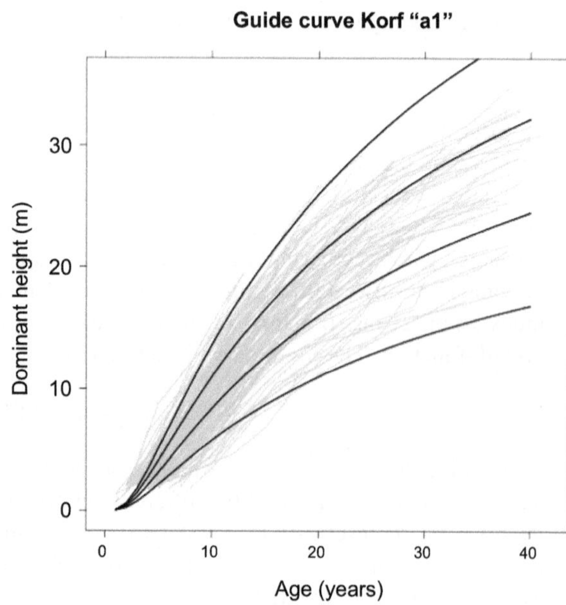

Fig. 5.5 Graph of the anamorphic site index system for *Pinus radiata*

```
library(lattice)
a1<-65.8462; a2<-7.6115; a3<-0.6036
t<-c(1:40)
alis.func <- function(IS){IS/exp(-a2*20^(-a3))}
curves.alis <- data.frame(t, IS11=alis.func(11)*exp(-a2*t^(-a3)),
                          IS16=alis.func(16)*exp(-a2*t^(-a3)),
                          IS21=alis.func(21)*exp(-a2*t^(-a3)),
                          IS26=alis.func(26)*exp(-a2*t^(-a3)))

xyplot(height~age, data=f00, groups=tree, type="l", col="grey", main="Guide curve Korf "a1"",
ylab="Dominant height (m)", xlab="Age (years)",
       panel=function(x, y, ...){
         panel.xyplot(x, y, ...)
         panel.lines(curves.alis$t, curves.alis$IS11, col=1, lwd=2)
         panel.lines(curves.alis$t, curves.alis$IS16, col=1, lwd=2)
         panel.lines(curves.alis$t, curves.alis$IS21, col=1, lwd=2)
         panel.lines(curves.alis$t, curves.alis$IS26, col=1, lwd=2)
       })
```

A different family is obtained by substituting the estimates of parameters a_1 and a_3 in the growth function for each site index and solving for a_2:

$$
\begin{aligned}
26 &= 65.8462 \cdot \exp\left(-a_2 \cdot 20^{-0.6036}\right) \Rightarrow a_2 = \frac{-\log\left(26/65.8462\right)}{20^{-0.6036}} = 5.6679 \\
21 &= 65.8462 \cdot \exp\left(-a_2 \cdot 20^{-0.6036}\right) \Rightarrow a_2 = \frac{-\log\left(21/65.8462\right)}{20^{-0.6036}} = 6.9706 \\
16 &= 65.8462 \cdot \exp\left(-a_2 \cdot 20^{-0.6036}\right) \Rightarrow a_2 = \frac{-\log\left(16/65.8462\right)}{20^{-0.6036}} = 8.6293 \\
11 &= 65.8462 \cdot \exp\left(-a_2 \cdot 20^{-0.6036}\right) \Rightarrow a_2 = \frac{-\log\left(11/65.8462\right)}{20^{-0.6036}} = 10.9148
\end{aligned}
\tag{5.7}
$$

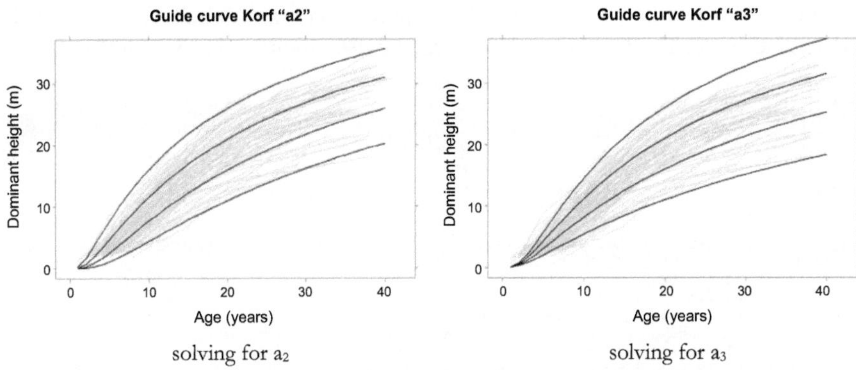

Fig. 5.6 Graph of two polymorphic site index systems. Left: Parameter a_2 is a function of the site index. Right: Parameter a_3 is a function of the site index

In this case, the family consists of polymorphic (nonproportional) curves with the same asymptote. The following R code can be used to draw the system (Fig. 5.6):

```
library(lattice)
t<-c(1:40)
a3is.func <- function(IS){-log(-log(IS/a1)/a2)/log(20)}
curves.a3is <- data.frame(t, IS11=a1*exp(-a2*t^(-a3is.func(11))),
                             IS16=a1*exp(-a2*t^(-a3is.func(16))),
                             IS21=a1*exp(-a2*t^(-a3is.func(21))),
                             IS26=a1*exp(-a2*t^(-a3is.func(26))))

xyplot(height~age, data=f00, groups=tree, type="l", col="grey", main="Guide curve Korf "a3"",
ylab="Dominant height (m)", xlab="Age (years)",
       panel=function(x, y, ...){
         panel.xyplot(x, y, ...)
         panel.lines(curves.a3is$t, curves.a3is$IS11, col=1, lwd=2)
         panel.lines(curves.a3is$t, curves.a3is$IS16, col=1, lwd=2)
         panel.lines(curves.a3is$t, curves.a3is$IS21, col=1, lwd=2)
         panel.lines(curves.a3is$t, curves.a3is$IS26, col=1, lwd=2)
       })
```

A third family can be obtained by substituting the estimates of parameters a_1 and a_2 in the growth function for each site index and solving for a_3:

$$
\begin{aligned}
26 &= 65.8462 \cdot \exp\left(-7.6115 \cdot 20^{-a_3}\right) \Rightarrow a_3 = \frac{-\log\left(-\log\left(26/65.8462\right)/7.6115\right)}{\log\left(20\right)} = 0.7020 \\
21 &= 65.8462 \cdot \exp\left(-7.6115 \cdot 20^{-a_3}\right) \Rightarrow a_3 = \frac{-\log\left(-\log\left(26/65.8462\right)/7.6115\right)}{\log\left(20\right)} = 0.6330 \\
16 &= 65.8462 \cdot \exp\left(-7.6115 \cdot 20^{-a_3}\right) \Rightarrow a_3 = \frac{-\log\left(-\log\left(26/65.8462\right)/7.6115\right)}{\log\left(20\right)} = 0.5617 \\
11 &= 65.8462 \cdot \exp\left(-7.6115 \cdot 20^{-a_3}\right) \Rightarrow a_3 = \frac{-\log\left(-\log\left(26/65.8462\right)/7.6115\right)}{\log\left(20\right)} = 0.4833
\end{aligned}
\tag{5.8}
$$

Again, the family consists of polymorphic (nonproportional) curves with the same asymptote. The following R code can be used to draw the system (Fig. 5.6):

```
library(lattice)
t<-c(1:40)
a3is.func <- function(IS){-log(-log(IS/a1)/a2)/log(20)}
curves.a3is <- data.frame(t, IS11=a1*exp(-a2*t^(-a3is.func(11))),
                          IS16=a1*exp(-a2*t^(-a3is.func(16))),
                          IS21=a1*exp(-a2*t^(-a3is.func(21))),
                          IS26=a1*exp(-a2*t^(-a3is.func(26))))

xyplot(height~age, data=f00, groups=tree, type="l", col="grey", main="Guide curve Korf "a3"",
ylab="Dominant height (m)", xlab="Age (years)",
       panel=function(x, y, ...){
         panel.xyplot(x, y, ...)
         panel.lines(curves.a3is$t, curves.a3is$IS11, col=1, lwd=2)
         panel.lines(curves.a3is$t, curves.a3is$IS16, col=1, lwd=2)
         panel.lines(curves.a3is$t, curves.a3is$IS21, col=1, lwd=2)
         panel.lines(curves.a3is$t, curves.a3is$IS26, col=1, lwd=2)
       })
```

An alternative to the guide curve method is the difference equation method.

5.1.1.2 A Site Index System Based on the Algebraic Difference Equation Method (ADA)

The difference equation method requires either permanent remeasured plots or stem analysis data and is based on the condition that observations of the same plot or dominant tree belong to the same site index curves system. The first step consists of developing a difference form of the growth function being fitted by considering two different measures of the same plot or dominant tree (H_1, t_1) and (H_2, t_2). When the Korf function is used, there are two sets of observations, H_1 at t_1 and H_2 at t_2:

$$\begin{aligned} H_1 &= a_1 \cdot \exp\left(-a_2 \cdot t_1^{-a_3}\right) \\ H_2 &= a_1 \cdot \exp\left(-a_2 \cdot t_2^{-a_3}\right) \end{aligned} \tag{5.9}$$

Because both points belong to the same site index curve, parameters a_1, a_2, and a_3 are the same. The first expression can therefore be solved for one parameter and then substituted in the second expression to obtain the required difference equation form. In this example, there are three difference equation forms, each one depending on the parameter solved. When we solve for a_1, the curves are anamorphic with a different horizontal asymptote. When we solve for a_2 and a_3, the curves are polymorphic with a common horizontal asymptote.

$$\begin{aligned} &\left.\begin{aligned} H_1 &= a_1 \cdot \exp\left(-a_2 \cdot t_1^{-a_3}\right) \\ H_2 &= a_1 \cdot \exp\left(-a_2 \cdot t_2^{-a_3}\right) \end{aligned}\right\} a_1 = \frac{H_1}{\exp\left(-a_2 \cdot t_1^{-a_3}\right)} \Rightarrow \\ &H_2 = H_1 \frac{\exp\left(-a_2 \cdot t_2^{-a_3}\right)}{\exp\left(-a_2 \cdot t_1^{-a_3}\right)} \end{aligned} \tag{5.10}$$

$$\left.\begin{array}{l} H_1 = a_1 \cdot \exp\left(-a_2 \cdot t_1^{-a_3}\right) \\ H_2 = a_1 \cdot \exp\left(-a_2 \cdot t_2^{-a_3}\right) \end{array}\right\} a_2 = \frac{-\log\left(H_1 / a_1\right)}{t_1^{-a_3}} \Rightarrow$$

$$H_2 = a_1 \cdot \left(\frac{H_1}{a_1}\right)^{\left(\frac{t_2}{t_1}\right)^{-a_3}} \tag{5.11}$$

$$\left.\begin{array}{l} H_1 = a_1 \cdot \exp\left(-a_2 \cdot t_1^{-a_3}\right) \\ H_2 = a_1 \cdot \exp\left(-a_2 \cdot t_2^{-a_3}\right) \end{array}\right\} a_3 = \frac{-\log\left(\frac{-\log\left(H_1 / a_1\right)}{a_2}\right)}{\log\left(t_1\right)} \Rightarrow$$

$$H_2 = a_1 \cdot \exp\left(-a_2 \cdot t_2^{\frac{\log\left(\frac{-\log(H_1/a_1)}{a_2}\right)}{\log(t_1)}}\right) \tag{5.12}$$

These difference equations can be fitted by means of nonlinear regression but the original database should be modified to combine all the pairs (*H*,*t*) of the same sample plot or the same dominant trees. After obtaining the parameter estimates, substitution of the reference age for t_1 and of site index (SI) for H_1 gives a specific site index curve. The following R code can be used to fit the difference equation form obtained by solving for a_1. The first three lines modify the original database to combine all the pairs (*H*,*t*) of the same sample plot or the same dominant tree to obtain suitable data for t_1, H_1, t_2, and H_2:

```
f01<-data.frame(tree1=f00$tree, t1=f00$age, h1=f00$height)
f02<-merge(f00,f01, by.x="tree", by.y="tree1")
f02<-subset(f02, f02$age!=f02$t1)

korf.ada.a1 <- function(h1, t1, age, a2, a3){
  yhat <- h1*exp(a2*t1^(-a3))/exp(a2*age^(-a3))
}

nlmod.korf.ada.a1  <-  nls(height~korf.ada.a1(h1,  t1,  age,  a2,  a3),  data=f02,
start=list(a2=7.6, a3=0.6))
summary(nlmod.korf.ada.a1)
```

The difference equation is:

$$H_2 = H_1 \frac{\exp\left(-7.9154 \cdot t_2^{-0.2124}\right)}{\exp\left(-7.9154 \cdot t_1^{-0.2124}\right)} \tag{5.13}$$

A graphic of the family with four different curves corresponding to site indices 26 m, 21 m, 16 m, and 11 m at age 20 years can be developed by using the following R code (Fig. 5.7):

```
library(lattice)
t<-c(1:40)
curve.korf.ada.a1 <- data.frame(t,
                         IS11=korf.ada.a1(11, 20, t, 7.9154, 0.2124),
                         IS16=korf.ada.a1(16, 20, t, 7.9154, 0.2124),
                         IS21=korf.ada.a1(21, 20, t, 7.9154, 0.2124),
                         IS26=korf.ada.a1(26, 20, t, 7.9154, 0.2124))

xyplot(height~age, data=f00, groups=tree, type="l", col="grey", main="ADA Korf "a1"",
ylab="Dominant height (m)", xlab="Age (years)",
       panel=function(x, y, ...){
         panel.xyplot(x, y, ...)
         panel.lines(curve.korf.ada.a1$t, curve.korf.ada.a1$IS11, col=1, lwd=2)
         panel.lines(curve.korf.ada.a1$t, curve.korf.ada.a1$IS16, col=1, lwd=2)
         panel.lines(curve.korf.ada.a1$t, curve.korf.ada.a1$IS21, col=1, lwd=2)
         panel.lines(curve.korf.ada.a1$t, curve.korf.ada.a1$IS26, col=1, lwd=2)
       })
```

One problem associated with the guide curve method and the algebraic difference equation method is the fact that it is not possible to develop a family of polymorphic curves with different asymptotes. A method called the "generalized algebraic difference approach" (GADA) and representing a generalization of the algebraic difference equation method has been proposed to solve this problem. The first step involves relating two parameters of the growth function with a specific index of site quality (X). To obtain curves with different asymptotes, one of the selected parameters has to be the parameter of the growth function related to the asymptote. When the Korf function is used, a GADA system can be obtained bysetting $a_1 = \exp(X)$; $a_2 = \frac{b_1}{X}$ and $a_3 = b_2$:

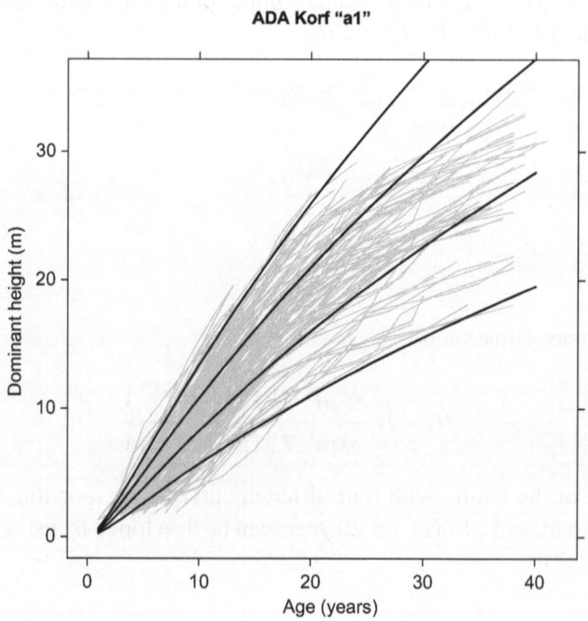

Fig. 5.7 Graph of the ADA site index system

$$
\begin{rcases}
H_1 = a_1 \cdot \exp\left(-a_2 \cdot t_1^{-a_3}\right) \\
H_2 = a_1 \cdot \exp\left(-a_2 \cdot t_2^{-a_3}\right)
\end{rcases} \Rightarrow H_1 = \exp(X) \cdot \exp\left(-\frac{b_1}{X} t_1^{-b_2}\right)
$$

$$
\log(H_1) = X - \frac{b_1}{X} t_1^{-b_2} \Rightarrow X^2 - X\log(H_1) - b_1 t_1^{-b_2} = 0
$$

$$
X = \frac{\log(H_1) + \sqrt{\log(H_1)^2 + 4 b_1 t_1^{-b_2}}}{2} \tag{5.14}
$$

$$
H_2 = \exp(X) \cdot \exp\left(-\frac{b_1}{X} t_2^{-b_2}\right)
$$

The following R code can be used to fit this GADA system:

```
korf.gada <- function(h1, t1, age, b1, b2){
  x<-(log(h1)+((log(h1))^2+4*b1*t1^-b2)^0.5)/2
  yhat <- exp(x)*exp(-(b1/x)*age^(-b2))
}
nlmod.korf.gada<-nls(height~korf.gada(h1, t1, age, b1, b2), data=f02, start=list(b1=40,
b2=0.6))
summary(nlmod.korf.gada)
```

The model finally obtained is:

$$
X = \frac{\log(H_1) + \sqrt{\log(H_1)^2 + 160.3424 \cdot t_1^{-0.3046}}}{2} \tag{5.15}
$$

$$
H_2 = \exp(X) \cdot \exp\left(-\frac{40.0856}{X} t_2^{-0.3046}\right)
$$

After the parameter estimates have been obtained, substitution of the reference age for t_1 and site index (SI) for H_1 gives a specific site index curve. The family with four different curves with site index 26 m, 21 m, 16 m, and 11 m at 20 years can be represented by means of the following R code (Fig. 5.8):

```
library(lattice)
t<-c(1:40)
curves.korf.gada <- data.frame(t,
                            IS11=korf.gada(11, 20, t, 40.0856, 0.3046),
                            IS16=korf.gada(16, 20, t, 40.0856, 0.3046),
                            IS21=korf.gada(21, 20, t, 40.0856, 0.3046),
                            IS26=korf.gada(26, 20, t, 40.0856, 0.3046))

xyplot(height~age, data=f00, groups=tree, type="l", col="grey", main="GADA Korf",
ylab="Dominant height (m)", xlab="Age (years)",
       panel=function(x, y, ...){
         panel.xyplot(x, y, ...)
         panel.lines(curves.korf.gada$t, curves.korf.gada$IS11, col=1, lwd=2)
         panel.lines(curves.korf.gada$t, curves.korf.gada$IS16, col=1, lwd=2)
         panel.lines(curves.korf.gada$t, curves.korf.gada$IS21, col=1, lwd=2)
         panel.lines(curves.korf.gada$t, curves.korf.gada$IS26, col=1, lwd=2)
       })
```

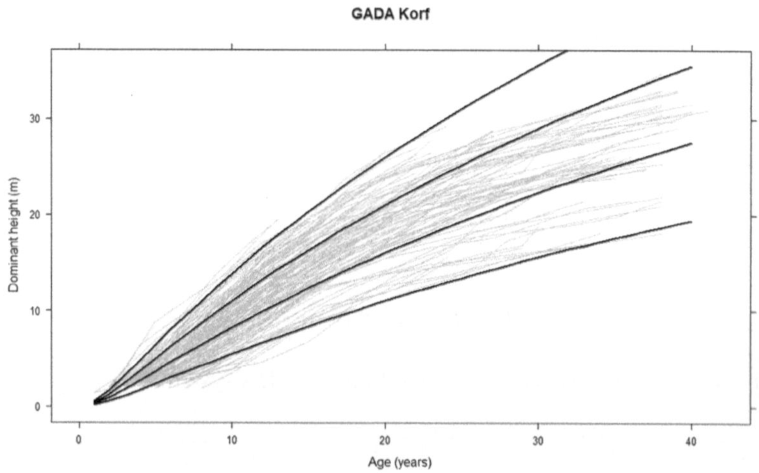

Fig. 5.8 Graph of the GADA site index system

Other examples of polymorph site index systems include one for *Pinus silvestris* forests in Spain (Rojo and Montero 1996) and a site index system for several species of *Pinus* in Mexico (Corral Rivas et al. 2004).

5.1.2 Phytocentric and Geocentric Measures of Site Productivity

The assessment of site quality in natural ecosystems where there is no common age, as is the case in a planted forest, has not received much attention. Among the first attempts to evaluate site quality in uneven-aged multi-species forests were the studies by Flury (1929) and Mitscherlich (1961). Flury found that the mean height of the trees in the biggest diameter class was a useful measure of site quality. Vanclay and Henry (1988) proposed the same measure of site productivity without reference to Flury and called it *site form* (to avoid confusion with *site index*, which is derived from the height–age relation).

5.1.2.1 Plant Communities

Estimating site productivity by means of the ground flora has a long tradition, especially in the boreal ecosystems of northern Europe. A well-known example is the system of vegetation types proposed by Cajander (1909). He identified five major plant associations to estimate the site productivity of forest ecosystems in Finland. Figure 5.9 shows the development of the growing stock of *Pinus sylvestris* forests for five vegetation types (see Kimmins 1987). Cajander's method has aroused much interest among European and North American scientists (Rennie 1963; Vanclay 1992). Another example of the application of natural plant associations to evaluate site productivity was published by Shafer (1989). He found that the association *Stoebe plumosa/Lachenalia* spp/*Tetraria cuspidata/Senecio juniperinus*

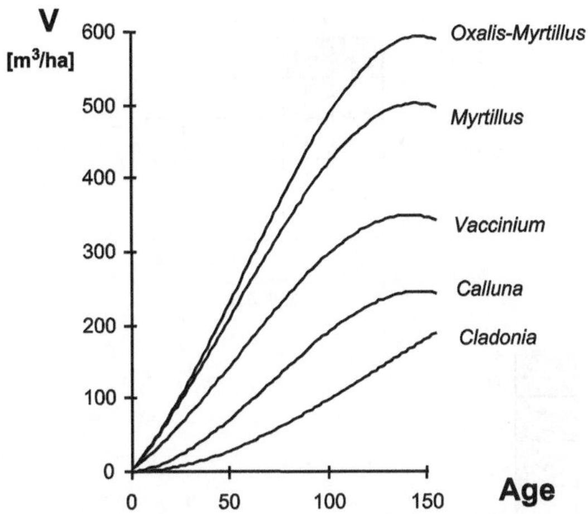

Fig. 5.9 Development of the growing stock of *Pinus sylvestris* forests for Cajander's (1926) five vegetation types (see Kimmins 1987:443)

characterized the best sites, while the association *Watsonia fourcadei/Berzelia intermedia/Cliffortia stricta/Pellaea chedrum* characterized the least productive sites.

Hakes (1994) made an attempt to use the vegetation classification approach in central European *Fagus* ecosystems. Applications in temperate regions are rare, however, especially where the ground flora does not reveal distinct patterns because of frequent anthropogenic disturbances and a high variability of site conditions.

5.1.2.2 Indicator Plant Species

An alternative approach is to use the presence of certain plant species as an indication of site productivity. The presence and abundance of specific indicator plants can indicate the presence of several of the site factors that are important for tree growth, and this approach seems to offer greater flexibility than the climax vegetation approach (Jones 1969). A number of examples are reported from regions with high species diversity and a small-scale distribution of different types of ground flora (Webb 1959; Marques 1991). Hodgkins (1960) and Corns and Pluth (1984) developed regression equations that incorporated the presence and abundance of several plant species to predict the site index of natural conifer stands. MacLean and Bolsinger (1973) and Wiant et al. (1975) presented equations to predict site index from the presence and absence of several indicator plants. Certain plant species allow an assessment of specific site conditions, such as the temperature, humidity, and nutritional conditions of the site (Ellenberg et al. 1991; Schönhar 1993; Fig. 5.10).

Most species are able to survive in a variety of conditions, while some are able to thrive on very specific sites that may be unfavorable for other species. These specialists have a high indicator value (Ellenberg 1986; Schröder 1998). Vanclay (1992:150) and presented a combined approach for a tropical rainforest in Queensland that included geological formation and presence/absence data of certain species.

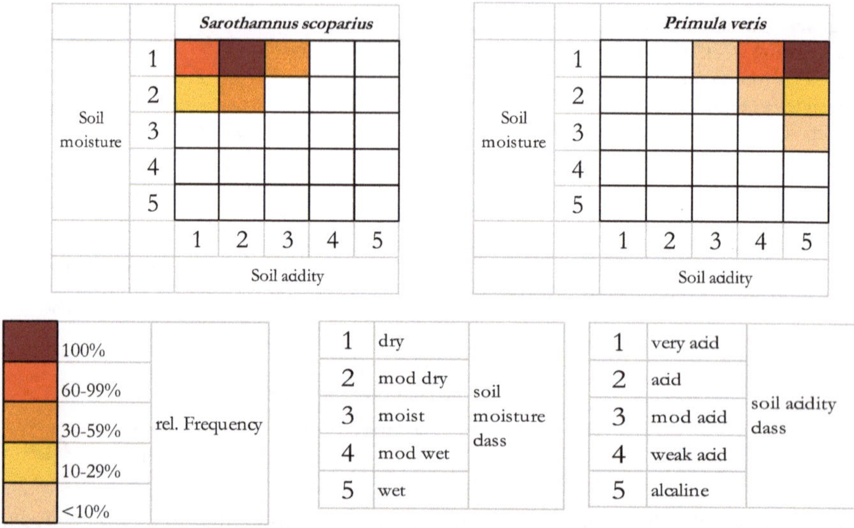

Fig. 5.10 Example of the preference of two indicator plant species for a particular combination of soil moisture and acidity (Schönhar 1993)

5.1.2.3 Climate Indices

The climate is a major influencing factor in tree growth and survival, and therefore in community assemblage. A number of climate indices that aggregate several individual factors have been proposed. Paterson (1962) proposed the climate, vegetation, and productivity (CVP) index that correlates the relation between climate and biosphere productivity:

$$CVP = \mathrm{N} \cdot \frac{TV}{TA} \cdot \frac{g}{360} \cdot \frac{Rp}{Rs} \tag{5.16}$$

where

N = Average annual rainfall (mm)

Tv = Average temperature of the warmest month (°C)

Ta = Difference: average temperature of the warmest and coldest month (°C)

g = Degree Days: number of days per year with a temperature of at least 7 °C

Rp/Rs = Coefficient of evapotranspiration (radiation at north pole/local radiation)

The CVP index was used with some success by Pardé and Bouchon (1990:207) to estimate the mean annual volume production of *Fagus sylvatica* forests. Climate often varies considerably from year to year. Short- and medium-term climate fluctuations, such as extended periods of drought, can influence productivity more than climate normals. Aggregate climate indices such as the CVP can therefore result in imprecise local estimates of site productivity (Weiskittel et al. 2011). In addition, variables are often highly correlated, which can make it difficult to identify the most influential response variable. Climate indices developed for large geographic scales may be interesting for comparative studies, but other factors will probably control productivity at the local scale.

5.1.2.4 Soil Parameters

The assessment of site productivity that is based on soil parameter is a field of research with a long tradition. Half a century ago, Carmean (1973) published a list of 793 publications that focused on this particular topic. Numerous authors found that rootable soil depth was a good indicator of site productivity (Turner et al. 1990; Turvey et al. 1990). One example is Shafer's research (1988) on the relation between soil depth and site index for *Pinus elliottii* ecosystems in the Cape Province, South Africa (see Fig. 5.11).

Various chemical and physical soil parameters have also been evaluated by means of a multiple regression approach (Baker and Broadfoot 1977; Schönau 1988; Henderson et al. 1990; Schönau and Aldworth 1991; Marques 1991).

5.1.2.5 Comparison of Methods

There is no straightforward answer to the question about the best method for assessing site productivity. The answer depends mostly on available observations, but also on the suitability of a particular approach within a specific geographical context. It is possible to answer the question, for instance, for a particular local ecosystem where geocentric and phytocentric observations are available. An example of this is a study published by Marques (1991) who presented a comparison of soil

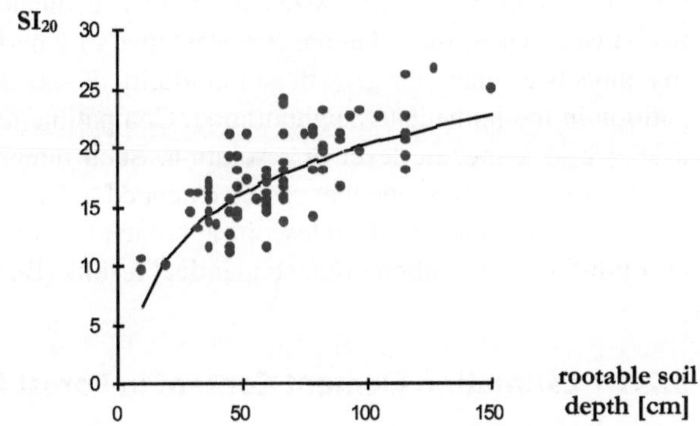

Fig. 5.11 Relation between rootable soil depth and site index for *Pinus elliottii* ecosystems in the Cape Province, South Africa (Shafer 1988)

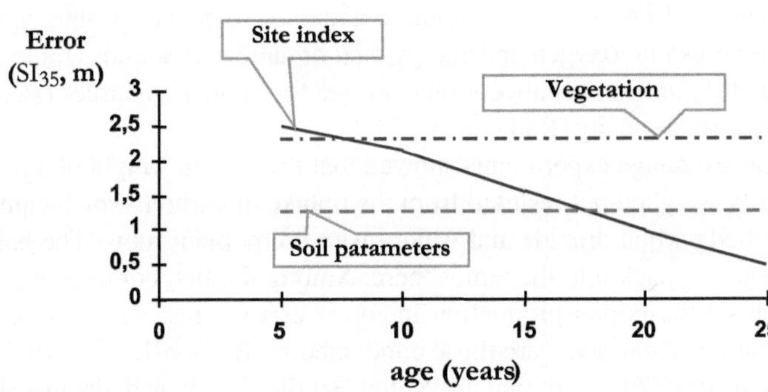

Fig. 5.12 Error of height estimate for *Pinus pinaster* forests in Portugal as a function of age and involving three methods (Marques 1991)

parameters, vegetation attributes, and site index (Fig. 5.12). Marques found that estimates based on soil parameters were more accurate than those derived from vegetation when both were compared with site index. The accuracy of site index increased as the real age approached the site index reference age.

Shugart (1984:52–61) used four growth reduction factors to model the effects of shading, temperature, nutrient status, and humidity on forest production. Similar approaches were followed by Kellomäki et al. (1993) and Kimmins et al. (1990). However, the multiple regression approach appears to have lost its popularity, and it now seems that better use can be made of observations that are assessed routinely in site surveys. The transformation of semantic-categorical data to continuous variables by means of the methods described above should be a promising line of research.

5.2 Estimating Forest Production

Forest production is concerned with the accumulation of tree volume or biomass per unit of time and area. According to convention, the standard unit of time is 1 year and the unit of area is 1 hectare. Highly aggregated regional production models are used to estimate biomass or volume production for large areas. Forest production models provide more detail, and usually include estimates of dominant height, basal area, and stems per hectare. Size class models estimate the growth of "representative" trees, that is, trees that are representative of a particular size class. Individual tree models estimate the growth and mortality of specific trees in response to competition in the immediate neighborhood. Compatible telescope models utilize data available at a specific level of resolution. Such integrated modeling systems are comparable to a telescope that can be extended to show varying levels of detail. An essential requirement of the telescope approach is compatibility, meaning that models of different resolution produce identical results (Burkhart 1987).

5.2.1 Estimating Element Content in Forest Production Models

Photosynthesis is the process by which green plants transform light energy into chemical energy. Light energy is captured and used to convert water, carbon dioxide, and minerals into oxygen and energy-rich organic compounds. Table 5.3 shows a typical pattern of biomass allocation over age for Scotch pine trees (*Pinus sylvestris*) in northern Germany (Rademacher 2002).

Early gas exchange experiments showed that the gain in weight of a plant grown in a carefully weighed pot resulted from the uptake of carbon, which came entirely from absorbed carbon dioxide and water taken up by plant roots. The balance was oxygen, released back into the atmosphere. Among the first comprehensive studies about photosynthesis and production in forest ecosystems, where light energy is stored as chemical energy, were those conducted by Boysen-Jensen (1918; see also Mitscherlich 1975:248). At that time, few studies contained details about age-related biomass allocation and element content in different tree species and tree compartments. The study of biomass allocation and element distribution in different

Table 5.3 A typical pattern of biomass partitioning over age for Scotch pine (*Pinus sylvestris*) stands in northern Germany, presented by Rademacher (2002)

young forest (25-years)

tree compartment	BM (t/ha)	N (kg/ha)	P (kg/ha)	K (kg/ha)	Ca (kg/ha)	Mg (kg/ha)
Needles	3,2	44,7	4,4	14,8	7,3	2,3
Branches < 7cm diameter	31,1	79,3	6,6	36,2	42,7	12,2
Bark > 7cm	3,3	11,1	1,1	3,5	21,3	1,1
Stemwood >7cm	16,3	14,9	0,5	10,6	30,3	3,3
Root stock	2,2	2,9	0,2	1,6	5,7	0,5
Coarse roots	4,9	7,1	0,9	5,4	3,7	1,3
Fine roots	4,1	24,1	2,7	7,2	3,4	1,3
Total	65,0	184,1	16,2	79,2	114,5	22,0

medium age forest (46-years)

tree compartment	BM (t/ha)	N (kg/ha)	P (kg/ha)	K (kg/ha)	Ca (kg/ha)	Mg (kg/ha)
Needles	6,5	99,2	8,9	35,6	19,9	5,4
Branches < 7cm diameter	27,5	126,2	10,6	46,7	99,2	14,3
Bark > 7cm	8,0	22,8	2,2	9,9	47,9	3,7
Stemwood >7cm	77,5	68,2	2,1	22,7	54,8	14,6
Root stock	9,5	10,1	0,5	3,6	11,4	2,0
Coarse roots	18,9	32,6	3,6	21,8	14,4	4,7
Fine roots	4,1	24,1	2,7	7,2	3,4	1,3
Total	152,0	383,2	30,5	147,4	251,1	46,1

old forest (115-years)

tree compartment	BM (t/ha)	N (kg/ha)	P (kg/ha)	K (kg/ha)	Ca (kg/ha)	Mg (kg/ha)
Needles	2,9	50,0	3,3	12,6	9,2	2,1
Branches < 7cm diameter	15,2	55,9	3,6	15,4	30,1	5,0
Bark > 7cm	7,9	31,7	2,4	10,6	66,2	3,8
Stemwood >7cm	104,3	47,9	2,7	27,3	71,4	16,7
Root stock	11,9	8,2	0,5	3,9	14,3	2,1
Coarse roots	22,2	39,7	2,3	19,0	18,8	5,4
Fine roots	4,1	24,1	2,6	7,2	3,4	1,3
Total	168,4	257,3	17,5	96,1	213,3	36,4

tree compartments began much later. The biomass allocation in needles, branches, bark, stem wood, root stock, coarse roots, and fine roots changes during the life of a forest tree. The proportion of stem wood biomass increases whereas the element content (milligram of the element per gram of biomass) decreases exponentially with increasing stand age (Fig. 5.13).

The rates of change of the biomass and element distributions cannot be measured directly because sampling is usually destructive. The effort required to collect this kind of information is considerable, and therefore such observations are scarce. Wirth et al. published their work on the Norway spruce (*Picea abies*) in 2004, showing the average partitioning of the total biomass of 17 test sites ranging in age from 16 to 172 years. Equations were proposed that allowed the prediction of biomass as a function of tree diameter for different individual tree compartments. Bartelink (1996) published a more general model for dry matter partitioning in trees.

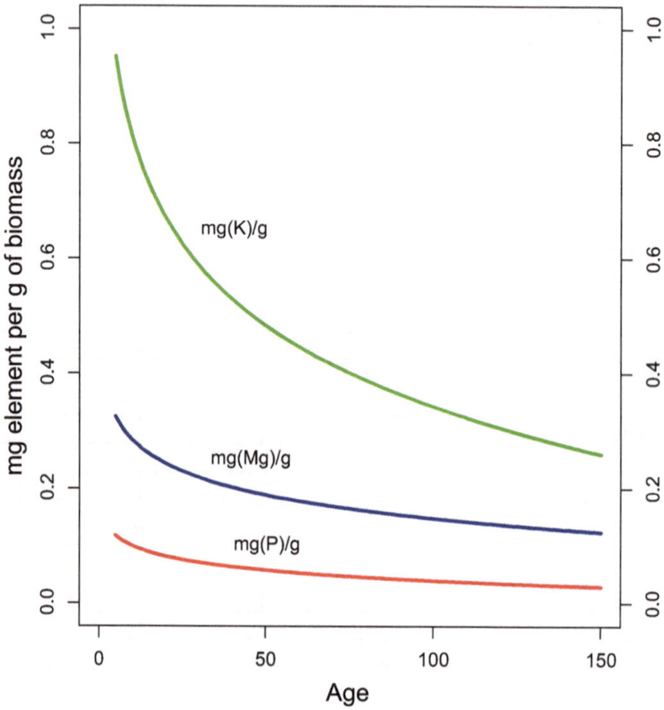

Fig. 5.13 Development of the relative content of three elements (K, Mg, and P) over time. (See Jacobsen et al. 2002)

5.2.2 Production in Planted Forests

Production in planted forests refers to total production, mean increment, and current increment per year and hectare. Total production (TP_t) at age t of a planted forest ecosystem is equal to the biomass or volume at age t (V_t) plus the sum of the removals from planting to t (r_t):

$$TP_t = V_t + r_t \quad (5.17)$$

Example*:* 80 years after planting, the volume of a *Cunninghamia lanceolata* forest (V80) is 550 m^3/ha; 250 m^3/ha had been harvested during thinning operations. TP80 is therefore equal to 800 m^3/ha. Accordingly, the *mean annual increment* at age t (MAI_t) is equal to TP_t divided by t:

$$MAI_t = TP_t / t \quad (5.18)$$

Example*:* MAI_{80} = 800/80 = 10 m^3/year/ha.

The *current annual increment* (CAI_t) is given by the slope of the tangent or the first derivative of a growth curve:

$$CAI_t = V_t' \tag{5.19}$$

Assignment

The following growth function applies to an even-aged *Pinus tabulaeformis* forest:

$$CAI_t = 1800 \cdot e^{-\left\{\frac{100}{t}\right\}}$$

Calculate the following:
the instantaneous growth rate at age 20 (CAI_{20}),
the mean annual increment at age 20 (MAI_{20}),
the age of maximum CAI, and
the age of maximum MAI.

Solution:

Maximum MAI:

$$DGZ_t' = 0 = \frac{1800}{t} \cdot e^{-\left(\frac{100}{t}\right)} \cdot \frac{100}{t^2} - \frac{1800}{t^2} \cdot e^{-\left(\frac{100}{t}\right)} \quad \rightarrow \frac{100}{t^2} = \frac{1}{t} \rightarrow t = 100$$

Maximum CAI:

$$GWL'' = \frac{180000}{t^2} \cdot e^{-\left(\frac{100}{t}\right)} \cdot \frac{100}{t^2} - \frac{360000}{t^3} \cdot e^{-\left(\frac{100}{t}\right)} = 0 \rightarrow \frac{50}{t^4} = \frac{1}{t^3} \rightarrow t = 50$$

The basic relations between the three classic quantities that define production in a planted forest (TP, MAI, and CAI) are shown in Fig. 5.14.

The following R-code was used to generate Fig. 5.14.

```
t=array(0);TP=array(0); MAI=array(0); CAI=array(0)
for(i in 1:150){t[i]=i;TP[i]=1800*exp(-100/i)                # total production
MAI[i]=TP[i]/i;CAI[i]=180000/(i^2)*exp(-100/i)}              # CAI and MAI
d=data.frame(t,TP,MAI,CAI)                                   # combine in data.frame
par(mar=c(5,5,2,5));
with(d,plot(t,TP, xlab="Age",ylab="Biomass (ton)",type="l",col="green",lwd=3,cex.lab=2))
par(new=T);
with(d, plot(t,MAI,type="l", axes=F, xlab=NA, ylab=NA, ylim=c(0,50),col="blue",lwd=3)); axis(side=4)
par(new=T);
with(d,plot(t,CAI,type="l",axes=F,xlab=NA, ylab=NA, ylim=c(0,50),col="red",lwd=3)); axis(side=4)
```

Maximum production per unit of time is defined by the age at which MAI = CAI.

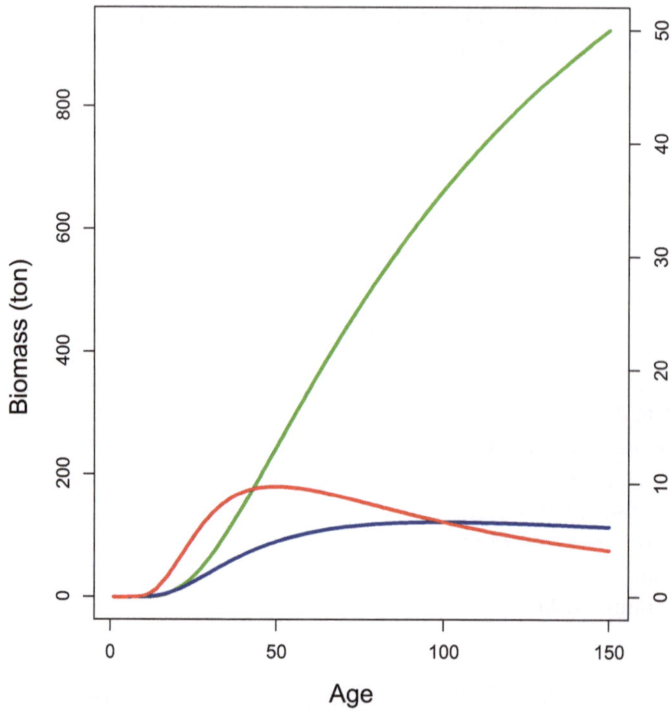

Fig. 5.14 The relation between the three classic quantities TP, MAI, and CAI

5.2.3 Production in Continuous Cover Forests

The main difference between a planted forest and a continuous cover forest is the fact that the latter is multi-aged. There is no common age. In continuous cover forests, production is therefore assessed for a specific *period* between two successive points in time t_1 and t_2. The average annual periodic production ($PP_{t1,t2}$) is then equal to the difference in total growing stock (V_2-V_1) divided by the number of years (t_2 and t_1):

$$PP_{t_1,t_2} = \frac{V_2 - V_1}{t_2 - t_1} = \frac{\Delta V}{\Delta t} \tag{5.20}$$

Total periodic production includes intermittent mortality and removals. Trees that died or were removed during the observation period contributed to production. Ecosystem production therefore often exceeds the difference between the observed growing stock volumes (V_2–V_1). This is an important consideration, and the effect can be illustrated by means of an example presented by Sterba (1991; Fig. 5.15).

The white core in Fig. 5.15 represents the stem volume of the three trees at time t_1. Tree growth between t_1 and t_2 is indicated by the lightly shaded areas. Tree number 2 died at time t_2 while the remaining two trees added more growth until t_3 (dark

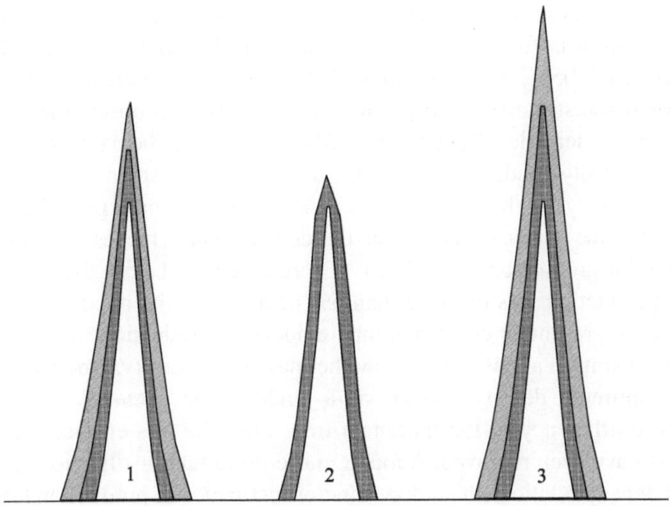

Fig. 5.15 A forest with three trees. The white core represents the stem volume of the three trees at time t_1; and their growth is indicated by the shaded areas (Sterba 1991)

shading). The total periodic production of the forest is equal to the sum of the light- and dark-shaded tree volumes, specifically:

$$\begin{aligned} PP_{t1,t3} &= (\mathrm{V}_{13} - \mathrm{V}_{11}) + (\mathrm{V}_{22} - \mathrm{V}_{21}) + (\mathrm{V}_{33} - \mathrm{V}_{31}) \\ &= (\mathrm{V}_{13} + \mathrm{V}_{33}) - (\mathrm{V}_{11} + \mathrm{V}_{21} + \mathrm{V}_{31}) + \mathrm{V}_{22} \\ &= \mathrm{V}_{\mathrm{E}} - \mathrm{V}_{\mathrm{A}} + \mathrm{N}, \end{aligned}$$

where VE = volume at end of period, VA = volume at beginning of period, N = mortality (or harvest volume).

5.2.3.1 Forest Density and Production

Bastin et al. (2017) determined the extent of forests and tree cover throughout the world's dryland biomes by means of a visual interpretation of Google Earth's very high spatial resolution images. It was only logical to extend their assessment to all the biomes of the earth. Using direct measurements of forest cover, Bastin et al. (2019) estimated that an additional 0.9 billion hectares of potential continuous tree cover exists outside existing forests and agricultural land. The greatest potential for restoration was found in six countries: Russia (151 million hectares); USA (103 million hectares); Canada (78.4 million hectares); Australia (58 million hectares); Brazil (49.7 million hectares); and China (40.2 million hectares). This kind of restoration, according to the authors would capture atmospheric carbon and mitigate climate change by reducing the atmospheric carbon pool by an estimated 25%. By storing the sequestered carbon dioxide in live woody tissues, forests are a major sink in the global carbon cycle, and such estimates, published in a reputable science journal, are sure to cause high expectations.

One of the most important factors affecting forest production and therefore carbon sequestration is forest density (Langsaeter 1941; Assmann 1961; Thomasius 1976; Pretzsch 2005; Corral Rivas et al. 2016). The first theories involving the relation between forest density and production are based on observations from field plots of varying densities. Langsaeter (1941) proposed a theory involving a much-cited forest density-production relation. According to Assmann (1961), the "optimum stand density" is that density which produces maximum periodic volume per hectare, while the "critical density" is the density that achieves 95% of the maximum potential production (Fig. 5.16). A forest density below the critical density implies a production loss of more than 5% of the potential maximum. Thomasius (1976, 1978) presented corresponding evidence: Production in spruce forests reached a maximum at densities below the maximum density. One justification for using the optimum density relation as a guide is the potential improvement of resource use efficiency of the dominant trees after the less efficient subdominant individuals have been removed. Another, more important justification is to develop empirical density relations that allow concrete estimates of production losses due to reduced stocking levels, for example, in Mexico (Torres-Rojo 2014; Corral Rivas et al. 2016).

Pretzsch (2005) evaluated the effect of forest density on volume production in *Picea abies* and *Fagus sylvatica* field plots. He found: (a) maximum production at below-maximum density in plots with small mean DBHs and (b) maximum production at maximum density in plots with large mean diameters. In both species, the relation was affected by site quality and mean diameter. According to Zeide (2004),

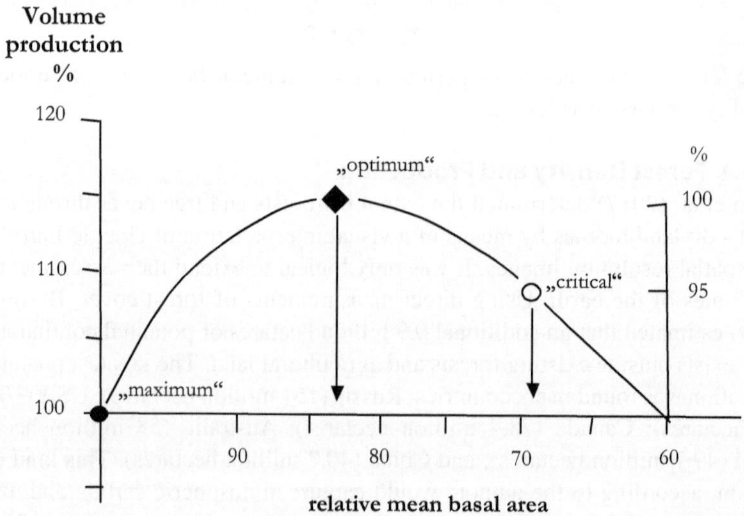

Fig. 5.16 Relation between forest density and production, according to Assmann (1961). Production (expressed as current annual volume increment, expressed in percent on the scale at right) increases as maximum stand density (100%) is reduced. Maximum production is attained at optimum basal area levels (well below-maximum density). Further reduction of density leads to loss of production. The density where 95% of maximum production is achieved is known as the critical basal area. (Graphic taken from Gadow 2004)

optimum density is equal to maximum density, because a complete crown closure intercepts more light than a broken canopy, and therefore the densest stands produce maximum volume growth.

The effect of density on production was studied by Nyland (2016) for the northern hardwoods. A summary of the results, which can be expressed by the following relation was presented by Schölch et al. (2014):

$$\Delta G = -0.8072 + 0.1423 \cdot G - 0.003512 \cdot G^2 \tag{5.21}$$

According to Eq. 5.21, the maximum annual basal area growth is 0.61 m²/ha and it can be attained at the comparatively low density of 20 m²/ha. This kind of growth response is evident in many density-production relations. Another example is an equation for stand basal area increment in cypress pine (*Callitris columellaris*) in southern Queensland, presented by Vanclay (1988):

$$\Delta G = e^{-3.071+1.094 \cdot \ln(G)+(0.007402 \cdot S-0.2258) \cdot G} \tag{5.22}$$

where ΔG is gross basal area increment (m^2ha^{-1} $year^{-1}$), G is stand basal area (m^2ha^{-1}), S is an index of site quality (Vanclay and Henry 1988), and ln is the natural logarithm to base e. For a site quality of S = 17, the optimum density, which produces 0.22 m^2ha^{-1} $year^{-1}$ is only 12 m^2ha^{-1}. Compared to such examples, the productivity of the natural forests of the Mexican Sierra Madre Occidental is very high with an estimated periodic annual increment of almost 20 m^3ha^{-1} $year^{-1}$ at densities of about 40 m^2ha^1 (Corral Rivas et al. 2016).

Most of the studies involving the production–density relation involve only a few state variables, and in some cases, just one variable (forest density). The omission of other variables that can also affect production, such as species composition and site conditions, is one of the reasons why many of the historical studies of the production–density relation have been criticized (Curtis et al. 1997).

Confusion arising from previous work in forest density-production relations is mostly caused by the fact that there can be considerable differences in forest density patterns. The same density can refer to poor as well as good sites. The same basal area can include a few large trees or many small ones (Pretzsch 2005). The biology, that is, the particular species mix, of a forest community is also relevant. The same forest density can include many fast-growing pioneers and a few slow-growing shade-tolerant species, or many sub-canopy species, and a few mature and dominant species. What further complicates the issue is the fact that different measures of forest production were used. Referring to previous work, Allan and Burkhart (2018) defined three theoretical relations between forest production and density, defining them as "increasing," "optimal," and "constant" (Fig. 5.17). Ambiguity can be caused by different interpretations of forest production terminology:

(a) *Net production*, that is, the periodic annual volume (or biomass) increment of all live trees, plus ingrowth, and the volume of trees removed in thinnings.
(b) *Gross production*, that is, net production plus the volume of trees lost to mortality.
(c) *Accretion*, that is, the volume increment of only those trees that survived during the growth period (Nyland 2016:185).

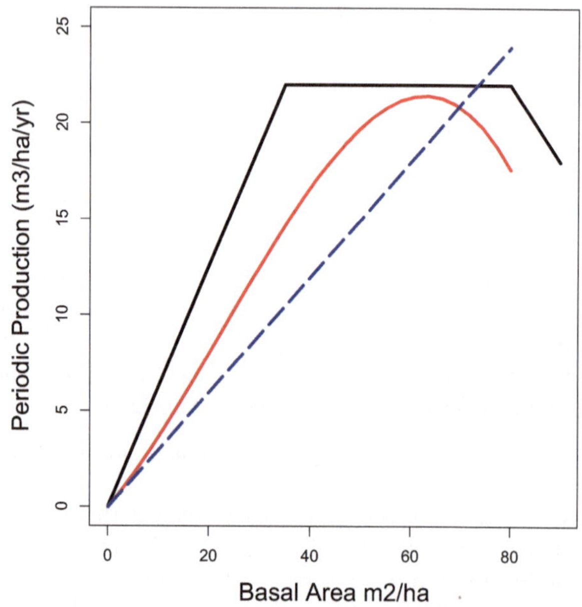

Fig. 5.17 Three growth-density patterns ("increasing," "optimal," and "constant"), as defined by Allan and Burkhart (2018). The red colored optimum relation refers to the empirical results of a study by Corral Rivas et al. (2018)

Differentiation between definitions of forest production is important, as each may result in different density-production patterns (Nyland 2016, p. 370).

The graph in Fig. 5.18 was developed by means of the following code:

```
BA=seq(0,80, length=30);PAI=0.3091*BA+0.006462*BA^2-0.00009467*BA^3      # Corral
# CAI=0.02152*BA^2-0.0003432*BA^3                                        # Jaime
par(mar=c(5,6,2,2) + 0.8)
plot(BA, PAI, xlim=c(0,90), ylim=c(0,25),type="l",xlab="Basal Area m2/ha", ylab="Periodic
Production (m3/ha/yr)", lwd=3,col="red",cex.lab=1.5)
x <- c(0,35); y <- c(0,22); lines(x, y,type="l",lwd=3)   #linear up
x <- c(35,80); y <- c(22,22); lines(x, y,type="l",lwd=3) #linear flat
x <- c(80,90); y <- c(22,18); lines(x, y,type="l",lwd=3) #linear down
x <- c(0,80); y <- c(0,24); lines(x, y,type="l",lty= "longdash",col="blue",lwd=3)
```

The models presented by Allan and Burkhart (2018) for estimating variations in gross production in loblolly pine plantations account for age, site quality, average tree size, and density. To compare the volume growth in thinned and unthinned plots, the volume of the thinned portion was included in the estimate of production. There is much uncertainty regarding the relation at very high densities, but at low densities the relation is often very obvious. An example of the linear relation between the basal area and the volume production was presented by Corral Rivas et al. (2016) for the municipalities San Dimas and Santiago Papasquiaro in the Mexican Sierra Madre Occidental (Fig. 5.18). The practical implications of these relations were illustrated by means of the management database of the Ejido San Diego de Tezains in the municipality of Santiago Papasquiaro. The database includes details of all the compartments on a forest property of 28596.95 ha. In total, 286 compartments, covering 1372.7 ha, have a basal area density less than 10 m^2/ha. The increase in annual production is estimated at 0.3735 m^3/ha if density increases by 1 m^2/ha. Assuming, for example, that all these compartments have a basal area of

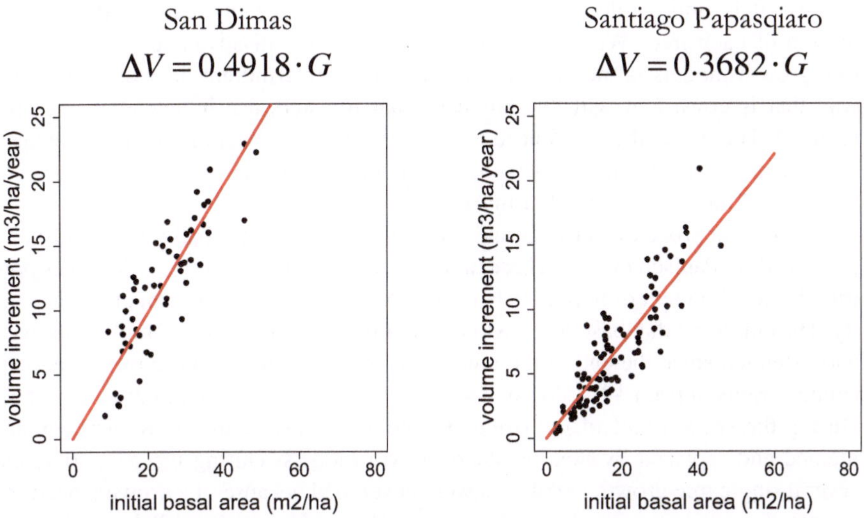

Fig. 5.18 Relations between forest density (basal area per hectare) and production (periodic annual volume increment per hectare) in the municipalities San Dimas and Santiago Papasquiaro

5 m^2/ha^{-1}, total production would be 1372.7 × (5 × 0.3735) = 2563.5 m^3 each year. A density of 30 m^2/ha would produce six times that volume, namely, 1372.7 × (30 × 0.3735) = 15381.1035 m^3 each year. This simple calculation shows that management should take great care in controlling minimum residual densities.

An important practical problem concerns the time required to raise the density of an overexploited ecosystem to a level that will ensure high and sustainable production. Corral Rivas et al. (2016) assumed that an ecosystem had a current unproductive density of 10 m^{2}ha^{-1}, and then estimated that it would take approximately 14 years to reach a productive density of 20 m^{2}ha^{-1}. During that time, there would be no timber harvests and the community would have to rely on other sources of income. An analysis such as this makes a contribution toward identifying and explaining the overriding significance of residual stocking on production.

Forest production on a single site is affected by density, but a specific relation is restricted to particular structural attributes and species compositions (Oliver and Larson 1996; Day 1998; Reyes-Hernandez et al. 2013). Very dense plots are rare, and the relation is often found to be linear and somewhere between low and medium densities. Because very high densities are very scarce, the complete relation covering the entire range of densities is often unknown.

5.2.4 Examples of Forest Production Models

The change from the traditional yield tables to tree growth models has been met with enthusiasm as well as skepticism. Yield tables are easy to use, and they are often found to be sufficiently reliable for practical application. Production estimates for whole forests based on individual tree models have elicited much criticism because it is impossible to capture many of the local factors that influence the growth of each tree (Windhager 1999). An additional disadvantage of individual tree growth models is the difficulty in describing intermittent harvest events in a way that is consistent with the language that foresters use in selecting trees for removal. Therefore, there is a need for models of forest dynamics that can be used to estimate (a) the changes caused by a particular harvest event and (b) the growth response of the community after that event.

The first approach to growth and yield modeling was probably the yield table published by Paulsen (1795). These and subsequent generations of yield tables portray the development of pure forests for a fixed silvicultural treatment and site quality. The empirical database of growth observations has been considerably improved and extended since the end of the nineteenth century and also during the twentieth century (Schwappach 1911; Wiedemann 1931; Schober 1972; Dittmar et al. 1986). During the second half of the twentieth century, many yield tables were supplemented and replaced by flexible stand growth models. During the past 50 years, several single tree growth simulators were developed because of a growing need for information and improved computer technology (Hasenauer 1994; Sterba and Monserud 1997; Pretzsch and Kahn 1998; Pretzsch et al. 2002; Nagel et al. 2002).

In a pioneering study, García (1994) presented his state-space approach for approximating the dynamic development of a forest ecosystem. García showed that the future state of a forest could be estimated by means of an accurate description of its current state. When the current state of a forest is known, estimates of future development do not depend on data collected in the past. In addition, it should be possible to estimate other variables of interest when a few selected state variables are known and specific "output functions" are used (García 2003). When the state variables are selected, the principle of parsimony should be taken into account. A parsimonious model is the simplest one that describes the biological phenomena while remaining consistent with the structure and function of the biological system (García 1994; Vanclay 1995).

Transition functions are used to update the state variables. These functions must have the following properties (García 1994): (a) path-invariance (the result of projecting the state first from t_0 to t_1, and then from t_1 to t_2, must be the same as the result of a one-step projection from t_0 to t_2) and (b) causality (a change in the state can only be influenced by inputs within the relevant time interval). Transition functions generated by the integration of differential equations (or the summation of difference equations when discrete time is used) satisfy these conditions and enable the computation of the future state trajectory. At least three state variables (dominant height, number of trees per hectare, and stand basal area per hectare) are needed to describe a forest ecosystem. Three transition functions are needed to estimate the future forest state. Once these state variables are known, additional output functions that are based on disaggregation can be used to estimate greater detail. Álvarez-González et al. (2012) proposed a basic structure of a comprehensive dynamic forest model. Their forest modeling system follows the *generalized algebraic difference approach* known as "GADA" (Cieszewski and Bailey 2000) and includes disaggregation involving diameter distributions, generalized height–diameter relations, tree level taper functions, and biomass fractions (Fig. 5.19).

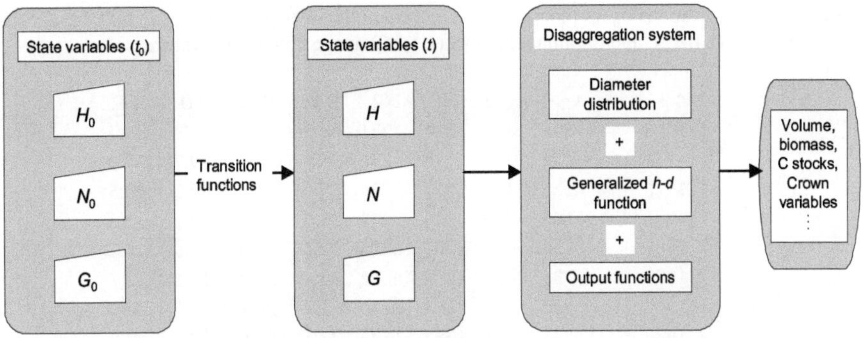

Fig. 5.19 Model of the basic structure of a forest presented by Álvarez-González et al. (2012). H_0, H, N_0, N, G_0, G are dominant height, number of trees per hectare and stand basal area at ages t_0 and t, respectively

Most of the dynamic equations have the general form of $Y = f(t, t_0, Y_0)$, where Y is the value of the function at age t, and Y_0 is the reference variable defined as the function value at age t_0. The structure allows us to expand the base equations according to specific theories about forest growth and production (e.g., asymptote or change of growth rate) thereby allowing more than one parameter to be site-specific and allowing the derivation of more flexible dynamic equations (Cieszewski 2003).

5.2.4.1 A *Fagus sylvatica* Ecosystem

The shade-tolerant European beech (*Fagus sylvatica* L.) is the dominant tree species in central Europe. The species prefers a maritime climate with moderate fluctuations of temperature and precipitation. Its geographic range has been described in numerous research papers. According to Bolte et al. (2007), more than 20 maps of its phytogeographical range have been published.[1] The data that were used to demonstrate the method represent a small subset of those used by Zhang et al. (2010b). Table 5.4 shows details of six plots with a wide range of ages, densities, and dominant heights.

The following function is used to predict dominant height (H_2) for age 2, assuming that age 1 and H_1 are known:

$$H_2 = H_1 \cdot \left(\frac{Age_2}{Age_1} \right)^p \tag{5.24}$$

```
rm(list = ls())                                                # clear the working environment
dat<-read.csv("/Users/kg/Desktop/DataBeech.csv", header=TRUE)    # import data
attach (dat)
H.f<-function(x1,x2,x3,p) { x1*(x2/x3)^p}                        # define the function
Hreg <- nls(H2 ~ H.f(H1,Age2,Age1,p), start=list(p=1),trace=TRUE)# nonlinear regression
summary(Hreg)                                                    # p=0.8672***
```

[1] See, for example, Peters 1997; Schröder1998; Otto 2002.

Table 5.4 Observations of six plots of *Fagus sylvatica*; the trees per hectare (N1,N2), basal areas per hectare (BA1,BA2), and dominant heights (H1,H2) at age 1 and 2 are known

Plot	Age1	Age2	N1	N2	BA1	BA2	H1	H2
1_1966	145	150	271	271	38.3	39.8	37.9	39.7
1_1971	150	155	246	246	36.8	38.6	39.7	39.0
1_1976	155	160	227	226	35.5	36.8	39.0	40.5
1_1981	160	178	216	215	35.9	41.9	40.5	42.3
2_1967	30	35	6007	6126	17.1	24.8	11.6	16.2
2_1972	35	40	4062	4072	19.4	26.2	16.2	18.7
2_1977	40	45	2561	2130	21.7	26.1	18.7	22.0
2_1982	45	58	1338	1295	22.4	34.6	22.0	27.9
3_1969	132	137	445	445	37.7	40.5	36.1	36.8
3_1974	137	142	309	315	30.4	33.2	36.8	37.5
3_1979	142	151	285	278	31.8	35.3	37.5	39.8
3_1988	151	157	200	193	29.7	31.2	39.8	40.4
4_1970	162	168	254	251	41.4	42.8	39.1	40.7
4_1976	168	172	220	221	38.4	40.0	40.7	41.0
4_1981	172	190	193	189	36.8	41.0	41.0	43.0
5_1966	65	70	1138	1114	27.1	30.3	23.4	26.1
5_1971	70	75	891	924	27.5	32.1	26.1	27.5
5_1976	75	80	782	778	29.5	33.0	27.5	28.6
5_1981	80	96	659	658	30.7	41.8	28.6	34.3
6_1966	94	99	297	293	26.6	29.3	27.5	29.1
6_1971	99	104	263	267	27.0	30.2	29.1	31.2
6_1976	104	109	264	263	29.9	32.9	31.2	32.1

Similarly, basal area at age 2 can be predicted by means of the known basal area at age 1 and assuming that the dominant heights H_1 and H_2 are known:

$$BA_2 = BA_1 \cdot \left(\frac{Age_2}{Age_1}\right)^{p1} \cdot \left(\frac{H_2}{H_1}\right)^{p2} \tag{5.25}$$

```
BA.f<-function(x1,x2,x3,x4,x5,p1,p2) { x1*(x2/x3)^p1* (x4/x5)^p2}
BAreg <- nls(BA2 ~ BA.f(BA1,Age2,Age1,H2,H1,p1,p2), start=list(p1=1.5, p2=0.5),trace=TRUE)
summary(BAreg)                                              # p1=1.26**; p2=0.4564***
```

Total volume per hectare (V) can be estimated by using a form factor (F), stand age, crowding (N), basal area, and dominant height:

$$F = e^{(-0.8295-0.00009916 \cdot N+0.001135 \cdot Age)} \tag{5.26}$$

and

$$V = BA \cdot H \cdot F \tag{5.27}$$

The general modeling approach was described in detail by Clutter et al. (1983). The specific application was reported by Zhang et al. (2010b).

5.2.4.2 A *Eucalyptus grandis* Ecosystem

Eucalyptus grandis is a native forest species in the Australian states of Queensland and New South Wales, where tree heights can exceed 70 m and diameters reach up to 183 cm (Carter 1945; Hall et al. 1970). The climate in the Australian native range is humid-subtropical and frost-free and rainfall ranges between 1000 mm and 1800 mm, mostly in the summer. In Australia, the species grows best on well-drained and deep loamy soils of alluvial or volcanic origin (Boland et al. 1984). The species has become one of the most important commercial eucalypts in tropical and subtropical regions. Planting programs have been established in South Africa, Brazil, Angola, Argentina, India, Sri Lanka, Uruguay, Zaire, Zambia, Zimbabwe, and southwest Florida in the USA (Meskimen and Franklin 1978).

This section presents an example of a set of equations for estimating forest production and product yields for a planted *Eucalyptus grandis* ecosystem on the Zululand coastal plain in South Africa, where the soils are of a light sandy texture with low clay and organic matter, and a low cation exchange and water holding capacity. Table 5.5 shows the relevant observations for each of four plots planted at different densities.[2]

Eight measurements are available for each plot. The experiment was observed for almost 39 years, and the interval between two consecutive measurements varies from 2 to 5 years. The site quality is uniform, and dominant plot heights are available for one plot.

Assignment

Estimate maximum density by using the data in Table 5.6 as well as the limiting relation between the quadratic mean diameter and number of trees. Estimate the parameters and present the result in a graph that shows the data and the maximum density for a 0.99 percentile.

Solution: The maximum number of trees per hectare is estimated with the equation (Fig. 5.20).

$$\text{Nmax} = 430591 \cdot \text{Dq}^{-1.812}$$

Sometimes it is useful to estimate tree survival by using the current stems per hectare (N1) and current basal area (BA1, m^2/ha). The following equation is based on the data in Table 5.6: $N2 = N1^{0.96} + BA1^{1.10}$.

Mathematical models for estimating forest production are usually developed by fitting a suitable set of equations to observed data. The model is then used to estimate production for conditions resembling those of the original data set but of

[2] The dataset was kindly provided by *Mr H. Kotzé, South Africa.*

Table 5.5 Four plots of a *Eucalyptus grandis* spacing experiment. N1 and N2 are number of trees per hectare at age 1 and age 2; H1 and H2 are dominant height (m); BA1 and BA2 are basal area (m^2/ha)

Plot	Age1	Age2	N1	N2	H1	H2	BA1	BA2
1	1.9	5.0	6003	4073	6.3	18.9	11.2	39.0
1	5.0	10.0	4073	3474	18.9	28.6	39.0	55.9
1	10.0	14.0	3474	2998	28.6	33.7	55.9	63.6
1	14.0	20.0	2998	2316	33.7	38.5	63.6	70.8
1	20.0	24.3	2316	1437	38.5	44.6	70.8	65.4
1	24.3	27.7	1437	1306	44.6	47.2	65.4	67.7
1	27.7	39.5	1306	731	47.2	52.0	67.7	66.5
3	1.9	5.0	2685	2487	6.4	19.4	7.6	33.8
3	5.0	10.0	2487	2084	19.4	29.5	33.8	48.9
3	10.0	14.0	2084	1903	29.5	34.8	48.9	56.8
3	14.0	20.0	1903	1837	34.8	39.4	56.8	67.9
3	20.0	24.3	1837	1334	39.4	45.6	67.9	68.2
3	24.3	27.7	1334	1301	45.6	47.7	68.2	71.1
3	27.7	39.5	1301	708	47.7	52.7	71.1	67.6
4	1.9	5.0	1474	1441	6.4	19.9	4.5	29.7
4	5.0	10.0	1441	1416	19.9	30.4	29.7	46.1
4	10.0	14.0	1416	1375	30.4	35.6	46.1	53.9
4	14.0	20.0	1375	1268	35.6	40.4	53.9	63.0
4	20.0	24.3	1268	1103	40.4	46.1	63.0	65.4
4	24.3	27.7	1103	1037	46.1	48.3	65.4	66.7
4	27.7	39.5	1037	609	48.3	52.6	66.7	62.5
5	1.9	5.0	988	988	6.4	20.1	2.8	25.5
5	5.0	10.0	988	938	20.1	30.9	25.5	40.7
5	10.0	14.0	938	856	30.9	36.4	40.7	46.1
5	14.0	20.0	856	856	36.4	41.2	46.1	56.3
5	20.0	24.3	856	799	41.2	47.0	56.3	60.2
5	24.3	27.7	799	790	47.0	49.2	60.2	64.0
5	27.7	39.5	790	544	49.2	53.3	64.0	62.4

Table 5.6 Parameter estimates for number of trees, basal area, and dominant height

Variable	Parameter	Variable	Independent fit	R^2
Trees per ha	α_0	Intercept	482.45	0.93
	α_1	N1	0.7100	
	α_2	Ln(BA1)	−72.289	
Basal area	β_0	Intercept	3.2858	0.96
	β_1	N2	0.0001241	
	β_2	Ln(H2/H1)	−0.1355	
	β_3	Age2	0.009837	
Height	γ_0	Intercept	0.06753	0.99
	γ_1	1/Age2–1/Age1	3.30675	

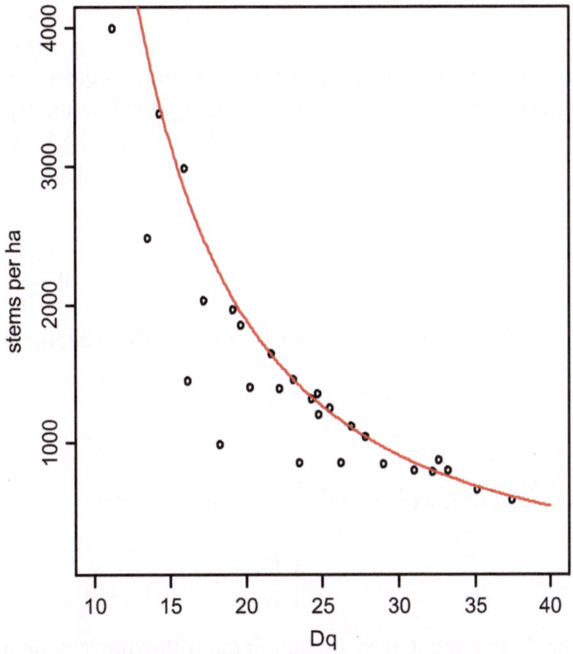

Fig. 5.20 Limiting line of maximum density where Dq is the quadratic mean diameter at breast height

unproven accuracy in other circumstances. As a result, empirical yield models tend to be specific rather than general. The production model should include at least three basic components: an estimate of maximum stand density, a model for estimating dominant height, and a model for estimating stand basal area. Firstly, we read the data of the file LangepanDiff.csv and then use the attach() function to attach the database to the R search path so that objects in the data.frame can be accessed by simply giving their names.

```
# reading the data
dat<-read.csv("/Users/kg/Desktop/LangepanDiff.csv")
attach(dat)
```

We select the following function to estimate the development of the number of trees per hectare:

$$N_2 = \alpha_0 + \alpha_1 \cdot N_1 + \alpha_2 \cdot \ell n(G_1)$$

which is implemented by using the following statements:

```
ln_BA1<-log(BA1)
n.model<- lm(N2~N1+ ln_BA1)
```

Summary details of n.model are provided when the following statement is used:

```
summary(n.model)
```

This is a convenient way to estimate the development of the number of trees per hectare. Strictly speaking, three conditions that apply in linear models are violated: (a) the variance is not constant (no homoscedasticity), which is important for tests of significance; (b) N2 is not normally (but binomially) distributed, and (c) N2 is not independent of N1. These conditions should be kept in mind and discussed when this model is used in a scientific publication. The next step involves the estimation of basal area at age 2:

$$\ln(BA2) = \beta_0 + \beta_1 \cdot N2 + \beta_2 \cdot \ln\left(\frac{H2}{H1}\right) + \beta_3 \cdot Age2 \quad (5.26)$$

It is implemented by using the following statements:

```
ln_BA2<-log(BA2); ln_H21<-log(H2/H1)
ba.model<-lm(ln_BA2~N2+ln_H21+Age2)
summary(ba.model)
```

The following height model can be used:

$$\ell n\left(\frac{H_2}{H_1}\right) = \gamma_0 + \gamma_1 \cdot \left(\frac{1}{Age_2} - \frac{1}{Age_1}\right). \quad (5.27)$$

and implemented by including the following statements:

```
ln_H21<-log(H2/H1); A2A1<-1/Age2-1/Age1
h.model<-lm(ln_H21~A2A1) # 0.99
summary(h.model)
```

The parameter estimates are presented in Table 5.6.

The development of trees per hectare, basal area, and height can be estimated simultaneously with the library *systemfit*:

```
library(systemfit)                         # only N and BA have left/right-variables
ln_BA1<-log(BA1); ln_BA2<-log(BA2); ln_H21<-log(H2/H1)
n<- N2~N1+ln_BA1                          # trees per ha
ba<- ln_BA2~N2+ln_H21+Age2                # basal area per ha
inst<-~N1+ln_BA1+N2+ ln_H21+Age2          # "instrumental" variables, right hand side, order
system<-list(n,ba)                        # list of model equations to be fitted
model<-systemfit(system, data=dat)
model                                     # print results
```

The residual standard error is 0.057896 on 24 degrees of freedom (SSR: 0.080447 MSE: 0.003352 Root MSE: 0.057896). The multiple R-squared is 0.960504 (adjusted R-squared 0.955567). The two remaining components, dominant stand height and basal area, are estimated by using the *nls* function of the ***R*** statistical package. All independent variables are highly significant.

$$\text{Hdom} = 59.0 \cdot \left(1 - e^{-0.073 \cdot Age}\right) \qquad H2 = H1 \cdot \left(\frac{1 - e^{-0.073 \cdot Age2}}{1 - e^{-0.073 \cdot Age1}}\right) \qquad 1.99\text{m} \quad (5.28)$$

The quadratic mean diameter can be derived from basal area and stems perhectare in the following way: $Dq = \sqrt{\frac{40000}{\pi} \cdot \frac{G}{N}}$. Stand volume per hectare canbe obtained from basal area and dominant height: V = 0.4215 · G · Hdom.

5.2.4.3 Estimating Production: No Observations

Regional forecasts of forest production are based on aggregate area information, typically involving a considerable lumping of detail. Such production models are often required for large-scale resource projections. In the absence of long-term observations there are few options for estimating production. Aggregate estimates of mean annual increment (MAI), based on experience, are sometimes available. However, forest production is not linear, because MAI changes with age, and such estimates have to be used in conjunction with a reference age. Murray and Gadow (1993) found that it is convenient to use the age at which the MAI culminates, designated as tmax, and the MAI at this age, MAImax. They proposed a general yield model that can be applied to situations where reasonable estimates of MAImax and tmax are available. A suitable basic model is the Chapman-Richards function:

$$V(t) = A\left(1 - e^{k\{t - to\}}\right)^m, \tag{5.29}$$

where V is the growing stock of a forest of age t, measured in m^3/ha. The parameter A is a scaling factor for the size of V, its units being the same as those of V. The parameter k scales the time axis. By varying k, the model can match the actual growth rate of the variable. The parameter t_0 indicates the lower bound age after which an increase of the biomass of the forest can be observed. The fourth parameter m offers further flexibility concerning the shape of the growth curve.

Specific values for A, k, t_0, and m are obtained if Eq. 5.29 is fitted to empirical yield data. Setting $t_0 = 1$ is quite satisfactory for fast-growing timber plantations, especially when the emphasis is on making good predictions in the vicinity of tmax. The parameter A calibrates the model, and its value is automatically determined whenever calibration data are provided, such as by specifying a value for MAI and a corresponding felling age. The form and orientation of the model with respect to the time axis is determined by the parameters k and m. The value of tmax is thus exactly defined by the values of k and m. For example, tmax is exactly equal to 25 years when k = 0.095 and m = 3.70. Consequently, k and m cannot be used in a model in which tmax is specified as an input variable.

In situations where data are not available, a practical solution can be to use local experience to define tmax. The values of MAImax and tmax (the age at which MAI culminates) are often known, and it would therefore be practical to define a yield function in such a way that the parameters are compatible with the assumed values of MAImax and tmax. A yield function based on MAI requires as inputs the known values of tmax and MAImax, the former specifying the MAI-culmination age and the latter calibrating the MAI curve. It is therefore convenient to introduce the relative MAI defined by the condition MAImax = 1. The relative MAI, now a dimensionless quantity per hectare per year, is given by:

$$relDGZ(t) = \frac{t_{max}}{t} \frac{\left(1 - e^{-k\{t - to\}}\right)^m}{\left(1 - e^{-k\{t_{max} - to\}}\right)^m} \tag{5.30}$$

The condition that MAI should culminate at a given tmax requires the derivative of MAI $(1-e^{-k(t-1)})$ m/t to vanish at tmax. The constant A calibrates the growth model and can therefore be set equal to 1. When we set the first derivative equal to zero, we obtain a relation between tmax and the parameters k and m. The multiplicative factor A can be ignored, and upon setting the derivative equal to zero, we obtain the condition:

$$1 = \left[1 + t_{max} \cdot m \cdot k\right] \cdot e^{-k \cdot \{t-1\}} \tag{5.31}$$

Murray and Gadow (1993) argued that 3 would be an appropriate value for m. As a result, the only unknown parameter that satisfies the condition is k, which needs to be determined iteratively. A first approach would be to use an Excel spreadsheet for implementing the system of models (Table 5.7).

Assignment
Use an Excel table to find the value of k for tmax = 50 years; m = 3 and MAImax = 5.3.

The estimated MAI values are obtained by multiplication with the given MAImax. The following steps can be used to implement the model development in R:

 # Step 1: Define tmax and MAImax and determine k (use R function uniroot).

```
tmax=50; MAImax=12                           # define tmax and MAImax; find the root, and k
f<-function (k) { (1+tmax*3*k)-(1/exp(-k*(tmax-1)))}; uniroot(f, lower=0, upper=0.3)$root     #=0
k                                             # k=0.0394868
```

 # Step 2: Calculate the relative MAI for a range of ages.

```
relMAI = array(0); MAI = array(0); age=array(0); V = array(0)
for (i in 1:100) {age[i]=i;relMAI[i]=(tmax*(1-exp(-k*(i-1)))^3)/(i*(1-exp(-k*(tmax-1)))^3)}
```

 # Step 3: Multiply the relative MAI with MAImax and calculate biomass (V/ha).

```
for (i in 1:100) {age[i]=i; MAI[i]=relMAI[i]*MAImax; V[i]= MAI[i]*i}; d=data.frame(age,MAI,V)
par(mar=c(5,5,2,5));              with(d,plot(d$age,d$V,                   xlab="Age",ylab="Biomass
(ton/ha)",type="l",col="green",lwd=3,cex.lab=2))
par(new=T);      with(d,      plot(d$age,d$MAI,type="l",      axes=F,      xlab=NA,      ylab=NA,
ylim=c(0,30),col="red",lwd=3)); axis(side=4)
```

The estimates of the production model are shown in Fig. 5.21.

Table 5.7 Iterative search for the unknown parameter k implemented in Excel

example					
tmax=	50	m=	3	MAI=	5.3
k	eq	age	rel mai	MAI	Vol
0.020	1.501	5	0.050	0.3	1.3
0.022	1.473	10	0.214	1.1	11.3
0.023	1.442	15	0.408	2.2	32.4
0.025	1.407	20	0.587	3.1	62.2
0.026	1.371	25	0.734	3.9	97.2
0.028	1.332	30	0.844	4.5	134.1
0.029	1.292	35	0.920	4.9	170.7
0.031	1.251	40	0.968	5.1	205.2
0.032	1.209	45	0.993	5.3	236.8
0.034	1.167	50	1.000	5.3	265.0
0.035	1.125	55	0.994	5.3	289.8
0.037	1.083	60	0.979	5.2	311.2
0.038	1.041	65	0.957	5.1	329.6
0.040	1.000	70	0.930	4.9	345.2
0.041	0.959	75	0.902	4.8	358.4
0.043	0.919	80	0.871	4.6	369.4
0.044	0.880	85	0.841	4.5	378.7
0.046	0.842	90	0.810	4.3	386.4
0.047	0.805	95	0.780	4.1	392.8
0.049	0.769	100	0.751	4.0	398.1
0.050	0.733	105	0.723	3.8	402.5
0.052	0.700	110	0.697	3.7	406.1

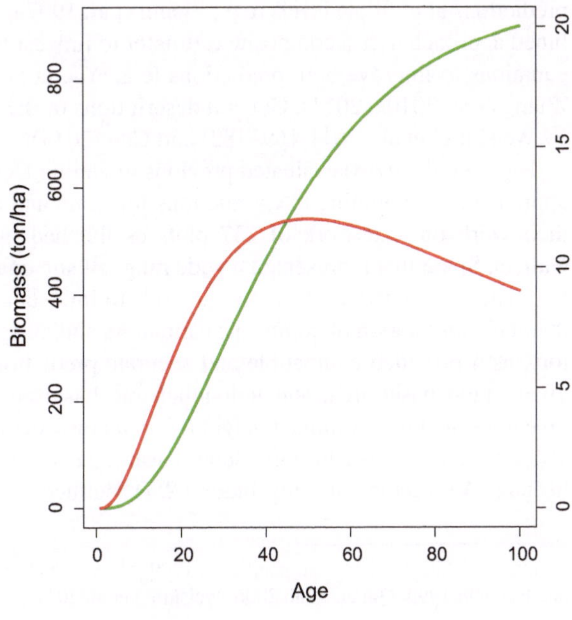

Fig. 5.21 Estimates of total biomass and mean annual increment

This approach for developing a model for estimating forest production has been applied extensively in areas where there is a lack of empirical observations (personal communication with Dr. GuoLiang Liu, Nanjing, China, 2016).

5.3 Compatible Models of Growth and Production

Growth model performance is highly dependent on the accuracy of the current system state. The development of the average height of a planted forest can be estimated with reasonable accuracy provided the current average height is known and the transition functions have been tested by means of long-term observations. Stand parameters like basal area or average height are more readily available than individual tree parameters. Increasing resolution can cause a decrease in prediction accuracy simply because knowledge about the current state is usually imperfect. However, if complete information is available, it makes sense to use that information for high-resolution estimates. Linking models with different levels of resolution is the key to simultaneous estimation of tree growth and survival and total forest production. The pioneering work by Cao, Zhang, and Ávarez González has provided a rich assortment of methods for achieving this important objective. Compatible tree-and-stand-level models represent state-of-the-art forest modeling. For this reason, we describe in detail the methods used by Hevia et al. (2015).

Hevia et al. (2015) used three approaches to link forest production and individual tree models: (i) disaggregation, (ii) constrained parameters, and (iii) a combination of the two methods.[3] In the disaggregation approach, growth and mortality are projected at the stand level and then adjusted to achieve compatibility with individual tree projections. In the constrained approach, a multi-response parameter estimation technique developed by Bates and Granger (1969) is used to optimize tree-level predictions at multiple levels (e.g., Zhang et al. 1997a, 1997b; Cao 2006). The combined approach uses a composite estimator to link estimates of tree- and stand-level equations to improve both predictions (e.g., Yue et al. 2008, Zhang and Lei 2010; Zhang et al. 2010a, 2011). General descriptions of the three approaches were given by Weiskittel et al. (2011:169–182) and Cao (2014).

Hevia et al. (2015) evaluated previous modeling techniques and presented a new approach for combining the predictions for stand and tree level. The authors based their work on a network of 137 plots established in birch-dominated forests in Galicia, Spain that represented a wide range of site qualities, ages, and stand densities. They evaluated 48 different methods to link stand and tree models and found that the combination of composite estimators and disaggregation was most satisfactory, as it provided compatible and accurate predictions of stand density, tree survival, stand basal area, and individual tree basal area. The observations at each inventory included dominant height (*H*, in meters) defined as the mean height of the 100 thickest trees per hectare, stand basal area (*G*, in m²/ha), number of trees per hectare (*N*), relative spacing index (*RSI*), defined as the ratio between the mean

[3] Clutter and Allison 1974; Campbell et al. 1979; Matney et al. 1990; Nepal and Somers 1992; Cao and Baldwin 1999; Qin and Cao 2006; Weiskittel et al. 2011.

distance between trees and dominant height $RSI = 100/\left(\sqrt{N}H\right)$, and site index (*SI*, in m), defined as the dominant height estimated with the model proposed by Diéguez-Aranda et al. (2006) at a reference age of 20 years.

When a forest is thinned, its growth characteristics and dynamics change. Studies involving thinning experiments in even-aged stands showed inconsistent results in terms of the effects of stand density variation on stand basal area growth (Clutter et al. 1983:68). It has been shown in several studies that there is no difference in the unit-area basal area growth in thinned and unthinned stands of the same age, site index, and stand basal area (e.g., Cao et al. 1982; Matney and Sullivan 1982). In contrast, it has been shown in other studies that basal area growth rates in thinned stands exceed those of unthinned stands with the same stand characteristics (e.g., Amateis et al. 1995; Hasenauer et al. 1997; Amateis 2000).

Two approaches have been used to consider the effect of thinning operations on dynamic stand growth models: (i) Development of different transition functions for basal area for different types of stands (unthinned and after the first, second, and subsequent thinning operations) that have the same mathematical structure but that have been provided with parameters by means of different datasets (e.g., Woollons and Hayward 1985; Zarnoch et al. 1991) and (ii) inclusion in the transition function of a thinning response function that expresses the basal area growth of a thinned stand as a product of a reference growth and the thinning response function (Hynynen 1995). The reference growth accounts for the factors affecting stand growth in unthinned stands while the thinning response function predicts the relative growth response after thinning.

The first approach has been commonly used in Spain to take into account the effect of thinning on basal area growth. Dummy variables associated with a different set of parameters for thinned and unthinned stands are employed. Data from thinning trials are required to include this effect on dynamic stand growth models.

5.3.1 Transition Function for Tree Number Reduction

Many functions have been used to model empirical mortality equations. Some of them are mathematical relations between stem number reduction and stand variables; others are biologically based functions derived from differential equations. These biologically based functions have properties that are essential in a mortality model but are not always present in a pure mathematical model (Clutter et al. 1983): consistency, path-invariance, and asymptotic limit of stocking approaching zero as age becomes very large. It is also reasonable to assume that ingrowth is negligible for even-aged stands, so if age at time 2 is greater than age at time 1, the density at time 2 will be less than density at time 1. Specification of a difference equation model that possesses these properties and properly expresses the population relation is often a difficult task, but a complete description of the approach to obtain difference equations with these desirable properties can be found in Álvarez-González et al. (2004).

The state-space approach has been widely used to develop stand growth models. Each transition function is usually fitted separately; however, the three transition functions together define a system of three regression equations and the random errors of the equations can be correlated. In order to improve the efficiency of the estimation, a method that takes into account the cross-equation correlations should be used (e.g., the iterative seemingly unrelated regression – ITSUR). An estimate of the cross-equation error covariance matrix to initiate the iterative procedure can be obtained by means of Ordinary Least Squares.

In the general formulation of the dynamic equations, the error terms e_{ij} are assumed to be independent and identically distributed with zero mean. Nevertheless, because of the longitudinal nature of the data sets used for model fitting, correlation between the residuals within the same plot can be expected. To overcome any possible autocorrelation, the error should be modeled (see, for example, Álvarez-González et al. 2010).

5.3.2 Disaggregation Involving Tree Diameters

The diameter distribution of the stand may be obtained by using a probability density function. The two-parameter Weibull function has frequently been used for describing the diameter distribution of even-age stands because of its flexibility and simplicity. The expression of the two-parameter Weibull density function is as follows:

$$f(x) = \left(\frac{c}{b}\right)\left(\frac{x}{b}\right)^{c-1} e^{-\left(\frac{x}{b}\right)^{c}}, \tag{5.32}$$

where x is the random variable, b the scale parameter, and c the shape parameter that controls skewness.

The Weibull parameters are obtained by means of the moments method. In this approach, the parameters of the Weibull function are recovered from the first two-order moments of the diameter distribution (i.e., the mean and variance, respectively). To recover parameters b and c, the following expressions are used:

$$\text{var} = \frac{\overline{d}^2}{\Gamma^2\left(1+\frac{1}{c}\right)}\left(\Gamma\left(1+\frac{2}{c}\right) - \Gamma^2\left(1+\frac{1}{c}\right)\right) \tag{5.33}$$

$$b = \frac{\overline{d}}{\Gamma\left(1+\frac{1}{c}\right)}, \tag{5.34}$$

where $\overline{d}$ is the arithmetic mean diameter of the observed distribution, var. its variance, and Γ the Gamma function.

Once the mean and the variance of the diameter distribution are known at any specific time, and the fact that Eq. (5.34) only depends on parameter c is taken into account, the latter can be obtained by means of iterative procedures. Parameter b can then be calculated directly from Eq. (5.34). Considering that the disaggregation system is developed for inclusion in a whole stand growth model, only the arithmetic mean diameter needs to be modeled, because the variance can be obtained directly from the arithmetic and the quadratic mean diameters (d_g).

The arithmetic mean diameter can be estimated at any point in time with the following expression, which ensures that predictions of $\bar{d}$ are lower than d_g for the ordinary range of stand conditions:

$$\bar{d} = d_g - e^{\mathbf{X}\beta}, \tag{5.35}$$

where X is a vector of explanatory variables (e.g., dominant height, number of trees per hectare, or age) that characterize the state of the stand at a specific time. X is obtained from the functions of the stand growth model, and β is a vector of parameters to be estimated. This procedure is based on the parameter recovery approach, and has been widely used in diameter distribution modeling (e.g., Cao et al. 1982; Burk and Burkhart 1984; Knoebel et al. 1986; Trincado et al. 2003). Figure 5.22 shows a diagram which explains the modular flow of the disaggregation system.

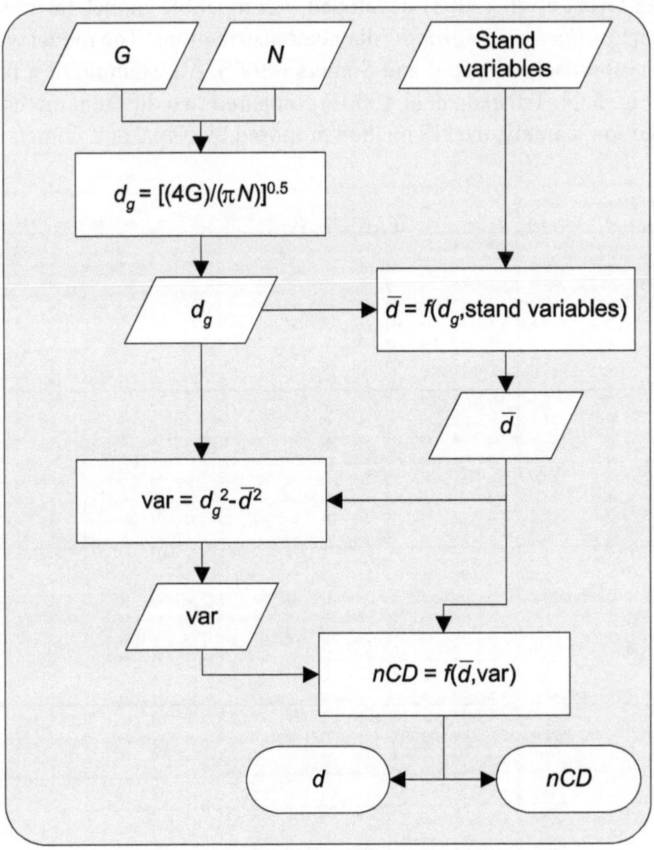

Fig. 5.22 Diagram presenting the flow of the disaggregation system. G = stand basal area; N = number of trees per hectare; d_g = quadratic mean diameter; $\bar{d}$ = arithmetic mean diameter; *var.* = diameter variance; nCD = number of trees per hectare in the diameter class; d = diameter at breast height of the average tree in the diameter class

If the diameter class frequencies are available for a range of study areas, each with a specific density and age, a more detailed prediction of stand structure is possible. The function proposed by Bailey (1980) can be applied to estimate the Weibull parameters of future diameter distributions:

$$DBH_2 = a_2 + b_2 \left(\frac{DBH_1 - a_1}{b_1} \right)^{\frac{c_1}{c_2}} \tag{5.36}$$

where:

DBHt = Diameter at time t,

a_t, b_t, c_t = Weibull parameters at time t.

This function can be used with the assumption that the diameter distribution is a Weibull distribution at both time periods and that each tree remains in the same relative rank with regard to all other individuals. An example of the resulting progression of the diameter class frequencies is shown in Fig. 5.23.

Sánchez Orois et al. (2003) developed a compatible model for predicting the development of *Eucalyptus grandis* diameter distributions. The model was based on the specific approach of Nepal and Somers (1992). An example of a projection is shown in Fig. 5.24. Trincado et al. (2003) compared two different methods of stand table projection, namely, the NS method proposed by Nepal and Somers (1992) and

Age (years)	N per ha	Hdom (m)	G (m2/ha)	2	6	10	14	18	22	26	30	34	38	42	46	50	54	58	62	66	N per ha
2.1	6091		13.9	1926	4114	51	0	0	0	0	0	0	0	0	0	0	0	0	0	0	6091
5.2	3998		40.1	74	1393	1187	849	462	33	0	0	0	0	0	0	0	0	0	0	0	3998
10.2	3379		55.0	52	864	941	542	374	232	271	103	0	0	0	0	0	0	0	0	0	3379
14.2	2984		60.0	0	553	887	463	424	193	180	167	77	39	0	0	0	0	0	0	0	2984
20.2	1846		56.3	0	51	278	443	341	240	114	114	139	51	63	13	0	0	0	0	0	1846
24.6	1451		60.8	0	8	141	298	307	207	133	75	91	116	50	17	8	0	0	0	0	1451
27.8	1352		64.8	0	0	77	264	315	230	94	43	111	77	77	51	9	9	0	0	0	1352
38.9	874		73.0	0	0	9	44	141	159	97	79	79	62	35	53	79	18	9	9	0	874
2.1	2734	6.5	8.5	391	2173	171	0	0	0	0	0	0	0	0	0	0	0	0	0	0	2734
5.2	2471	21.1	35.9	84	264	804	792	468	60	0	0	0	0	0	0	0	0	0	0	0	2471
10.2	2026	32.1	47.6	12	186	398	497	398	249	199	62	25	0	0	0	0	0	0	0	0	2026
14.2	1960	38.2	56.7	0	172	343	478	331	270	159	110	74	25	0	0	0	0	0	0	0	1960
20.2	1639	42.3	60.6	0	84	181	422	265	265	84	121	121	36	36	12	12	0	0	0	0	1639
24.6	1310	48.9	60.8	0	33	58	249	240	265	83	108	41	124	58	17	25	8	0	0	0	1310
27.8	1244	51.1	63.5	0	24	71	213	244	244	87	110	24	71	71	31	39	8	8	0	0	1244
38.9	799	57.1	69.2	0	0	0	41	124	198	66	99	41	16	82	25	49	33	16	8	0	799
2.1	1474		4.5	98	1253	123	0	0	0	0	0	0	0	0	0	0	0	0	0	0	1474
5.2	1441		29.7	0	12	197	431	640	148	12	0	0	0	0	0	0	0	0	0	0	1441
10.2	1400		45.3	0	13	113	325	300	263	225	125	25	13	0	0	0	0	0	0	0	1400
14.2	1391		53.8	0	13	88	238	338	238	188	138	113	13	13	13	0	0	0	0	0	1391
20.2	1202		58.1	0	12	12	168	252	252	120	132	120	84	12	24	0	12	0	0	0	1202
24.6	1112		63.2	0	8	33	108	216	216	158	108	58	100	41	41	17	0	8	0	0	1112
27.8	1037		62.9	0	0	41	82	156	230	148	107	58	82	66	8	49	0	0	8	0	1037
38.9	659		63.8	0	0	0	9	69	130	61	87	61	61	87	35	9	35	17	0	0	659
2.1	988		2.8	37	827	124	0	0	0	0	0	0	0	0	0	0	0	0	0	0	988
5.2	988		26.0	0	0	49	185	469	272	12	0	0	0	0	0	0	0	0	0	0	988
10.2	856		37.1	0	0	0	132	132	219	230	88	44	11	0	0	0	0	0	0	0	856
14.2	856		46.1	0	0	0	99	132	154	230	88	99	44	0	11	0	0	0	0	0	856
20.2	840		55.5	0	0	0	75	129	151	108	151	75	86	22	32	0	11	0	0	0	840
24.6	799		60.3	0	0	0	76	50	143	101	126	84	76	76	25	34	0	8	0	0	799
27.8	790		64.3	0	0	0	66	74	115	99	140	74	66	74	33	25	16	0	8	0	790
38.9	576		63.6	0	0	0	8	25	67	58	125	50	67	33	75	25	0	25	17	0	576

Fig. 5.23 An example of the progression of the diameter class frequencies. (See Sánchez Orois et al. (2003)

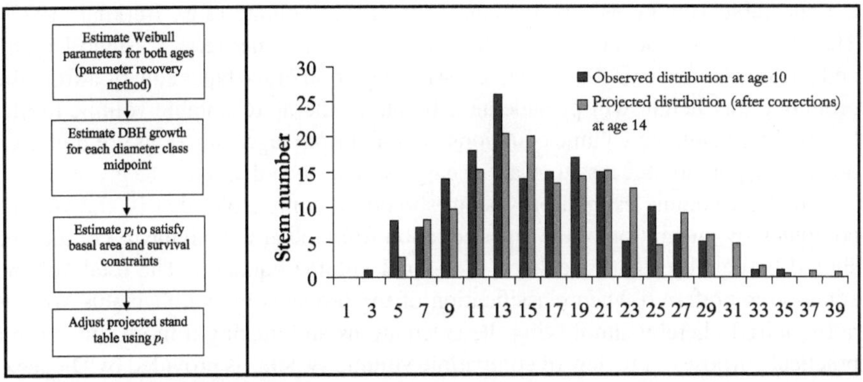

Fig. 5.24 Left: Algorithm proposed by Nepal and Somers (1992). Right: Example of a projection of a Langepan diameter distribution (Sánchez Orois et al. 2003)

the CB method of Cao and Baldwin (1999) and evaluated their suitability for use in *Eucalyptus nitens* plantations in Chile. Figure 5.24 presents one specific result.

Trincado et al. (2003) concluded that both stand table projection approaches are suitable. Moreover, the NS method can also be employed usefully in irregular stands with multimodal diameter distributions.

5.3.3 Disaggregation Involving Height Estimation

Once the diameter distribution is known, it is necessary to estimate the height of the average tree in each diameter class. A local height–diameter (h-d) relation can be used for this purpose; nevertheless, the height–diameter relation varies from stand to stand, and is not even constant over time within the same stand (Curtis 1967). Therefore, a single curve cannot be used to estimate all the possible relations that can be found within a forest. To minimize this level of variance, h-d relations can be improved by taking into account stand variables that introduce the dynamics of each stand into the model (e.g., Curtis 1967; López Sánchez et al. 2003; Sharma and Zhang 2004) and/or by using the mixed-model approach (e.g., Calama and Montero 2004; Castedo-Dorado et al. 2006; Vargas-Larreta et al. 2009).

5.3.4 Output Functions

5.3.4.1 Total and Merchantable Volume Estimation

Once the diameter and height of the average tree in each diameter class have been estimated, the total tree volume can be calculated directly by using a volume equation. If volume prediction to any merchantable limit is required, two methods are commonly applied. One is to develop volume ratio equations that predict merchantable volume as a percentage of total tree volume (e.g., Burkhart 1977). The other is to define an equation describing the stem taper (e.g., Castedo Dorado and Álvarez González 2000; Kozak 2004), integration of the taper equation from the ground to any height will provide an estimate of the merchantable volume to that height. Merchantable volume equations obtained from taper functions are preferred nowadays, perhaps because they allow easy estimation of diameter at a given height.

Ideally, a volume estimation system should be compatible, that is, the volume computed by integration of the taper equation from the ground to the top of the tree should be equal to that calculated by a total volume equation. The total volume equation is preferred when classification of the products by merchantable sizes is not required, thereby simplifying the calculations and making it more suitable for practical purposes. A review of compatible volume systems is provided by Diéguez-Aranda et al. (2006).

The following compatible volume system requires observations of tree diameters at different heights and the total stem volume from destructively sampled trees. Aggregation of total (v) or merchantable (v_i) tree volume times number of trees in each diameter class provides total (V_{total}) or merchantable stand volume (V_{clasif}), respectively. The diagram of the total and merchantable volume estimation system is shown in Fig. 5.25.

5.3.4.2 Biomass Distribution and Carbon Estimations

Tree biomass dry-weight equations are generally fitted in allometric form and are developed for the different tree fractions (stem, bark, branches of diverse sizes, foliage, roots, and total) according to their importance in the nutrient cycle or for their use as bioenergy. Probability density functions are also fitted to characterize the vertical crown fuel distributions (branches of diverse sizes and foliage) as a function of tree height. All models are fitted simultaneously to guarantee additivity among tree biomass components (the estimates of total biomass should be compatible with the sum of each fraction).

The dynamic forest growth models described are combined with these tree biomass equations and C content data to estimate biomass and C stocks by tree fractions. These tools allow assessment of the effects of forest type, age, site quality, silvicultural schedules, and size class on the allocation of tree biomass among components (e.g., Balboa Murias et al. 2006). Aggregation of biomass and C tree multiplied with the number of trees in each diameter class provides total stand biomass and C content, respectively. The fitting of the biomass equations and probability density functions requires data on biomass per fraction at different heights and the C contents from destructively sampled trees.

5.3.4.3 Forest Production and Individual Tree Growth

The simultaneous fitting combined with the variable growth rate (VGR) approach (Cao 2000) was found to be very satisfactory for fitting tree basal area growth and tree survival probabilities. Good results were reported for different species (e.g., Cao 2006; Nord-Larsen 2006; Cao and Strub 2008; Crecente-Campo et al. 2010) although only the last-named authors applied a simultaneous fitting of the components of the model.

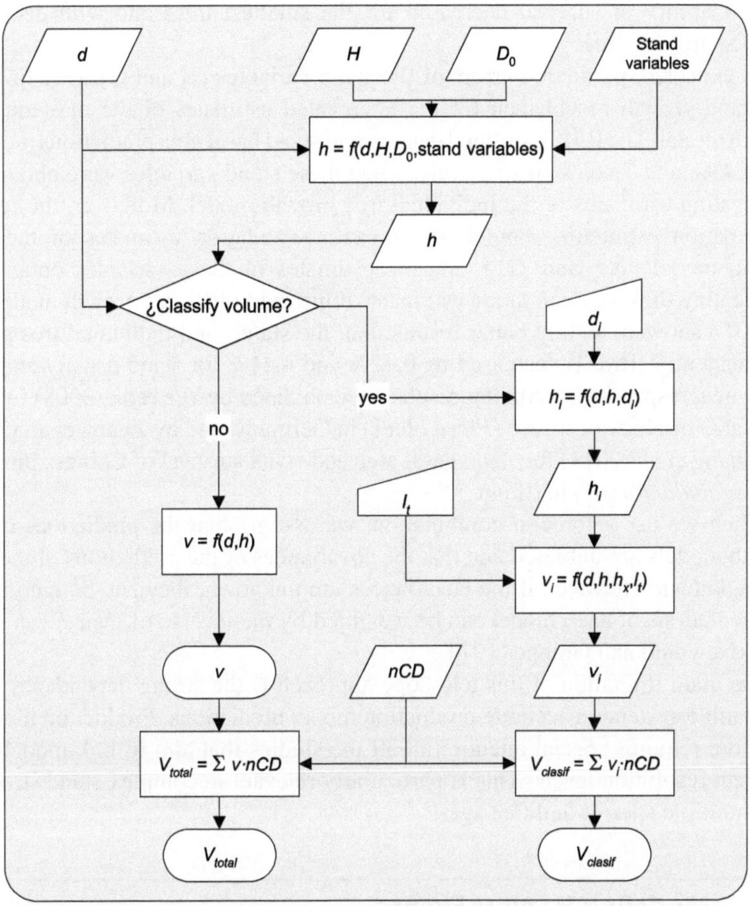

Fig. 5.25 Diagram of the total and merchantable volume estimation system. d = diameter at breast height of the average tree in the diameter class; H = dominant height; D_0 = dominant diameter; h = total tree height; d_i = top diameter at height h_i; h_{st} = stump height; h_i = height above ground level to top diameter d_i; l_t = log length; v = total tree volume above stump level; v_i = merchantable tree volume, the volume above stump level to a specified top diameter d_i; nCD = number of trees per hectare in the diameter class; V_{total} = total stand volume; V_{clasif} = merchantable stand volume

The individual tree growth model includes size variables (DBH and g) and competition variables (G and DBH/dg). Wykoff (1990) and Monserud and Sterba (1996) found that most of the explained variance was due to relative tree size while competition variables were not really important.

Site index did not have a significant effect in the basal area growth model of Hevia et al. (2015). The values and signs of the estimates in the individual tree growth model have biological meaning. The projected tree basal area decreases for lower initial diameters; for higher initial stand densities (represented by stand basal area); and for a lower social status of the tree (DBH/dg). With increasing basal area, the probability of survival decreased for the smallest trees and with decreasing social status of a tree.

As expected, prediction errors of the stand variables (N and G) were lower for the stand growth model than for the aggregated estimates of the individual tree growth model. The RMSE of stand density and stand basal area predictions increased by 13.45% and 73.65%, respectively, when these stand variables were obtained by aggregating estimates of the individual tree growth model. Moreover, the forecast combination estimators, combining the cross-validation estimates of the stand growth model (N_2^S and G_2^S), and the estimates of these variables obtained by aggregating the cross-validation estimates of the individual tree growth model (N_2^T and G_2^T) showed slightly better results than the stand- and individual tree models independently (RMSE decreased by 0.42% and 4.41% for stand density and stand basal area, respectively). Similar results were obtained by Yue et al. (2008) for stand basal area of Norway spruce (*Picea abies*) in Germany, and by Zhang et al. (2010a) and Zhang et al. (2011) for stand basal area and stand survival of Chinese pine trees (*Pinus tabulaeformis*) in China.

When we use a forecast combination, we assume that the predictions of both growth models are unbiased and that the covariances of the predictions of the models are known. However, if the covariances are unknown, they can be ignored and the predictions of each model can be weighted by the inverse of their mean square error (Newbold and Granger 1974).

The main limitation of this telescope approach is the strong dependency of the compatible system on accurate production model predictions. Production modeling therefore requires special attention in all the studies that aim to link models with different resolution levels. This is particularly relevant in complex stand structures with multiple species and tree ages.

5.4 Modeling Harvest Events

In a managed forest, thinnings usually have a far greater effect on forest development than natural growth. Although the prediction of the diameter distribution after thinning is an important problem, surprisingly, little has been published on this particular topic, especially in comparison to the wealth of literature covering tree growth. The problem has been addressed over the years by using tables or regressions involving various stand parameters and variables that describe the weight and type of thinning. For example, a table compiled by Craib (1939) was used by several authors to calculate thinning effects in South African plantations. A case in point is Grut (1970), who derived regression equations for predicting the change of the mean diameter after a thinning in *Pinus radiata* plantations. Similar methods were used by Kennel (1972), Preussner (1974), Alder (1979), Römisch (1983), Gadow (1987), Murray and Gadow (1991), Lemm (1991), and Kassier (1993). The most common approach is to describe a thinning in terms of the change of the diameter distribution. One of the methods for assessing the change is based on the estimation of the effect of a thinning on the parameters of the diameter distribution. An example of this approach is provided by Álvarez González (1997) who estimated the change of the Weibull parameters b and c after a thinning in *Pinus pinaster* forests in Spain:

$$b_{after} = -4.7067 + 1.0205 \times b_{before} + 85.35\frac{N_{removed}}{N_{total}} - 73.617\frac{G_{removed}}{G_{total}} \tag{5.37}$$

$$c_{after} = -1.059 + 1.178 \times c_{before} + 8.170\frac{N_{removed}}{N_{total}} - 5.255\frac{G_{removed}}{G_{total}} \tag{5.38}$$

with b_{before}, b_{after}, c_{before} and c_{after} = Weibull parameter b and c before and after a thinning,

$N_{removed}$, N_{total}, = number of stems per hectare, removed and before thinning,

$G_{removed}$, G_{total}, = basal area per hectare, removed and before thinning.

The removed stem number and basal area ratios are very suitable independent variables for modeling thinning effects.

An alternative approach was developed by Gadow and Hui (1997), who used data from the long-term *Cunninghamia lanceolata* thinning trials DGS and CX executed in China, and following Alder (1979). DGS was initiated in 1988 and low thinnings were carried out immediately after its establishment. Between 15% and 65% of the stems were removed per hectare. The thinning trial CX, comprising 12 plots was initiated in 1975, at age 11 years, by the Forestry Institute in Jiangxi. One low thinning was carried out in 1975 at age 11. The plots were enumerated again 1, 2, 5, 8, 10, and 14 years after the thinning.

According to Alder (1979), the survival rate of a representative tree in a diameter class is determined by its relative cumulative frequency. After a thinning, the survival rate of the remaining trees per hectare can be obtained according to:

$$n_{b_i} = N \int_{F(x_{i-1})}^{F(x_i)} l(F(x))dF \tag{5.39}$$

with nri = stems per hectare remaining after thinning in diameter class *i*,

N = stems per hectare before thinning,

F=F(x) = relative cumulative diameter frequency before thinning,

x = diameter [cm],

l = survival rate of a tree with diameter *x*.

The total survival rate *L*, that is, the relative proportion of the trees remaining after the thinning, is:

$$L = \int_{0}^{F=1} l(F(x))dF \tag{5.40}$$

The thinning weight in terms of stems per hectare is therefore *(1-L)*. Equation 5.42 can be used to calculate the diameter distribution after a thinning, if the distribution before the thinning and the survival rate as a function of the relative cumulative frequency are known. One way to model a thinning is therefore to estimate the survival rate as a function of the cumulative frequency of the diameter distribution.

5.4.1 Separation Parameters

Certain characteristics of a forest community, specifically the mean and the variance of the diameter distribution, are presumed to be known before a thinning. The objective is then to describe the diameter distributions of the removed as well as of the remaining trees when the thinning proportion, defined as the proportion of trees that are removed, is given. Intuitively, the degree of selectivity, that is, the preference for removing trees in certain diameter classes rather than at random, should be relevant when we describe the relation between the distributions. Several authors have found specific regression equations for predicting the changes in the distribution parameters after a thinning, but neither the values of the coefficients nor the particular form of the equation is necessarily applicable in other circumstances. In fact, there is no clear methodology of how such equations should be derived for other data sets.

In nonselective thinnings, such as when every n'th row is removed, the mean diameters before and after the thinning are expected to be the same. For modeling purposes, we can therefore make the assumption that the mean diameter does not change during a nonselective thinning. Clearly, no additional distributional information is required. In a selective thinning, however, the mean diameter does change. The spread in tree sizes permits the selection of trees with small (or large) diameters while the larger (or smaller) trees remain. Consequently, some additional information about the distribution is essential. At least, the variance must be known. As a result, Murray and Gadow (1991) focused their attention on the relations between the means and variances of the before thinning, the removed, and the remaining trees and the consistency that must be maintained between the three diameter distributions.

Let μ and σ^2 be the mean and variance of the diameter distribution, while the subscripts b, t, and r distinguish between the before-thinning distribution, the removed distribution, and the remaining distribution. The proportions of the number of trees removed and remaining are denoted by a_t and a_r, respectively. Obviously, the condition:

$$a_t + a_r = 1 \tag{5.41}$$

must hold and must be equivalent to

$$N_t + N_r = N_b \tag{5.42}$$

with N_t, N_r, and N_b denoting the number of trees. In terms of this notation, the before-thinning quantities μ_b and σ^2_b are assumed to be known, while the thinning proportion a_t is given. The problem is to determine μ_r and μ_t, and σ^2_r and σ^2_t from these known quantities.

A thinning splits the diameter frequency distribution into two parts, namely, the removed and the remaining part, and it is essential that the two parts add up to the total. The associated consistency conditions imply very specific forms for the equations, relating the quantities μ_r and μ_t, and σ^2_r and σ^2_t to their before-thinning counterparts μ_b and σ^2_b and the thinning proportion a_t. The consistency conditions will focus on the general structure of the model. In addition, an empirical element is required for describing the type of thinning encountered in practice. The selectivity of a thinning operation can be described by the difference between the means of the removed and the remaining distributions. This quantity can be made dimensionless by dividing it by a scale parameter, for example, the standard deviation of the original distribution. The *location separation parameter* S_1 is then defined as:

$$S_1 = \frac{\mu_r - \mu_t}{\sigma_b} \tag{5.43}$$

Equation (5.43) can be rewritten as

$$\mu_r - \mu_t = S_1 \cdot \sigma_b \tag{5.44}$$

The mean of the original distribution is the weighted average of the mean of the thinned and the remaining distribution, which is embodied in the condition[4]

$$a_r\mu_r + a_t\mu_t = \mu_b \tag{5.45}$$

Solving for μ_r and μ_t in (5.44) and (5.45) yields

$$\mu_r = \mu_b + a_t\sigma_b S_1 \tag{5.46a}$$

and

$$\mu_t = \mu_b - a_r\sigma_b S_1 \tag{5.46b}$$

Therefore, S_1 is the only empirical quantity required to determine the thinned and the remaining mean diameters in terms of the before-thinning parameters μ_b and σ^2_b and the thinning proportion a_t (actually a_r is also used, but a_r = 1-a_t). Analogous to the preceding analysis, a consistency condition can be derived that relates the variances of the three distributions. Firstly, we define the *scale separation parameter* S_2:

$$S_2 = \frac{\sigma^2_r - \sigma^2_t}{\sigma^2_b} \tag{5.47}$$

[4]An equivalent equation was used by Nagumo et al. (1981) to compute μ_r from the other variables, but this is only feasible when μ_t is already known.

The following consistency conditions based on classic sums of squares conditions in the analysis of variance can be derived[5]:

$$\sigma^2_r = \sigma^2_b \left(1 - a_r a_t S_1^2 + a_t S_2\right) \tag{5.48a}$$

and

$$\sigma^2_t = \sigma^2_b \left(1 - a_r a_t S_1^2 + a_r S_2\right) \tag{5.48b}$$

These equations express the variances of the remaining and of the removed distributions in terms of the before-thinning variance, the thinning proportions, and the separation parameters.

Note that no assumptions have been made regarding the form of the distributions. Should any particular distribution be necessary, the first two moments of the thinned and the remaining distributions are available from Eqs. (5.46) and (5.48). Third moments may be found by defining a skewness separation parameter and imposing specific consistency conditions. The Weibull distribution parameters can then be obtained by using methods based on moments (Burk and Newberry 1984; Shiver 1988).

The practice of silviculture is characterized by numerous semantic expressions that are used to describe a particular thinning operation. These descriptions are not very precise and so a particular expression, such as *high thinning,* can be associated with a range of different S_1 values. It is possible, though, to assign an S_1 parameter value to a semantic expression with different degrees of association. To this effect, we can use an association function, which defines the strength of association of a given S_1 value to one or more categories (Zimmermann 1991). Figure 5.26 shows an

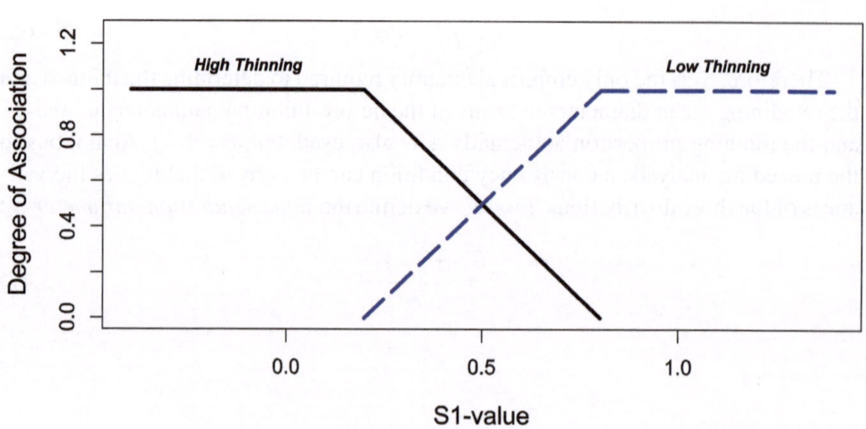

Fig. 5.26 S_1 association function for the categories high thinning and low thinning

[5] See Murray and Gadow (1991) for details of the derivation.

association function for the two categories *high thinning* and *low thinning*. The functions are based on an analysis of numerous long-term thinning experiments in beech forests in northern Germany and Denmark.

A thinning can be classified as definitely *low* when S_1 values are greater than 0.8. If the S_1 value is less than 0.2, we can be sure that it was a high thinning. However, with an S_1 value of 0.6, there is only a one-in-three degree of certainty (0.8–0.6)/0.8–0.2) that the thinning was actually *high,* while the degree of certainty that it was a *low thinning* is equal to (0.6–0.2)/0.8–0.2) = 2/3. This is another example of the problems associated with the use of semantic variables for describing silvicultural operations, and the opportunities for improving the descriptions, not only in the aftermath, but also at the predictive planning stage. The following R-code was used to develop Fig. 5.26:

```
dev.new(width=8, height=4); x<-c(-0.4,0.2); y<-c(1,1); par(mar=c(5,6,1,1) + 0.8)
plot(x,y,xlim=c(-0.4,1.4),  ylim=c(0,1.25),type="l",lwd=3,xlab="S1-value",  ylab="Degree  of
Association", cex.lab=1.2)
x <- c(0.2,0.8); y <- c(1,0); lines(x, y,type="l",lwd=3) #linear flat
x <- c(0.8,1.4); y <- c(1,1); lines(x, y,type="l",lty= "longdash",col="blue",lwd=3)
x <- c(0.2,0.8); y <- c(0,1); lines(x, y,type="l",lty= "longdash",col="blue",lwd=3)
```

text(−0.1,1.1,"High Thinning", font = 4,cex = 0.8); text(1.1,1.1,"Low Thinning", font = 4,cex = 0.8).

Forest development is the result of natural growth and modifications caused by silvicultural operations and other disturbances. One of the most important ecological events is a thinning operation. A model of forest development would be incomplete without a thinning component. Thinnings can be modeled either by imitating the way in which a forester normally selects trees for removal (*imitation*) or by using thinning rules which define the specific manner of selecting the trees to be harvested (*prescription*).

5.4.2 Thinning Rules

The normative character of a thinning decision is a basic precondition for a realistic modeling approach. It assumes that the selection of trees for removal is based on specific rules. This assumption is confirmed by empirical studies showing that a given thinning prescription can obviously be interpreted in a variety of ways, and can therefore produce different results (Zucchini and Gadow 1995). However, consistent decision-making can be observed when a thinning is repeated by the same person, an observation that quite naturally gives rise to the application of rule-based methods (Füldner et al. 1996).

Few areas of science have raised such high expectations and at the same time have met with so much skeptical resistance as the area of artificial intelligence and its most important derivative, the rule-based system. Rules of thumb that are based on practical experience have always played an important part in the practice of forest management, and it is only recently that they have become acceptable as a subject of scientific investigation.

Many phenomena can be described adequately by a relation between objects, commonly known as a mathematical model. For example, the following statement relates the relative spacing (RS), a measure of density in even-aged stands, to the objects N and H:

$$RS = \frac{\sqrt{10\ 000 / N}}{H} \tag{5.49}$$

with N = stems per hectare, and H = dominant tree height [m]. The stems per hectare corresponding to a given stand density is therefore defined by:

$$N = \frac{10\ 000}{(RS \cdot H)^2} \tag{5.50}$$

An equation is a quantitative statement, representing a relation between objects. An implication, commonly known as a rule, is a relation between statements and therefore a higher order concept. Consider the following rule which specifies that if the density exceeds an RS-value of 0.20 and the thinning is not risky, then the number of trees to be removed must be such that RS after the thinning will be equal to 0.3:

$$\left(\begin{array}{c} \frac{\sqrt{10\ 000 / N}}{H} < 0.20 \\ \& \\ \textit{thinning is not risky} \end{array} \right) \Rightarrow \left[\begin{array}{c} \textit{remove} \left\{ N - \frac{10000}{(0.30 \cdot H)^2} \right\} \textit{trees / ha} \\ \textit{in a thinning from below} \end{array} \right] \tag{5.51}$$

with N = stems per hectare and H = dominant tree height [m].

The world of silviculture is characterized by rules. Many rules prescribe some ideal state that should be attained and then maintained, or some ideal sequence of operations that should be followed. Examples of ideal states are balanced age class distributions and stable forest structures prescribed by silvicultural programs.

Ideal states and developments cannot always be attained, and therefore the most effective rules are those that recognize that conditions in the real world are usually suboptimal. Realistic rules that are oriented toward the current condition of the ecosystem have a greater chance of being followed than those that present an idealistic state. A detailed system of thinning rules for even-aged spruce stands in Austria that was based on dominant stand height was proposed by Johann and Pollanschütz (1981). Their rules can be used for planning specific developments for a forest with a given treatment history.

Silvicultural programs using the Z-tree concept are based on successive stem number reductions and aims to remove the Z-tree competitors. The number of trees removed in the vicinity of a Z-tree depends on the size of the Z-tree and the sizes and distances of the competitors. Threshold distances to neighboring trees can be calculated when the tree coordinates are known. The following critical distance was proposed by Johann (1982):

$$G_{iz} = \frac{h_z}{A} \cdot \frac{d_i}{d_z} \tag{5.52}$$

with G_{iz} = critical distance between a Z-tree and a competitor
h_z = height of the Z-tree
d_i, d_z = breast height diameter of competitor i and Z-tree
A = parameter which defines the thinning weight

A potential competitor is removed if its distance to the Z-tree E_{iz} is less than the critical distance G_{iz}. This simple rule can be implemented not only in a computer program but also in the forest. The critical distance to a Z-tree with a given height and diameter depends on the value of the thinning parameter A. The thinning weight decreases with decreasing critical distance, that is, with an increasing value of A.

Literature

Alder D (1979) A distance-independent tree model for exotic conifer plantations in East Africa. For Sci 25:59–71

Allen MG, Burkhart HE (2018) Growth-density relationships in Loblolly Pine plantations. For Sci XX(XX):1–15. https://doi.org/10.1093/forsci/fxy048

Álvarez González JG (1997) Análisis y caracterización de las distribuciones diamétricas de Pinus pinaster Ait. en Galicia. Ph D. Thesis. Universidad Politécnica de Madrid. Madrid, Spain. 270 p

Álvarez-González JG, Castedo-Dorado F, Ruiz-Gonzalez AD, López-Sanchez CA, Gadow Kv (2004) A two-step mortality model for even-aged stands of *Pinus radiata* D. Don in Galicia (Northwestern Spain). Ann For Sci 61:439–448

Álvarez-González JG, Zingg A, Gadow Kv (2010) Estimating growth in beech forests: a study based on long term experiments in Switzerland. Ann For Sci 67:307

Álvarez-González JG, Cañellas I, Alberdi I, Ruiz-González AD (2012) National Forest Inventory and Forest Observational Studies in Spain; Application to Forest Development Modeling. In: Zhao XiuHai (赵秀海), Zhang ChunYu (张春雨) and Klaus v. Gadow (eds) Forest observational studies. Proceedings of an international workshop at Beijing Forestry University (20/21 September 2012) Published by the Key Laboratory for Silviculture and Conservation, Ministry of Education, Beijing Forestry University: 3 pp 5–58

Amateis RL (2000) Modeling response to thinning in loblolly pine plantations. South J Appl For 24:17–22

Amateis RL, Radtke PJ, Burkhart HE (1995) TAUYIELD: A stand-level growth and yield model for thinned and unthinned loblolly pine plantations. Report No. 82. School of Forestry and Wildlife Resources. Virginia Polytechnic Institute and State University. Blacksburg, Virginia. 38 p

Assmann E (1961) Waldertragskunde. Organische Produktion, Struktur, Zuwachs und Ertrag von Waldbeständen. BLV Verlagsgesellschaft, München, Bonn, Wien, p 490

Bailey RL (1980) Individual tree growth derived from diameter distribution models. For Sci 26:626–632

Baker JB, Broadfoot WM (1977) A practical field method for site evaluation for eight important Southern hardwoods. USDA For. Serv., Southern Forest Experiment Station, New Orleans, LA, Gen. Tech. Rep. SO-14

Balboa Murias MA, Rodríguez Soalleiro R, Merino A, Álvarez González JG (2006) Temporal variations and distribution of carbon stocks in aboveground biomass of radiata and maritime pine. For Ecol Manag 237:29–38

Bartelink HH (1996) A model of dry matter partitioning in trees. Tree Physiol 18:91–101

Bastin J-F et al (2017) The extent of forest in dryland biomes. Science 356:635–638

Bastin J-F, Finegold Y, Garcia C, Mollicone D, Rezende M, Routh D, Zohner CM, Crowther TW (2019) The global tree restoration potential. Science 365(6448):76–79. https://doi.org/10.1126/science.aax0848

Bates JM, Granger CWJ (1969) The combination of forecasts. Oper Res Q 20(4):451–468

Boland DJ, Brooker MIH, Chippendale GM, Hall N, Hyland BPM, Johnston RD, Kleinig DA, Turner JD (1984) Forest trees of Australia. Nelson-CSIRO Melbourne, Australia. 687 p

Bolte A, Czajkowski T, Kompa T (2007) The north-eastern distribution range of European beech—a review. Forestry 80(4):413–429. https://doi.org/10.1093/forestry/cpm028

Boysen-Jensen P (1918) Studies on the production of matter in light- and shadow-plants. Dan Bot Tidsskrif 36:219–262

Burk TE, Burkhart HE (1984) Diameter distributions and yields of natural stands of loblolly pine. Publication No. FWS-1-84. School of Forestry and Wildlife Resources. Virginia Polytechnic Institute and State University. Blacksburg, Virginia, 46 p

Burk TE, Newberry JD (1984) A simple algorithm for moment-based recovery of Weibull distribution parameters. For Sci 30:329–332

Burkhart H (1977) Cubic-foot volume of loblolly pine to any merchantable top limit. South J Appl For 1(2):7–9

Burkhart HE (1987) Data collection and modelling approaches for forest growth and yield prediction. In: Predicting forest growth and yield – current issues, future prospects. Inst. of Forest Resources. Univ. of Washington. Contribution Nr 58:3–16

Cajander AK (1909) Über Waldtypen. Acta For Fenn 1(1):1–175. Helsingfors

Cajander AK (1926) The theory of forest types. Acta Forestalia Fennica 29(3):7193

Calama R, Montero G (2004) Interregional non-linear height-diameter model with random coefficients for stone pine in Spain. Can J For Res 34:150–163

Campbell RG, Ferguson IS, Opie JE (1979) Simulating growth and yield of mountain ash stands: a deterministic model. Aust For Res 9:189–202

Cao QV (2000) Prediction of annual diameter growth and survival for individual trees from periodic measurements. For Sci 46:127–131

Cao QV (2006) Predictions of individual-tree and whole-stand attributes for loblolly pine plantations. For Ecol Manage 236:342–347

Cao QV (2014) Linking individual-tree and whole-stand models for forest growth and yield prediction. For Ecosyst 1:18

Cao QV, Baldwin VC Jr (1999) A new algorithm for stand table projections models. For Sci 45:506–511

Cao QV, Strub M (2008) Evaluation of four methods to estimate parameters of an annual tree survival and diameter growth model. For Sci 54(6):617–624

Cao QV, Burkhart HE, Lemin RC (1982) Diameter distributions and yields of thinned loblolly pine plantations. VPI & SU, Sch. For. & Wildl. Resour. Publ FWS-1-82 62p

Carmean WH (1973) Forest Soils Bibliography for the North Central Region. USDA For. Serv. Tech. Rep. NC-5

Carter CE (1945) The distribution of the more important timber trees of the genus Eucalyptus. Atlas 1. Commonwealth Forestry Bureau, Canberra, Australia. 8 p., 34 plates

Castedo Dorado F, Álvarez González JG (2000) Construcción de una tarifa de cubicación con clasificación de productos para Pinus radiata D. Don en Galicia basada en una función de perfil del tronco. Investigación agraria: Sistemas y Recursos Forestales 9(2):253–268

Castedo-Dorado F, Diéguez-Aranda U, Barrio Anta M, Sánchez Rodríguez M, Gadow Kv (2006) A generalized height-diameter model including random components for radiata pine plantations in northeastern Spain. For Ecol Manag 229:202–213

Chen S-J, Hwang C-L (1992) Fuzzy multiple attribute decision making – methods and applications, Lecture notes in economics and mathematical systems, vol 375. Springer, Berlin

Chung D-J (1996) Konkurrenzverhältnisse und Struktur natürlicher *Pinus densiflora – Quercus variabilis* – Mischwälder in Korea. Diss. Fak. f. Forstw. u. Waldökologie, Universität Göttingen

Cieszewski CJ (2003) Comparing fixed-and variable-base-age site equations having single versus multiple asymptotes. For Sci 48(1):7–23

Cieszewski C, Bailey RL (2000) Generalized algebraic difference approach: theory based derivation of dynamic site equations with polymorphism and variable asymptotes. For Sci 46:116–126

Clutter JL, Allison BJ (1974) A growth and yield model for Pinus radiata in New Zealand. In: Fries J (ed) Growth models for tree and stand simulation, Research note, vol 30. Department of Forest Yield Research. Royal College of Forestry, Stockholm, pp 136–160

Clutter JL, Fortson JC, Pienaar LV, Brister GH, Bailey RL (1983) Timber management – a quantitative approach. Wiley, New York, 333 p

Corns IGW, Pluth DJ (1984) Vegetational indicators as independent variables in forest growth prediction in west-central Alberta, Canada. For Ecol Manage 9:13–25

Corral Rivas JJ, Álvarez González JG, Ruíz González AD, Gadow Kv (2004) Compatible height and site index models for five pine species in El Salto, Durango (Mexico). For Ecol Manag 201:145–160

Corral Rivas JS, Torres Rojo JM, Lujan Soto JE, Nava Miranda MG, Aguirre Calderón OA, Gadow K (2016) Density and production in the natural forests of Durango/Mexico. Allgemeine Forst- und Jagdzeitung 187(5/6):93–103

Corral-Rivas JJ, González-Elizondo MS, Lujan-Soto JE, von Gadow K (2018) Effects of density and structure on production in the communal forests of the Mexican Sierra Madre Occidental. Southern Forests: a Journal of Forest Science. https://doi.org/10.2989/20702620.2018.1463152

Craib IJ (1939) Thinning, pruning and management studies on the main exotic conifers grown in South Africa. Govt. Printer, Pretoria

Crecente-Campo F, Soares P, Tomé M, Diéguez-Aranda U (2010) Modelling annual individual-tree growth and mortality of Scots pine with data obtained at irregular measurement intervals and containing missing observations. For Ecol Manage 260:1965–1974

Curtis RO (1967) Height-diameter and height-diameter-age equations for second growth Douglas-fir. For Sci 13:365–375

Curtis RO, Marshall DD, Bell JF (1997) LOGS: a pio- neering example of silvicultural research in coast Douglas-Fir. J For 95(7):19–25

Day K (1998) Stocking standards for uneven aged interior Douglas-fir stands. In: Vyse A, Hollstedt C, Huggard D (eds) Managing the dry Douglas-Fir forests of the Southern Interior. Res. B.C. mim. For., Victoria B.C. Working Paper 34. pp 37–52

Diéguez-Aranda U, Grandas-Arias JA, Álvarez-González JG, Gadow Kv (2006) Site quality curves for birch stands in north-western Spain. Silva Fennica 40(4):631–644

Dittmar O, Knapp E, Lembcke G (1986) DDR-Buchenertragstafel. IFE-Berichte aus Forschung und Entwicklung Heft 4 1986

Ellenberg H (1986) Vegetation Mitteleuropas mit den Alpen in ökologischer Sicht. 6. Auflage. Ulmer, Stuttgart

Ellenberg H, Weber HE, Düll R, Wirth V, Werner W, Paulissen D (1991) Zeigerwerte von Pflanzen Mitteleuropas. Scripta Geobotanica XVIII. Goltze, Göttingen

Flury P (1929) Über den Aufbau des Plenterwaldes. Mitteilungen der schweizerischen Centralanstalt für das Versuchswesen, Band 15:305–357

Füldner K, Sattler S, Zucchini W, Gadow K v (1996) Modellierung personenabhängiger Auswahlwahrscheinlichkeiten bei der Durchforstung. AFJZ 167(8):159–162

Gadow Kv (1987) Untersuchungen zur Konstruktion von Wuchsmodellen für schnellwüchsige Plantagenbaumarten. Forstl. Forschungsberichte, Universität München Nr. 77, 147 S

Gadow K (2004) Waldstruktur und Wachstum. Universitätsdrucke Göttingen

Gadow Kv, Hui GY (1997) Modellierung forstucher Eingriffe in Plantagenwäldern. Forstarchiv 68:59–63

García O (1994) The state-space approach in growth modeling. Canadian Journal of Forest Research-Revue Canadienne de Recherche Forestiere 24:1894–1903

García O (2003) Dimensionality reduction in growth models: an example. For Bio Modell Inf Sci 1:1–15

Grut M (1970) Equations for calculating height increment and site index of Pinaster pine in the Cape province. S Afr For J 102:43–50

Hakes W (1994) On the predictive power of numerical and Braun-Blanquet classification: an example from beechwoods. J Veg Sci 5:153–160

Hall N, Johnston RD, Chippendale GM (1970) Forest trees of Australia. Forest and Timber Bureau, Canberra, 334 p

Hasenauer H (1994) Ein Einzelbaumwachstumssimulator für ungleichaltrige Fichten-, Kiefern- und Buchen-Fichtenmischbestände. Forstl. Schriftenreihe d. Universität f. Bodenkultur, Wien, Bd. 8: 152 S

Hasenauer H, Burkhart HE, Amateis RL (1997) Basal area development in thinned and unthinned loblolly pine plantations. Can J For Res 27:265–271

Henderson GS, Hammer RD, Grigal DF, (1990) Can measurable soil properties be in- tegrated into a framework for characterizing forest productivity? In: Gessel SP, Lacate DS, Weetman GF, Powers RF (eds) Sustained productivity of forest soils. Proc. 7th North Am. Forest soils conf. Vancouver, pp 137–154

Hevia A, Cao QV, Álvarez-GonzÁlez JG, Ruiz-GonzÁlez AD, von Gadow K (2015) Compatibility of whole-stand and individual-tree models using composite estimators and disaggregation. For Ecol Manage 348:46–56

Hodgkins EJ (1960) Forest site classification in the southeast: an evaluation in Southern Forest Soils. In: 8th Am. For. Symp. Proc. 1959. Lousiana State Univ. Press, pp 34-48

Hynynen J (1995) Predicting the growth response to thinning for Scots pine stands using individual-tree growth models. Silva Fennica 29:225–247

Jacobsen C, Rademacher P, Meesenburg H, Meiwes KJ (2002) Gehalte chemischer Elemente in Baumkompartimenten. Literaturstudie und Datensammlung im Auftrag des BMELF: 64p

Johann K (1982) Der „A-Wert“ – ein objektiver Parameter zur Bestimmung der Freistellungsstärke von Zentralbäumen. [The “*A*-thinning index” – an objective parameter for the determination of release intensity of frame trees]. Tagungsbericht der Jahrestagung 1982 der Sektion Ertragskunde im Deutschen Verband Forstlicher Forschungsanstalten in Weibersbrunn, 146–158

Johann K, Pollanschütz J (1981) Standraumregulierung bei Fichte und Buche, Mischwuchsregulierung. Seminarreihe der Forstlichen Bundesversuchsanstalt, Wien, 51 S

Jones JR (1969) Review and comparison of site evaluation methods. USDA For. Serv, Rocky

Kahn M (1994) Modellierung der Höhenentwicklung ausgewählter Baumarten in Abhängigkeit vom Standort. Dissertation, Forstwissenschaftliche Fakultät, Universität München

Kassier HW (1993) Dynamics of diameter and height distributions in commercial timber plantations, PhD dissertation, Faculty of Forestry, University of Stellenbosch, South Africa

Kellomäki S, Väisänen H. Strandman H (1993) FINNFOR: A model for calculating the response of a boreal forest ecosystem to climate change. Univ. of Joensuu. Faculty of Forestry. Research Notes 6

Kennel R (1972) Die Buchendurchforstungsversuche in Bayern von 1870 bis 1970. Forschungsberichte der forstl. Forschungsanstalt München Nr. 7

Kimmins JP (1987) Forest ecology. Macmillan Publishing Company, New York

Kimmins JP (1990) Modeling the sustainability of forest production and yield for a changing and uncertain future. For Chron 66:271–280

Kimmins JP, Comeau PG, Kurz W (1990) Modelling the interactions between moisture and nutrients in the control of forest growth. For Ecol Manage 30:361–379

Kiviste A, Álvarez González JG, Rojo A, Ruiz González AD (2002) Funciones de crecimiento de aplicaci6n en el ámbito forestal. Comunicaciones I.N.I.A. Serie: Recursos Naturales. No. 4

Knoebel BR, Burkhart HE, Beck DE (1986) A growth and yield model for thinned stands of yellow-poplar. Suppl For Sci 32(2)

Kozak A (2004) My last words on taper functions. For Chron 80(4):507–515

Langsæter A (1941) Om tynning I enaldret granog furuskog [Thinning of even-aged spruce and pine forests]. Medd Norsk Skogfor 8:131–216

Lemm R (1991) Ein dynamisches Forstbetriebs-Simulationsmodell. Dissertation, ETH Zürich: 235 p

López Sánchez CA, Gorgoso Varela J, Castedo Dorado F, Rojo Alboreca A, Rodríguez Soalleiro R, Álvarez González JG, Sánchez Rodríguez F (2003) A height-diameter models for Pinus radiata D. Don in Galicia (Northwest Spain). Ann For Sci 60:237–245

MacLean CD, Bolsinger CL (1973) Estimating dunning's site index from plant indicators. USDA For. Serv., Pacific Northwest Research Station, Portland, OR, Res. Note PNW-197, 10 pp

Marques P (1991) Evaluating site quality of even-aged stands in northern Portugal using direct and indirect methods. For Ecol Manage 41:193–204

Matney TG, Sullivan AD (1982) Compatible stand and stock tables for thinned and unthinned loblolly pine stands. For Sci 28:161–171

Matney TG, Belli KL, Farrar RM Jr (1990) Parameter free diameter-distribution recovery. In: Division 4 Proceedings of the IUFRO World Congress Meeting. School of Forestry and Wildlife Resources, Virginia Polytechnic Institute and State University. Blacksburg, Va. pp 95–107

Meskimen G, Franklin EC (1978) Spacing Eucalyptus grandis in southern Florida: a question of merchantable versus total volume. Southern J Appl For 1:3–5

Mischerlich G (1975) Waldwachstum und Umwelt. Volume 3: Boden, Luft und Produktion. J.D. Sauerländer's Verlag, Frankfurt am Main

Mitscherlich G (1961) Untersuchungen in Plenterwäldern des Schwarzwaldes. AFJZ 132:61–73 + 85–95

Monserud RA, Sterba H (1996) A basal area increment model for individual trees growing in even- and uneven-aged forest stands in Austria. For Ecol Manage 80(1–3):57–80

Montero G, Martinez F, Alía R, Candela JA, Ruiz-Peinado R, Cañellas I, Mutke S, Calama R, Gutierrez M, Pavon J, Alonso A, Del Rio M, Bachiller A (2004) El Pino Piñonero (Pinus pinea L.) en Andalucía. Consejería de Medio Ambiente, Junta de Andalucía: 261 p

Moosmayer H-U, Schöpfer W (1972) Beziehungen zwischen Standortsfaktoren und Wuchsleistung der Fichte. AFJZ 143(10):203–215

Murray DM, Gadow K (1991) Prioritizing mountain catchment areas. J Environ Manage 32(4):357–366

Murray DM, Gadow K (1993) A flexible yield model for regional timber forecasting. Southern J Appl For 17(2):112–115

Nagel J, Albert M, Schmidt M (2002) Das waldbauliche Prognose- und Entscheidungsmodell BWINPro 6.1. Forst und Holz 57(15/16):486–493

Nagumo H, Shiraishi N, Tanaka M, (1981) Computer programming for the construction of a sugi local yield-table - program TUSYCS. In: Nishizawa M (ed) Proceedings of forest resource inventory, growth models, management planning and remote sensing, XVII IUFRO World Congress. Kyoto, Japan, pp 103–114

Nepal SK, Somers GL (1992) A generalized approach to stand table projections. For Sci 38:120–133

Newbold P, Granger CWJ (1974) Experience with forecasting univariate time series and the combination of forecasts. J R Stat Soc Ser A 137:131–165

Nord-Larsen T (2006) Modeling individual-tree growth from data with highly irregular measurement intervals. For Sci 52:198–208

Nyland RD (2016) Silviculture: concepts and applications. Waveland Press, Long Grove

Oliver CD, Larson BC (1996) Forest stand dynamics. Wiley, New York

Otto HJ (2002) Antriebskräfte natürlicher Buchenwalddynamik in Europa. Forst u Holz 57(21):649–653

Pardé J, Bouchon J (1990) Dendrométrie. Ecole Nationale du Genie Rural des Eaux et des Forêts, Nancy

Paterson SS (1962) Introduction to Phytochorology of Norden. Medd Stat Skogsforskininginstitut 50:1–145

Paulsen JC (1795) Praktische Anweisung zum Forstwesen.. Detmold

Peters R (1997) Beech forests, Geobotany, vol 24. Kluwer Academic Publishers, Dordrecht, 187 p

Pienaar LV, Turnbull KJ (1973) The Chapman-Richards generalization of von Bertalanffy's growth model for basal area growth and yield in evenaged stands. For Sci 19(1):2–22

Pretzsch H (2005) Stand density and growth of Norway spruce (Picea abies (L.) Karst.) and European beech (Fagus sylvatica L.): evidence from long-term experimental plots. Eur J For Res 124:193–205. https://doi.org/10.1007/s10342-005-0068-4

Pretzsch H, Kahn M (1998) Forschungsvorhaben „ Konzeption und Konstruktion von Wuchs- und Prognosemodellen für Mischbestände in Bayern": Abschlußbericht Projekt W28 Teil 2. Konzeption und Konstruktion des Wuchsmodells SILVA 2.2 – Methodische Grundlagen. Lehrstuhl für Waldwachstumskunde der Ludwig-Maximilians-Universität München, Freising, 279 p

Pretzsch H, Biber P, Ďurský J, Gadow Kv, Hasenauer H, Kändler G, Kenk G, Kublin E, Nagel J, Pukkala T, Skovsgaard JP, Sodtke R, Sterba H (2002) Recommendations for standardized documentation and further development of forest growth simulators. Forstwissenschaftliches Centralblatt 121(3):138–151

Preussner K (1974) Aufstellung und experimentelle Überprüfung mathematischer Modelle für die Entwicklung der Durchmesserverteilung von Fichtenbeständen. Doctoral Dissertation. Univ. Tharandt: 123p

Qin J, Cao QV (2006) Using disaggregation to link individual-tree and whole-stand growth models. Can J For Res 36:953–960

Rademacher P (2002) Ermittlung der Ernährungssituation, der Biomasseproduktion und der Nährelementakkumulation mit Hilfe von Inventurverfahren sowie Quantifizierung der Entzugsgrößen auf Umtriebsebene in forstlich genutzten Beständen. Habilitationsschrift, Univ. Göttingen

Rennie PJ (1963) Methods of assessing site capability. Commonw For Rev 42:306–317

Reyes-Hernandez V, Comeau PG, Bokalo M (2013) Static and dynamic maximum size-density relationships for mixed trembling aspen and white spruce stands in western Canada. For Ecol Manage 289:300–311

Rojo A, Montero G (1996) El Pino Silvestre en la Sierra de Guadarrama. Ministerio de Agriculture, Pesca y Alimentacion, Madrid, 293 S

Römisch K (1983) Ein mathematisches Modell zur Simulation von Wachstum und Durchforstung gleichaltriger Reinbestände. TU Dresden, Sektion Forstwirtschaft, Tharandt, Diss. A

Sánchez Orois S, Kotzé H, Corral Rivas JJ (2003) Ein kompatibles Modell für die Fortschreibung von Durchmesserverteilungen. Vortrag anlässlich der Jahrestagung der Sektion Ertragskunde des Deutschen Verbandes Forstlicher Forschungsanstalten in Torgau am 2. – 4. Juni 2003

Schober R (1972) Die Rotbuche 1971. J. D. Sauerländer's Verlag, Frankfurt am Main, 333 S

Schölch M, Rodenkirchen H, Risse F-J (2014) ANW-Exkursion durch Mischwälder im Nordosten und mittleren Westen der USA – Teil II Waldbauexperimente und Urwälder. Der Dauerwald 50:45–53

Schönau APG (1988) A re-appraisal of a site evaluation study in Black Wattle with special reference to soil factors. In: Gadow K (ed) Research contributions to plantation forestry. Faculty of Forestry, Stellenbosch, pp 83–102

Schönau APG, Aldworth WJK (1991) Site evaluation in black wattle with special reference to soil factors. S Afr For J 156:35–43

Schönhar S (1993) Die Waldbodenvegetation als Standortsweiser. AFJZ 164:173–180

Schröder FG (1998) Lehrbuch der Pflanzengeographie. Quelle & Meyer Verlag, Wiesbaden, 457 p

Schwappach A (1911) Die Rotbuche. Neumann Verlag, Neudamm

Shafer GN (1988) A site growth model for *Pinus eliottii* in the Southern Cape. S Afr For J 146:12–17

Shafer GN (1989) Site indicator species for predicting productivity of pine plantations in the Southern Cape. S Afr For J 148:7–17

Sharma M, Zhang S (2004) Height–diameter models using stand characteristics for Pinus banksiana and Picea mariana. Scand J Fort Res 19: 442–451

Shiver BD (1988) Sample sizes and estimation methods for the Weibull distribution for unthinned slash pine plantation diameter distributions. For Sci 34(3):809–814

Shugart HH (1984) A theory of forest dynamics: the ecological implication of forest successsion models. Springer, New York, p 278 S

Sterba H (1991) Forstliche Ertragslehre. Lecture notes (Heft 4), Vienna, p 159

Sterba H, Monserud RA (1997) Applicability of the forest stand growth simulator Prognaus for the Austrian part of the Bohemian Massif. Ecol Model 98:23–34

Thomasius HO (1976) Anwendungsbeispiel eines Verfahrens zur Berechnung ertragskundlich optimaler Bestandesdichtewerte und Diskussion dieses Verfahrens. Als Manuskr. gedr. Techn. Univ. Dresden, Preprint Nr. 21-16-76

Thomasius HO (1978) Ableitung eines Verfahrens zur Berechnung der ertragskundlich optimalen Bestandesdichte. Beitr. Forstwirtsch 12, 83

Torres-Rojo JM (2014) Exploring volume growth-density of mixed multiaged stands in northern Mexico. Agrociencia 48:447–4613

Trincado G, Quezada R, Gadow Kv (2003) A comparison of two stand table projection methods for young *Eucalyptus nitens* (Maiden) plantations in Chile. For Ecol Manage 6193:1–9

Turner J, Thompson CH, Turvey ND, Hopmans P, Ryan PJ (1990) A soil technical classification for *Pinus radiata* (D. Don) plantations. I. Development. Aust J Soil Res 28:797–811

Turvey ND, Booth TH, Ryan PJ (1990) A soil technical classification for *Pinus radiata* (D. Don) plantations. II. A basis for estimation of crop yield. Aust J Soil Res 28:813–824

Vanclay JK (1988) A stand growth model for cypress pine. In Leech JW, McMurtrie RE, West PW, Spencer RD, Spencer BM (eds) Modelling Trees, Stands and Forests. Proc. Workshop, Aug. 1985, Univ. of Melbourne. School of Forestry, Univ. of Melbourne, Bulletin No. 5, pp 310–332

Vanclay JK (1992) Assessing site productivity in tropical moist forests: a review. For Ecol Manage 54:257–287

Vanclay JK (1995) Growth models for tropical forests: a synthesis of models and methods. For Sci 41:7–42

Vanclay JK, Henry NB (1988) Assessing site productivity of indigenous cypress pine forest in southern Queensland. Commonw For Rev 67:53–64

Vargas-Larreta B, Castedo-Dorado F, Álvarez González JG, Barrio-Anta M, Cruz-Cobos F (2009) A generalized height–diameter model with random coefficients for uneven-aged stands in El Salto, Durango (Mexico). Forestry 82:445–462

Webb LJ (1959) A physiognomic classification of Australian rainforests. J Ecol 47:551–570

Weiskittel AR, Hann DW, Kershaw JA (2011) Forest growth and yield modeling. Wiley, New York, p 415. ISBN-13: 9780470665008

Wiant HV, Ramirez MA, Barnard JE (1975) Predicting oak site index by species composition in West Virginia. J For 73:666–667

Wiedemann E (1931) Die Rotbuche. Mitteilung aus Forstwirtschaft und Forstwissenschaft, 96 S

Windhager M (1999) Evaluierung von vier verschiedenen Waldwachstumssimulatoren. Dissertation, Universität für Bodenkultur Wien, 217 S

Wirth C, Schumacher J, Schulze E-D (2004) Generic biomass functions for Norway spruce in Central Europe – a meta-analysis approach toward prediction and uncertainty estimation. Tree Physiol 24:121–139

Woollons R, Hayward W (1985) Revision of a growth and yield model for radiata pine in New Zealand. For Ecol Manage 11:191–202

Wykoff WR (1990) A basal area increment model for individual conifers in the northern Rocky mountains. For Sci 36(4):1077–1104

Yue C, Kohnle U, Hein S (2008) Combining tree- and stand-level models. A new approach to growth prediction. For Sci 54(5):553–566

Zarnoch SJ, Feduccia DP, Baldwin VC, Dell TR (1991) Growth and yield predictions for thinned and unthinned slash pine plantations on cutover sites in the West Gulf region. Res. Pap. SO–264. New Orleans, LA: U.S. Department of Agriculture, Forest Service, Southern Forest Experiment Station. 32 p

Zeide B (2004) Optimal stand density: A solution. Can J For Res 34:846–854

Zhang X, Lei Y (2010) A linkage among whole-stand model, individual-tree model and diameter-distribution model. J For Sci 56:600–608

Zhang S, Amateis RL, Burkhart HE (1997a) Constraining individual tree diameter increment and survival models for loblolly pine plantations. For Sci 43:414–423
Zhang S, Burkhart HE, Amateis RL (1997b) The influence of thinning on tree height and diameter relationships in loblolly pine plantations. South J Appl For 21:199–205
Zhang X, Lei Y, Cao QV (2010a) Compatibility of stand basal area predictions based on forecast combination. For Sci 56(6):552–557
Zhang CY, Petráš R, Zhao X, Gadow K (2010b) Estimating beech growth and survival. A study based on longterm experiments in Slovakia. Allgemeine Forst- und Jagdzeitung 181(3/4):44–53
Zhang X, Lei Y, Cao QV, Chen X, Liu X (2011) Improving tree survival prediction with forecast combination and disaggregation. Can J For Res 41:1928–1935
Zimmermann HJ (1991) Fuzzy set theory and its applications, 2nd edn. Kluwer Academic Publishers
Zucchini W, Gadow K (1995) Two indices of agreement among foresters selecting trees for thinning. Forest Landsc Res 1:199–206

Designing Forest Ecosystems

6

Abstract

Crutzen (2002) assigned the term "Anthropocene" to the current human-dominated geological epoch. He concluded that mankind would remain a major environmental force in the future. Ecologists and engineers should therefore guide society toward environmentally sustainable management. Humans are able to identify the causes and disastrous consequences of neglect and exploitation of the forest ecosystems of the earth. Nowadays, we have both the knowledge and the tools to prevent further damage, and to sustain the world's ecosystems, for our own benefit and for the survival of life on earth. Humans are capable of managing the environment to provide a range of ecosystem services and benefits to society. For this reason, the study of sustainable ecosystem design has become a central theme of forest ecology.

Pinus roxburghii in Uttarakhand, India

© Springer Nature Switzerland AG 2021

K. von Gadow et al., *Sustaining Forest Ecosystems*, Managing Forest Ecosystems 37,
https://doi.org/10.1007/978-3-030-58714-7_6

6.1 Risks and Preferences

Given the great variety of natural, economic, and operational hazards affecting forest ecosystems, uncertainty and multiple risks are important aspects of forest design. Risk analysis and the specification of preferences provide the essential basis for rational decisions, comparison of alternatives, and ecosystem design.

The scientific analysis of risk and uncertainty in the context of natural hazards had its beginning at the turn of the eighteenth century, when catastrophic events were no longer considered simple acts of destiny, but as cause–effect relations (Weiss 2001). Applications of vulnerability analyses are still surprisingly rare, in spite of the rich assortment of sophisticated modeling tools that are available today. Various empirical and ecophysiological models exist to predict potential hazards for forest ecosystems (e.g., Gadow 2001; Lindner et al. 2010). The objective is to identify potential hazards as well as the most vulnerable areas in a particular forest landscape. Specific hazards can also be expressed as economic risks when estimates of probabilities are available.

A potential hazard becomes a risk when its occurrence has a probability greater than zero and if its occurrence will cause damage to a valuable object. Although extensive experience about various kinds of damage to forest trees exists, models for predicting potential hazards are rarely applied in practice. If technical knowledge is a constraint to the implementation of risk analysis in forest ecosystem design, relevant publications can be consulted. A number of textbooks dealing with hazard and risk analysis have been published, with contributions from the natural, social, and engineering sciences (e.g., Gadow 2001; Vose 2008). Standard books include a range of methods, concepts, and specific case studies with worked examples that can be useful to ecologists, the legal and policy environment, natural resource administrators, and students.

Risk assessment is a formal procedure for quantifying risk with regard to the potential for damage, including all the possible threats within a given hazard domain. Risk is not the same as uncertainty. Uncertainty becomes a risk when the result of the uncertainty is an expected loss. It is important to describe the system in detail and to identify potential hazards. However, it is difficult to assess hazard factor probabilities because forest design, in contrast to industrial safety management, is affected by exogenous hazards that cannot be controlled. A practical approach is to estimate age-dependent cumulative survival rates for a given set of hazard factors. Expressions that are often used in the context of risk analysis and examples of definitions are provided in Table 6.1.

In silviculture, one of the most important hazards is drought. Trees react to drought with loss of vitality and higher vulnerability, for example, higher susceptibility to biotic hazards (Hanson and Weltzin 2000). What complicates the issue is the fact that individual species respond differently to varying thresholds of soil moisture deficit and these responses may or may not result in physiological damage. Valladares and Niinemets (2008) published a drought tolerance classification for trees in the temperate northern hemisphere, while Albert et al. (2015) published a comprehensive drought risk assessment to provide decision support for forest management.

The subject of a risk analysis is any valuable object, such as a forest ecosystem, that can be lost or damaged. The system needs to be defined in terms of time and space and the seriousness of the expected hazard. Within the scope of forest design, the scales of time and space can range from a year to a century and from an individual tree to an entire forest region. Short- to medium-term tree-based risk analysis is essential for the economic evaluation of removal decisions in selective thinning models. Medium-term stand-based risk analysis is concerned with specific abiotic or biotic hazards affecting particular types of stand structure, species, or age class. A typical problem in forest scenario planning is to predict major losses on a compartment or forest level. Long-term risk analysis for forest regions is concerned with the choice of tree species and the application of adaptive silviculture.

Forest ecosystems are facing new climate-induced hazards. Vulnerability analysis is therefore often applied in projections of ecosystem dynamics (Albert et al. 2015; Staupendahl and Möhring 2011). What is the response of a forest community in different management scenarios? What are the effects of a combination of risk factors such as windthrow and drought stress on forest community structure and dynamics? What are the uncertainties arising from climate projections, and how do these uncertainties affect estimates of forest production, tree growth, and mortality?

Table 6.1 Terms that are often used in the context of risk analysis and examples of definitions (see Albert et al. 2015)

Expression	Definition
Hazard	Threatening event within a given time period and area (United Nations Department of Humanitarian Affairs 1992).
Vulnerability	Degree to which a natural resource is susceptible to damage caused by a certain hazard (Adger 2006).
Risk	Probability of expected loss or damage caused by a particular hazard for a given area and reference period; mathematically, the product of hazard and vulnerability (United Nations Department of Humanitarian Affairs 1992). Economically, risk is expected loss (r), that is, the product of an expected damage (s) expressed in terms of monetary loss and its probability (p).
Uncertainty	Situation where the current state of knowledge is incomplete because of (1) lack of information, (2) imprecision in model projections, and (3) unknown preferences in decision-making (Ascough et al. 2008). Uncertainty can be caused by sampling errors, measurement errors, and prediction errors (Kangas and Kangas 2004).

6.1.1 Estimating Abiotic Risks

Schmidt et al. (2010) developed a model for estimating the risk of wind damage after the gale-force storm Lothar that occurred in the winter of 1999. Figure 6.1 shows a map of the Altdorf Forest in the south-west of the state of Baden-Württemberg in Germany. The study was based on observations of 63,452 trees (6639 trees broken or uprooted by the storm and 56,813 unharmed trees). They used a range of explanatory variables assessed within the plots of the first (1987) and subsequent (2002) German National Forest Inventory in Baden-Württemberg.

The particular storm damage inventory was conducted shortly after the storm damage on December 26, 1999. The field crews classified all downed trees as "storm-damaged." Trees with lesser or no damage were classified as "unharmed." A statistical model was developed to estimate the risk of storm damage for individual trees. The effects of tree-specific variables, topography, and site conditions were included in the model. The storm damage scenario assumes a gale-force wind of the strength of Lothar. The risk estimates are based on the model developed by Schmidt et al. (2010). Tree height proved to be the most important variable affecting the level of damage. Stands dominated by *Picea abies* had the highest potential for damage, followed by those dominated by *Abies alba*, *Pseudotsuga menziesii*, and *Pinus sylvestris*. Estimated damage probabilities were generally lower for deciduous trees

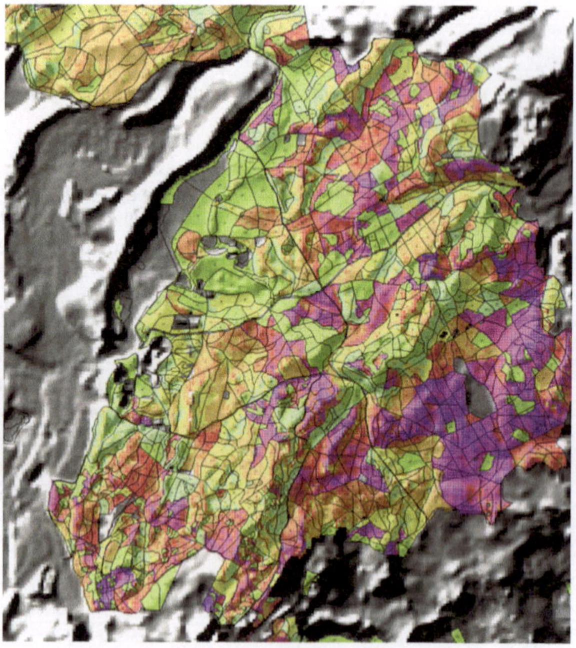

Fig. 6.1 Map of a storm damage scenario in the south-western region of Baden-Württemberg, Germany, assuming a gale-force wind identical to Lothar. Red indicates high, yellow moderate, and green low risk of storm damage

than for conifers. West- to south-exposed locations bore a considerably higher risk, and waterlogged soils showed a higher damage potential compared to slightly waterlogged or dry sites.

6.1.2 Estimating Biotic Risks

The spruce bark beetle *Ips typographus* is an essential component of *Picea abies* ecosystems and one of the most significant biotic risk agents threatening *Picea abies* communities in Europe. Available climate-change scenarios indicate that higher temperatures and longer growing seasons can be predicted. These scenarios are likely to produce more bark beetle generations per year, thus causing more frequent and more serious outbreaks of *Ips typographus* in the future. Periods of drought may occur more frequently. Droughts may also last for longer periods, resulting in the physiological weakening of the host species *Picea abies*. Overbeck and Schmidt (2012) developed the following binary linear regression model, including random effects, for estimating the bark beetle infestation risk for spruce stands in the Harz Mountains in Germany:

$$g\left(\pi_{ij}\right)=f_1\left(X_{1ij}\right)+f_2\left(X_{2ij}\right)+\ldots.+f_n\left(X_{nij}\right)+Z_j b_j \qquad (6.1)$$

where $\pi_{ij} = E(y_{ij})$ with y_{ij} ~ binomial $(1,\pi_{ij})$ with π_{ij}: Expectation value $E(y_{ij})$ is the probability of beetle infestation of stand i in forest subdistrict j within the next 10 years; g is the link function (logistic-link); f_1, f_2,. .. f_n are one-dimensional smoothing functions (penalized thin plate regression splines); x_{1ij}, x_{2ij},...,x_{nij} are 1...n predictor variables of stand i of forest subdistrict j; Z_j is a row in a model matrix including dummy variables for coding random effects for forest subdistrict j, where j = 1, …, b_j; and A is a vector of random effects with an unknown positive definite covariance matrix ω with b_j ~ N(0,ω), assuming the random effects are independently and identically normally distributed.

The model was based on observations that were collected during 10 years of detailed recording and covered 12,085 individual records of bark beetle damage in 4352 different stands. It was assumed that the damage was caused exclusively by *Ips typographus*. They found that infestation risk was significantly affected by two stand parameters (age and proportion of *Picea abies* within the species mix) and several site parameters (available soil-water capacity, thermal sum, and the topex-to-distance index). The response variable "degree of infestation" is binary, with the categories "not infested" and "infested." The model permitted a spatially explicit estimation of the risk intensity of bark beetle infestation, as shown in the map of the Harz mountain region in Germany (Fig. 6.2).

Only a sufficiently large dataset ensures that the various combinations of infestation rate and predictors are more or less equally affected by confounding variables to obtain unbiased estimates.

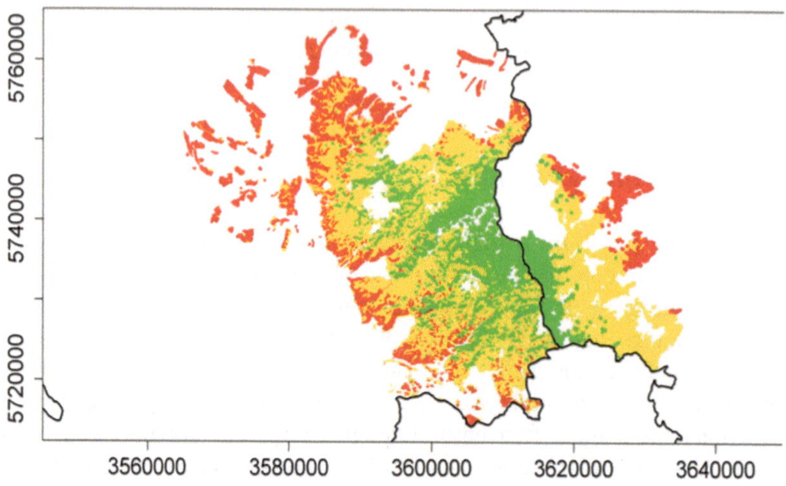

Fig. 6.2 Estimated risk intensity of bark beetle infestation for the Harz region in northern Germany. The y-axis shows northing, the x-axis easting

> **Assignment**
> Study the paper by Overbeck and Schmidt (2012), including the cited publications, and develop a similar model to estimate a biotic risk in a forest area of your choice.

6.1.3 Establishing a Preference Order

Forest ecosystems represent a scarce resource. The use of this resource is governed by diverse interests and affected by the expectations of a variety of stakeholders, that is, individuals who either care about or have a vested interest in the design. Preferences are expressed by goals, strategies, and objectives. A goal is a broad primary outcome, for example, greater biodiversity. A strategy represents a general approach that can be followed to achieve the goal (for example, "replace clear-felling silviculture with selective harvesting"). An objective is a measurable step that is taken to achieve a strategy. Objectives are operational if the formulation is unambiguous. They are vital when alternative plans of action are evaluated.

The "ideal" tree in a tree-breeding program is the straightest, the fastest growing, and the most resistant to disease. Such a tree does not exist. There is no tree that simultaneously performs best in all these categories. The best tree represents a compromise. It performs better than the other candidate trees when all the criteria are considered simultaneously, although it may be outperformed in any one criterion. When we want to determine the conservation value of several forests,

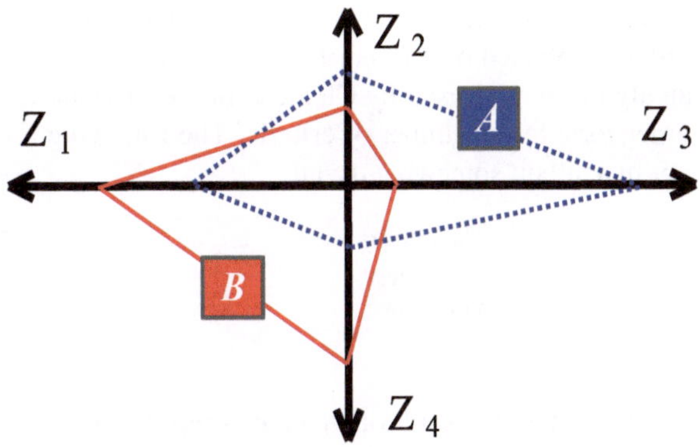

Fig. 6.3 Typical relationship between four objectives (Z1, Z2, Z3, Z4) and two alternative action plans (A and B)

we can assume that three criteria will be used: naturalness, diversity, and scarceness. These criteria need to be weighted to assess the conservation values of a specific forest. Objectives have no intrinsic value. Their value derives from the importance that people attach to them, relative to other objectives. Personal preferences are often considered intangible or irrelevant. Yet, many decisions seem to be made intuitively and based largely on personal preferences. It seems logical, therefore, to pay attention to the measurement of subjective and judgmental information. Figure 6.3 shows a typical relationship between four objectives (Z1, Z2, Z3, Z4) and two alternative plans of action (A and B). Alternative A meets Z3 very well, while B performs better in terms of Z1 and Z4. Obviously, the relative weights assigned to the four criteria determine the choice of the preferred alternative.

One of the most popular methods for assessing relative preferences among a set of objectives is the method of paired comparisons. The famous *law of comparative judgment* is based on the assumption that relative preferences among different stimuli can be established if the stimuli are lined up in pairs and then evaluated pairwise (Thurstone 1927). Many refinements have since been added to Thurstone's theory, for example, those of Guilford (1954) and Torgerson (1958). The method has been used to estimate the scenic beauty of different forestry landscapes (Hull et al. 1984). With the method of paired comparisons, n objectives are lined up in n(n-1)/2 pairs. This number is sometimes doubled if there are fewer than five criteria. Each pair is then evaluated to establish a measure of preference for one pair item over the other. The method of presenting the pairs in some optimum order has been the subject of extensive research work. Most noteworthy is the method proposed by Ross (1934). Ross defined an optimum order as an arrangement of pairs that are characterized by two conditions:

(a) *optimum balance*: any given pair-member, appearing m-1 times, is presented equally often as a first and second (left and right) member;
(b) *maximum spacing*: pairs containing the same item should be separated by the greatest possible number of pairs which do not contain that specific item.

The two conditions can only be met if n is an odd number. The tables of optimum orders published by Ross (1934) can be used for lining up the pairs. Alternatively, an algorithm proposed by Cloete et al. (1988) can be used to produce an optimum order for a large number of criteria. The following arrangement of three criteria is optimal, albeit somewhat trivial:

Naturalness	—	Diversity
Scarceness	—	Naturalness
Diversity	—	Scarceness

The numerical result of a paired comparison is defined as follows:

$$V_{ij} = \text{Value of } Z_i \text{ relative to } Z_j$$

V_{ij} is a measure of preference in a pair of criteria (Z_i, Z_j). An ordinal scale is used to establish a rank order. V_{ij} can therefore assume two values only

$$V_{ij} = \begin{cases} 1, Z_i \succ Z_j \\ 0, \text{ otherwise} \end{cases} \tag{6.2}$$

The ordinal scale does not allow equivalence (see Marazzi 1985). Absolute transitivity is an additional constraint of the ordinal scale that severely limits its use. A ratio scale is therefore often preferred in practical applications: V_{ij}, the numerical result of the paired comparison, is equal to the ratio of the Z:

$$V_{ij} = \frac{v(Z_i)}{v(Z_j)} \tag{6.3}$$

A value of 2 would imply, for example, that Z_i is twice as important as Z_j. Saaty (1980) used the ratio scale in his famous *Analytical Hierarchy Process* (AHP) concept. The general form of the pair matrix is shown in Fig. 6.4.

This concept implies that V_{ij} is reciprocal to V_{ji}, with the practical consequence that only *n(n-1)/2* pairs need to be evaluated. Saaty's AHP has been used in a number of studies with reference to natural resources (Schmoldt et al. 2001). Table 6.2 shows a hypothetical application where the relative preference of naturalness is twice as high as diversity, while the ratio naturalness–scarceness is assigned the value 8. Diversity is six times as important as scarceness.

The result of the paired comparison is presented in a pair matrix (Table 6.3). By normalizing the row sums, we obtain Saaty's eigenvector and the eigenvalues. To estimate the eigenvector for Saaty's λ_{max}, the row sums in Table 6.3 must be divided

Fig. 6.4 Pair matrix based on a ratio scale (see Saaty 1980)

$$\begin{bmatrix} 1 & v_{12} & \cdot & \cdot & v_{1n} \\ v_{21} & 1 & \cdot & \cdot & v_{2n} \\ \cdot & \cdot & \cdot & \cdot & \cdot \\ \cdot & \cdot & \cdot & \cdot & \cdot \\ v_{n1} & v_{n2} & \cdot & \cdot & 1 \end{bmatrix}$$

Table 6.2 Paired comparison involving the criteria naturalness, diversity, and scarceness and using the 9-point scale recommended by Saaty (1980)

a_i	8	6	4	2	1	1/2	1/4	1/6	1/8	a_j
Naturalness				x						Diversity
Diversity									x	Naturalness
Scarceness		x								Scarceness

Table 6.3 Pair matrix and normalized pair matrix for the assessment of a total biotope value

	Pair matrix			Normalized pair matrix			
Criterion	Naturalness	Diversity	Scarceness	Naturalness	Diversity	Scarceness	Sum
Naturalness	1	2	8	0.615	0.631	0.533	1.779
Diversity	1/2	1	6	0.308	0.316	0.400	1.024
Scarceness	1/8	1/6	1	0.077	0.053	0.067	0.197
Sum	1.625	3.167	15.00	1.000	1.000	1.000	3.000

by the number of elements in each row: 1.779/3 = 0.59 for naturalness; 0.34 for diversity; and 0.07 for scarceness.

The biotope values of different ecosystems can be determined according to the criteria naturalness, diversity, and scarceness. However, the specific aggregation is only valid for the forests included in the paired comparison. The preference ranking will change with each additional forest that is evaluated.

6.1.4 Constrained Optimization

Methods of mathematical programming as well as numerous applications of combinatorial optimization have been used in forest design. Complex decision-making and planning problems can be solved with applications of *linear programming* (LP). A linear-programming model includes a linear objective function (OF) that is maximized or minimized, subject to a set of linear constraints:

$$\text{linear objective function: max } (\text{min}) \quad c^T x$$

$$\text{linear constraints: } \mathbf{Ax} \begin{Bmatrix} = \\ \geq \\ \leq \end{Bmatrix} \mathbf{b}$$

where **A** is a m x n matrix of technological coefficients; **x** is an n x 1 vector of structure variables (the "unknown quantities"); **c** is an n x 1 vector of objective function coefficients; and **b** is a m x 1 vector of constants (also known as the "right-hand side").

6.1.4.1 A Pulp-mill with Boiler Constraint

A pulp-mill has secured contracts to ensure a sustained supply of roundwood from forests in the vicinity. Altogether, 300,000 tons of hardwood and 1.2 million tons of softwood are available each year. The mill produces bleached pulp from hardwoods (BH), bleached pulp from softwoods (BS), and unbleached kraft liner (KR). The production system is shown in Fig. 6.5.

The unknown tons of bleached pulp from softwood, hardwood, and unbleached pulp (known as kraft) are designated as BH, BS, and KR. The roundwood required and the profit for each ton of BH, BS, and KR are listed in Table 6.4.

The variables are the unknown quantities of BH, BS, and KR. The objective is maximum profit, subject to the roundwood supply limitations, which requires two linear constraint equations. An additional constraint refers to the recovery boiler where unwanted residues are burnt and certain chemicals are recovered for reuse. The boiler operates almost continuously for 350 days per year and 24 h per day, a total of 8400 h per year. For 1000 tons of bleached pulp (BH or BS), 23.9 boiler hours are required. For 1000 tons of unbleached kraft liner (KR), 20.2 boiler hours are required. One ton of bleached pulp therefore "consumes" 23.9/1000 = 0.0239 boiler hours, while one ton of kraft uses 20.2/1000 = 0.0202 boiler hours. This requires a boiler time constraint. The model formulation involves five steps:

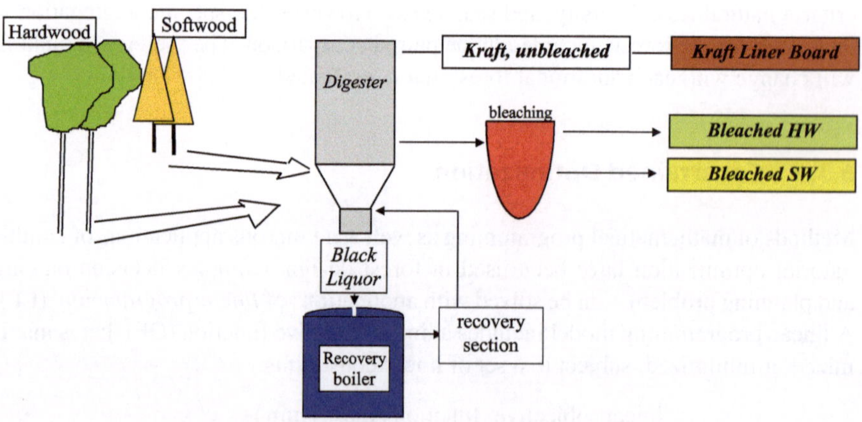

Fig. 6.5 Simplified representation of the pulp-mill production system

Table 6.4 Roundwood required and profit for each ton of BH, BS, and KR

Product	Profit: *Euro* per ton of product	Tons of roundwood required for each ton of product
BH	10	2.5
BS	8	2.0
KR	5	1.3 hardwood +0.7 softwood

1. defining the variables: BH, BS, and KR are the unknown quantities of bleached hardwood, bleached softwood, and kraft liner pulp that should be produced by the mill in one year;
2. defining the objective function: Z = 10 (BH) + 8 (BS) + 5 (KR) max.;
3. defining the hardwood constraint: 2.5 (BH) + 1.3 (KR) ≤ 300,000 t (round hardwood);
4. defining the softwood constraint: 2.0 (BS) + 0.7 (KR) ≤ 1,200,000 t (round softwood);
5. defining the boiler constraint: 0.0239 (BH) + 0.0239 (BS) + 0.0202 (KR) ≤ 8400 h.

Assignment

Formulate the problem as a linear-programming (LP) model and determine the optimum product mix. Use the R library Rglpk to solve the model.

```
library(Rglpk)
obj<-c(10,8,5)
mat<-rbind(c(2.5,0,1.3), c(0,2,0.7),c(0.0239,0.0239,0.0202))
# Each element must be one of "<", "<=", ">", ">=", or "==".
sign <- c("<=","<=","<=")                    # constraint signs
rhs<-c(300000,1200000,8400)                  # vector of rhs coeffs
Rglpk_solve_LP(obj, mat, sign, rhs, max = TRUE)

# Solution:
$optimum                             # [1] 3051715
$solution                            # [1] 120000.0   231464.4   0.0
$auxiliary$dual
[1] 0.800    0.000    334.728        # increase per 1 additional boiler hour
```

Only part of the roundwood is used because of the boiler constraint. Relaxing the boiler hour constraint would increase profit by $334.728. This information is important when a decision has to be made as to whether or not to invest in an additional boiler.

6.1.4.2 Afforestation and Water Yield

In many regions of the world, it has become necessary to regulate afforestation, because of forest plantations' high demand for scarce water resources. The problem of regulating afforestation can be addressed with methods of optimization, provided relevant models for estimating water yields before and after afforestation are available. In addition, evidence from catchment studies regarding the spatial variation of evapotranspiration in catchments, critical low flow constraints, and the influence of forest rotation age on water use is indispensable. Bosch and Gadow (1990) published a review of the procedures for regulating afforestation of catchment areas and made some suggestions for using methods of optimization for regulating afforestation. They developed a simple model to illustrate the method and the required information. The relation between the water used by vegetation in a mountain catchment and water yield from the catchment can be expressed as

$$Q = Pg - Et \tag{6.4}$$

where Q is the mean annual stream-flow; Pg is the mean annual gross rainfall (mm); and Et is the mean annual evapotranspiration (mm). The simultaneous optimization of timber and water yield can therefore result in the maximization of timber production, subject to a limited reduction of Et. We assume that the mean annual production of the plantation ecosystems can be calculated, given the site quality and environmental conditions.

Evapotranspiration increases progressively after afforestation (see Bosch and Gadow 1990). An assumption underlying this basic trend is that the annual increase in Et over an entire catchment causes an equivalent decrease in annual stream-flow from the catchment. The general shape of the time trend in annual Et is sigmoidal. A Chapman-Richards equation was therefore fitted to five catchments in South Africa that were deemed important after long-term observations (Wicht 1949; Nänni 1970; Van Lill et al. 1980; Scott and Lesch 1997; Scott and Smith 1997). The afforestation and water-use problem that affects the design of forest ecosystems is quite common. We will use a simple example to demonstrate an approach for the simultaneous optimization of timber and water yield by considering a large hypothetical mountain catchment area that includes three subcatchments (Fig. 6.6).

The objective is to maximize timber yield (mean annual increment ,MAI)), subject to the constraint that the total water yield may not be reduced by more than 10% after afforestation. Details of the catchment area are shown in Table 6.5.

The objective is to achieve a compromise between water yield and timber yield. The total runoff (Q) is estimated to equal precipitation (Pg. mm/year) minus evapotranspiration (Et. mm/year). An example of two models for estimating mean annual reduction of runoff over age of a *Eucalyptus grandis* and a *Pinus patula* ecosystem is presented in Fig. 6.7 (see Bosch and Gadow 1990).

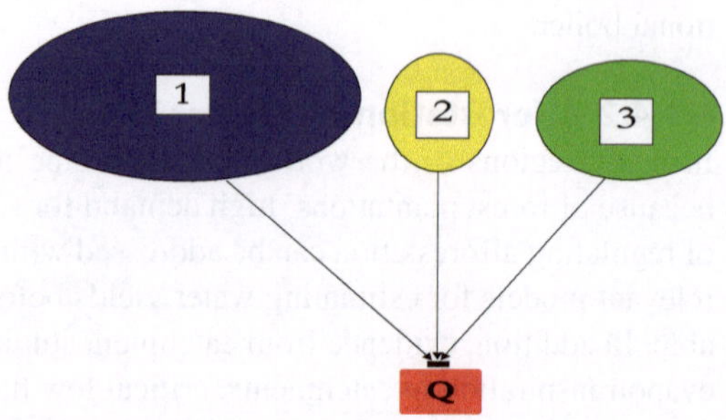

Fig. 6.6 Hypothetical mountain catchment with three subcatchments. Qu is the total water yield before afforestation (m³/ha)

Table 6.5 Details of the catchment area presented in Fig. 6.6. Qp is the water yield after afforestation (m³/ha). MAI is the expected timber yield (m³/ha/year)

Subcatchment	Area (ha)	Qu (m3/ha)	Qp (m3/ha)	MAI (m3/ha)
1	5000	7000	4500	20
2	2500	3400	1700	16
3	3000	3000	2000	12

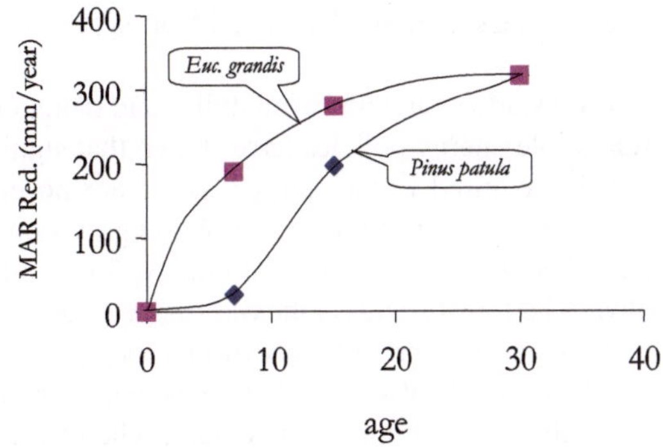

Fig. 6.7 Estimated reduction of mean annual runoff (MAR Red.) from the catchment over stand age for Eucalyptus grandis and Pinus patula (see Bosch and Gadow 1990)

The first important step in the development of an optimization model is to define the variables, that is, the unknown hectares to be afforested in the three subcatchments:

P1, P2, P3, unknown hectares

Next, the objective function needs to be specified:

$$\mathbf{Z = 20(P1) + 16(P2) + 12(P3)}\,\text{max}.$$

The three area constraints are:

$$\mathbf{P1 \leq 5000; P2 \leq 2500; P3 \leq 3000}$$

The minimum water yield constraint is:

$$\sum_{i=1}^{3} P_i \cdot Q_{pi} + \sum_{i=1}^{3} (F_i - P_i) \cdot Q_{ui} \geq 0.9 \cdot \sum_{i=1}^{3} F_i \cdot Q_{ui}$$

which can be simplified to: **25 (P1) + 17 (P2) + 10 (P3) ≤ 52,500.**

The R function Rglpk can be used to solve the problem. The formulation is:

```
library(Rglpk)
obj<-c(20,16,12)
mat<-rbind(c(1,0,0), c(0,1,0),c(0, 0,1), c(25,17,10))
# Each element must be one of "<", "<=", ">", ">=", or "==".
sign <- c("<=","<=","<=","<=")              # constraint signs
rhs<-c(5000,2500,3000,52500)                # vector of rhs coeffs
Rglpk_solve_LP(obj, mat, sign, rhs, max = TRUE)
# Solution:
$optimum                                # [1] 57176.47
$solution                               # [1] 0.000 1323.529 3000.000
$auxiliary$dual                         # [1] 0.0000000 0.0000000 2.5882353 0.9411765
```

The annual total timber yield is estimated at 57176.47 m^3, which is equivalent to an average MAI of 13.22 m^3/ha/year.[1] After rounding to P2 = 1323, we get 4,725,090 mm water per year and a timber yield of 57,168 m^3 per year in the entire catchment area.

[1] This is a very simplified assumption because MAI is age-dependent. The main purpose of the model is to illustrate the general decision structure.

6.2 Sustaining Planted Forests

Several studies on wood availability and demand and the resulting implications for renewable energy policies have shown that significant additional amounts of wood will be required in the future.[2] There are no energy-efficient or environmentally acceptable substitutes for wood. Wood grows on trees, and to successfully maintain an industry based on timber products, forest planning must ensure that there will always be forest stands at the right stage of development and in sufficient number to yield the desired product mix from a forest.

The principle of sustainable forest management is now as relevant as it was at the beginning of the eighteenth century, when Carlowitz (1713) emphasized the need for careful planning to ensure that future wood supplies would be secure. A traditional view of a managed forest is that its development follows a perpetual series of growth cycles that are defined by the rotation. One cycle begins with the establishment of the young trees and ends with the harvest of the mature ones. Intermediate thinnings may improve the value of the final crop. Productivity is measured in terms of the total volume produced or the mean annual increment at rotation age (MAI_R). The present value of a specific silvicultural program is equal to the sum of the annual net incomes, discounted to the age of establishment.[3] Sustainable yield control is based on the model of the "normal forest."

6.2.1 Classical Methods

A so-called normal forest is a theoretical model where timber-growing stock, harvest, and production are in perfect equilibrium (Hundeshagen 1826). The growing conditions in this idealized forest are uniform and all age classes are represented by the same areas. A yield table is used to estimate the development of the growing stock over age (Fig. 6.8).

Assignment
A rotation (R) of 120 years has been set for a forest with a total area (F) of 2 400 ha. Calculate the normal area for each 20-year age class.

Solution: The normal area for each age is equal to f = F/R = 20 ha. The normal area of a 20-year age class is f x 20 = 400 ha, which equals the normal area harvested during a 20-year planning period.

[2] http://www.fao.org/docrep/003/X8423E/X8423E04.htm (accessed on 2 January, 2019); www.unece.org/trade/timber/docs/tc-sessions/tc-65/policyforum/

[3] See, for example, Clutter et al. (1983) and Chang (1998) for details concerning the economic criteria.

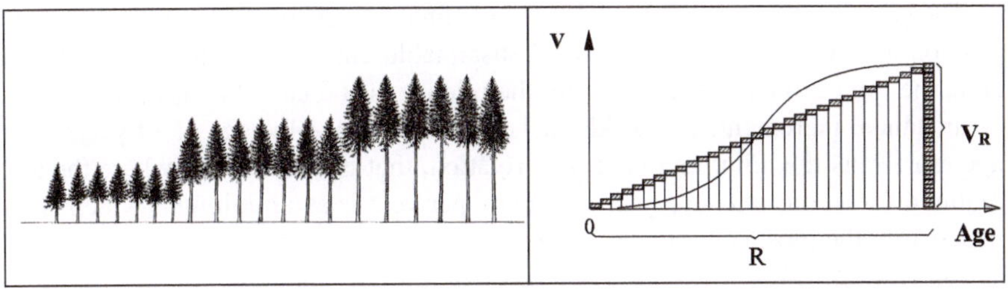

Fig. 6.8 The model of a normal forest is a model of an idealized forest where all age classes are represented in equal proportion (Speidel 1971)

Table 6.6 Effects of the rotation on the area harvested, the mean annual increment, the growing stock volume, and the annual sustainable harvest are illustrated by means of a yield model of Alnus incana for site index 20 m (Lockow 2004)

Rotation (years)	Volume (m3/ha)	area harvested each year	total forest area (ha)	total growing stock (m³)	harvest (% of growing stock)	MAI (m³/ha/yr)
A	B	C = 10 000/B	D =A*C	E =2*(summeB)*D	F =100* B*C/E	G = B/A
2	1.2	8505.9	17011.8	40000.0	25.0	0.59
4	15.2	656.0	2624.0	43084.9	23.2	3.81
6	42.5	235.1	1410.4	54333.7	18.4	7.09
8	76.1	131.4	1051.4	62364.3	16.0	9.51
10	110.3	90.6	906.3	67581.9	14.8	11.03
12	142.1	70.4	844.7	71067.1	14.1	11.84
14	169.8	58.9	824.6	99467.3	10.1	12.13
16	193.1	51.8	828.7	127462.3	7.8	12.07
18	212.2	47.1	848.3	155989.6	6.4	11.79
20	227.6	43.9	878.9	185452.7	5.4	11.38
22	239.8	41.7	917.6	216007.6	4.6	10.90
24	249.4	40.1	962.4	247683.8	4.0	10.39
26	256.9	38.9	1012.1	280443.3	3.6	9.88
28	262.7	38.1	1065.8	314211.8	3.2	9.38
30	267.2	37.4	1122.6	348896.9	2.9	8.91

The normal forest is defined by two conditions: (a) a yield model that estimates the development of the growing stock over age and (b) a rotation that defines the expected volume per hectare at felling age.

The effect of the rotation on the area harvested, the mean annual increment, the growing stock volume, and the sustainable harvest can be illustrated with a yield model of *Alnus incana* for site index (SI) 20 m[4] (Lockow 2004; Table 6.6).

We assume that a biomass power plant requires 10,000 m^3 of roundwood per year for its operation. The wood is obtained from *Alnus incana* energy plantations that are located in the vicinity of the power plant. The various effects of the rotation assuming the annual sustainable harvest are illustrated in Table 6.6.

[4] Site index is defined here as the expected top height at age 50 years.

The growing stock that is available for the final harvest increases with increasing rotation age, which affects the annual sustainable cutting area, the required total area, the total growing stock volume, the relative sustainable harvest rate, and the mean annual increment. The MAI reaches a maximum at the age of 14 years. That age represents the minimum land-base rotation, that is, the rotation where the area required for a constant supply of 10,000 m^3 per year is at a minimum (which is not necessarily the most economical rotation).

6.2.2 Age-Class Simulation

The model of the normal forest is a standard that can be used for comparisons of current and normal growing stock volumes. Its value as a scenario model is limited. Methods based on age-class simulation can be used in situations where information about forest areas by age class is available. In the simplest simulation approach, the forest area is subdivided into *m* age classes, each covering an area of a_{ij} ha in the *j*th felling period (I = 1 .. m; j = 0 .. n). The available timber volume in the *i*th age class is equal to v_{ij}, and the planned total harvest volume for the *j*th felling period is h_j. The algorithm presented in Fig. 6.9 can be used to simulate the effect of a given harvest level on the development of the age-class distribution of a given forest resource.

Example The following empirical yield model for a fully stocked *Pinus silvestris* forest growing on a uniform site of quality class III in northern Russia is available:

$$V(t) = 205.3 \cdot \left(1 - e^{-0.0231 \cdot t}\right)^{2.93} \tag{6.5}$$

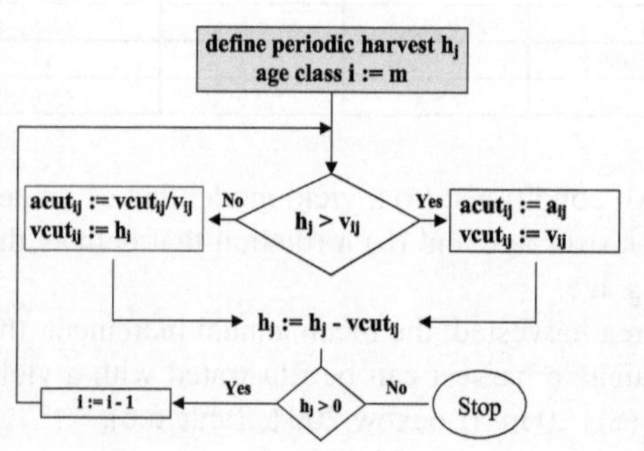

For each harvest period *j*, do:

For each age class i, starting with the oldest one, and while h_j>0, do:

a) calculate the available growing stock volume v_{ij} ;

b) if $h_j \geq v_{ij}$, then harvest the entire growing stock available in age class *i*;

else harvest only h_j m^3;

c) subtract the volume harvested from h_j

Fig. 6.9 Algorithm for age-class simulation with flowchart (left) and abbreviated pseudocode (right). a_{ij} = forest area available in age class i (I = 1...m) and period j (j = 1...n); v_{ij} = timber volume available in age class i and period j (m^3); $vcut_{ij}$ = timber volume harvested in age class i and period j (m^3); $acut_{ij}$ = harvested area in age class i and period j (ha); h_j = specified harvest volume for period j (m^3)

Table 6.7 Yield table for a fully stocked Pinus silvestris forest growing on a uniform site of quality class III in northern Russia

	Age-class midpoint									
	10	30	50	70	90	110	130	150	170	190
m^3/ha	2.0	26.9	67.7	107.4	138.8	161.4	176.8	187.1	193.7	197.9

Table 6.8 Simulated development of the age-class distribution of a Pinus silvestris forest growing on a uniform site of quality class III in northern Russia and assuming a harvest rate of 30,000 m^3 per 20-year cutting period

	Age-class midpoint										Total
	10	30	50	70	90	110	130	150	170	190	Area
a_{i0} (ha)	20.0	300.0	100.0	100.0	200.0	50.0	50.0	100.0	10.0	70.0	1000
a_{i1} (ha)	156.0	20.0	300.0	100.0	100.0	200.0	50.0	50.0	24.0	0.0	1000
a_{i2} (ha)	168.3	156.0	20.0	300.0	100.0	100.0	155.7	0.0	0.0	0.0	1000
a_{i3}(ha)	171.0	168.3	156.0	20.0	300.0	100.0	84.7	0.0	0.0	0.0	1000
a_{i4} (ha)	177.7	171.0	168.3	156.0	20.0	300.0	7.0	0.0	0.0	0.0	1000
a_{i5} (ha)	185.2	177.7	171.0	168.3	156.0	20.0	121.8	0.0	0.0	0.0	1000

where V(age) = growing stock (m^3/ha) at the specified forest age (years).

The model, based on the work by Shvidenko et al. (1995),[5] was used to compile the yield table presented in Table 6.7.

The forest covers an area of 1000 ha. Assuming a given initial age-class distribution (a_{i0} in Table 6.8), a constant periodic harvest of 30,000 m^3 per 20-year cutting period, and immediate regeneration of the felling areas, the age-class distributions can be calculated for a series of future cutting periods.

The development of the resource presented in Table 6.8 shows the effect of a given harvest level on the age-class distribution. In the specific example, a periodic harvest of 30,000 m^3 would result in a wipeout of the three oldest age classes already in the second cutting period. The normal age-class area would amount to 200 hectares if the rotation of a maximum MAI of 100 years is selected. With a mean annual increment of 1.54 m^3/ha/year at 90 years, a harvest level of 30,000 m^3 per year could be sustained indefinitely, assuming that the risks of fire and other disturbances are insignificant.

The method obviously involves considerable aggregation over growing sites, forest types, and management regimes. Therefore, predictions have to be interpreted with caution. However, an age-class simulation is often the only feasible way to predict the dynamic development of a forest resource for large timber-growing regions.

[5]Empirical models are not always available and it is often necessary to construct a plausible yield model based on local experience or other evidence. Murray and Gadow (1993) developed a yield model that is suitable for highly aggregated applications. The model is derived from (a) estimates of the maximum mean annual increment (MAImax) and (b) the age corresponding to MAImax.

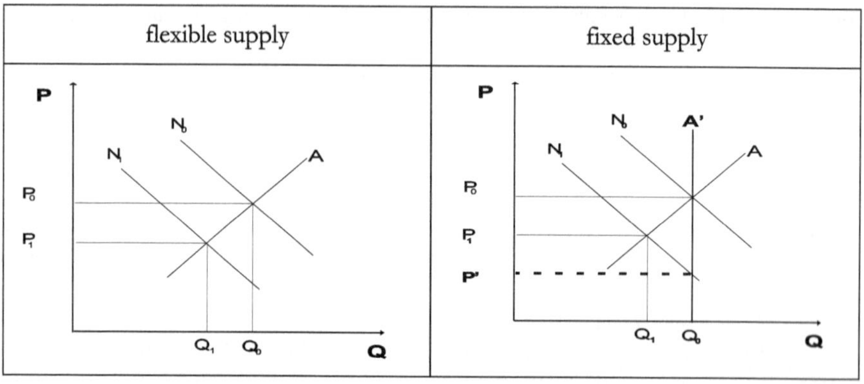

Fig. 6.10 Demand and supply in the market model. The quantity Q can be sold at price P

The original idea of the normal forest model was to prescribe strict sustainable timber yields for each management unit, and it was achieved by means of the periodic block system developed by Hartig (1795) and Cotta (1804). A periodic block is a forest area scheduled for harvest in a given time period. The economic disadvantages of the strict application of the sustainability concept are obvious. To ensure an even flow (EF) of roundwood to the market, it may be necessary to harvest young stands prematurely or to leave overmature ones unutilized when there is an abnormally high proportion of old stands. Strict sustainability, as defined by the normal forest concept, can only be realized in a supply-oriented, controlled economy. High demand cannot be met because the cutting volume is limited to normal levels. In times of low demand, the normal amounts are supplied, irrespective of the financial losses (Fig. 6.10).

The market equilibrium applies for goods, such as timber, that are purchased and sold in a competitive market. The quantity Q can be sold at price P. The actual willingness to pay for the quantity of timber Q_0 is the price P_0 at the demand curve N_0. When the demand, that is, the current willingness to pay, changes as illustrated by the demand curve N_1, a new market equilibrium is established with the supply quantity Q_1 and the price P_1 (Bergen 1997). Even if the supply is fixed, as implied by the strictly sustained timber yield concept, demand may rise or fall depending on the market. When the demand falls, the price will drop more drastically than in the flexible supply scenario (see Waggener 1969).

6.2.3 Multiperiod Harvest Scheduling

Numerous applications of timber harvest planning based on mathematical programming have been reported. The models include an objective function that must be maximized or minimized and a set of constraints. The objective may be to maximize the harvest or the discounted cash flows, or to minimize the cost of supplying the mill. The constraints are typically formulated, so that a smoothing-out of production levels over time and a specified level of sustainability can be achieved simultaneously. A complete survey of mathematical programming applications to timber harvest scheduling is beyond the scope of this chapter.[6]

In multiperiod harvest scheduling models, the state of the forest at the start of period t + 1 is a linear function of the forest state at the start of period t and the areas cut in period t. The application of linear programming to multiperiod harvest planning can be illustrated with an example adapted from Buongiorno and Gilless (1987).[7] Consider a *Eucalyptus grandis* plantation covering an area of 150 hectares. The trees on 100 hectares are seven years old and the site index is 22.0.[8] The remaining 50 hectares of site index 24.5 are three years old. The owner wishes to maximize the volume yield during the next five years. The following model can be used to predict total volume yield (V), measured in m³/ha for a given site index (SI) and age (t):

$$V = e^{2.83+0.151\cdot SI} \cdot \left[1 - e^{-0.055\cdot(t-1)}\right]^{1.19} \tag{6.6}$$

Two yield tables can be developed with Eq. (6.2) (Table 6.9). The silvicultural regime is clear-felling followed by coppicing. A strict area control by means of the rotation of maximum volume production is necessary to ensure sustained yields. An additional objective is to increase the total cutting volume by 2% each year.

Table 6.9 Two yield tables generated by Eucalyptus volume–yield Eq. (6.2)

	Site index = 22.0		Site index = 24.5	
Age	V (m³/ha)	MAI (m³/ha/year)	V (m³/ha)	MAI (m³/ha/year)
1	0.0	0.00	0.0	0.00
2	14.4	7.20	21.0	10.51
3	31.8	10.61	46.4	15.48
4	49.9	12.49	72.9	18.21
5	68.1	13.63	99.4	19.88
6	86.1	14.35	125.6	20.94
7	103.7	14.82	151.3	21.61
8	120.8	15.10	176.2	22.03
9	137.4	15.26	200.4	22.27
10	153.4	15.34	223.7	22.37
11	168.7	15.34	246.1	22.37
12	183.5	15.29	267.6	22.30
13	197.6	15.20	288.3	22.17
14	211.2	15.08	308.0	22.00
15	224.1	14.94	326.9	21.79

[6] See García (1990); Buongiorno and Gilless (1987); Leuschner (1990); Clutter et al. (1983).

[7] See similar applications by Johnson and Scheurman (1977); Hobbelstad and Hofstad (1988); García (1990); Peyron (1993).

[8] Site index is defined here as the stand top height at a reference age of 20 years.

The first step is to define the model variables. Let X_{ij} be the area cut from the initial age class i (I = 1…n) in the cutting period j (j = 1…m). Given two initial age classes and five cutting periods, there are 10 variables, namely X_{31}, X_{32}, ..., X_{75}. Let v_{ij} be the volume per hectare in initial age class i at planning period j. The total volume cut during a specific cutting period is a linear function of the areas cut:

$$\text{Volume cut} = 72.9X_{31} + 99.4X_{32} + \ldots + 183.5X_{75}.$$

This volume is to be maximized. There are two area constraints, one for each initial age class:

$$X_{31} + X_{32} + X_{33} + X_{34} + X_{35} = 50$$
$$X_{71} + X_{72} + X_{73} + X_{74} + X_{75} = 100.$$

To ensure sustainability, it is decided that the harvest volume in period j + 1 must be 2% more than the volume harvested in period j. For period 2, for example, we can say [volume cut in 2] = [1.02(volume cut in 1)]; or, more specifically:

$$99.4X_{32} + 137.4X_{72} = 1.02\left(72.9X_{31} + 120.8X_{71}\right)$$

The rotation of maximum volume production is 11 years for both age classes, by a slight margin. When the potential economic benefits of lowering the rotation age without jeopardizing the objective are considered, it is decided to set the rotation at 10 years. The normal annual felling area is thus 150/10 = 15.0 ha/year. Consequently, the area harvested may not exceed 75 hectares during the five-year period. The appropriate constraint equation reads as follows:

$$X_{31} + X_{32} + \ldots + X_{75} = 75.$$

The main reason for regulating production in a multiperiod optimization model is to maintain sustained levels of timber supply to processing plants. A useful approach is to model relevant aspects of the processing plant subsystem and its interactions with the forestry system (see applications by Kallio et al. 1981; Barros and Weintraub 1982). Newnham (1976, 1991) presented an example of a linear-programming model that determines the timber resources available to meet mill demands and schedules the necessary harvesting, processing, transportation, and regeneration operations in such a way that the cost of meeting mill demands is minimized.

Assignment
Formulate the problem as a linear-programming (LP) model and solve the model by using the R library Rglpk.

6.3 Sustaining Continuous-Cover Forests

Forest ecosystems constitute an almost continuous gradient from uneven-aged multispecies forests to fiber crops such as fast-growing willow or poplar plantations (Angelstam et al. 2004). Multispecies forests are often believed to be superior to plantation monocultures regarding a wide range of expectations. Despite higher harvesting costs, uneven-aged multispecies forests have important advantages, such as greater acceptance by the public and greater flexibility in responding to changing markets (Dedrick et al. 2007). For these reasons, there is wide and increasing interest in developing management methods to rehabilitate disturbed natural forests (Geldenhuys 2009), to re-convert industrial plantations to natural ecosystems (Oppili 2013), and to manage existing multispecies forests more effectively.

Schütz (1757) in the Ore Mountains and Dralet (1820) in the French Pyrenees were among the first to find that clear-felling was detrimental to soil conservation. They proposed that a forest should not be regenerated by planting, but be allowed to regenerate naturally under the forest canopy. Two and a half centuries ago, the French used the term *jardinage* (forest gardening) to describe specific methods of selective harvesting and natural regeneration (Schütz et al. 2012). Subsequently, Biolley (1901) introduced a new method of sustainability control in the communal forest of Couvet and in the Canton Neuchâtel. Their approach became known as the Plenterwald system. Schütz (2002) reports that about 370,000 hectares of classic *Abies alba/Fagus sylvatica/Picea abies* forests still exist in temperate western Europe. These forests are found only in the montane regions where *Abies alba* occurs naturally.

6.3.1 The Inverse J-Shaped Liocourt Curve

Many uneven-aged forests exhibit a negative exponential diameter at breast height (DBH)-class distribution, which is also known in the literature as an inverse J-shaped distribution. The ideal DBH-class distribution in an uneven-aged forest should be "balanced," which means that it should be possible to maintain a specific diameter distribution by applying a given harvest rate in perpetuity. In a balanced inverse J-shaped distribution, a constant quotient exists between successive DBH classes (Buongiorno and Mitchie 1980; Virgilietti and Buongiorno 1997; Rautiainen 1998). This quotient defines the curve's shape and permits the calculation of an ideal number of trees for each DBH class in a forest with a given basal area. In a forest that is managed by defining a balanced inverse J-shaped diameter distribution, the assumption is made that periodic growth can be removed intermittently while maintaining an ideal DBH distribution and density. The DBH distribution of an uneven-aged forest can be represented by the equation (Cancino and Gadow 2001):

$$N_i = k_0 e^{-k_1 d_i} \tag{6.7}$$

The quotient q that defines the ratio of the trees per hectare between successive DBH classes is

$$q = N_{i+1} / N_i \tag{6.8}$$

where d_i (cm) is the DBH-class midpoint, N_i is the number of trees in the ith DBH class, N_{i+1} is the number of trees in the next smaller class, k_0 is the intercept, and k_1 is the rate at which trees diminish in successive DBH classes. Substituting N_i of Eqs. [6.1] in [6.2], we obtain

$$q = k_0 e^{-k_1(d_i - h)} / k_0 e^{-k_1 d_i} \text{ so that } q = e^{k_1 h} \tag{6.9}$$

Therefore, q is dependent on h, the diameter-class width. For example, assuming that $q^1 = 1.3$ for $h = 5$ cm, this would be equivalent to $q^2 = 1.69$ for $h = 2*5 = 10$, and both are equivalent to $q = 2.197$ for $\boldsymbol{h} = 3*5 = 15$ or $h = 1.5*10 = 15$ ($q^{15/5}$ or $q^{15/10}$). Given a specific value of q and with $N_{i+1} = qN_i$ (from Eq. [6.2]), we can calculate the number of trees in each DBH class. Thus, the number of trees in the 2nd, 3rd, 4th, ... DBH class is, respectively, $N_2 = qN_1 = q^1N_1, N_3 = qN_2 = q^2N_1, N_4 = qN_3 = q^3N_1,...$ where N_1 is the number of trees in the 1st (largest) class. In general

$$N_i = q^{i-1} \cdot N_1 \tag{6.10}$$

that is, the entire DBH distribution is dependent on q and N_1. We can now determine the growing stock for an uneven-aged stand with a residual basal area (RBA) B (m^2/ha), which is

$$B = k_2 \sum_{i=1}^{c} N_i d_i^2 \tag{6.11}$$

Substituting Eqs. [6.4] in [6.5], we obtain $B = k_2 \sum_{i=1}^{c} N_1 q^{i-1} d_i^2 = N_1 k_2 \sum_{i=1}^{c} q^{i-1} d_i^2$, or $B = N_1 \cdot k_3$ and

$$N_1 = B / k_3 \tag{6.12}$$

where $\boldsymbol{c}$ is the number of DBH classes; $k_2 = \pi/40000$ and

$$k_3 = k_2 \sum_{i=1}^{c} q^{i-1} d_i^2 \tag{6.13}$$

Thus, k_3 is the key parameter for calculating N_1 and the entire DBH distribution.

Assignment 1

Develop a table of k_3 values for three values of q (q = 1; q = 1.3; q = 1.6), given a DBH distribution with a class width of h = 5 cm and maximum DBHs of 20 cm, 30 cm, 40 cm, and 50 cm.

Solution: I have the table.

Assignment 2
Draw a graph in R which shows the values of $N1$ (the number of trees in the largest residual DBH class), for different values of **q** (1, …, 1.6) and **D** (20, …, 50; h = 5 cm) and a target basal area of $B = 30$ (m²/ha).

The ideal residual DBH distribution is determined by using the B, D, and q parameters. Table 6.10 shows details of a real forest with 1490 trees/ha and a basal area of 57 m²/ha, and an ideal forest defined by a basal area of 30 m²/ha, a maximum DBH of 35 cm, and $q = 1.3$.

Management requires a residual structure with basal area $B = 30$ (m²/ha), a maximum DBH of 35 cm, and $q = 1.3$. A balanced ideal structure requires $N1 = 30/0.429 = 70$ trees/ha in the largest residual DBH class (Eq. [6]), $N2 = 70*1.3 = 91$ trees/ha in the next smaller class, etc.

Assignment 3
Draw a graph in R which shows the values of $N1$ (the number of trees in the largest residual DBH class), for different values of q (1, …, 1.6) and D (25, …, 65; h = 5 cm) and a target basal area of $B = 30$ (m²/ha).

Although the ideal DBH distribution produces the wished-for target basal area (B-ideal in Table 6.10), it cannot be established in practice because of the deficit of trees in three DBH classes. The basal area of the deficit can, however, be

Table 6.10 Details of a real forest with 1490 trees/ha and a basal area (B) of 57 m²/ha, and an ideal forest defined by a basal area of 30 m²/ha, a maximum DBH of 35 cm, and and q = 1.3 (example taken from Cancino and Gadow 2001)

DBH class	N-real	N-ideal	B-real	B-ideal
5	370	338	0.7	0.7
10	230	260	1.8	2.0
15	145	200	2.6	3.5
20	240	154	7.5	4.8
25	210	118	10.3	5.8
30	150	91	10.6	6.4
35	30	70	2.9	6.7
40	36		4.5	
45	24		3.8	
50	32		6.3	
55	20		4.7	
60	3		0.8	
Total	1490	1230	57.0	30.0

redistributed into the DBH classes where the real number of trees exceeds the ideal number. Management may decide to harvest all the trees in DBH classes exceeding 35 cm, and to ignore the surplus in the smaller DBH classes.

An Example from North America

The development of uneven-aged forest management in the USA began in the early 1930s as a way to manage cutover, understocked stands of loblolly and shortleaf pine stands in southern Arkansas (Baker et al. 1996). Guldin (1991) published an example of an uneven-aged mixed conifer forest that included *Pinus ponderosa* and *P. lambertiana* in the overstory and *Pseudotsuga menziessi*, *Abies concolor*, and *Calocedrus decurrens* in the understory. *Lithocarpus densiflorus*, *Arbutus menziesii*, and *Quercus kelloggii* also occurred in the forest. Guldin (1991) calculated the ideal DBH distribution for the forest by using a target basal area of B = 180 ft^2/acre, a maximum residual DBH of 36 inches, a q parameter of 1.21, and a DBH-class width of 2 inches. Guldin's example illustrates the method very well. It would be interesting to evaluate the potential of alternative harvest options.

Examples from Europe

Rois (2000) analyzed an uneven-aged mixed hardwood stand consisting of generally small-sized trees of *Fagus sylvatica, Fraxinus excelsior, Acer platanoides, A. pseudoplatanus*, and *Sorbus torminalis*. The forest, located near Göttingen in Germany, demonstrates an almost perfect ideal negative exponential DBH-class distribution with a very high q value. That particular *Fagus* community originates from natural regeneration after the removal of the overstory and is an example of a very early stage of development. This early stage is only temporary and does not represent a desirable structure. However, although a perfect fit of the negative exponential distribution was obtained, the model proved to be unsuitable as an ideal guide curve for management.

Natural forests can display a great variety of diameter distributions, depending on the successional stage (Košir 1966; Korpel 1992). This observation was confirmed in a comparative study of nine virgin beech forests by Westphal et al. (2006). Several of the forests showed a bimodal diameter distribution that seemed to be typical for beech forests, confirming the results of other studies (Wenk 1996; Condés 1997; Hessenmöller and Gadow 2001). One example is shown in Fig. 6.11.

The bimodal Weibull distribution is given by:

$$f(x) = g \cdot f_u(x) + (1 - g) \cdot f_o(x) \tag{6.14}$$

where $f_u(x)$ and $f_o(x)$ are the functions for the canopy and subcanopy cohorts; g is a parameter which estimates the number-of-tree proportions in the canopy and subcanopy cohorts and also links the two functions. The density of the bimodal Weibull function is given by:

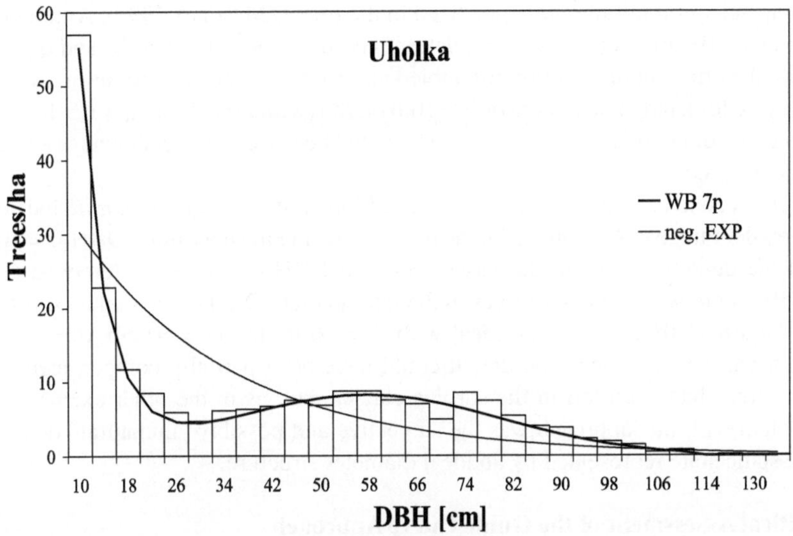

Fig. 6.11 Observed diameter frequencies in the Uholka virgin forest, and two models fitted: WB 7p refers to the 7-parameter bimodal Weibull distribution, and neg EXP to the negative exponential distribution (Westphal et al. 2006)

$$f(x)=\begin{cases} 0 & \text{for } x \le a_u \\ g\cdot\left[\dfrac{c_u}{b_u}\cdot\left(\dfrac{x-a_u}{b_u}\right)^{c_u-1}\cdot e^{-\left(\frac{x-a_u}{b_u}\right)^{c_u}}\right] & \text{for } a_u < x \le a_o \\ g\cdot\left[\dfrac{c_u}{b_u}\cdot\left(\dfrac{x-a_u}{b_u}\right)^{c_u-1}\cdot e^{-\left(\frac{x-a_u}{b_u}\right)^{c_u}}\right]+(1-g)\cdot\left[\dfrac{c_o}{b_o}\cdot\left(\dfrac{x-a_o}{b_o}\right)^{c_o-1}\cdot e^{-\left(\frac{x-a_o}{b_o}\right)^{c_o}}\right] & \text{for } a_o < x \end{cases} \tag{6.15}$$

where a_u; a_o are the location parameters for subcanopy (u) and canopy (o) cohorts; b_u; b_o are the scale parameters for subcanopy (u) and canopy (o) cohorts; c_u; c_o are the shape parameters for subcanopy (u) and canopy (o) cohorts; and g is the joining (link) parameter.

Examples from Chile and Argentina

Cancino and Gadow (2001) presented two examples of uneven-aged forest management in hardwoods in South America. The first, an uneven-aged mixed hardwood stand located in the region of Valdivia in Chile, consists of *Nothofagus obliqua*, *Laurelia sempervirens*, *Persea lingue*, and several other minor species.

Group-selection fellings were practiced in the forest (Martínez 1999). Although the residual DBH distribution was not determined by means of the B, D, and q parameters, the after-cut distribution resembled an inverse J-shaped distribution. An ideal structure for a target basal area of B = 16.6 m2/ha (with D = 35 cm, q = 1.5, h = 5 cm) was defined, but this ideal could not be reached because of the deficiency of trees in the largest DBH classes.

The second example was an uneven-aged forest of *Nothofagus pumilio* located at Tierra del Fuego in Argentina. All the trees exceeding the maximum DBH D = 50 cm and the defective trees in the largest residual DBH classes were removed (Bava 1999). There was a deficit of trees in the intermediate DBH classes and a surplus in the largest DBH classes if an ideal with B = 30 m^2/ha, D = 50 cm, q = 1.3, and h = 5 cm was assumed. The deficit could have been partially compensated for if more trees had been left in the surplus classes, but, as in the other examples, the problem with this approach was the subjective and possibly "unnatural" definition of a static state represented by an ideal diameter structure.

Critical Assessment of the Guide Curve Approach

It is possible to represent the empirical diameter distribution of uneven-aged forests by means of a negative exponential guide curve. This approach has been practiced sporadically for many years. But its practical use for determining periodic harvest levels in uneven-aged forests is limited. The assumption that there is a unique and possibly ideal diameter distribution is not substantiated by empirical evidence. Uneven-aged forest management in Europe has a long tradition and has been common practice in a number of regions. Mitscherlich (1952), in a much-cited study, could not confirm the universal applicability of the negative exponential distribution in selectively managed forests in south-western Germany. Instead, he found a wide range of balanced uneven-aged forest structures, and not just one, as is implied by the plenter principle (Cancino and Gadow 2001). In his assessment, which he based on numerous observations, he concluded that the concept of an ideal diameter distribution has no biological foundation.

It appears that there is no ecological rationale for maintaining a forest in some fixed state of equilibrium, and the economic benefits of marketing a constant harvest of a specific mix of log dimensions are doubtful. It appears therefore that the guide curve approach for managing forest ecosystems is nothing but an interesting historical relic. Models of tree growth and specific analytical tools, such as methods of simulation or optimization, are required for generating and evaluating alternative harvest scenarios where a variety of medium-term options and constraints for a given point in time can be considered. Markets, social and economic demands as well as environmental conditions are changing, and they require a continuous adaptation and adjustment of objectives.

6.3.2 Optimum Residual Structures

Many of the world's forest ecosystems are used for several purposes, and these often include the potentially conflicting goals of timber production and biodiversity conservation. Franklin et al. (1997; see also Mitchel and Beese 2002) proposed *variable retention* as a silvicultural concept that retains forest structural elements in order to preserve environmental values associated with structurally complex forests. Retention forestry emerged some 25 years ago in north-western USA under the labels "new forestry" and "green-tree retention" (Gustafsson et al. 2012). Both live and dead trees should be retained after harvesting, with a preference for large specimens and rare species. In retention forestry, a portion of the original stand is left unlogged to maintain the continuity of structural and compositional diversity. This practice increases public acceptance of large-scale clear-felling by preserving key habitat components with benefits for tree-dwelling plants, wood-inhabiting insects, and fungi. The methods were developed specifically for the vast forest resources in the Pacific Northwest in the United States, and some of the basic principles may be applicable in other forest biomes (Roberge et al. 2013).

Puettmann et al. (2015) put forward convincing arguments in favor of the global acceptance of alternative silvicultures. They propose a set of principles that include avoidance of clear-cutting, emphasis on structural diversity and small-scale variability, deployment of mixed species with natural regeneration, and avoidance of intensive site-preparation methods. Bauhus and Schmerbek (2010) described various methods for creating and maintaining structural variability in commercial plantation forests by retaining patches of native vegetation. These patches are sometimes called understory islands if they are deliberately excluded from harvesting and site preparation (Ough and Murphy 1998; Hartley 2002). To advance the acceptance of retention approaches such as these in commercial timber plantations, it is necessary to quantify the associated losses in productivity and the increase in operational costs. Bi et al. (2002) and Palik et al. (2003) provided examples for assessing the influence of retained mature trees on plantation productivity and seedling regeneration, respectively.

A particularly challenging objective of continuous-cover silviculture is to derive economic benefits from the timber without modifying the key features of the undisturbed ecosystem. To achieve this objective, a range of practical silvicultural systems have been developed in various regions of Europe, North America, and in some tropical and subtropical forests of South Africa, Asia, and South America. In all these applications, forest development oscillates around some level of residual stocking, which is assumed to be optimal or "near natural." Unfortunately, as pointed out before, the "natural state" is something that cannot be defined. It is not possible to measure and objectively evaluate an ecosystem's degree of "naturalness." It is possible, however, to describe some aspects of the ecosystem that appear to be natural, such as the distribution of tree species that have evolved in a particular environment. While the advantages of maintaining particular species mixtures and structural variability in forest ecosystems are obvious, the translation of those specific structures and ideal mixtures into management can be a major challenge. Acceptance by management is vital for successful implementation, and therefore there is a need for simplification without sacrificing essential characteristics of forest structure.

Most forests display well-defined vertical layers of vegetation. It is therefore possible to identify canopy, subcanopy, and shrub species by consulting standard ecological textbooks (Korpel 1992; Adams 1993). Some tree species are fast-growing pioneers, while others are known to be shade tolerant; some species become dominant members of the canopy when mature, while others, including shrub species, have adapted to remain permanently in the subcanopy layer even when they are mature. While this grouping may seem to be largely experienced based, certain statistical methods can be used to identify particular cohorts, especially the mature and immature individuals, and so simplify a complex multispecies forest.

Mature and Immature Canopy Species

The forest canopy includes the mature individuals of the canopy species and also forms a habitat with unique flora and fauna. Prominent canopy species in a European mixed beech forest are *Fagus sylvatica*, *Fraxinus excelsior*, and *Picea abies* (Kramer 1988), while *Pinus koraiensis, Fraxinus mandshurica*, and *Acer mandshuricum* are prominent canopy species in the mixed broadleaf-conifer forests in Jilin province in north-eastern China (Zhang et al. 2013a, b). *Podocarpus falcatus*, *Olea capensis*, and *Ocotea bullata* are examples of canopy species in the subtropical Knysna forests in South Africa (Breitenbach 1974), while *Pinus durangensis*, *P. cooperi*, and *P. teocote* are among the canopy species in the unique multispecies forests of Pueblo Nuevo in Durango, Mexico (Corral-Rivas et al. 2012).

The population of canopy trees is composed of a mixture of mature and immature subpopulations. Zucchini et al. (2001) presented a model for diameter–height distribution that is specifically designed to differentiate between the two subpopulations. A mixture of two bivariate normal distributions was found useful for differentiating between the mature and immature canopy species (Fig. 6.12).

The parameters have familiar interpretations and can be estimated with the R function *mclust*. If $f(d,h)$ denotes the bivariate probability density function of diameter and height, the proposed model is

$$f(d,h) = \alpha \cdot n_1(d,h) + (1-\alpha) \cdot n_2(d,h) \tag{6.16}$$

where α, a parameter in the interval (0, 1), determines the proportion of trees belonging to each of the two-component bivariate normal distributions $n_1(d,h)$ and $n_2(d,h)$. The parameters of $n_j(d,h)$ are the expectations u_{dj}, u_{hj}; the variances σ^2_{dj} and σ^2_{hj}, and the correlation coefficients, ρ_j(j = 1, 2). The examples in Fig. 6.12 show that the height–diameter ratios of the mature and immature canopy species are quite different. The ratio tends to be lower for the trees that have already reached the canopy and higher in those that still grow in the shade. Inspection of the bivariate diameter–height distribution therefore helps to separate mature and immature canopy species.

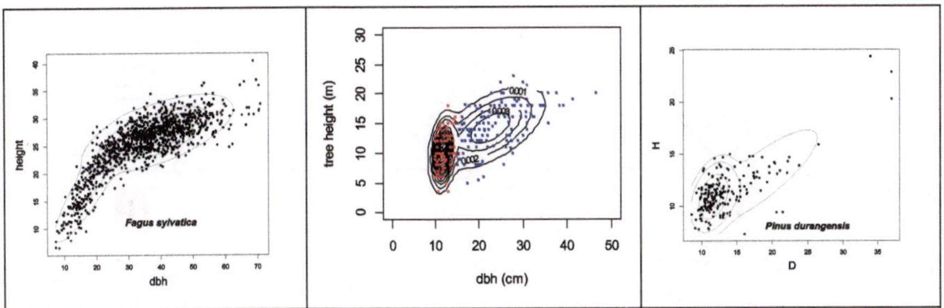

Fagus sylvatica, Germany	*Acer mandshuricum*, China	*Pinus durangensis*, Mexico

Fig. 6.12 Examples of bivariate distributions of three canopy species that are dominant in specific ecosystems. Obviously, the diameters separating the two clusters are different in each species

The separating DBHs are 13 cm, 16 cm, and 16 cm for *Fagus sylvatica, Acer mandshuricum*, and *Pinus durangensis*, respectively.

```
# bivariates for Quercus sideroxyla
Qsid=read.csv('/Users/kg/Desktop/Quercus_sid.csv',header=T)
library(mclust) # at least R.15 needed
dbh_cm= Qsid$DBH; height_m= Qsid$H
x<-as.matrix(data.frame(dbh_cm,height_m))
dens = densityMclust(x,G=2,Selection=2)
# quantiles plot
par(mar=c(5,5,1,1) + 0.2)          # must be drawlabels = F, when prompted: use
Selection=2
plot(dens, x, col = "red", drawlabels = F, cex.axis=1.5,cex.lab=1.7,xlim=c(5,40),
ylim=c(0,15), levels =quantile(dens$density, probs = c(0.3,0.5,0.7)))
 text(25,12,'Quercus_sideroxyla',cex=1.5,font=4)
# perspective plot
plot(dens, type = "persp",)
```

Assignment
Use the R function mclust to identify the separating dbh of *Quercus sideroxyla*.

Beneath the canopy there is another layer, the subcanopy or understory which is formed by low woody plants, sometimes with multiple stems from the base that attain a height at maturity which is usually less than one third of the canopy height. Accordingly, we can subdivide canopy trees into dominant and subdominant individuals, using Roman numerals to identify three cohorts, I: mature canopy trees; II: immature canopy trees; and III: subcanopy species of any size. Figure 6.13 shows a hypothetical example of different species cohorts in a forest.

The distribution of basal area by species type reveals the relative importance of the different species cohorts in a particular forest community. This information can be used to design optimum structures and differentiated harvest levels.

Fig. 6.13 Schematic representation of different species cohorts in a multispecies forest, including I: mature canopy trees; II: immature canopy trees; and III: subcanopy species. The shading separates the canopy and understory

Residual Basal Areas

The residual basal area concept is based on the assumption that sustainable use in multi-aged forests is not achieved by prescribing *how much to harvest*, but by defining a desirable forest structure that *should remain after* the harvest. What complicates the implementation of practical harvesting guidelines in multispecies forests is the need to determine appropriate thresholds of residual structure and density. The Liocourt curve prescribes the residual structure in terms of the number of trees in different diameter classes. The guide curve is therefore "species-blind," and this is certainly not desirable in most multispecies systems (Cancino and Gadow 2001). The principle of the residual basal area (RBA) concept is explained in Fig. 6.14. The first step involves the structuring of the community, for example, by identifying canopy and subcanopy species and finding the separating DBHs for each of the canopy species based on a bivariate analysis of the DBH–height relations. When an objective basis for defining three tree classes has been established, the basal area distribution over tree classes is assessed, usually by sampling the population before the harvest. Then an RBA normality must be defined, usually based on studies relating forest productivity to different levels of density. Finally, the RBA normality is superimposed over the observed distribution of basal areas. Figure 6.14 explains the principle.

The development of three tree-class basal areas in the communal forests of Durango in Mexico is based on the theoretical concept explained in Fig. 6.14 and illustrated in Fig. 6.15. The basal area distributions of the three species cohorts are

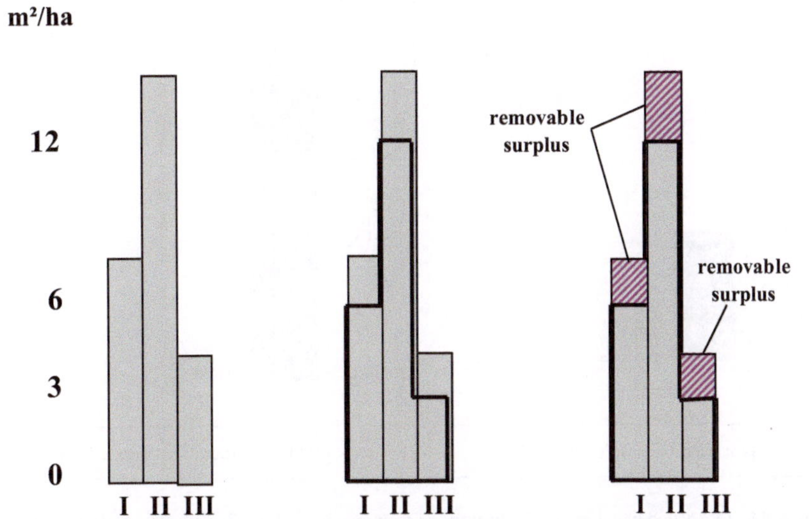

Fig. 6.14 Principle of the residual basal area (RBA) concept. Left: Observed basal area distribution over tree classes before the harvest. Center: RBA normality, the remaining basal area distribution superimposed over the observed one. Right: The removable surplus

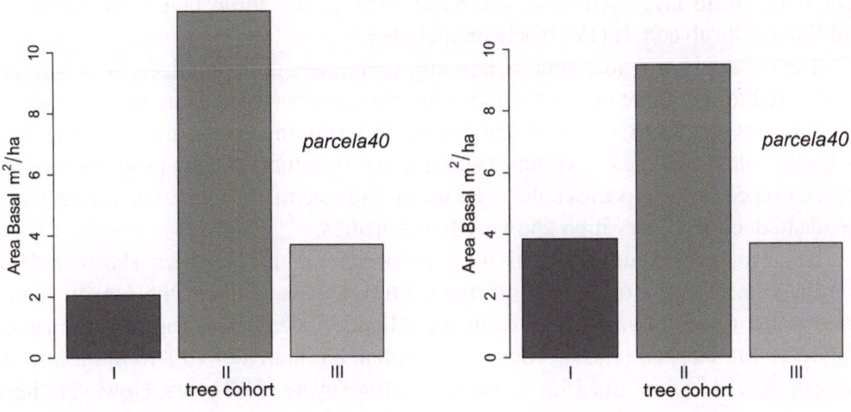

Fig. 6.15 Left: Observed basal areas of three species cohorts in a permanent field plot in Durango that were measured in 2007 and remeasured in 2012. There was considerable ingrowth from cohort II into cohort I

shown for the year 2007 for one field plot known as Parcela40. The same plot was assessed five years later. In terms of basal area, the ingrowth from cohort II (immature canopy species) into cohort I (mature canopy species) during the five-year observation period slightly exceeds the loss in cohort II. The reason for the increase in basal area is growth. The basal area of the cohort of subcanopy species shows almost no change.

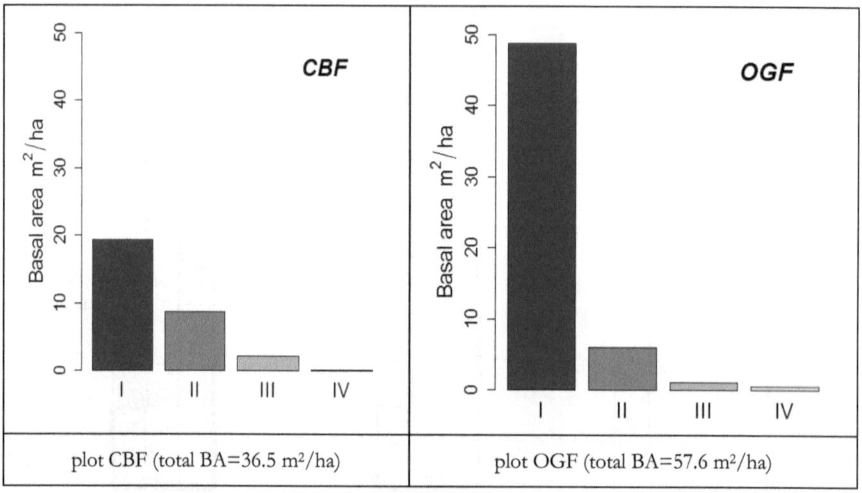

Fig. 6.16 Observed basal area distributions of four species cohorts in two observational field plots (CBF and OGF) in north-eastern China

The basal area distributions for four species cohorts in two multispecies temperate forests (known as conifer and broad-leaved forest (CBF) and old-growth forest (OGF)) in north-eastern China are shown in Fig. 6.16. The two forests have a characteristic shrub layer. Although the basal areas of the shrub layers are small, an additional shrub cohort (IV) has been included.

The residual basal area approach is easy to implement in practice, and when it is compared to the guide curve method it has the considerable advantage that natural tree species distributions can be maintained. Determining an optimum residual density (the total basal area or volume) and residual structure (density proportions allocated to specific tree species cohorts) is the task of scientists. Ecosystem productivity tends to decline at very high and very low densities.

Basal area distribution norms were proposed for the Northern Hardwoods, a multispecies forest extending from the central Appalachians in the United States northward toward Canada and from the Atlantic Ocean westward to Minnesota (Eyre 1980). Hansen (1987) proposed a residual basal area of 19.8 m^2/ha for a cutting cycle of 15 years and 19.8 m^2/ha for a cutting cycle of 30 years. However, these norms apply to size classes: the total basal area is distributed over tree diameter classes while the RBA method is based on species cohorts.

The objective of the RBA concept is to maintain the most (or sufficiently) productive forest density, and at the same time to preserve the natural structure and functioning of the ecosystem. When permanent observational plots are available, as is the case in many regions of the world, it is relatively easy to identify optimum densities and developing technical prescriptions for distributing the total density over particular species and/or tree size cohorts.

6.4 Sustaining Green Landscapes

To successfully maintain an industry based on timber products, traditional forest planning had to ensure that there would always be stands at the right stage of development and in sufficient numbers to yield the desired product mix from the forest. For planning purposes, it was necessary a) to prescribe future management on a long-term basis using standard silvicultural prescriptions and b) to provide long-term estimates of forest development based on yield tables or growth models (Schober 1978).

Standards are agreements on technical specifications that are used as rules or definitions to facilitate manufacturing, trade, and communication. In forestry, standard units are used to facilitate data assessment, analysis, and planning. Standard units of length, temperature, mass, and time have been adopted and are used to measure trees. It was also proposed that the development and documentation of tree growth modeling systems be standardized (Pretzsch et al. 2002). A silvicultural standard is an agreed-upon set of instructions for managing a forest. Determining standard treatment schedules (including planting, pruning, thinning, and clear-felling) was an important research topic from the 1960s to the 1980s (Brodie et al. 1978).

Standardized silviculture creates uniform habitats that may reduce both the local and regional diversity and fail to generate desirable conditions for species that require specific habitat conditions. Furthermore, tree species choice and silviculture are influenced by changing policies and changing economic and environmental conditions. A new standard silviculture may be implemented, but its acceptance may only be temporary. Amling (2005) identified a succession of changing silvicultural standards for the period 1945–2005 in the state forests of Nordrhein-Westfalen in Germany. Similar changes in silvicultural standards were reported from East Germany (Koch 2005). In both cases, the phase-lengths of the cycles of policy changes, which affected the types of harvesting practices (clear-felling vs selective harvesting), species selection (deciduous species vs conifers), and preferred forest structures (even-aged monocultures vs uneven-aged multispecies forests), were much shorter than the life of a tree. Many forests seem to be in constant transition from one standard silviculture to another. This is a paradox, because forest management has always been committed to long-term strategy.

6.4.1 Traditional Methods of Forest Design

The strategic significance and economic benefits of industrial standardization are generally accepted. Traditional forest designs were therefore based on standard silvicultural programs that were prescribed for an entire rotation. Since economic, environmental, and political conditions are constantly changing, however, intermittent adaptation is required. Consequently, the negative effects of using fixed silvicultural regimes meant to be followed for several decades outweigh the advantages of those fixed regimes. In addition, spatial differentiation of forest management activities can contribute to a mix of habitat conditions. For example, the provision of recreation amenities, timber production, or maintenance of deadwood may be more effective if these activities are distributed within the landscape. Standardization can reduce the capacity of an ecosystem to provide multiple services if it is applied too rigidly.

Forest biomes are found in a variety of environmental conditions and have been subdivided into distinct types. The simplest classification distinguishes three types (Gadow et al. 2008): *tropical forests* which are characterized by a great diversity of species and occur between the latitudes of 23.5 degrees N and 23.5 degrees S; *temperate forests* which have a growing season of 140–200 days and occur in eastern North America, north-eastern Asia, and western and central Europe; and the *boreal forests* which represent the largest terrestrial biome, occur between 50 and 60 degrees N in Eurasia and North America, and have a growing season of about 130 days. Forests grow under a great variety of environmental conditions, and it is not surprising that such a simple classification cannot portray the great diversity of forest ecosystem formations. Numerous attempts have been made by geographers, botanists, ecologists, and foresters to capture this diversity by means of ever finer classifications (Folch and Camarasa 2000; Matthews et al. 2000; Hendrick 2001). These ecosystems are not virgin forests, but have been subject to human use. They are managed to satisfy a great variety of needs. Natural forests have often been replaced by fast-growing commercial timber plantations (Cossalter and Pye-Smith 2003). Plantations were established to stabilize domestic wood supplies, to provide jobs, to combat soil erosion, and to reduce pressure on natural forest resources.

Global surveys of human impact indicate that only a very small part of the world's land area has escaped direct human influence (Sanderson et al. 2002; Kareiva et al. 2007). There is evidence of both indiscriminate exploitation to satisfy short-term human needs and systematic domestication to provide a variety of goods and services on a sustainable basis (Tewari and Arya 2005; Kumar and Nair 2006; Tewari and Singh 2006; Ruppert 1972; Ebert and Langer 1977). Forests play a significant role in the climate system. Wood is one of the main sources of renewable energy. Forests are harvested for timber, but silvicultural systems are also sometimes adapted to increase water yields (Bosch and Gadow 1990), reduce fire hazard (González et al. 2006), improve nitrogen processes (Rusanen et al. 2004; Laurén et al. 2005), and enhance scenic beauty (Liao and Nogami 1999). Trees are killed and left standing to increase the amount of deadwood habitat (Humphrey et al. 2002), while species compositions and tree size distributions are modified to create and maintain "near-natural" forest structures (Emborg et al. 2000; Pommerening 2006). Ecosystem services are increasingly determined by human design and management, not by nature. The design of green landscapes,[9] constrained by changing demands, changing environmental conditions, and increasing population pressures, has therefore become a key area of ecological research, more relevant than the

[9]A green landscape, in this context, is a rural expanse that may comprise a variety of features and usually, but not necessarily, includes woody vegetation.

Operation	Site Index: 8		Site Index: 13		Site Index: 18	
	Age	N/ha	Age	N/ha	Age	N/ha
Planting	0	2000	0	2000	0	2000
Brush control	2	-	2	-	2	-
Brush control	4	-	4	-	4	-
Low pruning & brush control	32	1670	22	1645	14	1588
First thinning (semi-systematic)	58	497	40	539	30	575
High pruning	-	-	40	450	30	450
Second thinning (from below)	84	224	56	253	42	259
Third thinning (from below)	94	157	66	177	52	181
Clearfelling	104	366	76	415	62	423

Fig. 6.17 A cradle-to-grave silvicultural program for Pinus sylvestris in north-western Spain (see Rojo et al. 2005). Site index is dominant height (m) at age 50; N/ha is trees per hectare

traditional focus on the "few remaining pristine ecosystems," as suggested by Cushing et al. (1997).

Trees are long-living organisms, and foresters have been educated to think long term, often in terms of several decades or even a century. Accordingly, traditional systems of silviculture are characterized by long planning horizons that may extend from planting to the final harvest, or by some residual structure that is thought to be optimal.

6.4.1.1 "Cradle-to-Grave" Design

Two examples of long-term silvicultural programs are shown in Fig. 6.17. The first example involves a detailed schedule proposed by Rojo et al. (2005) for *Pinus sylvestris* forests in north-western Spain. The system includes details of specific operations (brush control, pruning, thinning, and clear-felling) for three site quality types. The planning horizon covers the entire life span of a forest from its planting to the final harvest (from the cradle to the grave) and includes intermediate ages at which specific operations should take place.

The second example in Fig. 6.17 is the silvicultural program proposed by Burschel and Huss (1987:263). It covers the entire life span from the juvenile stage to maturity of a *Fagus sylvatica* forest in Europe. That particular schedule is divided into three distinct phases. The juvenile phase is characterized by high crowding to encourage competition and reduce branchiness. Selective thinning is practiced during an intermediate phase at midrotation when the dominant height is around 20 m. Heavy thinnings and eventual final fellings are proposed when the forest has reached maturity, and when there is sufficient regeneration to initiate a new development cycle. Such cradle-to-grave designs have dominated silvicultural thinking for a long time.

6.4.2 Multiple-Path Design

Such long-term schedules may be optimal in theory, but silvicultural programs developed for single products, single services, and specific markets can turn out to be undesirable when demands, technical processing facilities, and environmental policies change. Continual changes in markets, climate, and public opinion require continual adaptation. A long-term schedule that was optimal 10 years ago may be outdated now. This requires constant reorientation, taking into account realistic time windows and following recommendations such as those by Messier et al. (2019), to overcome the command and control approach, and to manage forests as complex adaptive systems. New modeling tools enable us to specify a range of possible silvicultural alternatives. This range of alternatives, which we call "multiple paths," is defined by specific "guard railings," such as maximum or minimum densities. An example involving a *Pinus pinaster* ecosystem in north-western Spain illustrates the approach. The following equation is used to estimate the basal area production for a specific age, site index, and density:

$$\Delta G = 17.53 \cdot G^{0.387} \cdot t^{-1.238} \tag{6.17}$$

where ΔG = basal area production (m²/ha/year) and G = current basal area (m²/ha); t = age (years). The equation can be used to generate two alternative development paths for a ten-year-old forest. Each path is defined by a 25-year time window, and by a specific succession of harvest events. Path I is defined by four heavy thinnings and a final basal area of about 35 m²/ha at age 35; Path II by four light thinnings and a final basal area of about 45 m²/ha at age 35 (Fig. 6.18).

The following additional equation for dominant height can be used to estimate the difference in total production and the terminal growing stock between the two alternative paths (Sánchez and Rodríguez 2002):

$$H_0 = 0.229 \cdot SI^{0.943}\left[1 - \frac{1}{e^{0.055 \cdot t}}\right] \tag{6.18}$$

where H_0 = dominant height (m) and SI = site index (dm); t is age (years). Hinrichs (2006) generated multiple paths for multispecies forests dominated by *Fagus sylvatica*. He defined the range of alternatives by specific silvicultural guard railings, such as maximum forest density (G_{max}), specific thinning weights (rG), and a time window (t0, t1). The time of a harvest event is determined by forest density. An example is shown in Fig. 6.19 where the guard railings of the multiple-path (MP) model were defined by the maximum forest density, specific thinning weights, and a specific time window.

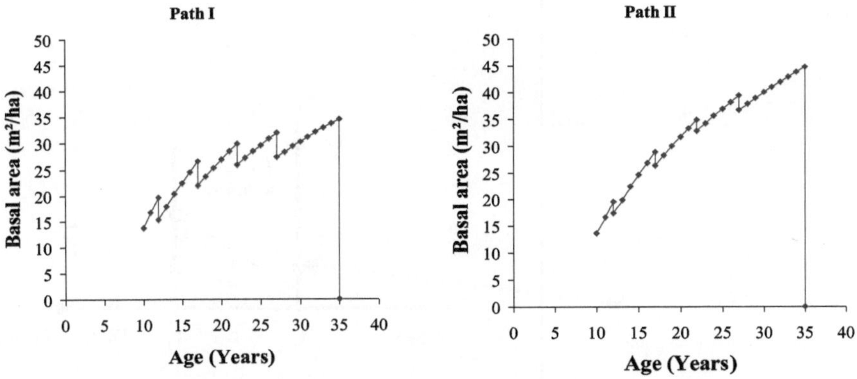

Fig. 6.18 Two alternative development paths for a ten-year-old Pinus pinaster forest. Path I is defined by four heavy thinnings and a final basal area of about 35 m²/ha at age 35; Path II by four light thinnings and a final basal area of about 45 m²/ha at age 35

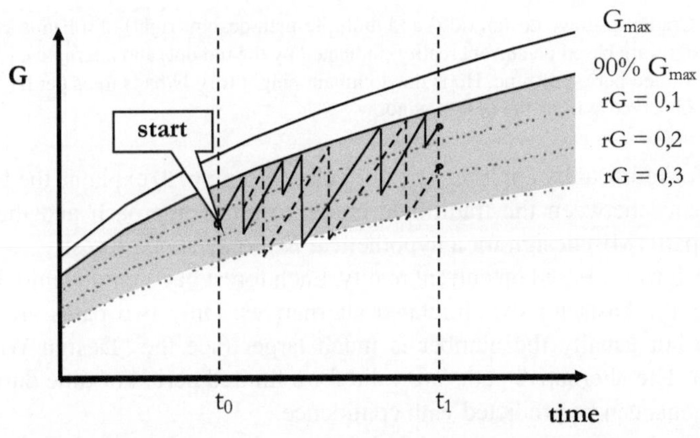

Fig. 6.19 Hinrichs (2006) defined the guard railings of his multiple-path approach by the maximum forest density (G_{max}), specific thinning weights (rG), and a time window (t0, t1)

Assuming six harvest events, each with three different thinning weights[10] and two maximum densities within the specified time window, we obtain 55,986 alternative paths:

$$\sum_{i=0}^{6}(3\cdot 2)^{i} = 55986$$

Implementation in R confirms the result: k = rep(NA); for(i in 0:6){k[i] = (3*2)^i}; sum(k). In a multispecies community, the basal area removed in a selective thinning is distributed among the different tree species. In a second step, individual trees are

[10] The weight of a thinning is rG = removed basal area/total basal area

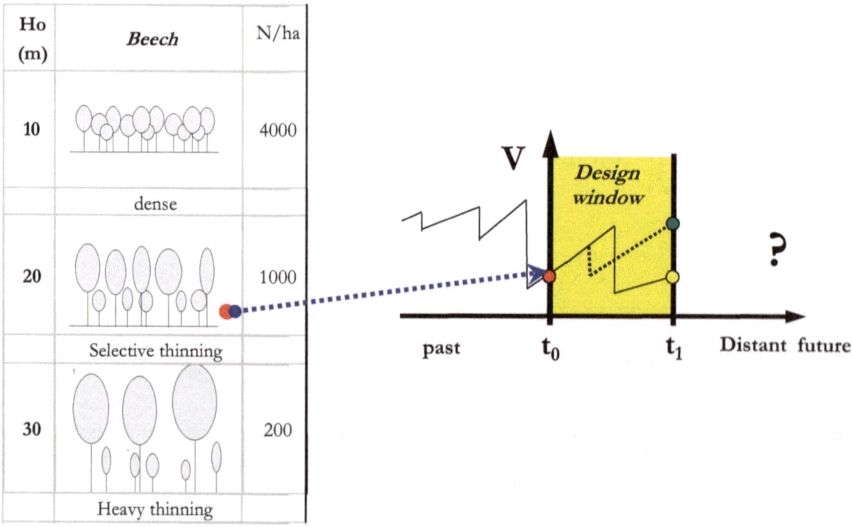

Fig. 6.20 Cradle-to-grave design (left) and multiple-path design (right) of a Fagus sylvatica forest. MP designs are based on current reality (indicated by the red dot) and alternative paths that are valid for a limited period of time. Ho is the dominant height (m); N/ha is trees per ha. The red dot defines the observed start of the design window

selected for removal by applying specific rules. Figure 6.20 explains the fundamental difference between the traditional cradle-to-grave approach and the adaptive multiple-path (MP) design for a hypothetical *Fagus sylvatica* forest.

MP designs are based on current reality. Each forest has a unique initial state that represents the basis for the simulated alternatives. Only two paths are shown in Fig. 6.20 but usually the number is much larger (see the "Design Window" in Fig. 6.19). The alternative paths are valid for a limited period of time during which developments can be predicted with confidence.

People's perceptions are modified by new research and changing societal demands, and therefore subject to continual change. Ecosystem designs have to consider a mix of objectives that must be satisfied simultaneously (Pukkala 2002a, b). This requires expertise in many areas and a need to bridge gaps between related disciplines and to incorporate discipline-specific know-how (Sayer and Campbell 2004). Pending environmental changes call for strategies that allow us to respond in ways that limit their negative effects. A tree must be able to obtain water and nutrients from the soil, prevent excessive water loss through transpiration, transport water and minerals from one part of the plant to another, maintain leaves for photosynthesis, and reproduce. Particular tree species have adapted to survive under extremely dry conditions, while others depend on sufficient moisture for survival. Their existence may be threatened when these conditions change and therefore active management is required.

A forested landscape represents a mosaic of spatial units which are called stands, compartments, or subcompartments. Forest ecosystems are spatially structured, and

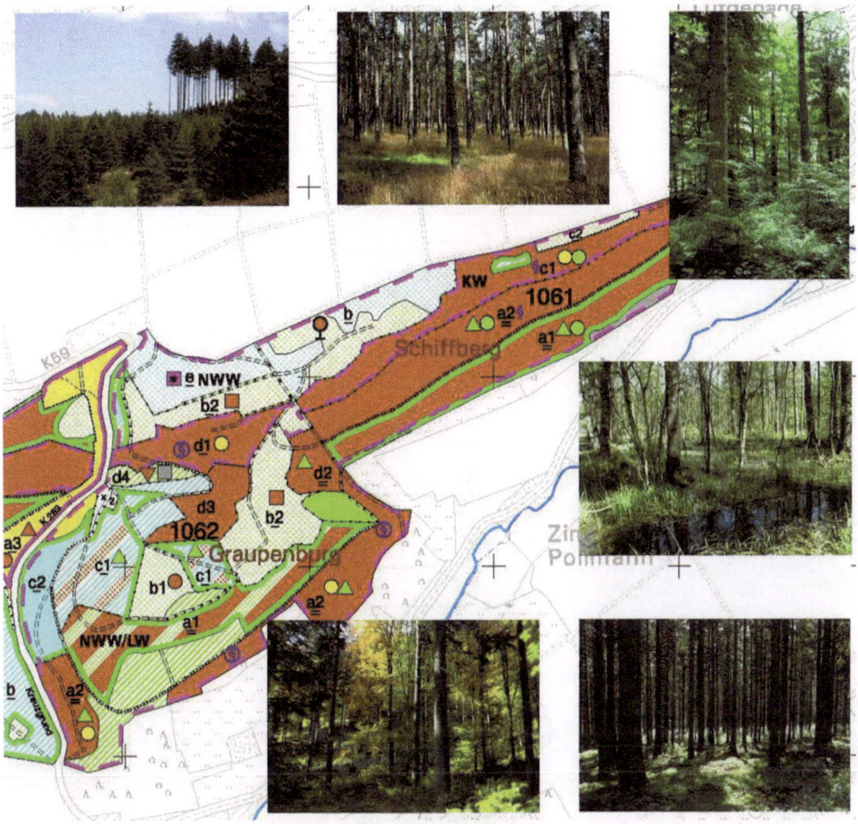

Fig. 6.21 Example of a forested landscape in northern Germany shaped by a long history of sustainable human use. This "domesticated anthrome" (Hui and Richardson 2017) represents a spatial mix and abrupt changes between neighboring ecosystems which are so characteristic of much of the world's forests

the spatial structure of a forested landscape forms the basis of ecosystem design and dynamics. To be effective, sustainable designs consider both the landscape as a whole and each individual stand. Linking the different levels of the spatial hierarchy is therefore an essential part of ecosystem design. Figure 6.21 shows an example of a forested landscape in northern Germany, a *domesticated anthrome* (Hui and Richardson 2017). The landscape represents a spatial mix of neighboring stands. Such patterns, which have evolved during a long history of human use, show a surprisingly high diversity of plant, bird, and animal species. The high diversity of small-scale habitats, which often exceeds the diversity of a protected landscape, is directly attributable to the great variety of habitats (Schulze et al. 2014, 2019).

The multiple-path design facilitates the linking of the different levels of the spatial hierarchy. The principle is illustrated with the simplified example of a forested landscape that consists of three stands, A, B, and C (Fig. 6.22). The initial condition

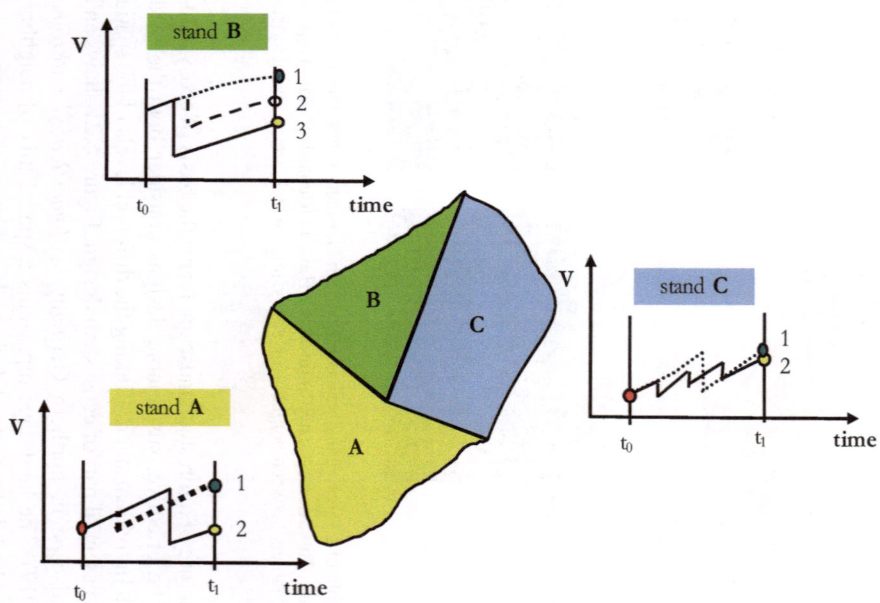

Fig. 6.22 Simplified example to explain the spatial relevance of the multiple-path theory: A forest consists of stands A, B, and C. Two development paths are defined for stands A and C and three for stand B during the time window $t_0 - t_1$, resulting in $2 \times 3 \times 2 = 12$ possible development paths for the forest as a whole

of each individual stand is defined by the biomass at time t_0 indicated by the red dot in each of the three time windows.

Two development paths are defined for stands A and C, three for stand B. There are 2 × 3 × 2 = 12 possible development paths for the forest as a whole. Stand B starts with a high biomass, and path B1 remains undisturbed throughout the time window. Path B2 involves a moderate thinning, B3 an early heavy thinning. Accordingly, B1 has the highest terminal biomass at the end of the time window, followed by B2 and then B3. The design involves 2 × 3 × 2 = 12 possible development paths for the forest as a whole, namely A1B1C1; A1B1C2; A1B2C1; A1B2C2; A1B3C1, A1B3C2, A2B1C1; A2B1C2; A2B2C1, A2B2C2, A2B3C1, A2B3C2. If the overall objective is to maintain a high growing stock, then the combination A1B1C1 is the most desirable one. The optimal combination depends on the existing preference structure. The multiple-path approach to forest design therefore refers to the spatiotemporal organization of a forest and involves the search for a combination of management options that provides a desirable mix of products and services. The basic concept is straightforward and can be summarized according to a proposal by Pukkala and Kangas (1993) who defined the utility of a specific forest design as follows:

$$Maximize\ \mathbf{U} = \sum_{i=1}^{n} w_i \cdot u_i\left(q_i\right) \tag{6.19}$$

where U is the utility of a given path combination; n refers to the number of services; w_i is the relative weight of service i $\left(0 \le w_i \le 1; \sum_{j=1}^{m} w_i = 1\right)$; q_i is the realized amount of service i for a given path combination; and $u_i(q_i)$ is the partial utility function for service i ($0 \le u_i(q_i) \le 1$).

Alternative formulations and refinements of the basic concept are possible, including period weighting or formulations as a linear-programming model. Applications can be found in Gadow and Pukkala (2008). The multiple-path (MP) model can be expressed in different ways, one being a linear-programming (LP) structure. The general LP formulation is defined by a linear objective function

$$Z = \sum_j C_j x_j \to \max, \min \tag{6.20}$$

where c_j are objective function coefficients and x_j are variables (unknown quantities), subject to

$$\sum_j a_{ij} x_j \begin{Bmatrix} = \\ \le \\ \ge \end{Bmatrix} b_i, \forall i \tag{6.21}$$

where the a_{ij} are technical coefficients (certain products or services) and b_i are constants (the required total periodic products or services). The general model structure, depicted as a block diagram in Fig. 6.23, includes the three basic elements of

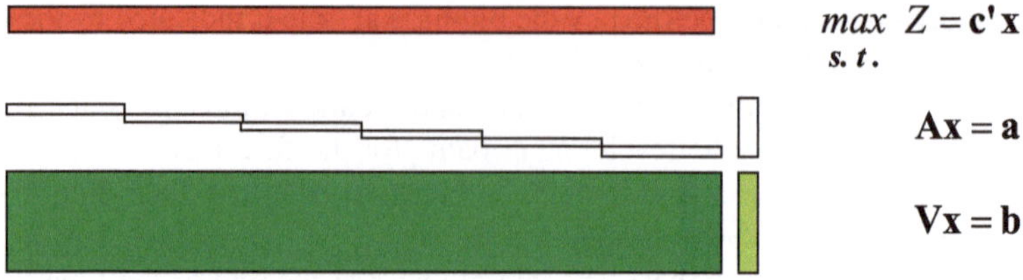

Fig. 6.23 Diagram illustrating the three elements of the multiple-path model: the objective function (red), area constraints (white), and technical constraints (green)

the Multiple-Path model: an objective function (red), a series of area constraints (white) and technical constraints (green).

In planted forests, the objective function coefficients are usually calculated by means of the Faustmann method of cash flow analysis. The Faustmann method is based on a long-term vision of a forest with a known initial and final state. Hille et al. (1999) presented a critical analysis of the traditional Faustmann formula when applied to continuous-cover forests where there is no beginning or end. In their approach, the net present value (NPV) of a silvicultural alternative is calculated as the sum of the discounted periodic net returns plus the discounted terminal value of the stand at the end of the silvicultural "time window." The idea is to buy a forest stand and sell it again at the end of the time window. We can generate income from intermediate harvest events resulting in a low terminal value at the end of the time window, or alternatively, invest in costly tending treatments that will generate a high terminal value. The time window method appears to be a realistic alternative to the classic cradle-to-grave approach of discounted cash flow analysis in planted forests.

6.4.3 Heuristic Combinatorial Optimization

Management-oriented forest planning is often conducted in two steps. First, alternative treatment schedules are simulated for the calculation units. Second, the optimal combination of the simulated schedules is found by using combinatorial optimization. The traditional calculation unit in management-oriented planning is the stand (sometimes called compartment or subcompartment). However, it could also be inventory plot, grid cell, micro segment, or stratum, depending on the purpose and geographical scale of the planning task.

The methods available for combinatorial optimization can be divided into two categories: mathematical programming and heuristic methods (also called meta-heuristics). Examples of mathematical programming methods are linear programming and goal programming. Mathematical programming finds the optimal solution for the planning problem with certainty while heuristics usually find very good solutions, but it is not known whether the solution found is the global optimum. The advantage of heuristic methods is that more complex and extensive planning problems can be solved than would be possible using mathematical programming.

Heuristic methods used in forest-level planning can be classified into two categories: global or centralized methods, and local or decentralized methods. Examples of global heuristics are simulated annealing (SA), tabu search (TS), and genetic algorithm (GA). These are the most frequently used heuristics in forest planning, although several other algorithms such as threshold accepting, great deluge, and ant colony optimization (ACO) have also been tested. All heuristic methods change the combination of treatment schedules simulated for stands and evaluate the effect of each change on the objective function value. Net present value used to be the most important variable to be maximized. Utility functions are increasingly used in multifunctional designs where several objectives have to be achieved.

6.4.3.1 Simulated Annealing

Simulated annealing (SA), which belongs to a category known as cooling methods, is probably the most used centralized metaheuristic in forest planning. Other cooling methods are threshold accepting and great deluge. All cooling methods begin with an initial solution (an initial forest design or management plan), which is often a random combination of the treatment schedules simulated for the stands before optimization. Then, *moves* (changes) are made in the initial solution by selecting first a random stand and then a random schedule for this stand from those produced beforehand. This schedule replaces the current schedule (the schedule that is currently in the plan), resulting in a new candidate solution. The replacement (move) is maintained if it improves the objective function (OF) value. Otherwise (if the candidate is worse than the current solution), the move is maintained with the following probability:

$$p = \exp\left[\frac{\left(OF_{\text{Candidate}} - OF_{\text{Current}}\right)}{T}\right] \tag{6.22}$$

where OF_{Current} is the OF value of the solution before implementing the move, $OF_{\text{Candidate}}$ is the OF value after the move, and T is the "temperature," which determines the probability of accepting inferior moves. The temperature is decreased toward the end of the search, which means that the probability to accept inferior moves also decreases. Several candidate moves are produced at each temperature. The search is stopped when the temperature reaches a predefined low value, which is called freezing temperature. A new temperature is obtained by multiplying the current temperature by a multiplier smaller than 1. It is possible to increase or decrease the number of candidate moves per temperature when the process cools.

Accepting inferior moves (candidates are worse than the current solution) is a way to get out from local optima. Acceptance of inferior moves leads to optimization processes where the OF value does not increase monotonously during the optimization run (Fig. 6.24).

6.4.3.2 Tabu Search

Tabu search (TS) is somewhat different from the cooling methods. The first difference is that several candidate moves are produced at each iteration and only one of

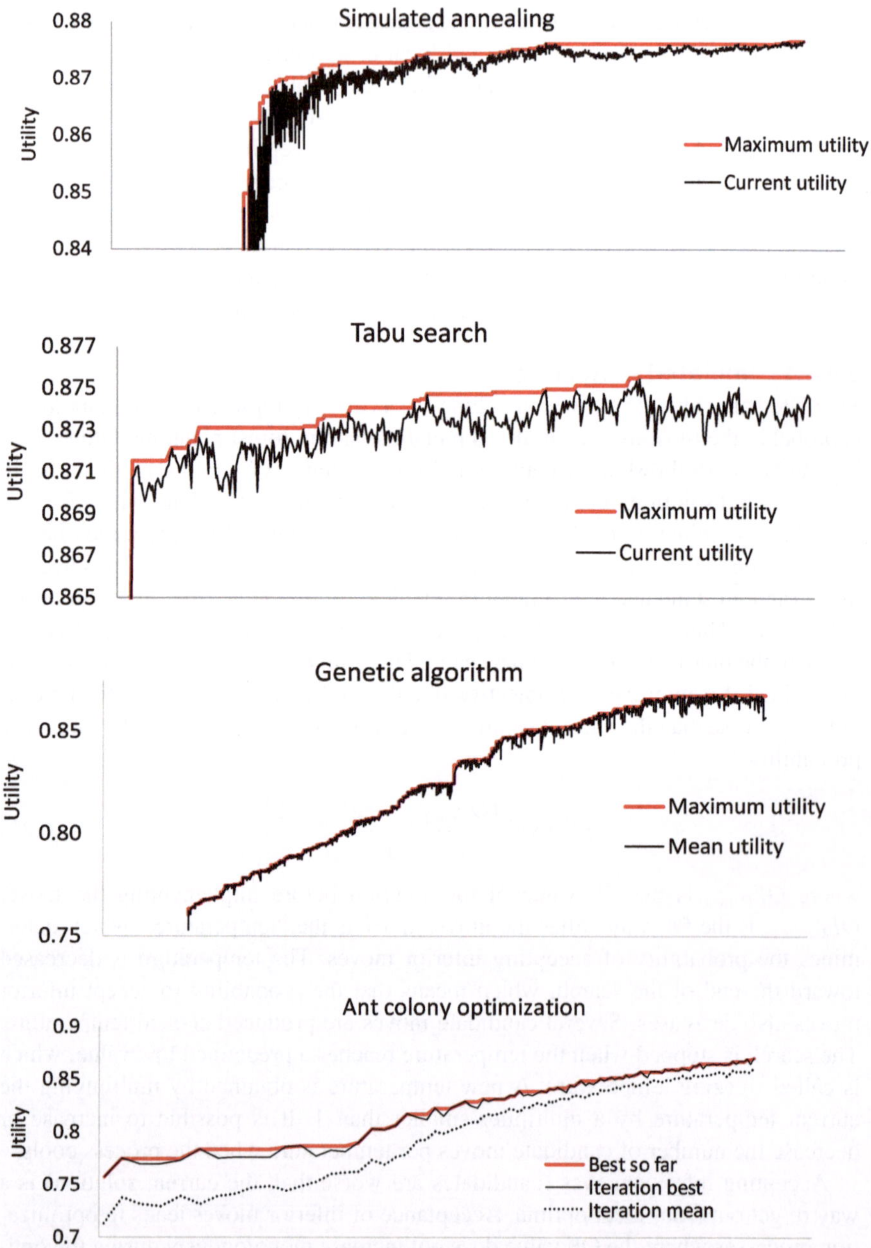

Fig. 6.24 Development of objective function value (utility) in different heuristic methods when they are used in the same forest planning problem (Jin et al. 2016). The red line shows the best solution found so far. In simulated annealing and tabu search, the black line shows the utility of the current solution. In genetic algorithm, the black line is the mean objective function value of the population (chromosomes) and in ant colony optimization, it is the objective function value of the best ant of the iteration. The dotted line is the mean value of all ants (mean utility value of the solutions found by the ants)

them is accepted. The second difference is that tabu lists are used. A tabu list means that a recent change of the solution is forbidden (tabu) for some iterations. Tabu lists prevent the search algorithm from repeating the same changes back and forth, forcing the algorithm to find new combinations of treatment schedules.

The number of candidates produced during iteration can be anything from two to the highest possible value (total number of schedules minus number of stands). The best "non-tabu" candidate is accepted, even if it produces an OF value lower than the current value. If all candidate moves are in the tabu list (forbidden), the move that has the shortest tabu tenure is accepted (shortest tabu tenure means that the move will the first to be removed from the tabu list). If a move would produce a better OF value than the best found so far, the move is accepted even if it is tabu (and has a long tabu tenure). A solution better than the best found so far is called elite solution.

Most forest planning applications use only one tabu list. The two schedules that participate in a move go to a tabu list where they stay for a certain number of iterations. Iteration refers to the generation of a set of candidate moves and selecting the best of them. For example, an initial tabu tenure 10 means that the schedule which goes to the tabu list cannot take part in a move during the next 10 iterations. Every iteration reduces the remaining tabu tenures of all tabu-listed schedules by 1. The idea of a tabu list is illustrated in the following example, where the forest consists of eight stands, and 3–8 different treatment schedules have been simulated for the stands:

Stand	**1**	**2**	**3**	**4**	**5**	**6**	**7**	**8**
Number of treatment schedules	4	4	6	8	8	8	3	5
Current solution	3	1	**2**	4	8	1	1	2
The best candidate solution	3	1	**5**	4	8	1	1	2
Schedules with full tabu tenure			**2, 5**					

One candidate solution is produced by replacing schedule 2 of stand 3 by schedule 5 of the same stand. This candidate happens to be the best non-tabu candidate of the iteration and becomes therefore the new current solution. Those two schedules (2 and 5 of stand 3) that participated in the move cannot be used to produce new candidate solutions during the coming iterations (for example, 10 next iterations). These schedules are given a full tabu tenure (10), which is decreased by 1 at each new iteration.

6.4.3.3 Genetic Algorithm

The genetic algorithm (GA) is based on principles of evolution where natural selection favors the fittest individuals (more exactly: the fittest chromosomes). In GA, alternative solutions of the planning problem (forest plans) are called chromosomes, and fitness is measured with the objective function value (for instance, NPV). A chromosome is a list of schedules included in the solution. A gene of a chromosome corresponds to a stand, and different treatment schedules of the stand are the alleles of the gene.

GA can be implemented in several different ways. Common to all implementations is that, opposite to all previous heuristics, they operate with several solutions. New chromosomes (offspring) are produced by combining two parent chromosomes using crossing-over. The new chromosome may have mutations, that is, small changes somewhere in the chromosome. A common way to use GA in forest planning is to produce only one new offspring during one generation (iteration). Two parent chromosomes are selected from the current population. The probability of selection depends on the fitness value (OF value) of the chromosome, so that the best chromosomes have the highest probability to become selected. The second parent is often selected randomly, using the same probability for all chromosomes.

Then, two crossing-over places are selected randomly. The genes located in the chromosome before the first and after the second point are taken from the first parent and the remaining genes are taken from the second parent. In the example shown below, two parents are selected from the population of solutions (Parents 1 and 2). Each parent consists of the numbers of the treatment schedules selected for 10 stands. The randomly selected crossing-over places are 3/4 and 7/8. The schedules for stands 1–3 and 8–10 are taken from the first parent and the schedules of stands 4–7 from the second parent.

Stand	1	2	3	4	5	6	7	8	9	10
Parent 1	3	2	11	13	7	15	1	9	4	6
Parent 2	12	5	10	4	5	7	7	8	11	1
Off-spring after crossing-over	3	2	11	4	**5**	7	7	9	4	6
Off-spring after mutation	3	2	11	4	**11**	7	7	9	4	6

Mutations can be applied to the new offspring. A mutation means that a random gene (stand) is selected, and its schedule is replaced by another, randomly selected schedule simulated for the same stand (blue shading in the above example). If mutations are made, their probability is usually low. Mutation rate may also change during the GA search process.

After producing a new offspring, one chromosome is removed from the population. The probability of removal is inversely proportional to the ranking of the chromosome (worst chromosomes have the highest probability of removal). The search is stopped when a predefined number of generations are completed.

6.4.3.4 Ant Colony Optimization

Ant colony optimization (ACO) represents swarm intelligence. Another swarm intelligence method is particle swarm optimization, which has been used in the optimization of stand management. Typical of swarm intelligence is a social component, which means that the behavior of an individual depends on the other individuals of the swarm. In ACO, the social component is implemented by using pheromone, which measures how good solutions have been found earlier, and how recently the good solutions were found.

There are a few different versions of ant colony optimization. In the Max–Min Ant System, all treatment schedules are given the same initial pheromone level. Each ant of the first ant iteration constructs a solution by visiting every stand and selecting a random treatment schedule for each stand. Then, the objective function value is calculated for each ant. Good solutions get additional dosage of pheromone. The pheromone also evaporates, which means that the memory of the ant colony is limited. When a new set of ants start constructing solutions, the probability to select schedules with high pheromone level is elevated. Those treatment schedules belong to solutions that were evaluated as good during previous ant generations.

The pheromone levels of all treatment schedules are updated at the end of each iteration. In the Max–Min Ant System, only the best ant of the iteration (iteration-best) or the best of all iterations completed so far (best-so-far) updates the pheromones. As a consequence, when new ants generate solutions during the next iterations, they are attracted toward the best solutions found during previous iterations.

The process of constructing solutions and updating pheromones is repeated for a certain number of ant iterations. The number of ants (number of solutions) is the same at each iteration. The best solution constructed during the whole search process (best so far) is the solution found by the ACO heuristic.

Of the four metaheuristics described above, SA has often been identified as the best in forest planning, followed by tabu search. However, compared to SA, the other methods offer more options and variants. They can be used in several different ways, although it seems that not all possible ways have been tested in forest planning. Therefore, the other heuristics (Genetic Algorithm and Ant Colony Optimization) are candidates for further development and testing.

6.4.3.5 Decentralized Heuristics

Two decentralized heuristic methods, namely the reduced cost method and cellular automata, have been used in forest management planning. Cellular automata can be regarded as multipurpose tools in forest planning, because they can also be used for stand delineation and to simulate spatial processes such as forest fire and pest outbreak.

Decentralized heuristics differ from the centralized ones in that the effect of changing the treatment schedule of a stand is evaluated locally, that is, at the stand level. This property greatly reduces the potential number of alternatives from which the best one should be selected. For example, if the forest consists of 100 stands and each stand has 10 alternative treatment schedules, the number of evaluations is 1000 (ten per stand). In centralized methods, all combinations of treatment schedules are evaluated. The 10 alternatives simulated for each of the 100 stands can be combined in 10^{100} different ways, which is such a huge number that it is impossible to evaluate all possible combinations separately. Therefore, heuristic algorithms and other methods developed for combinatorial optimization (for instance, the simplex method in linear programming) have been developed in such a way that they can find a very good or the best combination by inspecting only a small fraction of all potential combinations.

Based on the fact that the potential number of alternatives (decision space) is much larger in centralized solution methods, it can be assumed that decentralized heuristics should be preferred in planning problems where the number of stands or other calculation units is very large. However, forest-level planning usually has constraints or objectives that cannot be evaluated at the stand level. A typical forest-level constraint is the need for an even flow of incomes or harvested wood. These constraints make stand-level decisions dependent on each other, complicating the use of decentralized heuristics. This fact is dealt with in different ways in cellular automata and the reduced cost method. The method of reduced costs employs so-called dual prices (shadow prices) of the forest-level constraints to penalize the stand-level objective function if the constraints are not met. For example, if net present value (NPV) is maximized at the stand level subject to certain total harvest from the whole forest during three 10-year periods, the stand-level objective function is as follows (Pukkala et al. 2008):

$$\max z_j = NPV_j - v_1 H_{1j} - v_2 H_{2j} - v_3 H_{3j} \tag{6.23}$$

where NPV_j is the net present value of stand j (€); H_{1j}, H_{2j}, and H_{3j} are the volumes harvested from the stand during 10-year periods 1, 2, and 3 (m^3); and v_2, v_2, and v_3 are the dual prices of the harvest volume constraints. They can be zero, positive, or negative. In this example, the unit of the dual prices is €/m^3 (NPV is calculated in euros).

Dual prices depend on the difference between the realized harvested volume and its target value in such a way that the absolute value of dual price increases with increasing deviation from the target value. The first round (selecting, for each stand, the schedule that maximizes NPV) can be done by using zero dual prices. Then, the total harvest is calculated and the dual prices are updated based on the deviation between the total harvested volume of the selected schedules and the target volume, and the process is repeated. This is repeated until the harvested volumes are close enough to their target values. The so-called subgradient method can be used to update the dual prices.

In the cellular automata approaches that have been used so far in forest planning, the usual approach is to ignore the forest-level constraint at first. After finding the best (unconstrained) schedules for the stands, a global part is added to the objective function, and the best schedules are selected again. The global objective function measures deviations from the global constraints. It has a weight (b in Eq. 6.24 below), which is very small at first, and gradually incremented until deviations from the global constraints are acceptable (e.g., less than 1% of the value of the constraint). In the same planning problem as shown above for the reduced cost method, the formula for selecting the treatment schedule for stand j is Eq. 6.24 (Heinonen and Pukkala 2007):

$$\max z_j = \frac{a_j}{A} NPV_j + b\left[\left(w_1 p_1\left(H_1\right) + w_2 p_2\left(H_2\right) + w_3 p_3(H_3)\right)\right] \tag{6.24}$$

where a_j is the area of stand j, A is the area of the whole forest, w_k is the weight of the global objective k, p_k is the subpriority function for the harvested volume of period k, and H_k is the total harvested volume of period k. The subpriority functions are formulated in such a way that an increasing deviation from the target harvest decreases the subpriority.

6.4.3.6 Using Stand-Level Optimization in Forest-Level Planning

The two decentralized heuristics described above can be used in a stand-level optimization, which takes into account forest-level objectives and constraints. In the normal optimization-based forest planning, a limited number of alternative treatment schedules are simulated for the stands, after which their optimal combination is found by using combinatorial optimization.

The approach is different in stand-level optimization: the time points, intensities, and types of cuttings are optimized by using methods developed for continuous decision variables (thinning years, thinning intensities, etc.). In this optimization, the potential number of alternative schedules is unlimited, which means that the optimal schedules found for the stands are better (for example, in terms of NPV) than in the case where the best schedule is selected from among a limited number of schedules simulated before the optimization step.

The formulas presented above for the reduced cost method and cellular automata offer a means to develop forest-level management plans by using stand-level optimization. It can be expected that stand-level optimization, conducted in such a way that forest-level constraints are respected, increases the efficiency of sustainable forest management, compared to plans developed by using the traditional two-step approach (first simulation, then optimization).

6.4.3.7 Spatial Heuristic Optimization

The reduced cost method and cellular automata can also be used in spatially explicit problems, for example, for aggregating harvesting operations or protecting old-growth forests. In these problems, the local part of the objective function is augmented with a neighborhood-relation function, which measures the similarity of the stand or its treatment with adjacent stands. For example, if the aim is to aggregate harvests, the local objective function may be formulated as follows:

$$\max z = w_1\left(NPV / NPV_{\max}\right) + w_2\left(CCB / TB\right) - w_3\left(CuCB / TB\right) \tag{6.25}$$

where CCB is the length of the common border of the stand with those adjacent stands that are harvested at the same time, $CuCB$ is the common border with those adjacent stands that have a different harvest prescription (only one of the adjacent stands is harvested within a particular year), and TB is the total border length of the stand. For example, in Fig. 6.25, the CCB of stand 1 is its common boundary with

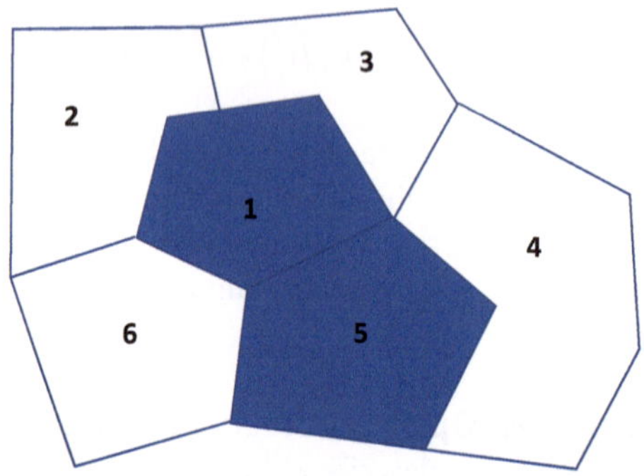

Fig. 6.25 Visualization of cut–cut boundary (CCB) and cut–uncut boundary (CuCB). The shaded stands (1 and 5) have a cutting prescription for a certain year, while the other stands do not have a cutting prescription. The CCB of stand 1 is its common boundary with stand 5, and the CuCB is its common boundary with stands 2, 3, and 6

stand 5 (both stands have cutting prescription for the same year) and the *CuCB* is the length of common boundary with stands 2, 3, and 6.

Maximizing *CCB* leads to aggregated cuttings and a large total cutting area while minimization of *CuCB* leads to compact shapes (short perimeters) of cut areas and a small total area of cutting prescriptions. Instead of cuttings or other treatments, it is possible to aggregate old-growth forests or other habitat types, by maximizing the boundary between adjacent stands that both represent the wanted habitat and minimizing habitat–nonhabitat boundary.

In the above formulation, *CCB* and *CuCB* were divided by the total border length of the stand (*TB*), and NPV was divided by the highest possible NPV of the stand (NPV_{max}). In this way, all three objective variables were converted to relative values ("priorities," "sub-utilities") which made it possible to calculate their sum. The weighted sum of the relative values of objective variables can be understood as a priority index or utility index.

The same types of spatial variables (*CCB* and *CuCB*) can also be used as objective variables in global heuristics (Fig. 6.26). In global heuristics, the objective variables may be the total cut–cut border, calculated over all stands, divided by the total border length of all stands. The cut–uncut border or the habitat–habitat border is calculated in the corresponding way. By using these simple variables, both local (decentralized) and global (centralized) heuristics can be used to aggregate or disaggregate harvests or certain types of forest (for instance, old-growth forest).

Several other types of forest variables can also be used in spatial optimization. An example is shown in Fig. 6.27, where the aim was to minimize differences in tree height between adjacent stands. For comparison, another plan was developed where height differences were maximized.

The total harvest removal was constrained to be the same in both cases. Simulated annealing was used to find the optimal combination of the stand-level treatment schedules.

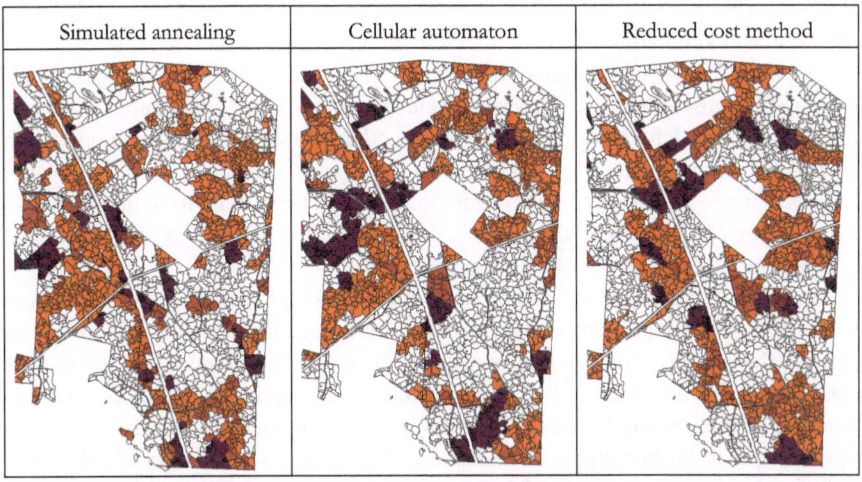

Fig. 6.26 Cutting prescriptions in solutions obtained with three different heuristic optimization methods (Pukkala et al. 2014). Dark purple tones indicate regenerative cuttings and brown tones indicate thinning. NPV is maximized with cutting aggregation target and with a fixed total harvest

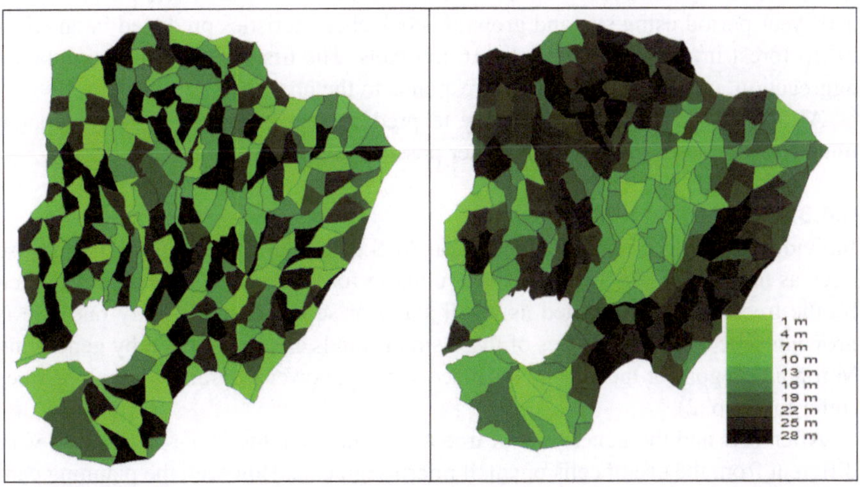

Fig. 6.27 Maximizing tree height differences between adjacent stands (left) leads to fragmented forest landscape where the risk of wind damage is high and habitat patches are frequently disconnected. Minimization of height differences (right) leads to "smoothed" landscape characterized by gradual change where the risk of wind damage is low and similar habitats, such as instance mature stands, form large contiguous areas

6.4.3.8 Use of Small Spatial Units in Forest Planning

The use of airborne laser scanning (ALS) has opened completely new avenues to develop forest management plans. Through the use of ALS, growing stock characteristics can be estimated for small spatial units, for instance small grid cells (their size is currently 16 m × 16 m in Finland), micro segments (see Chap. 2: the interpretation and simulation units are micro segments), or even individual trees.

The use of small spatial units in forest planning leads to the need for spatial optimization because otherwise the treatment areas would be too small and irregular for the practical implementation of the plan. Another problem related to small spatial units is the increasing number of simulations and greatly increasing size of the planning problems. For these reasons, heuristic methods need to be used more often. If the number of spatial units is very large, decentralized methods might be most appropriate for solving the planning problem.

A way to decrease the number of simulation units and avoid spatial optimization is to aggregate the grid cells or other small units into large enough stands before starting the optimization calculations, using the methods of automated stand delineation described in Chap. 2. It is also possible to use segmentation to create homogeneous segments larger than cells but smaller than stands, and then use spatial optimization to further aggregate treatments. This reduces the number of calculation and simulation units (compared to small cells) but guarantees that the simulation units are homogeneous, which is not necessarily the case when very large stands are used as calculation and simulation units.

The differences between the use of cells, small segments, and large stands are illustrated in Fig. 6.28 (adapted from Pukkala 2019). The plans were developed for a 10-year period using site and growing stock characteristics predicted in an ALS-based forest inventory for 16 m by 16 m pixels. The first approach (presimulation aggregation, use of large stands) corresponds to the prevailing practices.

ALS inventory and the possibility to predict stand variables for small spatial units lead to increasing use of the other possibilities.

6.4.3.9 Tree Level Planning

Individual-tree detection (ITD) based on ALS pulse data makes it possible to use trees as the calculation and simulation units in forest planning. The area occupied by the tree can be interpreted as small stand or segment, where only one tree is growing. The exact boundaries of these small stands can be obtained by generating Voronoi polygons or their more advanced version, power diagrams around the trees (refer to Chap. 2).

After ITD and the generation of tree areas, the planning is not necessarily very different from the use of cells or small micro segments. However, the planning cannot be done using the two separate steps of the traditional systems, namely first simulating alternative treatment schedules for the spatial units (in this case, trees) and then finding the optimal combination of the simulated treatment schedules. This is because the growth of a tree depends on the competition by trees growing in adjacent polygons. Therefore, it is impossible to predict the growth of an individual tree without knowing the growth and treatments of adjacent trees.

A suitable planning approach with tree level data is stand-level optimization conducted in such a way that the selection of the trees is also optimized or follows a rule that results in near-optimal removal of trees. For example, the development of trees growing in the forest is simulated using individual-tree growth models, and the time

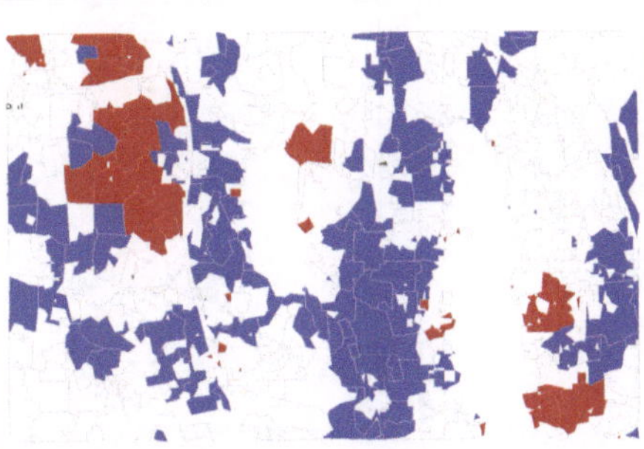	Pre-simulation aggregation (large stands used) • Growth overestimated • Area of rare fertility classed under-estimated • Very fast simulation • Easy and fast optimization • Additional segmentation step required • Total time consumption 21.7 min in a 2500-ha forest
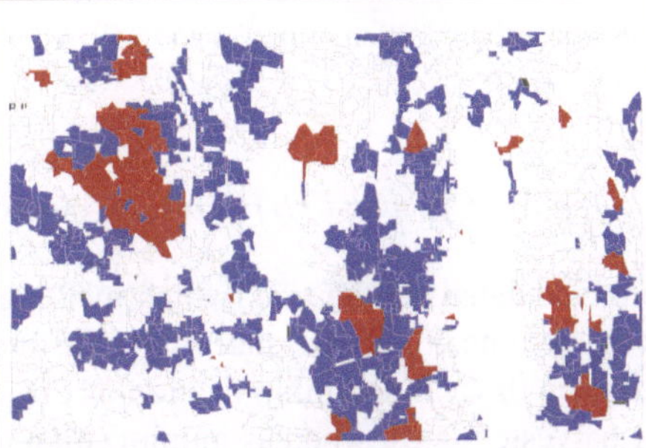	Pre- and post-simulation aggregation (small segments used) • Growth may be overestimated • Area of rare fertility classes may be underestimated • Fast simulation • Additional segmentation step required • Spatial optimization required • Total time consumption 90.4 min in a 2500-ha forest
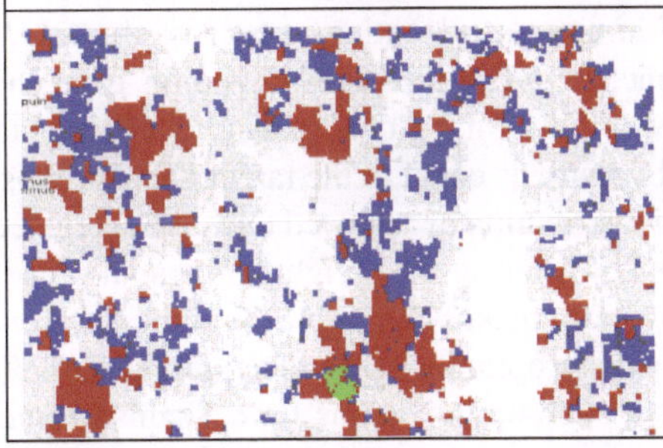	Post-simulation aggregation (cells used) • High time consumption in simulation and optimization • Highest number of small treatment units • Smallest biases in results • Total time consumption 433.6 min in a 2500-ha forest

Fig. 6.28 Summary of the results for different approaches to deal with raster data in forest planning

points of cuttings and cutting intensities are optimized in the same way as is normally done in stand-level optimization. The individual trees that are removed during a partial harvest operation are selected based on a meaningful rule, for instance based on relative value increment. Those trees that have the lowest value increment are financially the most mature, and they should therefore be removed if the aim is to maximize the economic profitability of a forest. The rule, which is used to select cut trees, may also be optimized (Fig. 6.29).

The example explained above and shown in Fig. 6.29 does not offer possibilities to control the spatial distribution of harvested trees. Such a control becomes possible through the use of decentralized heuristics where a tree level objective function is maximized using, for instance, cellular automaton. In a recent study (Packalen et al. 2019), the following objective function was maximized separately for each tree

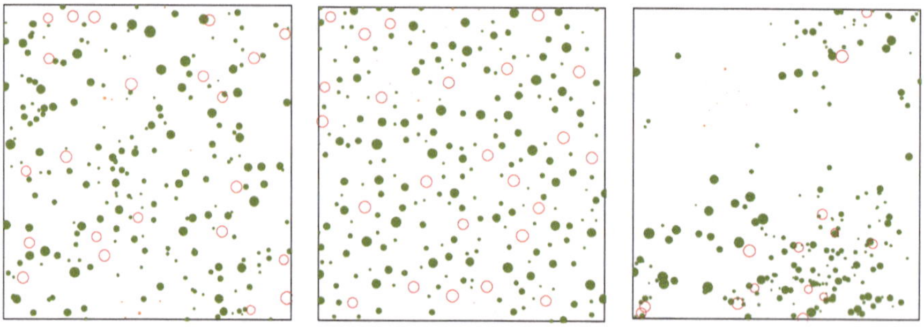

Fig. 6.29 Examples of stand-level optimization where the selection of harvested trees (red circles) is also optimized (Pukkala et al. 2016)

$$\max z = \frac{V_{\text{Tree}}}{V_{\text{Total}}}\left[w_1 p_1 (RVI) + w_2 p_2 (CCB/TB) + w_3 p_3 (CuCB/TB)\right] + b p_4 (H) \quad (6.26)$$

where *RVI* is the relative value increment calculated for the tree (which measures the financial maturity of the tree), V_{Tree} is the volume of the tree, and V_{Total} is the total volume of all trees in the forest. Maximization of *CCB* and minimization of *CuCB* lead to aggregated spatial distribution of harvested trees, whereas minimization of *CCB* and maximization on *CuCB* result in dispersed location of harvested trees. *H* is the total harvest from the forest. Function $p_4(H)$ measures deviation from the target harvest. Gradual increment in the weight of the global objective (*b*) leads to tree selections where the total harvested volume is equal to the target harvest, most mature trees are removed, and the spatial distribution of removed trees is aggregated or dispersed, depending on what is wanted. Altering the weights of spatial objectives (negative weight is equal to minimization) makes it possible to mimic different disturbance regimes and silvicultural systems in optimized tree selection. Figure 6.30 (Packalen et al. 2019) shows examples of such optimizations for a Spanish *Pinus pinea* forest.

The use of individual trees as the basic calculation unit in forest management planning is a new field. Therefore, a new methodological development is required. The approaches described above are just examples of the first attempts to optimize the treatment prescription separately for each tree. In addition, in the same way as in cell-, segment-, or stand-based spatial forest planning, it is also possible to use global metaheuristics for selecting the harvest event in which different trees are removed from the stand. The difference, compared to the use on larger spatial units, is the need to simulate each combination of harvest events of trees. This means that planning can no longer be divided into separate simulation and optimization steps.

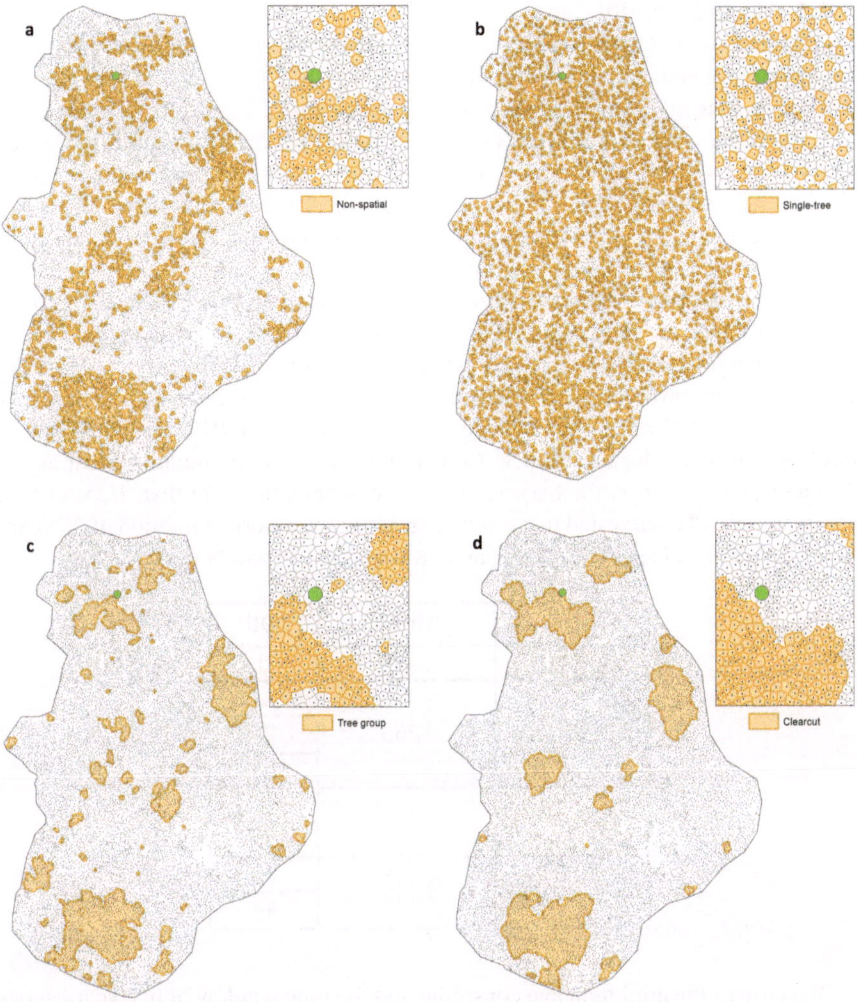

Fig. 6.30 Harvest block maps obtained by altering the weights of the spatial objective variables of tree level objective function. Case **a** is a nonspatial plan, case **b** aims at dispersed distribution of harvested trees (corresponds to single-tree selection), case **c** aggregates harvested trees with low weights on spatial objectives (group selection), and case **d** aggregates harvested trees with high weight on spatial objective (corresponds to clear-cuts)

6.4.4 Application Examples

Having discussed the basic theory and a range of methods for implementation, we will now use several examples to illustrate the range of possible applications of the multiple-path (MP) concept of sustaining ecosystems.

6.4.4.1 Indira Gandhi Canal

The Indira Gandhi Canal, one of the biggest canal projects in India, provides water for agriculture and human use in remote areas. Drought-resistant tree species, such as *Acacia senegal, Acacia nilotica, Prosopis cineraria, Tecomella undulata, and Salvadora persica,* were planted on either side of the canal to protect the watercourse from shifting sand dunes. The main canal as well as subordinate canals serves small villages. The villagers own cattle and sheep, and grow vegetables. Additional income is provided by woodcarving. Wood from trees harvested in the canal plantations provides the raw material.

The tree plantings on either side of the watercourse are subdivided into blocks. Individual blocks usually cover an area of three hectares (300 m along the canal x 100 m extending into the desert). Each block has a unique identifier (e.g., 1365RD). In a pilot study involving four selected blocks planted with *Tecomella undulata* of varying density (N = trees/ha) and age, Tewari and Gadow (2010) generated several development paths for each block. Constraints include a minimum harvest age of 21 years and a limit on the harvest area, for example, no more than 1/25th of the block area may be harvested in a given year, assuming a normal rotation of 25 years. The block IDs and ages (16, 20 17, and 18 years) are shown below:

age	block	path
16	1333	1b
20	1365	2a
		2b
		2c
17	1406	3a
		3b
18	1438	4a
		4b

Because of the minimum age constraint and the time window of five years, block 1333 does not qualify for a harvest. Three development paths are specified for block 1365 and two each for blocks 1406 and 1438. Typical objectives are age-related wood quality attributes (wood density, wood energy content) and even flow of harvest yields for the woodcarving community. The even flow (EF), for example, is defined as follows:

$$EF = \sqrt{\frac{1}{M}\sum_{j=1}^{M}\left(V_j - \bar{V}\right)^2} \tag{6.27}$$

where M is the number of years in the design period; V_j is the harvest yield in each year; and $\bar{V}$ is the average yearly volume yield during the entire period (3.6 m^3). Table 6.12 shows the 12 (2 × 3 × 2) path combinations, the periodic wood volume yields, and the deviations.

Table 6.11 The 12 path combinations, periodic wood volume yields, and the deviations from even flow

	volume yield (m³)						deviations					
combination	year 1	year 2	year 3	year 4	year 5	total (m³)	1	2	3	4	5	penalty
2a3a4a	0.0	0.6	0.0	3.1	8.0	11.7	13.0	8.9	13.0	0.3	19.7	3.3
2a3b4a	0.0	0.6	2.2	3.1	10.6	16.4	13.0	8.9	2.0	0.3	48.8	3.8
2a3a4b	0.0	0.6	2.2	3.1	10.6	16.4	13.0	8.9	2.0	0.3	48.8	3.8
2a3b4b	0.0	0.6	2.2	8.9	10.6	22.2	13.0	8.9	2.0	27.7	48.8	4.5
2b3a4a	0.6	0.0	0.6	2.4	8.7	12.3	9.1	13.0	8.7	1.5	26.5	3.4
2b3b4a	0.0	0.6	2.2	8.9	10.6	22.2	13.0	8.9	2.0	27.7	48.8	4.5
2b3a4b	0.6	0.0	2.8	2.4	11.3	17.1	9.1	13.0	0.6	1.5	59.3	4.1
2b3b4b	0.6	0.0	2.8	8.2	11.3	22.9	9.1	13.0	0.6	21.0	59.3	4.5
2c3a4a	0.6	0.6	0.6	3.1	8.7	13.6	9.1	8.9	8.7	0.3	26.5	3.3
2c3b4a	0.6	0.6	2.8	3.1	11.3	18.4	9.1	8.9	0.6	0.3	59.3	4.0
2c3a4b	0.6	0.6	2.8	3.1	11.3	18.4	9.1	8.9	0.6	0.3	59.3	4.0
2c3b4b	0.6	0.6	2.8	8.9	11.3	24.2	9.1	8.9	0.6	27.7	59.3	4.6
average	0.3	0.5	2.0	4.8	10.4	3.6					average	4.0

For example, the deviation in the first year is equal to the volume yield in year 1 (0.0) minus the average total volume yield (3.6). This difference (0.0–3.6) is squared to give 12.96 or 13.0 when rounded to one decimal (first line, column 8). The even-flow penalty in the last column of Table 6.11 is equal to the average deviations during the 5 years in the design period. For example, $\sqrt[2]{(13.0+8.9+13.0+0.3+19.7)/5} = 3.3$ for combination 2a3a4a.

The optimum design represents the best combination of development paths in terms of the specified objectives and subject to the given constraints. The example shows the trivial truth that ecosystem dynamics is determined to a large extent by the impact of human activity and to a much lesser extent by natural processes. For this reason, it is essential that ecological research which focuses on the processes and functions of wild systems (which hardly exist anymore) needs to be complemented by studies dealing with novel ecosystems.

6.4.4.2 A Forest in Germany

Implementing the most important principles outlined in the previous sections, Chen and Gadow (2008) applied the multiple-path approach to the Winnefeld forest in north-western Germany. The method of simulated annealing was used to identify the optimum combination of management paths. The objective function included three components:

$$a \times EC - b \times EF + c \times FSV \rightarrow \max \tag{6.28}$$

where a, b, and c are constant factors which are used to balance the relative importance of the three components. ***EC*** is an economic criterion that represents the discounted cash flow derived from selective harvests during the planning period plus the terminal stumpage value at the end of the planning period. ***EF*** is an even-flow penalty function that quantifies the degree of output fluctuation during the planning period. It is defined as follows:

$$EF = \sqrt{\frac{1}{M}\sum_{j=1}^{M}\left(V_j - \bar{V}\right)^2} \tag{6.29}$$

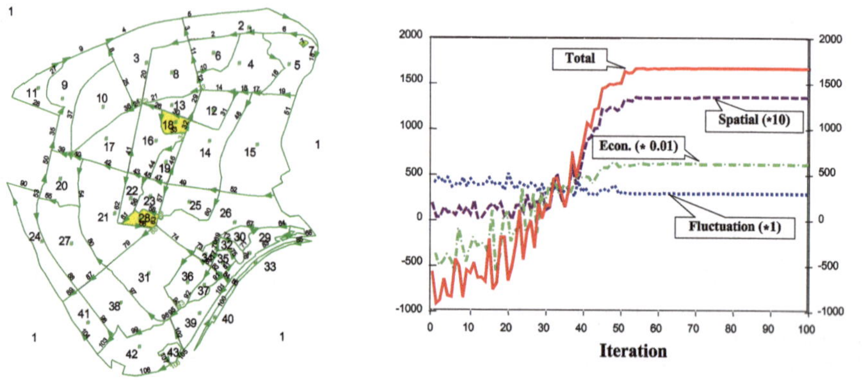

Fig. 6.31 Left: Map of the Winnefeld South forest with three themes: polygon number (large digit), polyline number (small digit), and centroid position (#). There are 108 polylines in this coverage. The CBL is derived from the polyline length. The distance between every two centroids is calculated by means of the Euclidian distance formula. Right: Example of the convergence when a simulated annealing process with three objective function components (economic, even-flow, and spatial) is used

where M is the planning horizon (the number of years in the planning period); V_j is the timber yield in each year; and $\overline{V}$ is the average yearly timber yield during the entire design period. **FSV** (forest stock volume) is a forest spatial criterion that can be used to achieve spatial aggregation or dispersion of forestry operations during the design period. Aggregation of operations is important as it reduces the cost of moving harvesting machines and equipment at a given point in time. Dispersion is important if conservation objectives are to be reached (Öhman 2000). The variable FSV for a forest with known geographical detail was defined as follows:

$$FSV = \sum_{i=1}^{N} NV_i = \sum_{i=1}^{N}\sum_{k=1}^{N} \frac{R_{ik} L_{ik}}{D_{ik}} \quad (i \neq k) \tag{6.30}$$

where N is the number of land parcels (commonly known as stands); NV_i is the neighbor value for stand i; R_{ik} is a value which quantifies the relation between two treatment options of two neighboring stands i and k; L_{ik} is the common boundary length (CBL) of stands i and k; and D_{ik} is the distance between the two centroids of stands i and k. The variable R_{ik} is used to encourage or discourage pairs of neighboring stands to have the same, similar, or different management paths. R_{ik} defines a value for management path i and management path k, whenever i and k refer to a pair of neighboring stands. If a spatial concentration of management activities at a given point in time is desired, the pair of adjacent stands will contribute a positive value to the objective function ($R_{ik} > 0$); otherwise, R_{ik} is assigned a negative value. The magnitude of the positive or negative contribution is determined by the value of R_{ik}. The entire matrix of R_{ik} among all the management paths is determined beforehand. A detailed description of the variable R_{ik} and the method of aggregation or dispersion of management activities in the landscape can be found in Chen and Gadow (2002).

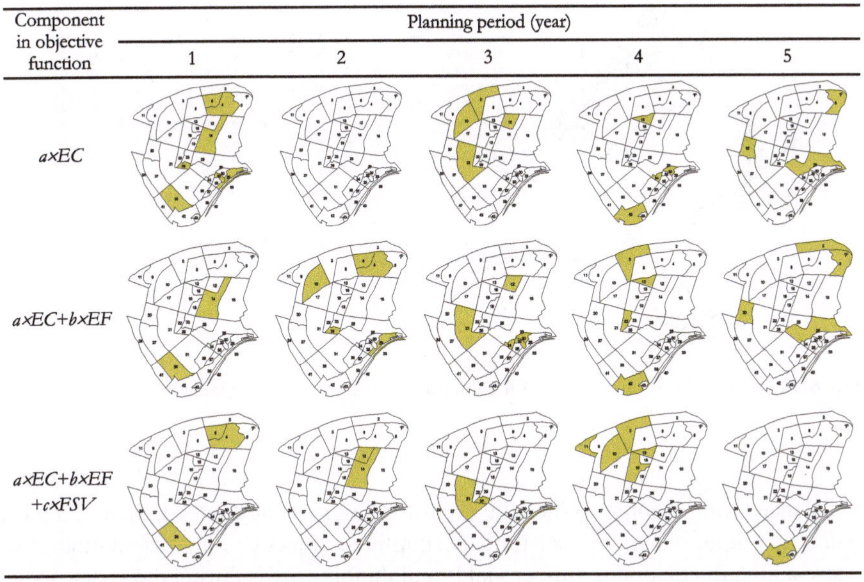

Fig. 6.32 Comparison of thinning areas resulting from the use of a one-, two-, and three-component objective function. Shadings indicate compartments where thinnings take place. The solution changes after the introduction of even-flow and spatial components

Winnefeld South includes 41 stands that differ in terms of area, age, site index, dominant height, number of trees, and basal area. A typical forest map in ArcView GIS can provide the required geographic information about the forest stands, such as the area and the centroid coordinates. Details about obtaining the CBL are provided by Chen and Gadow (2008). Figure 6.31 is a map of the forest and an example of a simulated annealing process. All the objective function components converged after fewer than 100 iterations. If the economic criterion is the only component in the objective function, the value of *EC* is the highest, as expected, while the other components are ignored. Adding the even-flow component improves the corresponding desirable property of the harvest schedule while decreasing the economic component value to some extent.

When compared to the results of the one-component objective function, the yearly operational areas of the two-component objective function show some similarities (Fig. 6.32, second row). The result of the optimization can be expressed in terms of annual operational plans. The operational schedules for the first five years are shown in Fig. 6.32. The maps in Fig. 6.32 indicate the (shaded) compartments where thinnings are scheduled to take place in the different years. As a result, the period-maps facilitate the interpretation of the spatial effect of adding specific components to the objective function. Thinnings were not scheduled in period 2 when only EC featured in the objective function (Fig. 6.32, first row, second map). There are no timber harvests during the second year in the planning period, resulting in an

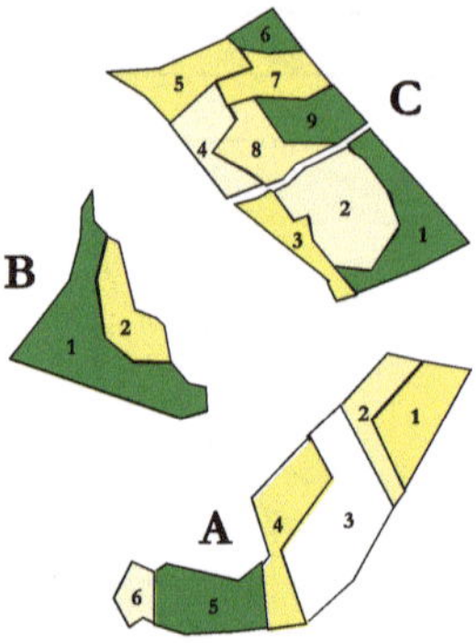

stand	SI	ha	age	0	1	2	3	4	5	6	7	8	9	10
									age					
A1	24.5	3.4	9	0	0	0	0	0	0	0	0	0	3.4	0
A2	24.5	2.8	8	0	0	0	0	0	0	0	0	2.8	0	0
A3	24.5	5.9	5	0	0	0	0	0	5.9	0	0	0	0	0
A4	26.1	4.2	9	0	0	0	0	0	0	0	0	0	4.2	0
A5	26.1	3.8	2	0	0	3.8	0	0	0	0	0	0	0	0
A6	26.1	0.9	0	0.9	0	0	0	0	0	0	0	0	0	0
B1	22.8	6.7	5	0	0	0	0	0	6.7	0	0	0	0	0
B2	22.8	2.9	9	0	0	0	0	0	0	0	0	0	2.9	0
C1	26.1	4.0	6	0	0	0	0	0	0	4	0	0	0	0
C2	26.1	5.8	6	0	0	0	0	0	0	5.8	0	0	0	0
C3	26.1	2.7	6	0	0	0	0	0	0	2.7	0	0	0	0
C4	26.1	3.3	6	0	0	0	0	0	0	3.3	0	0	0	0
C5	25.0	3.5	2	0	0	3.5	0	0	0	0	0	0	0	0
C6	25.0	0.9	2	0	0	0.9	0	0	0	0	0	0	0	0
C7	25.0	3.2	9	0	0	0	0	0	0	0	0	0	3.2	0
C8	24.5	3.3	7	0	0	0	0	0	0	0	3.3	0	0	0
C9	25.0	3.4	3	0	0	0	3.4	0	0	0	0	0	0	0
		60.7	Sum	0.9	0.0	8.2	3.4	0.0	12.6	15.8	3.3	2.8	13.7	0.0

Fig. 6.33 Map and relevant details of the the Eucalyptus grandis ecosystem

unbalanced timber output. When the even-flow component is added, there is a more balanced output of timber over time. A simplified objective function can therefore cause a high fluctuation in timber yields and thinning areas during the years, something that is not often desirable.

When the spatial component is added into the objective function, the spatial aggregation of thinnings improves notably (Fig. 6.32: maps in third row). The economic component decreased somewhat while the yield fluctuation showed only a slight increase. To realize spatial aggregation, it is necessary to ensure that neighboring stands will be thinned during the same time period. Ways in which this can be achieved are explained by Chen and Gadow (2002).

6.4.4.3 A Planted Eucalyptus Ecosystem

To illustrate the most relevant details of the multiple-path theory, we will use the example of a forest farm with 60.7 hectares of *Eucalyptus grandis*. The farm consists of three blocks subdivided into compartments. A map of the forest is shown in Fig. 6.33.

The following yield model is used to estimate timber yields from site index and age:

$$V = e^{(2.83+0.151\cdot SI)} \cdot \left(1 - e^{-0.055\cdot(age-1)}\right)^{1.19} \tag{6.31}$$

where V is the volume (m^3/ha) and SI is the site index (dominant stand height in m, at age 10 years). The model structure is shown in Table 6.12.

The third block shows the area constraints that specify that the sum of the path areas in each compartment must be exactly equal to the compartment area. Finally, the line at the bottom shows the solution: the entire area of both compartments A1 and A4 is subject to one path (one in both cases), while B1 is partly treated with path B11 (3.8 ha) and B12 (2.9 ha).

> **Assignment 2**
> Use the R function Rglp and the data in Table 6.12 to formulate the model. You may use the formulation below, but the coefficients need to be changed. Run the model and interpret the results.

Table 6.12 Details of the multiple-path model for the Eucalyptus grandis ecosystem

	A1			A2	A3	A4			A5	A6	B1		B2		
	A11	A12	A13	A21	A31	A41	A42	A43	A51	A61	B11	B12	B21		
age year 0	9	9	9	8	5	9	9	9	2	0	5	5	9		
year 1	223.7					284.8							173.1	>	400
year 2		246.1		223.7			313.4							>	400
year 3			267.6					340.8						>	400
year 4												136.3		>	400
m³ yr 4	46.4	21.0	0.0	21.02	200.4	59.1	26.8	0.0	160.0	92.8	155.0	0.0	35.9		
vol tot	270.1	267.1	267.6	244.7	200.4	344.0	340.1	340.8	160.0	92.8	155.0	136.3	209.0		
A1	1	1	1											=	3.4
A2				1										=	2.8
A3					1									=	5.9
A4						1	1	1						=	4.2
A5									1					=	3.8
A6										1				=	0.9
B1											1	1		=	6.7
B2													1	=	2.9
solution (ha)	3.4	0	0	2.8	5.9	4.2	0	0	3.8	0.9	3.8	2.9	2.9	6511.9	m³

```
library(Rglpk)
# names<-c("A11","A12","A13","A21","A31","A41","A42","A51","A61","B11","B12","B21")
obj<-c(61.3,66.1,68.6,53.5,45.8,61.3,66.1,73.9,45.8,81.2,73.9,53.5)
Year1<-c(30.7,0,30.7,0,0,30.7,0,43.2,0,43.2,43.2,30.7)
Year2<-c(0,43.2,0,30.7,0,0,43.2,0,0,0,0,0)
Year3<-c(0,0,22.9,0,30.7,0,0,0,30.7,22.9,0,22.9)
Year4<-c(30.7,0,0,0,0,30.7,0,30.7,0,0,30.7,0)
A1<-c(1,1,1,0,0,0,0,0,0,0,0,0); A2<-c(0,0,0,1,0,0,0,0,0,0,0,0)
A3<- c(0,0,0,0,1,0,0,0,0,0,0,0); A4<-c(0,0,0,0,0,1,1,0,0,0,0,0)
A5<- c(0,0,0,0,0,0,0,1,0,0,0,0); A6<-c(0,0,0,0,0,0,0,0,1,0,0,0)
B1<- c(0,0,0,0,0,0,0,0,0,1,1,0); B2<-c(0,0,0,0,0,0,0,0,0,0,0,1)
mat<-rbind(Year1, Year2, Year3, Year4,A1,A2,A3,A4,A5,A6,B1,B2)    # constr coeffs
# dir is a character vector with the directions of the constraints.
# Each element must be one of "<", "<=", ">", ">=", or "==". We have 12 constraints:
dir <- c(">=",">=",">=",">=","==","==","==","==","==","==","==","==")
rhs<-c(300, 300, 300, 300,3.4,2.8,5.9,4.2,3.8,0.9,6.7,2.9)          # vector of rhs coeffs
Rglpk_solve_LP(obj, mat, dir, rhs, max = TRUE)

# $optimum
# [1] 1906.628
# [1] 0.000000 3.400000 0.000000 2.800000 5.900000 2.645370 1.554630  3.800000 0.900000
3.373383
# [11] 3.326617 2.900000
```

Designing green landscapes demands a continuous adaptation to changing environmental and economic conditions by searching for a combination of management paths that provide a desirable mix of services to the landowner (or village community) within a realistic time horizon. The multiple-path theory is a helpful general concept for a great variety of specific ecological applications.

6.4.4.4 A Coastal Dune Ecosystem

Coastal dune areas are important ecosystems because of their role in protecting agricultural land. Dunes are also important resources as they provide sand for construction as well as a variety of minerals, but coastal dune systems are vulnerable to

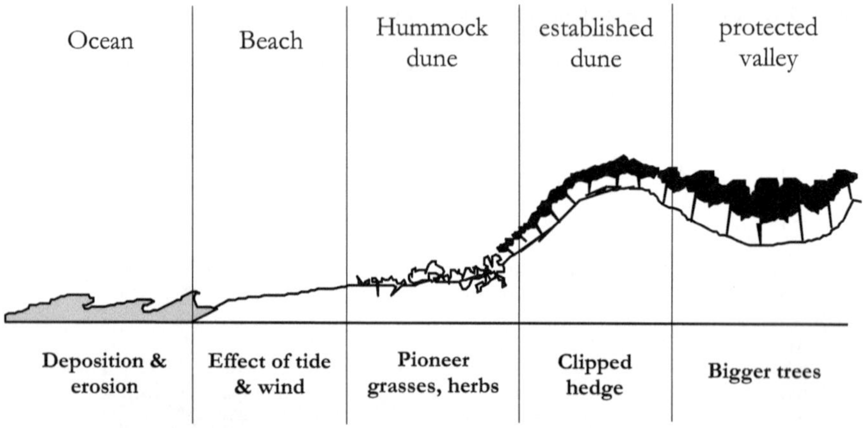

Fig. 6.34 Schematic classification of a coastal dune ecosystem at Bulugha, South Africa

the demands of recreation, urban development, and industry. As management policies change, sand dunes are either vegetated with invasive species such as *Acacia cyclops* or devegetated. Some exotic species, such as *Ammophila arenaria*, do not displace indigenous species and are good sand dune stabilizers (Hertling and Lubke 1999). Figure 6.34 is a schematic impression of a coastal dune ecosystem at Bulugha in the Eastern Cape province of South Africa. The development of a multiple-path model for the Bulugha area that is based on the specific vegetation structure of a given dune ecosystem can involve several steps:

- Spatial subdivision of the area into well-defined geographical units
- Stratification of dune types (e.g., beach; hummock dune; established dune, protected dune valley)
- Definition of alternative management paths for each type, including required input and expected output during the design period
- Establishing an optimum design by identifying the optimum path combination. Both linear programming (LP) and simulated annealing (SA) can be used to identify an optimum path combination

Alternative management activities can include the regular packing of branches to reduce sand movement; seeding of hummock dunes; planting of *Ammophila arenaria*; special protective measures to maintain established dunes; fencing of dune valleys; and control of sand mining.

Assignment 3
Develop an MP model for a coastal dune area. Use the R function Rglp to run the model and interpret the results.

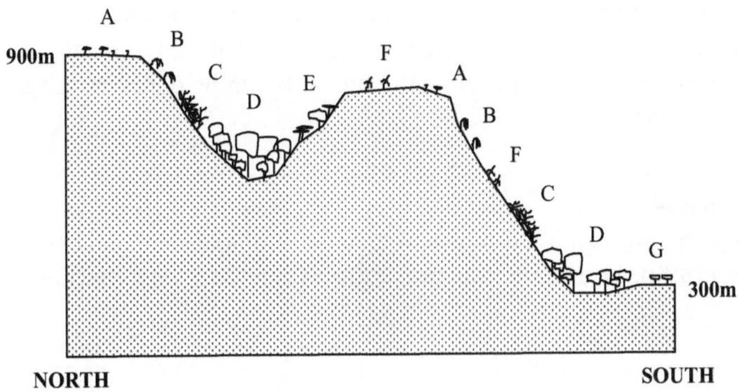

Fig. 6.35 Technical classification of vegetation types along the profile of a mountain ecosystem. A: Tristachya – Themeda; B: Festuca – Myrsine; C: Cliffortia; D: dry scrub forest; E: scattered Vachellia karroo; F: Bobartia – Trachypogon; G: very dry scrub

6.4.4.5 A Mountain Ecosystem

Criticism of the effect of planted forests on water supply has been the cause for establishing long-term paired experiments involving planted and unplanted mountain catchments (Hewlett and Pienaar 1973; Bosch and Gadow 1990; Versfeld 1994; Scott et al. 1999). In many mountain ecosystems where water resources are increasingly stressed, successful management requires a detailed understanding of the system hydrology as well as the integration of a range of environmental, social, and economic factors (McCartney et al. 2004). As there is also an awareness that financial resources are not unlimited, priorities must be set regarding management activities in different mountain ecosystems. Murray and Gadow (1991) showed how this can be achieved for a number of catchments, while keeping in mind their conservation status and their socioeconomic importance.

When considering a spatially explicit MP design of a mountain ecosystem, two types of information, namely a technical classification of the vegetation and a spatial subdivision, are useful. Stratification of vegetation is useful because it simplifies and organizes complex resource data, allows the harmonization and integration of different objectives, and facilitates descriptive reporting and the setting up of work schedules. A simplified profile of a mountain ecosystem is shown in Fig. 6.35.

Based on the specific vegetation structure of the mountain ecosystem, the development of a multiple-path (MP) model typically involves the following steps:

- Spatial subdivision of the area into well-defined geographical units (blocks)
- Classification of vegetation types (e.g., Tristachya – Themeda; B: Festuca – Myrsine; C: Cliffortia; dry scrub forest; Bobartia – Trachypogon)
- Definition of alternative management paths for each vegetation type (fire rotation; burning season; and eradication of alien invaders), including required input and expected output during the design period
- Establishing an optimum spatiotemporal design with an optimum path combination

A typical objective may be to maximize water yields subject to a limit on labor cost, or to set a maximum water yield reduction based on the results of long-term catchment studies. An example showing how this can be done was given by Bosch and Gadow (1990).

Assignment 4
Develop an optimization model based on the example provided by Bosch and Gadow (1990). Use the R function Rglp to run the model and interpret the results.

6.4.4.6 A Novel Ecosystem

The degradation or invasion of a natural ecosystem by alien vegetation often results in the formation of so-called novel ecosystems with new species combinations. As more of the earth's land surface becomes transformed by human use, novel ecosystems are increasing in importance (Richardson et al. 1997; Hobbs et al. 2006; Lindenmayer et al. 2008). Table 6.13 gives some details of the Bot River farm and its invasive tree and shrub species.

The Bot River farm covers a total area of 22.9 hectares and has been divided into seven blocks. *Acacia cyclops* is an aggressive alien invader species. The cluster pine (*Pinus pinaster*), introduced from the Mediterranean region in Europe, has also been classified as a Category 2 alien invader species in South Africa. Long-leaf wattle (*Acacia longifolia*) and Australian myrtle (*Leptospermum laevigatum*) have been classified as a Category 1 alien invader species. All of these invaders are controlled with mechanical, chemical, and biological methods.

Financial resources are often insufficient for investing in capital intensive crops, and a common objective is to improve the land by eliminating exotic invaders, and by using part of the land to grow flowers for export and producing firewood for local use on the remaining areas. Given these objectives, the first step is to develop alternative design paths for each spatial unit and to assign a value (an objective function coefficient) to each path. Typical constraints involve the smoothing over time of periodic levels of input (salaries and other expenditure) and output (income from the sale of firewood; measurable success in the eradication of undesirable vegetation). As mentioned before, both linear programming (LP) and heuristics such as simulated annealing (SA) can be used to identify an optimum path combination by solving this particular multiple-path model.

Assignment 5
Use the example provided by Bosch and Gadow (1990) to develop an optimization model. Use the R function Rglp to run the model and interpret the results.

Table 6.13 Relevant details of the alien vegetation in the seven compartments of the Bot River farm

Map	Compt	Area (ha)	Description	Photo
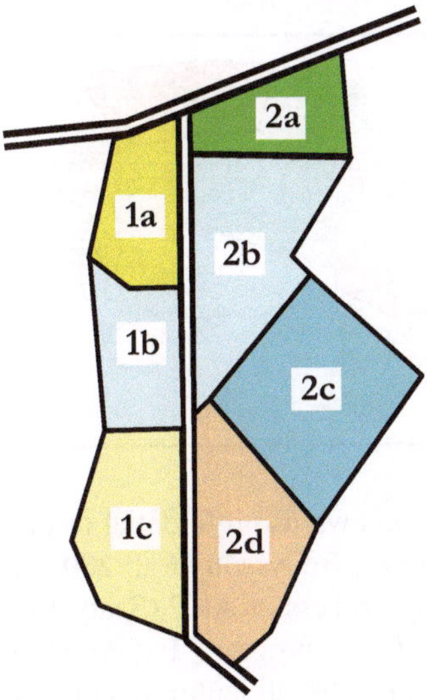	1a	2.5	Fynbos with few *Pinus pinaster* trees	
	1b	2.8	*Acacia cyclops* recently harvested for firewood	
	1c	3.6	Harvestable *Acacia cyclops*	
	2a	2.3	Dense *Pinus pinaster*, few tall trees	
	2b	4.4	Mix of small *Leptospermum laevigatum* and *Acacia cyclops*	
	2c	3.8	Mix of tall *Leptospermum laevigatum* and *Acacia cyclops*	
	2d	3.5	Mix of *Acacia longifolia* and *Acacia cyclops*	
Total area		**22.9**	ha	

6.4.4.7 A Forest Farm

A hypothetical forest farm covers 127.1 hectares and has been divided into 12 spatial units, known as compartments. The owner has defined two working circles.[11] The working circle *pine* includes all compartments planted with *Pinus taeda*; the working circle *wattle* includes all compartments planted with *Acacia mearnsii*. The relevant information of the forested parts, as well as a map, is shown in Table 6.14. The objective is to ensure that input/output fluctuations are smoothed, that silvicultural operations are carried out in accordance with specific preferences, and that the chosen design is feasible when available resources are considered.

Operational volume must be reduced, for example, if man-days are insufficient in a particular year to carry out all the work, then the reduction must affect the least important work. Ranking the operations is a typical multicriteria problem where the common assumption is that the value of each criterion can be assessed

[11] "Working circle" is a traditional term used by foresters to indicate an area of land where specific silvicultural treatments apply.

Table 6.14 Relevant data of the farm, including a map with compartment details. The code for each compartment in the illustration is the compartment number (A, B, ..L), the age (enumerator), and the area in hectares (denominator)

Compt	area ha	working circle	trees per hectare	age	map
A	10.8	Pine	400	25	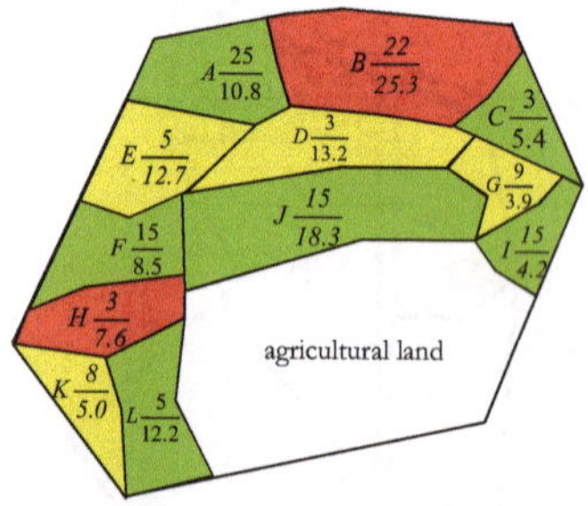
B	25.3	Pine	400	22	
C	5.4	Pine	1 100	3	
D	13.2	Wattle	1 500	3	
E	12.7	Wattle	1 461	5	
F	8.5	Pine	1 050	15	
G	3.9	Wattle	1 356	9	
H	7.6	Wattle	2 000	3	
I	4.2	Pine	400	15	
J	18.3	Pine	400	15	
K	5.0	Wattle	1 380	8	
L	12.2	Pine	650	5	

independently of all the others. This assumption is known as preferential or utility independence (Keeney and Raiffa 1976; Vincke 1992; Bouyssou et al. 2000) and forms the theoretical basis of the analytic hierarchy process or AHP (Saaty 1980). The coefficients can be determined by means of the program Expert Choice (McGinley 2012). In our example, the utility represents the silvicultural preference, but it can also be one of several other objectives, or a mix of these. Both linear programming (LP) and heuristics can be used to solve this particular multiple-path model (Chen and Gadow 2002, 2003).

The main objective of plantation forestry is to produce high quantities of timber obtained from market-driven species. Near-natural forest management with selective harvesting wants to maintain stable ecosystems that provide timber as well as other important products and services (Dedrick et al. 2007). The design of multispecies forests that exist in a natural or near-natural state is particularly demanding. Grouping tree species into a small number of cohorts facilitates management. The resulting simplification creates a sound basis for a design concept that does not prescribe the harvest amount but the residual forest structure after the harvest. The general concept is straightforward, but researchers also need to take into account specific details and adapt for specific ecosystems. This includes, in particular, the definition of residual densities and the identification of cohort specifications.

Many forest maps and aerial photographs of forested landscapes show a rather fragmented pattern, the result of property boundaries on the one hand and spatially organized human activity on the other. Forested landscapes can include open and closed forests, savannah woodlands, and commercial timber plantations. These landscapes provide multiple services, and their value is often only recognized when these services are lost after deforestation. Some of the common results of deforestation are soil erosion, flooding, reduced high-quality water, loss of species diversity, and increased greenhouse gas emissions.

Most landscapes, especially those where sustainable forest management is practiced, are subdivided into well-defined geographical units. These units have different names, depending on the type of land use. They are usually referred to as blocks, stands, or compartments. Each management activity has an influence on the mix of services, not only within these spatial units, but also within the landscape as a whole. The basic challenge of forest design is to continue to adapt to changing environmental and economic conditions. This is not a trivial task, because forest development is slow and rapid change is not possible. It takes a long time to change the species composition, the tree size distribution, and other features influencing the desirable quantity and quality of forest products and services.

Transdisciplinary systems approaches to landscape design have been proposed where the role of humans in the people–landscape interaction proceeds in mutually influencing loops of actions and reactions. The multiple-path theory of forest design, which involves a search for a combination of management paths with the aim of providing a desirable mix of products and services, is an example. The approach involves a continuous reassessment of changes within a limited planning horizon. The multiple-path design of forest ecosystems is preferable to the traditional cradle-to-grave design because numerous constraints related to forest ecosystem dynamics are considered simultaneously. Small-scale spatial mixing of different options, based on new pragmatic concepts and tools and advanced models of ecosystem dynamics, introduced in this book, is generating a continuously evolving landscape in response to human demand (the preferences of bona fide stakeholders) and changing environmental conditions.

Literature

Adams MD (1993) Genotypic and phenotypic variation as stress adaptations in temperate tree species: a review of several case studies. Tree Physiol 14(7–8–9):833–842

Adger WN (2006) Vulnerability. Glob Environ Chang 16:268–281. https://doi.org/10.1016/j.gloenvcha.2006.02.006

Albert M, Hansen J, Nagel J, Schmidt M, Spellmann H (2015) Assessing risks and uncertainties in forest dynamics under different management scenarios and climate change. For Ecosyst 2:14. https://doi.org/10.1186/s40663-015-0036-5

Amling C (2005) Waldbaurichtlinien in NRW, 1945 – heute. Unpublished thesis to the Seminar "Steuerung und Analyse der Waldentwicklung." Institut für Waldinventur und Waldwachstum d. Univ. Göttingen

Angelstam P, Boutin S, Schmiegelow F, Villard M-A, Drapeau P, Host G, Innes J, Isachenko G, Kuuluvainen T, Mönkkönen M, Niemelä J, Niemi G, Roberge J-M, Spence J, Stone D (2004) Targets for boreal forest biodiversity conservation – a rationale for macroecological research and adaptive management. Ecol Bul 51:487–509

Ascough IIJC, Maier HR, Ravalico JK, Strudley MW (2008) Future research challenges for incorporation of uncertainty in environmental and ecological decision-making. Ecol Model 219:383–399. https://doi.org/10.1016/j.ecolmodel.2008.07.015

Baker JB, Cain MD, Guldin JM, Murphy PA, Shelton MG (1996) Uneven-aged Silviculture for the loblolly and shortleaf pine forest cover types. Gen. Tech. Rep. SO-1 18. Asheville, NC: U.S. Department of Agriculture, Forest Service, Southern Station. 65 p

Barros O, Weintraub A (1982) Planning for a vertically integrated forest industry. Oper Res 30:1168–1182

Bauhus J, Schmerbek J (2010) Silvicultural options to enhance and use forest plantation biodiversity. In: Bauhus J, Van der Meer P, Kanninen M (eds) Ecosystem goods and services from plantation forests. Earthscan, London, pp 96–139

Bava J (1999) Los bosques de lenga en Argentina. In: Donoso C, Lara A (eds) Silvicultura de los bosques nativos de Chile. Universitaria, Santiago, pp 273–296

Bergen V (1997) Volkswirtschaftslehre, Teil I. Lecture Notes, University of Göttingen: 62 pp

Bi HQ, Bruskin S, Smith RGB (2002) The zone of influence of paddock trees and the consequent loss in volume growth in young Eucalyptus dunnii plantations. For Ecol Manage 165:305–315

Biolley H (1901) Oeuvre écrite. Supplement to the Zeitschrift des Schweizerischen Forstvereins, No. 66: 458 p

Bosch JM, Gadow K v (1990) Regulating afforestation for water conservation in South Africa. South AfFor J 153:41–54

Bouyssou D, Marchant T, Pirlot M, Perny P, Tsoukiàs A, Vincke P (2000) Evaluation and decision models: a critical perspective. Kluwer Academic Publishers, Boston

Breitenbach F (1974) Southern Cape forests and trees. The Government Printer, Pretoria. 328 p

Brodie JD, Adams DM, Kao C (1978) Analysis of economic impacts on thinning and rotation for Douglas-fir, using dynamic programming. For Sci 24(4):513–522

Buongiorno J, Gilless JK (1987) Forest management and economics. Macmillan, New York

Buongiorno J, Mitchie BR (1980) A matrix model of uneven-aged forest management. For Sci 26(4):609–625

Burschel P, Huss J (1987) Grundriß des Waldbaus. Ein Leitfaden für Studium und Praxis. Verlag Paul Parey. Hamburg und Berlin

Cancino J, Gadow Kv (2001) Stem number guide curves for uneven-aged forests – development and limitations. In: Gadow, Kv, Nagel J, Saborowski J (eds) Continuous cover forestry – assessment, analysis, scenarios. Kluwer Academic Publishers, pp 163–175

Chang SJ (1998) A generalized Faustmann model for the determination of optimal harvest age. Can J For Res 28:652–659

Chen BW, Gadow K (2002) Timber harvest planning with spatial objectives using the method of simulated annealing. Forstwiss Centralblatt 121:25–34

Chen BW, Gadow Kv (2003) A case study of simulated annealing, genetic algorithm, tabu search and linear programming for solving forest planning problems. Forstachiv 74(1):26–31

Chen BW, Gadow K (2008) Combining spatial and other objectives in forest design. For StudlMetsanduslikud Uurimused 48:30–40

Cloete WG, Cloete I, Von Gadow K (1988) An algorithm for presenting pairs in optimum orders in the method of paired comparisons. EDV in Medizin und Biologie 19(2–3):75–77

Clutter JL, Fortson JC, Pienaar LV, Brister GH, Bailey RL (1983) Timber management – a quantitative approach. Wiley, New York, 333 p

Condés S (1997) Simulación de parcelas arboladas con datos del 2 Inventario Forestal Nacional. Tesis Doctoral. Tesis Doctoral Escuela Técnica Superior de Ingenieros de Montes de Madrid. 616 pp

Corral-Rivas JJ, Reyes RI, Wehenkel C, Aguirre-Calderón OA, Gadow Kv (2012) A network of forest observational studies in Durango (Mexico). In: Zhao XiuHai, Zhang ChunYu, Gadow Kv (eds) Forest observational studies. Proceedings of an international workshop at Beijing Forestry University, which convened on 20/21 September 2012: 125–138

Cossalter C, Pye-Smith C (2003) Fast-wood forestry – myths and realities. Center for International Forestry Research (CIFOR): 50 p

Cotta H (1804) Systematische Anleitung zur Taxation der Waldungen. Sander, Berlin

Crutzen PJ (2002) Geology of mankind. Nature 415:23–23

Cushing CE, Cummins KW, Minshall GW (1997) Ecosystems of the world: river and stream ecosystems. J North Am Soc 16(3):717–719

Dedrick S, Spiecker H, Orazio C, Tomé M, Martinez I (2007) Plantation or conversion – the debate. European Forest Institute, Discussion paper 13: 98 p

Dralet M (1820) Traité des forêts des arbres résineux et des terrains adjacent, sur les montagnes de la France. Vieussieux, Toulouse

Ebert W, Langer R (1977) Frankfurter Stadtwald. Published by Forstamt Frankfurt, 64 pp

Emborg J, Christensen M, Heilmann-Clausen J (2000) The structural dynamics of Suserup Skov, a near-natural temperate deciduous forest in Denmark. For Ecol Manage 126(2):173–189

Eyre FH (1980) Forest cover types of the United States and Canada. Society of American Foresters, Bethesda, Maryland, 148 p (1 map sheet)

Folch R., Camarasa JM (2000) Encyclopaedia of the Biosphere Vol. 1-11; Gale Group

Franklin JF, Berg DR, Thornburgh DA, Tappeiner JC (1997) Alternative silvicultural approaches to timber harvesting: variable retention harvest systems. In: Kohn KA, Franklin JF (eds) Creating a forestry for the 21st century: the science of ecosystem management. Island Press, Washington, DC, 111p

Gadow K (ed) (2001) Risk analysis in forest management, Managing forest ecosystems, vol 2. Springer, Dordrecht

Gadow Kv, Pukkala T (2008) Designing green landscapes, Vol. 15 of the book series "Managing forest ecosystems". Springer, Dordrecht, 322 p

Gadow Kv, Nuutinen T, Kellomäki S (2008) Adaptive design of forested landscapes. In: Gadow K, Pukkala T (eds) Designing green landscapes, Vol. 15 of the Book series "Managing forest ecosystems". Springer, Dordrecht, pp 3–30

Garcia O (1990) Linear programming and related approaches in forest planning. N Z J For Sci 20(3):307–331

Geldenhuys CJ (2009) Managing forest complexity through application of disturbance-recovery knowledge in development of silvicultural systems and ecological rehabilitation in natural forest systems in Africa. J For Res 15:3–13

González JR, Palahí M, Trasobares A, Pukkala T (2006) A fire risk model for forest stands in Catalonia (north-east of Spain). Ann For Sci 63:169–173

Guilford JP (1954) Psychometric methods. Mc Graw-Hill, New York

Guldin JM (1991) Uneven-aged BDq regulation of Sierra Nevada mixed conifers. Western J Appl For 6(2):27–32

Gustafsson L, Baker SC, Bauhus J, Beese WJ, Brodie A, Kouki J, Lindenmayer DB, Lõhmus A, Pastur GM, Messier C, Neyland M, Palik B, Sverdrup-Thygeson A, Volney WJA, Wayne A, Franklin JF (2012) Retention forestry to maintain multifunctional forests: a world perspective. BioScience 62(7):633–645

Hansen GD (1987) Effects of diameter distribution on the growth of northern hardwoods. Can J For Res 17(1)

Hanson PJ, Weltzin JF (2000) Drought disturbance from climate change: response of United States forests. Sci Total Environ 262:205–220

Hartig GL (1795) Anweisung zur Taxation der Forste, oder zur Bestimmung des Holzertrags der Wälder. Heyer, Gießen

Hartley MJ (2002) Rationale and methods for conserving biodiversity in plantation forests. For Ecol Manage 155:81–95

Heinonen T, Pukkala T (2007) The use of cellular automaton approach in forest planning. Can J For Res 37:2188–2200

Hendrick RL (2001) Forest types and classification. In: Evans J (ed) The forest handbook, vol 1. Blackwell Science Ltd, Malden

Hertling UM, Lubke RA (1999) Use of Ammophila arenaria for dune stabilization in South Africa and its current distribution – perceptions and problems. Environ Manag 24:467–482

Hessenmöller D, Gadow K (2001) Beschreibung der Durchmesserverteilung von Buchenbeständen mit Hilfe der bimodalen Weibullfunktion. Allg Forst- u J-Ztg 172(3):46–50

Hewlett JD, Pienaar L (1973) Design and analysis of the catchment experiment. In: White EH (ed) Proceedings of the Symposium on use of small watersheds in determining effects of forest land use on water quality. University of Kentucky, Lexington, pp 88–106

Hille M, Hessenmöller D Möhring B, Gadow Kv (1999) Evaluating alternative silvicultural regimes in a 110-year old beech forest. In: Proc IUFRO conference from theory to practice – gaps and solutions in managerial economics and accounting in forestry. Prag, pp 94–100

Hinrichs L (2006) Automatic generation of treatment paths for mixed forest stands of spruce (Picea abies) and beech (Fagus sylvatica). In: Proceeding of the IUFRO Conf. Precision Forestry, 5–10 March, 2006; Univ. Stellenbosch, South Africa, pp 261–276

Hobbelstad K, Hofstad O (1988) AVVIRK-3, a model for long-term forest planning. Med Norsk Institutt for Skogsforskning 41(35):504–516

Hobbs RJ, Arico S, Aronson J, Baron JS, Bridgewater P, Cramer VA, Epstein PR, Ewel JJ, Klink CA, Lugo AE, Norton D, Ojima D, Richardson DM, Sanderson EW, Valladares F, Vilà M, Zamora R, Zobel M (2006) Novel ecosystems: theoretical and management aspects of the new ecological world order. Glob Ecol Biogeogr 15:1–7

Hui C, Richardson DM (2017) Invasion dynamics. Oxford University Press, Oxford, 322 p

Hull RB, Buhyoff CJ, Daniel TC (1984) Measurement of scenic beauty – the law of comparative judgement and scenic beauty estimation procedures. For Sci 30(4):1084–1096

Humphrey J, Stevenson A, Whitfield P, Swailes J, (2002) Life in the deadwood. Forestry Commission: 18 pp

Hundeshagen JC (1826) Die Forstabschätzung auf neuen wissenschaftlichen Grundlagen. H. Laupp, Tübingen

Jin X, Pukkala T, Li F (2016) Fine-tuning heuristic methods for combinatorial optimization in forest planning. Eur J For Res 135(4):765–779

Johnson KN, Scheurman HL (1977) Techniques for prescribing optimal timber harvest investment under different objectives – discussion and synthesis. For Sci Monogr 18

Kallio M, Propol A, Seppälä R (1981) A model for the forest sector. In: Dantzig GB, Dempster MAH, Kallio MJ (eds) Large scale linear programming. Internat. Inst. for Applied Systems Analysis, Laxenburg, pp 1055–1100

Kangas AS, Kangas J (2004) Probability, possibility and evidence: approaches to consider risk and uncertainty in forestry decision analysis. For Pol Econ 6:169–188

Kareiva P, Watts S, McDonald R, Boucher T (2007) Domesticated nature: shaping landscapes and ecosystems for human welfare. Science 316(5833):1866–1869

Keeney RL, Raiffa H (1976) Decisions with multiple objectives: preferences and value tradeoffs. Wiley

Koch F (2005) Entwicklung der Waldbaurichtlinien in der DDR. Unpublished contribution to the Seminar Steuerung und Analyse der Waldentwicklung. Institut für Waldinventur und Waldwachstum Univerity Göttingen

Korpel S (1992) Ergebnisse der Urwaldforschung für die Waldwirtschaft im Buchen-Ökosystem. Allgemeine Forstzeitschrift 47(21):1148–1152

Košir Z (1966) Beitrag zur Erforschung der Urwaldstruktur reiner Buchenwälder. In: Gesellschaftsmorphologie, Ber. Int. Symp. Int. Ver. Veg.-Kunde, Rinteln: 4.–7

Kramer H (1988) Waldwachstumslehre. Verlag Paul Parey, Hamburg/Berlin, 374 p

Kumar BM, Nair PKR (2006) Tropical homegardens: a time-tested example of sustainable agroforestry. Springer, 370 pp

Laurén A, Finér L, Koivusalo H, Kokkonen T, Karvonen T, Kellomäki S, Mannerkoski H, Ahtiainen M (2005) Water and nitrogen processes along a typical water flowpath and streamwater exports from a forested catchment and changes after clearcutting: a modelling study. Hydrol Earth Syst Sci 9(2):1–17

Leuschner A (1990) Forest regulation, harvest scheduling and planning techniques. John Wiley, 281 p

Liao W, Nogami K (1999) Prediction of near-view scenic beauty in artificial stands of Hinoki (Chamaecyparis obtusa S. et Z.). J For Res 4:93–98

Lindenmayer DB, Fischer J, Felton A, Crane M, Michael D, Macgregor C, Montague-Drake R, Manning A, Hobbs RJ (2008) Novel ecosystems resulting from landscape transformation create dilemmas for modern conservation practice. Conserv Lett 1:129–135

Lindner M, Maroschek M, Netherer S, Kremer A, Barbati A, Garcia-Gonzalo J, Seidl R, Delzon S, Corona P, Kolström M, Lexer MJ, Marchetti M (2010) Climate change impacts, adaptive capacity, and vulnerability of European forest ecosystems. For Ecol Manage 259:698–709. https://doi.org/10.1016/j.foreco.2009.09.023

Lockow K.-W (2004) Weißerle im Norddeutschen Tiefland.. Landesforstanstalt Eberswalde

Marazzi CA (1985) A value analysis method for the evaluation of telecommunication systems bid proposals. IEEE Trans Eng Manag EM-32(2):55–62

Martínez A (1999) Silvicultura práctica en renovales puros y mixtos, y bosques originales remanentes del tipo forestal Roble-Raulí-Coigüe. In: Donoso C, Lara A (eds) Silvicultura de los bosques nativos de Chile, pp 145-176

Matthews E, Payne R, Rohweder M, Murray S (2000) Forest ecosystems: pilot analysis of global ecosystems. World Resource Institute, Washinton, DC

McCartney MP, Yawson DK, Magagula TF, Seshoka J (2004) Hydrology and water resources development in the Olifants River Catchment, Working Paper 76. International Water Management Institute (IWMI), Colombo

McGinley P (2012) Decision analysis software survey. OR/MS Today 39

Messier C, Bauhus J, Doyon F, Maure F, Sousa-Silva R, Nolet P, Mina M, Aquilue N, Fortin M-J, Puettmann K (2019) The functional complex network approach to foster forest resilience to global changes. For Ecosyst 6:21. https://doi.org/10.1186/s40663-019-0166-2

Mitchel SJ, Beese WJ (2002) The retention system: reconciling variable retention with the principles of silvicul- tural systems. For Chronicle 78(3):397–403

Mitscherlich G (1952) Der Tannen-Fichten-(Buchen)-Plenterwald. Heft 8 d. Schriftenreihe d. Bad. Forstl. Vers. Anst., Freiburg: 42 p

Murray DM, Gadow K (1991) Prioritizing mountain catchment areas. J Environ Manage 32(4):357–366

Murray DM, Gadow K (1993) A flexible yield model for regional timber forecasting. Southern J Appl For 17(2):112–115

Nänni UW (1970) Trees, water and perspective. S Afr For J 75:9–17

Newnham RM (1976) Computer may cut costs $1-2/cord. Pulp Pap Can 77(9):73–79

Newnham RM (1991) LOGPLAN II – a model for planning logging and regeneration activities. For. Can. Petawawa Nat. For. Inst. Inf. Rep. PI-X-102

Öhman K (2000) Creating continuous areas of old forest in long-term forest planning. Can J For Res 30:1817–1823

Oppili P (2013) Plantations to turn into natural forests. The Hindu, September 5, 2013

Ough K, Murphy A (1998) 'Understorey islands: A method of protecting understorey flora during clearfelling operations', VSP Internal Report 29, Department of Natural Resources and Environment, Victoria

Overbeck M, Schmidt M (2012) Modelling infestation risk of Norway spruce by Ips typographus (L.) in the Lower Saxon Harz Mountains (Germany). For Ecol Manage 266 (2012) 115–125

Packalen P, Pukkala T, Pascual A (2019). Combining spatial and economic criteria in tree-level harvest planning. Manuscript submitted to Forest Ecosystems

Palik B, Mitchell RJ, Pecot S, Battaglia M, Pu M (2003) Spatial distribution of overstory retention influences resources and growth of longleaf pineseedlings. Ecol Appl 13:674–686

Peyron J-L (1993) Présentation illustreé d'une méthode de planification de la gestion forestière et de détermination de l'effort de régeneration. Rev For Fr XLV(1):59–73

Pommerening A (2006) Evaluating structural indices by reversing forest structural analysis. For Ecol Manage 224:266–277

Pretzsch H, Biber P, Ďurský J, Gadow Kv, Hasenauer H, Kändler G, Kenk G, Kublin E, Nagel J, Pukkala T, Skovsgaard JP, Sodtke R, Sterba H (2002) Recommendations for standardized documentation and further development of forest growth simulators. Forstwissenschaftliches Centralblatt 121(3):138–151

Puettmann KJ, Wilson SMG, Baker SC, Donoso PJ, Drössler L, Amente G, Harvey BD, Knoke T, Lu Y, Nocentini S, Putz FE, Yoshida T, Bauhus J (2015) Silvicultural alternatives to conventional even-aged forest management – what limits global adoption. For Ecosyst 2:8. https://doi.org/10.1186/s40663-015-0031-x

Pukkala T (2002a) Multi-objective forest planning, Vol. 6 of the book series Managing forest ecosystems. Kluwer Academic Publishers, 264 p.

Pukkala T (2002b) Measuring non-wood forest outputs in numerical forest planning. In: Pukkala T (ed) Multi-objective forest planning. Kluwer Academic Publishers, Dordrecht, pp 173–207

Pukkala T (2019) Using ALS raster data in forest planning. J For Res 30(5):1581–1593

Pukkala T, Kangas J (1993) A heuristic optimization method for forest planning and decision making. Scand J For Res 8:560–570

Pukkala T, Heinonen T, Kurttila M (2008) An application of the reduced cost approach to spatial forest planning. For Sci 55(1):13–22

Pukkala T, Packalen P, Heinonen T (2014) Dynamic treatment units in forest management planning. Manag For Ecosyst 33:373–392

Pukkala T, Lähde E, Laiho O (2016) Which trees should be removed in thinning treatments? For Ecosyst 2(23):2–12

Rautiainen O (1998) Modelling the yield and growth of uneven-aged Shorea robusta stands. Ph.D thesis, University of Joensuu

Richardson DM, Macdonald IAW, Hoffmann JH, Henderson L (1997) Alien plant invasions. In: Cowling RM, Richardson DM, Pierce SM (eds) Vegetation of Southern Africa. Cambridge University Press, Cambridge, pp 535–570

Roberge JM, Mönkkönen M, Toivanen T, Kotiaho J (2013) Retention forestry and biodiversity conservation: a parallel with agroforestry. Nat Conserv 4(0):29. https://doi.org/10.3897/natureconservation.4.5116

Rois Diaz M (2000) Analyse eines Buchen-Edellaubholz-Mischbestandes im Göttinger Wald. Master thesis, Faculty of Forestry and Forest Ecology, Georg-August-University, Göttingen

Rojo A, Ulises D-A, Rodríguez-Soalleiro R, Gadow Kv (2005) Modelling silvicultural and economic alternatives for Scots pine (Pinus sylvestris L.) plantations in Northwestern Spain. Forestry 78(3) 17pp

Ross RT (1934) Optimum orders for the presentation of pairs in the method of paired comparisons. J Educ Psychol 25:375–382

Ruppert K (1972) 600 Jahre Frankfurter Stadtwald. Published by Forstamt Frankfurt

Rusanen K, Finér L, Antikainen M, Korkka-Niemi K, Backman B, Britschgi R (2004) The effect of forest cutting on the quality of groundwater in large aquifers in Finland. Boreal Environ Res 9:253–261

Saaty TL (1980) The analytic hierarchy process. McGraw-Hill

Sánchez S, Rodríguez R (2002) Modelling the growth and management of uneven-aged maritime pine-broadleved species forests in Galicia (NW Spain). Scand J For Res 17:538–547

Sanderson EW, Jaiteh M, Levy MA, Redfrod KH, Wannebo AV, Woolmer G (2002) The human footprint and the last of the wild. BioScience 52:891–904

Sayer J, Campbell B (2004) The science of sustainable development. Cambridge University Press, Cambridge

Schmidt M, Hanewinkel M, Kändler G, Kublin E, Kohnle U (2010) An inventory-based approach for modeling single- tree storm damage — experiences with the winter storm of 1999 in southwestern Germany. Can J For Res 40:1636–1652

Schmoldt DL, Kangas J, Mendoza GA, Pesonen M (eds) (2001) The analytic hierarchy process in natural resource and environmental decision making. Vol 5 of the book series Managing forest ecosystems, Kluwer Academic Publishers, 305 p

Schober R (1978) Ertragstafeln wichtiger Baumarten. Sauerländer's Verlag

Schulze E-D, Bouriaud L, Bussler H, Gossner M, Walentowski H, Hessenmöller D, Bouriaud O, Kv G (2014) Opinion paper: forest management and biodiversity. Web Ecol 14(1):3

Schulze ED, Craven D, Durso AM, Reif J, Guderle M, Kroiher F, Hennig P, Weiserbs A, Schall P, Ammer C, Eisenhauer N (2019) Positive association between forest management, environmental change, and forest bird abundance. For Ecosyst 6:3

Schütz J-P (2002) Silvicultural tools to develop irregular and diverse forest structures. Forestry 75:329–338

Schütz J-Ph, Pukkala T, Donoso P, Gadow Kv (2012) Historical emergence and current application of CCF. In: Pukkala T, Gadow K.v (eds) Continuous cover forestry. book series managing forest ecosystems. vol 24, © Springer, pp 1–28

Scott DF, Lesch W (1997) Streamflow responses to afforestation with Eucalyptus grandis and Pinus patula and to felling in the Mokobulaan experimental catchments, Mpumalanga Province, South Africa. J Hydrol 199:360–377

Scott DF, Smith RE (1997) Preliminary empirical models to predict reductions in total and low flows resulting from afforestation. Water SA 23:135–140

Scott DF, Le Maitre DC, van Wyk E (1999) The effects of forestry on low flows: further studies on low flow in wet and dry cycles and in relation to catchments of increasing size. Contract report No. ENV/S-C 99059, CSIR, Pretoria, South Africa

Shvidenko A, Venevsky S, Raille G, Nilsson S (1995) A system for evaluation of growth and mortality in Russian forests. Water Air Soil Pollut 82:333–348

Speidel G (1971) Planung im Forstbetrieb. Paul Parey, Hamburg

Staupendahl K, Möhring B (2011) Integrating natural risks into silvicultural decision models: a survival function approach. For Pol Econ (accepted)

Tewari VP, Arya R (2005) Degradation of arid rangelands in Thar Desert, India: a review. Arid Land Res Manag 19(1):1–12

Tewari VP, Gadow Kv (2010) Domesticated ecosystems – analysis and design – an example involving Indira Gandhi Canal Plantations in Rajasthan, India. Poster presented at the IUFRO World Congress in Seoul/Korea; August 2010

Tewari VP, Singh M (2006) Tree-crop interaction in the Thar desert of Rajasthan, India. Sѐcheresse 17(1-2):326–332

Thurstone LL (1927) A law of comparative judgement. Psychol Rev 34:273–286

Torgerson WS (1958) Theory and methods of scaling. Wiley, New York

United Nations Department of Humanitarian Affairs (1992) Internationally agreed glossary of basic terms related to disaster management. Geneva. http://reliefweb.int/sites/reliefweb.int/files/resources/004DFD3E15B69A67C1256C4C006225C2-dha-glossary-1992.pdf

Valladares F, Niinemets Ü (2008) Shade tolerance, a key plant feature of complex nature and consequences. Annu Rev Ecol Evol Syst 39:237–257

Van Lill WS, Kruger FJ, Van Wyk DB (1980) The effect of afforestation wilh Eucalyptus grandis Hill exMaiden and Pinus patula Schlecht. et Cham. on streamflow from experimental catchments at Mokobulaan' Transvaal. J Hydrol 48:107–118

Versfeld DB (1994) The forest industry and management for water conservation. In: Southern African forestry handbook 1994. South African Institute of Forestry, pp 657–674

Vincke P (1992) Multicriteria decision-aid. Wiley, New York

Virgilietti P, Buongiorno J (1997) Modelling forest growth with management data – a matrix approach for the Italian Alps. Silva Fennica 31(1):27–42

Vose D (2008) Risk analysis: a quantitative guide, 3rd edn. Wiley

von Schütz JE (1757) Ökonomisches Bedenken von dem Nachteil deren Gehaue in den sogenanten Schwartzen Hölzer. In: Ökonomisches Bedenken über allerhand in die Hauswirtschaft einschlagende Sachen, vol 1. Stössel, Chemnitz, pp 1–18

Waggener TR (1969) Some economic implications of sustained yield as a forest regulation model. Contribution No. 6, University of Washington, 22 pp

Weiss G (2001) Administrating risk – a social science perspective on natural hazard prevention based on an Austrian case study. In: Gadow Kv (ed) 2001: risk analysis in forest management. Vol 2 in the book series managing forest ecosystems. Kluwer Academic Publishers, New York, pp 171–200

Wenk G (1996) Durchmesserverteilungen im Buchenplenterwald (als Bestandteil des Forschungsprojektes „Sortenorientiertes Wachstumsmodell"). Tagungsband des Deutschen Verbandes Forstlicher Versuchs-anstalten. Sektion Ertragskunde, Mai 1996 in Neresheim

Westphal C, Tremer N, Oheimb Gv, Hansen J, Gadow K, Härdtle W (2006) Is the reverse J-shaped diameter distribution universally applicable in European virgin beech forests? For Ecol Manag 223:75–83

Wicht CL (1949) Forestry and water supplies in South Africa. Department of Forestry Bulletin No. 33, Pretoria

Zhang C, Wei Y, Zhao X, Gadow Kv (2013a) Spatial characteristics of tree diameter distributions in a temperate old-growth forest. PLoS One 8:e58983

Zhang CY, Zhao XH, Gadow Kv (2013b) An analysis of three selective harvest events based on large forest observational studies in North Eastern China. For Ecol Manage 316:100–109

Zucchini W, Schmidt M, Gadow K (2001) A model for the diameter-height distribution in an uneven-aged beech forest and a method to assess the fit of such models. Silva Fennica 35(2):168–183

Literature

Abetz P (1976) Kann und soll die Standraumregulierung in Fichtenbeständen programmiert werden? FuH 31:117–119

Ackerman S, Ackerman PA, Seifert T (2013) Effects of irregular stand structure on tree growth, crown extension and branchiness of plantation grown Pinus patula. Southern For 75(4):247–256

Adams MD (1993) Genotypic and phenotypic variation as stress adaptations in temperate tree species: a review of several case studies. Tree Physiol 14(7–8–9):833–842

Adger WN (2006) Vulnerability. Glob Environ Chang 16:268–281. https://doi.org/10.1016/j.gloenvcha.2006.02.006

Affleck DLR (2001) A comparative study of spatial analysis methods for forestry Nelder experiments. MSc thesis, The University of British Columbia, 70 pp

Affleck DLR, Diéguez-Aranda U (2016) Additive nonlinear biomass equations: a likelihood-based approach. For Sci 62(2):129–140

Aguirre O, Hui GY, Gadow KV, Jimenez J (2003) Comparative analysis of natural forest sites in Durango, Mexico. For Ecol Manage 183:137–145

Akaike H (1974) A new look at the statistical model identification. IEEE Trans Autom Contr 19(6):716–723

Akça A, Gadow Kv, Mench A, Mann P, Pahl A Setje-Eilers U (1994) Überprüfung des Formquotienten q7 (d7/BHD) in der Bundeswaldinventur für die Hauptbaumarten Buche, Eiche, Fichte und Kiefer im landeseigenen Wald Nordrhein-Westfalens. Project report, 23 p

Albert M (1994) Erhebung von Strukturinformationen in einem Buchen-Edellaubholz Mischbestand. Diplomarbeit am Institut für Forsteinrichtung und Ertragskunde, Universität Göttingen. 109 p

Albert M (1998) Analyse der eingriffsbedingten Strukturveränderung und Durchforstungsmodellierung in Mischbeständen. Dissertation, Fak. f. Forstw. u. Waldökologie., University of Göttingen, 178 S

Albert M (1999) Analyse der eingriffsbedingten Strukturveränderung und Durchforstungsmodellierung in Mischbeständen. Dissertation Universität Göttingen. Hainholz Verlag, 201 p

Albert M (2001) Predicting the selection of elite trees in mixed- species stands – a rule-based algorithm for silvicultural decision support systems. Allg Forst- u J-Ztg 173(9):153–161

Albert M (2003) Analyse und multikriterielle Bewertung alternativer Waldentwicklungsszenarien mit Hilfe des Analytischen Hierarchieprozesses. In: Proceedings of the annual meeting of the section Ertragskunde of the German Union of Forest Research Stations in Torgau, 2–4 June 2003

Albert M (2004) Modifikationen des BAL-Konkurrenzindex zur Prognose des Durchmesserzuwachses in Mischbeständen. Jahrestagung der Sektion Ertragskunde des DVFFA in Schlägl vom 24–26 Mai 2004

© Springer Nature Switzerland AG 2021

K. von Gadow et al., *Sustaining Forest Ecosystems*, Managing Forest Ecosystems 37, https://doi.org/10.1007/978-3-030-58714-7

Albert M, Hansen J, Nagel J, Schmidt M, Spellmann H (2015) Assessing risks and uncertainties in forest dynamics under different management scenarios and climate change. For Ecosyst 2:14. https://doi.org/10.1186/s40663-015-0036-5

Alder D (1979) A distance-independent tree model for exotic conifer plantations in East Africa. For Sci 25:59–71

Alfaro Reyna T, Martínez-Vilalta J, Retana J (2019) Regeneration patterns in Mexican pine-oakforests. For Ecosyst 6:50. https://doi.org/10.1186/s40663-019-0209-8

Alekseev A, Tomppo E, McRoberts RE, Gadow K v (2019) A constructive review of the State Forest Inventory in the Russian Federation. For Ecosyst 2019:6

Alho JM, Kolehmainen O, Leskinen P (2001) Regression methods for pairwise comparisons data. In: Schmoldt DL, Kangas J, Mendoza GA, Pesonen M (eds) The analytic hierarchy process in natural resource and environmental decision making. Kluwer Academic Publishers, Dordrecht, pp 235–251

Allen MG, Burkhart HE (2018) Growth-density relationships in Loblolly Pine plantations. For Sci XX(XX):1–15. https://doi.org/10.1093/forsci/fxy048

Álvarez González JG (1997) Análisis y caracterización de las distribuciones diamétricas de Pinus pinaster Ait. en Galicia. Ph D. Thesis. Universidad Politécnica de Madrid. Madrid, Spain. 270 p

Álvarez-González JG, Rodríguez Soalleiro R, Vega Alonso G (1999) Elaboración de un modelo de crecimiento dinámico para rodales regulares de *Pinus pinaster* Ait. en Galicia. Investigación Agraria: Sistemas y Recursos Forestales 8:319–334

Álvarez-González JG, Castedo-Dorado F, Ruiz-Gonzalez AD, López-Sanchez CA, Gadow Kv (2004) A two-step mortality model for even-aged stands of *Pinus radiata* D. Don in Galicia (Northwestern Spain). Ann For Sci 61:439–448

Álvarez-González JG, Zingg A, Gadow K (2009) Estimating growth in Beech Forests – a study based on longterm experiments in Switzerland. Ann For Sci 67:307

Álvarez-González JG, Zingg A, Gadow Kv (2010) Estimating growth in beech forests: a study based on long term experiments in Switzerland. Ann For Sci 67:307

Álvarez-González JG, Cañellas I, Alberdi I, Ruiz-González AD (2012) National Forest Inventory and Forest Observational Studies in Spain; Application to Forest Development Modeling. In: Zhao XiuHai (赵秀海), Zhang ChunYu (张春雨) and Klaus v. Gadow (eds) Forest observational studies. Proceedings of an international workshop at Beijing Forestry University (20/21 September 2012) Published by the Key Laboratory for Silviculture and Conservation, Ministry of Education, Beijing Forestry University: 3 pp 5–58

Álvarez-González JG, Cañellas I, Alberdi I, Gadow K, Ruiz-González AD (2014) National Forest Inventory and forest observational studies in Spain: applications to forest modeling. For Ecol Manage 316:54–64

Amaro A, Reed DD, Themido I, Tomé M (1997) Stand growth modelling for first rotation Eucalyptus globulus Labill. in Portugal. In: Amaro A, Tomé M (eds) Empirical and process-based models for forest tree and stand growth simulation. Oeiras, Lisbon (Portugal), 21–27 September 1997. Edições Salamandra. pp 99–110

Amateis RL (2000) Modeling response to thinning in loblolly pine plantations. South J Appl For 24:17–22

Amateis RL, Burkhart HE, Burk TE (1986) A ratio approach to predicting merchantable yields of unthinned loblolly pine plantations. For Sci 32:187–296

Amateis RL, Radtke PJ, Burkhart HE (1995) TAUYIELD: A stand-level growth and yield model for thinned and unthinned loblolly pine plantations. Report No. 82. School of Forestry and Wildlife Resources. Virginia Polytechnic Institute and State University. Blacksburg, Virginia. 38 p

Amling C (2005) Waldbaurichtlinien in NRW, 1945 – heute. Unpublished thesis to the Seminar "Steuerung und Analyse der Waldentwicklung." Institut für Waldinventur und Waldwachstum d. Univ. Göttingen

Anderson MC (1966) Stand structure and light penetration. II. A theoretical analysis. J Appl Ecol 3:41–54

Anderson MJ, Ellingsen KE, McArdle BH (2006) Multivariate dispersion as a measure of beta diversity. Ecol Lett 9(6):683–693

Angelstam P, Boutin S, Schmiegelow F, Villard M-A, Drapeau P, Host G, Innes J, Isachenko G, Kuuluvainen T, Mönkkönen M, Niemelä J, Niemi G, Roberge J-M, Spence J, Stone D (2004) Targets for boreal forest biodiversity conservation – a rationale for macroecological research and adaptive management. Ecol Bul 51:487–509

Angiosperm Phylogeny Group (2016) An update of the Angiosperm Phylogeny Group classification for the orders and families of flowering plants: APG III. Bot J Linn Soc 161:105–121

Aphalo PJ, Rikala R (2006) Spacing of silver birch seedlings grown in containers of equal size affects their morphology and its variability. Tree Physiol 26(9):1227–1237

Aravind NA, Manjunath J, Rao D, Ganeshaiah KN, Shaanker RU, Vanaraj G (2005) Are red-listed species threatened? A comparative analysis of red-listed and non-red-listed plant species in the Western Ghats, India. Curr Sci 88(02):258–265

Arias-Rodil M, Diéguez-Aranda U, Rodríguez Puerta F, López-Sánchez CA, Canga Líbano E, Cámara Obregón A, Castedo-Dorado F (2015) Modelling and localizing a stem taper function for *Pinus radiata* in Spain. Can J For Res 45(6):647–658

Arnswaldt HJ v (1950) Wertkontrolle in Laubholzrevieren. Forstarchiv:130–135

Ascough IIJC, Maier HR, Ravalico JK, Strudley MW (2008) Future research challenges for incorporation of uncertainty in environmental and ecological decision-making. Ecol Model 219:383–399. https://doi.org/10.1016/j.ecolmodel.2008.07.015

Assmann E (1953) Zur Bonitierung süddeutscher Fichtenbestände. AFZ 10:61–64

Assmann E (1961) Waldertragskunde. Organische Produktion, Struktur, Zuwachs und Ertrag von Waldbeständen. BLV Verlagsgesellschaft, München, Bonn, Wien, p 490

Assunção R (1994) Testing spatial randomness by means of angles. Biometrics 50:513–537

Asunta A, Hiltunen V, Väisänen M (2004) Metsähallituksen luonnonvarasuunnittelu. Forestry Publications of Metsähallitus 47. (In Finnish)

Atkinson D (2000) Root characteristics: why and what to measure. In: Smit AL, Bengough AG, Engels C, Van Noordwijk M, Pellerin S, Van de Geijn SC (eds) Root methods: a hand book. Springer, Berlin/Heidelberg, pp 2–32

Ausburg T (2006) Becoming interdisciplinary – an introduction to interdisciplinary studies, 2nd edn. Kendall/Hunt Publishing, New York

Avis AM (1992) Coastal dune ecology and management in the Eastern Cape. Ph.D. dissertation, Rhodes University, Grahamstown: 300 pp

Ayrey E, Hayesa DJ, Kilbrideb JB, Weiskittel AR (2019) Synthesizing disparate LiDAR and satellite datasets through deep learning to generate wall-to-wall forest inventories of New England. Posted online Mar. 17, 2019. https://doi.org/10.1101/580514

Bachmann P (1990) Produktionssteigerung im Wald durch vermehrte Berücksichtigung des Wertzuwachses. Eidgenössische Forschungsanstalt für Wald, Schnee und Landschaft, Birmensdorf

Bachmann M (1998) Indizes zur Erfassung der Konkurrenz von Einzelbäumen – Methodische Untersuchung in Bergmischwäldern. LMU/LWF, München

Baddeley A (2008) Analysing spatial point patterns in R. Workshop Notes February 2008. CSIRO

Baddeley A, Rubak E, Turner R (2015) Spatial point patterns: methodology and applications with R. Chapman and Hall/CRC Press, London.

Baddeley A et al (2019) Spatial point pattern analysis, model-fitting, simulation, tests. https://cran.r-project.org/web/packages/spatstat/spatstat.pdf. Accessed on 4 Mar 2019

Baeten L, Verheyen K, Wirth C, Bruelheide H, Bussotti F, Finér L, Jaroszewicz B, Selvi F, Valladares F, Allan E, Ampoorter E, Auge H, Avacariei D, Barbaro L, Barnoaiea I, Bastias CC, Bauhus J, Beinhoff C, Benavides R, Benneter A, Berger S, Berthold F, Boberg J, Bonal D, Brüggemann W, Carnol M, Castagneyrol B, Charbonnier Y, Chécko E, Coomes D, Coppi A, Dalmaris E, Dañila G, Dawud SM, Wd V, Wandeler HD, Deconchat M, Domisch T, Duduman G, Fischer M, Fotelli M, Gessler A, Gimeno TE, Granier A, Grossiord C, Guyot V, Hantsch L, Hättenschwiler S, Hector A, Hermy M, Holland V, Jactel H, Joly FX, Juckers T, Kolb S, Koricheva J, Lexer MJ, Liebergesell M, Milligan H, Müller S, Muys B, Nguyen D, Nichiforel L, Pollastrini M, Proulx R, Rabasa S, Radoglou K, Ratcliffe S, Raulund-Rasmussen K, Seiferling I, Stenlid J, Vesterdal L, Kv W, Zavala MA, Zielinski D, Scherer-Lorenzen M (2013) A novel comparative research platform designed to determine the functional significance of tree species diversity in European forests. Persp Plant Ecol Evol Syst 15:281–291

Bailey RL (1980) Individual tree growth derived from diameter distribution models. For Sci 26:626–632

Bailey RL, Clutter JL (1974) Base-age invariant polymorphic site curves. For Sci 20:155–159

Baker JB, Broadfoot WM (1977) A practical field method for site evaluation for eight important Southern hardwoods. USDA For. Serv., Southern Forest Experiment Station, New Orleans, LA, Gen. Tech. Rep. SO-14

Baker JB, Cain MD, Guldin JM, Murphy PA, Shelton MG (1996) Uneven-aged Silviculture for the loblolly and shortleaf pine forest cover types. Gen. Tech. Rep. SO-1 18. Asheville, NC: U.S. Department of Agriculture, Forest Service, Southern Station. 65 p

Balboa Murias MA, Rodríguez Soalleiro R, Merino A, Álvarez González JG (2006) Temporal variations and distribution of carbon stocks in aboveground biomass of radiata and maritime pine. For Ecol Manag 237:29–38

Barkman JJ (1989) A Critical evaluation of minimum area concept. Vegetatio 85:89–104

Barros O, Weintraub A (1982) Planning for a vertically integrated forest industry. Oper Res 30:1168–1182

Barry RP (2013) Nelder plots. Encyclopedia of environmetrics 4

Bartelink HH (1996) A model of dry matter partitioning in trees. Tree Physiol 18:91–101

Baskent EZ, Keles S (2005) Spatial forest planning: a review. Ecol Model 188:145–173

Basson MS, van Niekerk PH, van Rooyen JA (1997) Overview of water resources availability and utilization in South Africa. Department of Water Affairs and Forestry and BKS (Pty) Limited

Bastin J-F et al (2017) The extent of forest in dryland biomes. Science 356:635–638

Bastin J-F, Finegold Y, Garcia C, Mollicone D, Rezende M, Routh D, Zohner CM, Crowther TW (2019) The global tree restoration potential. Science 365(6448):76–79. https://doi.org/10.1126/science.aax0848

Bates JM, Granger CWJ (1969) The combination of forecasts. Oper Res Q 20(4):451–468

Bauhus J, Schmerbek J (2010) Silvicultural options to enhance and use forest plantation biodiversity. In: Bauhus J, Van der Meer P, Kanninen M (eds) Ecosystem goods and services from plantation forests. Earthscan, London, pp 96–139

Bauhus J, Khanna PK, Menden N (2000) Aboveground and belowground interactions in mixed plantations of Eucalyptus globulus and Acacia mearnsii. Can J For Res 30(12):1886–1894

Bauhus J, Pokorny B, Van der Meer P, Kanowski PJ, Kanninen M (2010) Ecosystem goods and services – the key for sustainable plantations. In: Bauhus J, Van der Meer P, Kanninen M (eds) Ecosystem goods and services from plantation forests. Earthscan, London, pp 205–227

Bava J (1999) Los bosques de lenga en Argentina. In: Donoso C, Lara A (eds) Silvicultura de los bosques nativos de Chile. Universitaria, Santiago, pp 273–296

Beck HE, Zimmermann NE, McVicar TR, Vergopolan N, Berg A, Wood EF (2018) Present and future Köppen-Geiger climate classification maps at 1-km resolution. Sci Data 5:180214

Beese F (1996) Indikatoren für eine multifunktionelle Waldnutzung. Forstw Cbl 115:65–79

Begon M, Harper JL, Townsend CR (1996) Ecology: individuals, populations and communities. Blackwell Science, Oxford

Behre CE (1923) Preliminary notes on studies of tree form. J For 21:507–511

Bella IE (1971) A new competition model for individual trees. For Sci 17:364–372

Bellman RE (1954) The theory of dynamic programming. Bull Am Math Soc 60:503–516

Bennett B (2015) Plantations and protected areas: a global history of forest management. MIT Press, Cambridge, MA

Bergen V (1997) Volkswirtschaftslehre, Teil I. Lecture Notes, University of Göttingen: 62 pp

Bernstorff A v, Kurth A, Grilc J, Günter R (1988) Betriebswerk Gartow. Unpublished management plan: 163p

Besag J (1977) Contribution to the discussion of Dr Ripley's paper. J R Statist Soc Ser B 39:193–195

Besag JE, Glaves JT (1973) On the detection of spatial pattern in plant communities. Bulletin of the International Statistical Institute 45:153–158

Bettinger P, Sessions J, Boston K (1997) Using Tabu search to schedule Timber harvests subject to spatial Wildlife goals for Big Game. Ecol Model 94:111–123

Bey A et al (2016) Collect earth: land use and land cover assessment through augmented visual interpretation. Remote Sens (Basel) 8(10):807

Bi HQ, Bruskin S, Smith RGB (2002) The zone of influence of paddock trees and the consequent loss in volume growth in young Eucalyptus dunnii plantations. For Ecol Manage 165:305–315

Biau G, Scornet E (2016) A random forest guided tour. TEST 25(2):197–227

Biber P (1996) Konstruktion eines einzelbaumorientierten Wachstumssimulators für Fichten-Buchen-Mischbestände im Solling. Diss. Forstw. Fak. d. Ludwig-Maximilians-Univ. München: 239 S

Biber P (1997) Analyse verschiedener Strukturaspekte von Waldbeständen mit dem Wachstumssimulator SILVA 2. Vortrag anläßlich der Jahrestagung 1997 der Sektion Ertragskunde im Deutschen Verband Forstlicher For- schungsanstalten. Tagungsbericht: 100–120.

Bigelow NH, Brubaker LB, Edwards ME, Harrison SP, Prentice IC, Anderson PM, Andreev AA, Bartlein PJ, Christensen TR, Cramer W, Kaplan JO, Lozhkin AV, Matveyeva NV, Murray DF, McGuire AD, Razzhivin VY, Ritchie JC, Smith B, Walker DA, Gajewski K, Wolf V, Holmqvist Björn H, Igarashi Y, Kremenetskii K, Paus A, Pisaric MFJ, Volkova VS (2003) Climate change and Arctic ecosystems: 1. Vegetation changes north of 55 N between the last glacial maximum, mid-Holocene, and present. J Geophys Res Atmos 108(D19)

Biging GS, Dobbertin M (1992) A comparison of distance-dependent competition measures for height and basal area growth of individual conifer trees. For Sci 38(3):695–720

Biging GS, Dobbertin M (1995) Evaluation of competition indices in individual tree growth models. For Sci 41(2):360–337

Biging GS, Gill SJ (1997) Stochastic models for conifer tree crown profiles. For Sci 43(1):25–33

Biging GS, Wensel LC (1990) Estimation of crown form for six conifer species of northern California. Can J For Res 20(8):1137–1142

Biolley H (1901) Oeuvre écrite. Supplement to the Zeitschrift des Schweizerischen Forstvereins, No. 66: 458 p

Bischl B, Lang M, Kotthoff L, Schiffner J, Richter J, Studerus E, Casalicchio G, Jones ZM (2016) Machine learning in R. J Mach Learn Res 17(170):1–5

Blandon P (1985) Forest economics and the Gentan probability distribution. J Jap For Soc 67(12):478–485

Blavascunas E (2012) Reversing orders: foresters and the local in Poland's Bialowieza Forest. NCEEER Working Paper, National Council for Eurasian and East European Research, University of Washington, Seattle, WA. Available online at: http://ifmlab.for.unb.ca/People/Kershaw/PDF_Library/B/BlavascunasE2012a.pdf

Blomberg SP, Garland T, Ives AR (2003) Testing for phylogenetic signal in comparative data: behavioral traits are more labile. Evolution 57:717–745

Boardman R (1984) Fast-growing species-pattern, process and ageing. Proceedings: IUFRO symposium on site and productivity of fast growing plantations, Pretoria: p 49

Bohn U, Zazanashvili N, Nakhutsrishvili G (2007) The map of the natural vegetation of europe and its application in the caucasus ecoregion. Bulletin of the Georgian National Academy of Sciences 175:112–121

Boland DJ, Brooker MIH, Chippendale GM, Hall N, Hyland BPM, Johnston RD, Kleinig DA, Turner JD (1984) Forest trees of Australia. Nelson-CSIRO Melbourne, Australia. 687 p

Bolte A, Rahmann T, Kuhr M, Pogoda P, Murach D, Gadow K (2004) Relationships between tree dimensions and coarse root biomass in mixed stands of European beech (Fagus sylvatica L.) and Norway Spruce (Picea abies [L.] Karst.). Plant and Soil 0:1–11

Bolte A, Czajkowski T, Kompa T (2007) The north-eastern distribution range of European beech—a review. Forestry 80(4):413–429. https://doi.org/10.1093/forestry/cpm028

Bolton D, Coops NC, Hermosilla T, Wsulder M, White JC, Ferster C (2019) Uncovering regional variability in disturbance trends between parks and greater park ecosystems across Canada (1985–2015). Sci Rep 9(1). https://doi.org/10.1038/s41598-018-37265-4

Bond-Lamberty B, Wang C, Gower ST (2002) Aboveground and below-ground biomass and sapwood area allometric equations for six boreal tree species of northern Manitoba. Can J For Res 32:1441–1450

Bongers F, Popma J, Del Castillo JM, Carabias J (1988) Structure and floristic composition of the lowland rain forest of Los Tuxtlas, Mexico. Vegetatio 74:55–80

Borcard D, Legendre P (2002) All-scale spatial analysis of ecological data by means of principal coordinates of neighbour matrices. Ecol Model 153:51e68

Borda M (2011) Fundamentals in information theory and coding. Springer, Berlin/Heidelberg

Borggreve B (1891) Die Holzzucht, 2nd edn. Parey, Berlin

Bosch JM, Gadow K v (1990) Regulating afforestation for water conservation in South Africa. South AfFor J 153:41–54

Bouyssou D, Marchant T, Pirlot M, Perny P, Tsoukiàs A, Vincke P (2000) Evaluation and decision models: a critical perspective. Kluwer Academic Publishers, Boston

Bowman DMJS, Balch J, Artaxo P, Bond WJ, Cochrane MA, D'Antonio CM, DeFries R, Johnston FH, Keeley JE, Krawchuk MA, Kull CA, Mack M, Moritz MA, Pyne S, Roos CI, Scott AC, Sodhi NS, Swetnam TW (2011) The human dimension of fire regimes on earth. J Biogeogr 38:2223–2236. https://doi.org/10.1111/j.1365-2699.2011.02595

Box GP, Jenkins GM (1976) Time series analysis. Forecasting and control. Rev edn. Holden-Day, San Francisco, 575 pp. ISBN 0-8162-1104-3.

Boyd D, Crawford K (2012) Critical questions for big data. Inf Commun Soc 15(5):662–679

Boyden S, Binkley D, Senock R (2005) Competition and facilitation between Eucalyptus and nitrogen-fixing Falcataria in relation to soil fertility. Ecology 86:992–1001

Boysen-Jensen P (1918) Studies on the production of matter in light- and shadow-plants. Dan Bot Tidsskrif 36:219–262

Bray JR, Curtis JT (1957) An ordination of the upland forest communities of southern Wisconsin. Ecol Monogr 27:325–349

Bredenkamp BV (1984) The CCT concept in spacing research – a review. In: Proc., IUFRO Symposium on site and productivity of fast growing plantations. Pretoria and Pietermaritzburg, South Africa, pp 313–332

Bredenkamp BV (1988) An analysis of correlated curve trend experiments in E. grandis. PhD thesis, Virginia Polytechnic Institute and State University

Bredenkamp BV (1990) The triple-S CCT design. In: Gadow K, Bredenkamp BV (eds) Proceedings of a symposium arranged by the Forest mensuration and modelling working group in collaboration with the Southern African Institute of Forestry and the *Eucalyptus grandis* Research Network on "Management of *Eucalyptus grandis* in South Africa". pp 198–205

Bredenkamp BV, Kotze H, Kassier H (2000) The langepan CCT. A guide to a correlated curve trend spacing experiment in *Eucalyptus grandis* at Kwambonambi, Kwazulu-Natal, South Africa. 32 pp

Breiman L (2001) Random forests. Mach Lear 45(1):5–32

Breitenbach F (1974) Southern Cape forests and trees. The Government Printer, Pretoria. 328 p

Brenning, A. 2012. Spatial cross-validation and bootstrap for the assessment of prediction rules in remote sensing: The R package sperrorest," 2012 IEEE International Geoscience and Remote Sensing Symposium, 2012, pp. 5372-5375, doi: 10.1109/IGARSS.2012.6352393.

Breymann (1868) Anleitung zur Holzmesskunst, Waldertragsbestimmung und Waldwertberechnung. Wien

Brincken Jv (1826) Mémoire descriptif sur la forêt impériale de Białowieża en Lithuanie" (A Brief Description of the Imperial Białowieża Forest in Lithuania). N. Glücksberg, Imprimeur-Libraire de l'Université Royale:. 127 pages

Brink C u, Gadow K v (1986) On the use of growth and decay functions for modelling stem profiles. EDV in Medizin und Biologie 17(1/2):20–27

Brodie JD, Adams DM, Kao C (1978) Analysis of economic impacts on thinning and rotation for Douglas-fir, using dynamic programming. For Sci 24(4):513–522

Bronstein IN, Semendjajew KA (1991) Taschenbuch der Mathematik. 25., durchgesehene Auflage herausgegeben von G. Grosche, V. Ziegler und D. Ziegler. B.G.Teubner Verlagsgesellschaft Stuttgart, Verlag Nauka Moskau. 824 S

Brouillet L, Depierre A, Demolis C, Verdot M, Preney S, Latreille C, Giovanini J-P, Herbert I, Nicot Pu, Allegrini C (1991) La Silviculture des Peuplements Réguliers de Hêtre en Franche-Comté – de la Régénération Naturelle à la Première Éclaircie. ONF Bulletin Technique Nr 22:9–42

Brown GS (1965) Point density in stems per acre. Forestry Research Notes No. 38, New Zealand, p 11

Bruelheide H, Nadrowski K, Assmann T, Bauhus J, Both S, Buscot F, Chen XY, Ding B, Durka W, Erfmeier A, Gutknecht JLM, Guo D, Guo LD, Härdtle W, He JS, Klein AM, Kühn P, Liang Y, Liu X, Michalski S, Niklaus PA, Pei K, Scherer-Lorenzen M, Scholten T, Schuldt A, Seidler G, Trogisch S, Gv O, Welk E, Wirth C, Wubet T, Yang X, Yu M, Zhang S, Zhou H, Fischer M, Ma K, Schmid B (2014) Designing forest biodiversity experiments: general considerations illustrated by a new large experiment in subtropical China. Method Ecol Evol 5:74–89

Brus D, Hengeveld G, Walvoort D, Goedhart P, Heidema A, Nabuurs G, Gunia K (2012) Statistical mapping of tree species over europe. Eur J For Res 131(1):145–157

Bryant JB, Lamanna C, Morlon H, Kerkhoff AJ, Enquist BJ, Green JL (2008) Microbes on mountainsides: contrasting elevational patterns of bacterial and plant diversity. Proc Natl Acad Sci U S A 105:7774–7778

Bungart R, Hüttl RF (2004) Growth dynamics and biomass accumulation of 8-year-old hybrid poplar clones in a short-rotation plantation on a clayey-sandy mining substrate with respect to plant nutrition and water budget. Eur J For Res 123(2):105–115

Buongiorno J, Gilless JK (1987) Forest management and economics. Macmillan, New York

Buongiorno J, Gilless J (2002) Decision methods for forest resource management. Elsevier, Burlington

Buongiorno J, Mitchie BR (1980) A matrix model of uneven-aged forest management. For Sci 26(4):609–625

Burden RL, Faires JD (1989) Numerical analysis, 4th edn. PWS-KENT, Boston, 729 p

Burger H (1939) Kronenaufbau gleichaltriger Nadelholzbestände. Mitt Schweiz Anst f d forstl Vers Wesen 21:5–58

Burgers TF (1976) Management graphs derived from the correlated (CCT) projects: South Africa. For Res Inst Bull 1/54, Dept. Forestry, Pretoria, South Africa

Burgos MF, Villa SAB (1974) La silvicultura en el desarrollo económico-social de México. U.I.E.F. San Rafael. S.A.G.-S.F.F. México, D.F. Bol. Núm. 5. 144 p

Burk TE, Burkhart HE (1984) Diameter distributions and yields of natural stands of loblolly pine. Publication No. FWS-1-84. School of Forestry and Wildlife Resources. Virginia Polytechnic Institute and State University. Blacksburg, Virginia, 46 p

Burk TE, Newberry JD (1984) A simple algorithm for moment-based recovery of Weibull distribution parameters. For Sci 30:329–332

Burkhart H (1977) Cubic-foot volume of loblolly pine to any merchantable top limit. South J Appl For 1(2):7–9

Burkhart HE (1987) Data collection and modelling approaches for forest growth and yield prediction. In: Predicting forest growth and yield – current issues, future prospects. Inst. of Forest Resources. Univ. of Washington. Contribution Nr. 58: 3-16

Burkhart HE, Tomé M (2012) Modeling forest trees and stands. Springer, Dordrecht. 457 p

Buys MH, Maritz JS, Boucher C, Van Der Walt JJA (1994) A model for species-area relationships in plant communities. J Veg Sci 5:63–66

Byrd DM, Cothern CR (2000) Introduction to risk analysis. Government Institutes: 433p

Byth K, Ripley BD (1980) On sampling spatial patterns by distance methods. Biometrics 36:279–284

Cáceres MD, Legendre P, He F (2013) Dissimilarity measurements and the size structure of ecological communities. Method Ecol Evol 4(12):1167–1177

Cadotte MW, Cardinale BJ, Oakley TH (2008) Evolutionary history and the effect of biodiversity on plant productivity. Proc Natl Acad Sci 105:17012–17017

Cadotte M, Cecile H, Albert CH, Walker SC (2013) The ecology of differences: assessing community assembly with trait and evolutionary distances. Ecol Lett 16:1234–1244

Cahill JF (1999) Fertilization effects on interactions between above- and below- ground competition in an old field. Ecology 80:466–480

Cahill JF (2002) Interactions between root and shoot competition vary among species. Oikos 99:101–112

Cajander AK (1909) Über Waldtypen. Acta For Fenn 1(1):1–175. Helsingfors

Cajander AK (1926) The theory of forest types. Acta Forestalia Fennica 29(3):7193

Calama R, Montero G (2004) Interregional non-linear height-diameter model with random coefficients for stone pine in Spain. Can J For Res 34:150–163

Campbell D, Stanley J (1963) Experimental and Quasi-experimental designs for research and teaching. In: Gage (ed) Handbook on research on teaching. Rand McNally, Chicago

Campbell RG, Ferguson IS, Opie JE (1979) Simulating growth and yield of mountain ash stands: a deterministic model. Aust For Res 9:189–202

Cancino J, Gadow Kv (2001) Stem number guide curves for uneven-aged forests – development and limitations. In: Gadow, Kv, Nagel J, Saborowski J (eds) Continuous cover forestry – assessment, analysis, scenarios. Kluwer Academic Publishers, pp 163–175

Canham CD, Coates KD, Bartemucci P, Quaglia S (1999) Measurement and mdoelling of spatially explicit variation in light transmission through interior cedar-hemlock forests of Bristish Columbia. Can J For Res 29:1775–1783

Canham CD, lePage PT, Coates KD (2004) A neighborhood analysis of canopy tree competition: effects of shading versus crowding. Can J For Res 34:778–787

Cao QV (2000) Prediction of annual diameter growth and survival for individual trees from periodic measurements. For Sci 46:127–131

Cao QV (2006) Predictions of individual-tree and whole-stand attributes for loblolly pine plantations. For Ecol Manage 236:342–347

Cao QV (2010) Adjustment of individual-tree survival and diameter-growth equations to match whole-stand attributes. In: Stanturf JA, editor. Proceedings of the 14th biennial southern silvicultural research conference. USDA, Forest Service, Southern Research Station. Gen. Tech. Rep. SRS-121. pp 369–373

Cao QV (2014) Linking individual-tree and whole-stand models for forest growth and yield prediction. For Ecosyst 1:18

Cao QV, Baldwin VC Jr (1999) A new algorithm for stand table projections models. For Sci 45:506–511

Cao QV, Strub M (2008) Evaluation of four methods to estimate parameters of an annual tree survival and diameter growth model. For Sci 54(6):617–624

Cao QV, Burkhart HE, Max TA (1980) Evaluation of two methods of cubic volume prediction of loblolly pine to any merchantable limit. For Sci 26:71–80

Cao QV, Burkhart HE, Lemin RC (1982) Diameter distributions and yields of thinned loblolly pine plantations. VPI & SU, Sch. For. & Wildl. Resour. Publ FWS-1-82 62p

Carcea F (1986) Determination of the forest's allowable cut in various countries of the world. IUFRO document, Inst. de Cercetari si amenajari silvice, Romania: 268 p

Carmean WH (1973) Forest Soils Bibliography for the North Central Region. USDA For. Serv. Tech. Rep. NC-5

Carnahan J (1989) Australia natural vegetation: Australia's vegetation in the 1780's. Australian Surveying and Land Information Group, Dept. of Administrative Services, Queensland, Australia

Carter CE (1945) The distribution of the more important timber trees of the genus Eucalyptus. Atlas 1. Commonwealth Forestry Bureau, Canberra, Australia. 8 p., 34 plates

Casper BB, Jackson RB (1997) Plant competition underground. Annu Rev Ecol Syst 28:545–570

Casper BB, Cahill JF, Hyatt LA (1998) Above-ground competition does not alter biomass allocated to roots in Abutilon theophrasti. New Phytol 140:231–238

Castedo Dorado F, Álvarez González JG (2000) Construcción de una tarifa de cubicación con clasificación de productos para Pinus radiata D. Don en Galicia basada en una función de perfil del tronco. Investigación agraria: Sistemas y Recursos Forestales 9(2):253–268

Castedo-Dorado F, Diéguez-Aranda U, Barrio Anta M, Sánchez Rodríguez M, Gadow Kv (2006) A generalized height-diameter model including random components for radiata pine plantations in northeastern Spain. For Ecol Manag 229:202–213

Castedo-Dorado F, Diéguez-Aranda U, Álvarez-González JG (2007) A growth model for Pinus radiata D. Don stands in north-western Spain. Ann For Sci 64:453–465

Cavender-Bares J, Kozak KH, Fine PV, Kembel SW (2009) The merging of community ecology and phylogenetic biology. Ecol Lett 12:693–715

Champion HG, Seth SK (1968) A revised survey of the forest types of India. Govt. of India Press, Delhi, 404 p

Chan D, Wu Q (2015) Significant anthropogenic-induced changes of climate classes since 1950. Sci Rep 5(13487):13487. Bibcode:2015NatSR...513487C. https://doi.org/10.1038/srep13487. PMC4551970.PMID26316255

Chang SJ (1998) A generalized Faustmann model for the determination of optimal harvest age. Can J For Res 28:652–659

Chao A, Chazdon RL, Colwell RK, Shen TJ (2005) A new statistical approach for assessing similarity of species composition with incidence and abundance data. Ecol Lett 8:148–159

Chao A, Chiu C-H, Jost L (2014) Unifying species diversity, phylogenetic diversity, functional diversity, and related similarity and differentiation measures through hill numbers. Annu Rev Ecol Evol Syst 45:297–324

Chatfield C (1992) The analysis of time series. An introduction, 4th edn. Chapman and Hall, London, 241 pp. ISBN 0-412-31820-2

Chazdon RL, Brancalion PHS, Laestadius L, Bennett-Curry A, Buckingham K, Kumar C, Moll-Rocek J, Vieira ICG, Wilson SJ (2016) When is a forest a forest? Forest concepts and definitions in the era of forest and landscape restoration. Conserv Lett 10(1):125–132. https://doi.org/10.1111/conl.12220

Chen D, Chen HW (2013) Using the Köppen classification to quantify climate variation and change: an example for 1901–2010. Environ Dev 6:69–79

Chen BW, Gadow K (2002) Timber harvest planning with spatial objectives using the method of simulated annealing. Forstwiss Centralblatt 121:25–34

Chen BW, Gadow Kv (2003) A case study of simulated annealing, genetic algorithm, tabu search and linear programming for solving forest planning problems. Forstachiv 74(1):26–31

Chen BW, Gadow K (2008) Combining spatial and other objectives in forest design. For Stud|Metsanduslikud Uurimused 48:30–40

Chen T, Guestrin C (2016) XGBoost: a scalable tree boosting system.. In: Proceedings of the 22nd ACM SIGKDD International Conference on Knowledge Discovery and Data Mining

Chen S-J, Hwang C-L (1992) Fuzzy multiple attribute decision making – methods and applications, Lecture notes in economics and mathematical systems, vol 375. Springer, Berlin

Chen HY, Luo Y (2015) Net aboveground biomass declines of four major forest types with forest ageing and climate change in western Canada's boreal forests. Glob Chang Biol 21:3675–3684

Chen L, Wang L, Yeerjiang B, Zhang C, Zhao X, Gadow K (2013) Seed dispersal and seedling recruitment of trees at different successional stages in a temperate forest in northeastern China. J Plant Eco 7:337–346. https://doi.org/10.1093/jpe/rtt024

Cheng Y, Zhang C, Zhao X, Gadow K (2018) Biomass-dominant species shape the productivity-diversity relationship in two temperate forests. Ann For Sci 75:97. https://doi.org/10.1007/s13595-018-0780-0

Chisholm RA, Muller-Landau HC, Rahman KA, Bebber DP, Bin Y, Bohlman SA et al (2013) Scale-dependent relationships between tree species richness and ecosystem function in forests. J Ecol 101:1214–1224

Chown NS, Froneman PW (2008) The Prince Edward islands – land-sea interactions in a changing ecosystem. SUN PReSS, Stellenbosch

Christenhusz MJ, Reveal JL, Farjon A, Gardner MF, Mill RR, Chase MW (2011) A new classification and linear sequence of extant gymnosperms. Phytotaxa 19(1):55–70

Christin S, Hervet É, Lecomte N (2018) Applications for deep learning in ecology. bioRxiv preprint first posted online May. 30, 2018; https://doi.org/10.1101/334854

Chung D-J (1996) Konkurrenzverhältnisse und Struktur natürlicher *Pinus densiflora – Quercus variabilis* – Mischwälder in Korea. Diss. Fak. f. Forstw. u. Waldökologie, Universität Göttingen

Cieszewski CJ (2001) Three methods of deriving advanced dynamic site equations demonstrated on inland Douglas-fir site curves. Can J For Res 31:165–173

Cieszewski CJ (2003) Comparing fixed-and variable-base-age site equations having single versus multiple asymptotes. For Sci 48(1):7–23

Cieszewski C, Bailey RL (2000) Generalized algebraic difference approach: theory based derivation of dynamic site equations with polymorphism and variable asymptotes. For Sci 46:116–126

Clark PJ, Evans FC (1954) Distance to nearest neighbor as a measure of spatial relationships in populations. Ecology 35:445–453

Clarke KR, Warwick RM (1998) A taxonomic distinctness index and its statistical properties. J Appl Ecol 35:523–531

Clarke KR, Warwick RM (1999) The taxonomic distinctness measure of biodiversity: weighting of step lengths between hierarchical levels. Mar Ecol Prog Ser 184:21–29

Clarke KR, Warwick RM (2001) A further biodiversity index applicable to species lists: variation in taxonomic distinctness. Mar Ecol Prog Ser 216:265–278

Clements FE, Weaver JE (1938) Plant ecology, 2nd edn. McGraw-Hill Book Company, New York

Cloete WG, Cloete I, Von Gadow K (1988) An algorithm for presenting pairs in optimum orders in the method of paired comparisons. EDV in Medizin und Biologie 19(2–3):75–77

Clutter JL, Allison BJ (1974) A growth and yield model for Pinus radiata in New Zealand. In: Fries J (ed) Growth models for tree and stand simulation, Research note, vol 30. Department of Forest Yield Research. Royal College of Forestry, Stockholm, pp 136–160

Clutter JL, Jones EP (1980) Prediction of growth after thinning in old field slash pine plantations. USDA For. Serv. Res. Paper SE-217

Clutter JL, Fortson JC, Pienaar LV, Brister GH, Bailey RL (1983) Timber management – a quantitative approach. Wiley, New York, 333 p

Coates KD, Canham CD, LePage PT (2009) Above- versus below-ground competitive effects and responses of a guild of temperate tree species. J Ecol 97:118–130

Cochran WG (1977) Sampling techniques, 3rd edn. Wiley, New York

Coetzee J (1986) Growth prediction equations for unthinned, short rotation E. grandis coppice stands using data obtained from working plan surveys. Institute for Commercial Forestry Research, Annual Report: 140–147

Coetzee P (2015) Fighting invasive alien plants in the Kleinmond Coastal and Mountain Nature Reserve. Unpublished report, Kleinmond Nature Conservation Society

Condés S (1997) Simulación de parcelas arboladas con datos del 2 Inventario Forestal Nacional. Tesis Doctoral. Tesis Doctoral Escuela Técnica Superior de Ingenieros de Montes de Madrid. 616 pp

Condit R (1998) Tropical forest census plots: methods and results from Barro Colorado Island, Panama and a comparison with other plots. Springer, 223 p

Condit R, Pitman N, Leigh EG Jr, Chave J, Terborgh J, Foster RB (2002) Beta-diversity in tropical forest trees. Science 295(5555):666–669

Condit R, Engelbrecht BM, Pino D, Pérez R, Turner BL (2013) Species distributions in response to individual soil nutrients and seasonal drought across a community of tropical trees. Proc Natl Acad Sci U S A 110:5064–5068

Conrad O, Bechtel B, Bock M, Dietrich H, Fischer E, Gerlitz L, Wehberg J, Wichmann V, Böhner, J. (2015) System for automated geoscientific analyses (saga) v. 2.1.4. Geosci Model Dev 8(7):1991–2007

Convention on Biological Diversity (2017) Definitions: indicative definitions taken from the report of the ad hoc technical expert group on forest biological diversity. https://www.cbd.int/forest/definitions.shtml. Retrieved 29 March 2017

Cook T, Campbell D (1979) Quasi-experimental design. Rand McNally, Chicago

Coomes DA, Allen RB (2007) Effects of size, competition and altitude on tree growth. J Ecol 95:1084–1097

Cornelissen JHC, Lavorel S, Garnier E, Díaz S, Buchmann N, Gurvich DE, Reich PB, ter Steege H, Morgan HD, van der Heijden MGA, Pausas JG, Poorter H (2003) Handbook of protocols for standardised and easy measurements of plant functional traits worldwide. Aust J Bot 51:335–380

Corns IGW, Pluth DJ (1984) Vegetational indicators as independent variables in forest growth prediction in west-central Alberta, Canada. For Ecol Manage 9:13–25

Corral Rivas JJ, Álvarez González JG, Ruíz González AD, Gadow Kv (2003) Compatible height and site index models for five pine species in El Salto, Durango (Mexico)

Corral Rivas JJ, Álvarez González JG, Ruíz González AD, Gadow Kv (2004) Compatible height and site index models for five pine species in El Salto, Durango (Mexico). For Ecol Manag 201:145–160

Corral Rivas JS, Torres Rojo JM, Lujan Soto JE, Nava Miranda MG, Aguirre Calderón OA, Gadow K (2016) Density and production in the natural forests of Durango/Mexico. Allgemeine Forst- und Jagdzeitung 187(5/6):93–103

Corral-Rivas J, Sánchez Orois S, Kotzé H, Gadow K (2009) Testing the suitability of a stand table projection method for Eucalyptus grandis plantations in South Africa. Southern For 71(3):207

Corral-Rivas JJ, Wehenkel C, Castellanos BH, Vargas LB, Diéguez-Aranda U (2010) A permutation test of spatial randomness: application to nearest neighbour indices in forest stands. J For Res 15:218–225

Corral-Rivas JJ, Reyes RI, Wehenkel C, Aguirre-Calderón OA, Gadow Kv (2012) A network of forest observational studies in Durango (Mexico). In: Zhao XiuHai, Zhang ChunYu, Gadow Kv (eds) Forest observational studies. Proceedings of an international workshop at Beijing Forestry University, which convened on 20/21 September 2012: 125–138

Corral-Rivas JJ, Hernández-Díaz JC, López-Sánchez CA, Luján-Soto JE, Gadow Kv (2015a) Ejido Borbollones, Durango, Mexico.Chapter 9; pag. In: Siry JP, Bettinger P, Merry K, Grebner DL, Boston K, Cieszewski Ch (eds) Forest plans of North America. Academic Press; Elsevier, pp 61–68

Corral-Rivas S, Álvarez-González JG, Corral-Rivas JJ, López-Sánchez CA (2015b) Characterization of diameter structures of natural forests of northwestern Durango, Mexico. Revista Chapingo Serie Ciencias Forestales y del Ambiente 21(2):221–236

Corral-Rivas JS, Torres Rojo JM, Lujan Soto JE, Nava Miranda MG, Aguirre Calderón OA, Gadow Kv (2016) Density and production in the natural forests of Durango/Mexico. Allgemeine Forstund Jagdzeitung 187(5/6):93–103

Corral-Rivas JJ, González-Elizondo MS, Lujan-Soto JE, von Gadow K (2018) Effects of density and structure on production in the communal forests of the Mexican Sierra Madre Occidental. Southern Forests: a Journal of Forest Science. https://doi.org/10.2989/20702620.2018.1463152

Cossalter C, Pye-Smith C (2003) Fast-wood forestry – myths and realities. Center for International Forestry Research (CIFOR): 50 p

Costanza R, d'Arge R, de Groot R, Farber S, Grasso M, Hannon B, Limburg K, Naeem S, O'Neill RV, Paruelo J, Raskin RG, Sutton P, van den Belt M (1997) The value of the world's ecosystem services and natural capital. Nature 387:253–260

Cotta H (1804) Systematische Anleitung zur Taxation der Waldungen. Sander, Berlin

Courbaud B (1995) Modélisation de la croissance en forêt irrégulière, – perspectives pour lespessières irrégulières de montagne. Rev For Fr XLVII:173

Cox DR (1958) Planning of experiments. Wiley, New York

Cox DR, Oakes D (1984) Analysis of survival data. Chapman & Hall, London

Craib IJ (1939) Thinning, pruning and management studies on the main exotic conifers grown in South Africa. Govt. Printer, Pretoria

Crawford G, Williams C (1985) A note on the analysis of subjective judgment matrices. J Mathl Psychol 29:387–405

Crawley MJ (2005) Statistics – an introduction using R. Wiley, West Sussex, 327 p

Crecente-Campo F, Soares P, Tomé M, Diéguez-Aranda U (2010) Modelling annual individual-tree growth and mortality of Scots pine with data obtained at irregular measurement intervals and containing missing observations. For Ecol Manage 260:1965–1974

Crecente-Campo F, Corral-Rivas JJ, Vargas-Larreta B, Wehenkel C (2014) Can random components explain differences in the height-diameter relationship in mixed uneven-aged stands? Ann For Sci 71:51–70

Crutzen PJ (2002) Geology of mankind. Nature 415:23. https://doi.org/10.1038/415023a

Cunia T (1974) On the objectives and methodology of national forest inventories. In: Forest resource inventory. Joint Meeting of IUFRO Groups S4.02 and S4.04, 18–26 June 1978, Institutul de Cerctari si Amenageri Silvice, XI-XXIX. Bucharest

Curt T, Prévosto B (2003) Root biomass and rooting profile of naturally regenerated beech in mid-elevation Scots pine woodlands. Plant Ecol 167(2):269–282

Curtis RO, Marshall DD, Bell JF (1997) LOGS: a pio- neering example of silvicultural research in coast Douglas-Fir. J For 95(7):19–25

Curtis RO (1967) Height-diameter and height-diameter-age equations for second growth Douglas-fir. For Sci 13:365–375

Cushing CE, Cummins KW, Minshall GW (1997) Ecosystems of the world: river and stream ecosystems. J North Am Soc 16(3):717–719

Cutler DR, Edwards TC, Beard KH, Cutler A, Hess KT, Gibson J, Lawler JJ (2007) Random forests for classification in ecology. Ecology 88(11):2783–2792

Czaplewski RL (2003) Accuracy assessment of maps of forest condition: statistical design and methodological considerations. In: Wulder MA, Franklin SE (eds) Remote sensing of forest environments: concepts and case studies. Boston, Kluwer Academic Publishers, pp 115–140

Daily GC (1997) Nature's services: societal dependence on natural ecosystems. Island Press, Washington, DC

Daniels RF (1976) Simple competition indices and their correlation with annual loblolly pine tree growth. For Sci 22:454–456

Danneyrolles V, Dupuis S, Fortin G, Leroyer M, de Römer A, Terrail R, Vellend M, Boucher Y, Laflamme J, Bergeron Y, Arseneault D (2019) Stronger influence of anthropogenic disturbance than climate change on century-scale compositional changes in northern forests. Nat Commun 10:1265. https://doi.org/10.1038/s41467-019-09265-z

Daume S (1995) Durchforstungssimulation in einem Buchen-Edellaubholz Mischbestand. Diplomarbeit, Fachbereich Forstwissenschaft, Universität Göttingen

Daume S, Füldner K, Gadow K (1997) Zur Modellierung personenspezifischer Durchforstungenin ungleichaltrigen Mischbeständen. AFJZ 169(2):21–26

Daume S, Albert M, Gadow K (2014a) Assessing citizen science opportunities in forest monitoring using probabilistic topic modelling. For Ecosyst 1:11. (available for download at http://www.forestecosyst.com/content/1/1/11)

Daume S, Albert M, Gadow K v (2014b) Forest monitoring and social media – complementary data sources for ecosystem surveillance? For Ecol Manage 316(2014):9–20

Day K (1998) Stocking standards for uneven aged interior Douglas-fir stands. In: Vyse A, Hollstedt C, Huggard D (eds) Managing the dry Douglas-Fir forests of the Southern Interior. Res. B.C. mim. For., Victoria B.C. Working Paper 34. pp 37–52

de Caprariis P, Lindemann RH, Collins CM (1976) A method for determining optimum sample size in species diversity studies. Math Geol 8(5):575–581

de Miguel S et al (2010) Models for simulating the development of even-aged Pinus brutia stands in Middle East. For Syst 19:449–457

de Miguel S, Pukkala T, Morales M (2014) Using optimization to solve tree misidentification and uneven measurement interval problems in individual-tree modeling of Balsa stand dynamics. Ecol Eng 69:232–236

Dedrick S, Spiecker H, Orazio C, Tomé M, Martinez I (2007) Plantation or conversion – the debate. European Forest Institute, Discussion paper 13: 98 p

Demaerschalk JP (1973) Integrated systems for the estimation of tree taper and volume. Can J For Res 3(1):90–94

Demaerschalk JP, Kozak A (1977) The whole bole system: a conditioned dual-equation system for precise prediction of tree profiles. Can J For Res 7:488–497

Deo RC, S ṃ ahin, M. (2015) Application of the extreme learning machine algorithm for the prediction of monthly effective drought index in Eastern Australia. Atmos Res 153:512–525

Devictor V, Mouillot D, Meynard C, Jiguet F, Thuiller W, Mouquet N (2010) Spatial mismatch and congruence between taxonomic, phylogenetic and functional diversity: the need for integrative conservation strategies in a changing world. Ecol Lett 13:1030–1040

DGCN. Tercer Inventario Forestal Nacional. (Ministerio de Medio Ambiente, Madrid, Spain 2008)

Dickel M, Kotze H, Gadow K, Zucchini W (2010) Growth and survival of *Eucalyptus grandis* – a study based on modeling lifetime distributions. Int J Math Comput For Nat-Resour Sci 2(2):20–30. (available online at http://mcfns.com)

Diéguez-Aranda U, Grandas-Arias JA, Álvarez-González JG, Kv G (2006) Site quality curves for birch stands in north-western Spain. Silva Fennica 40(4):631–644

Diggle PJ (1983) Statistical analysis of spatial point patterns. Academic, London, p 148

Diggle PJ (2003) Statistical analysis of spatial point patterns, 2nd edn. Arnold, London, 159p

Diggle PJ, Besag JE, Glaves JT (1976) Statistical analysis of spatial point patterns by means of distance methods. Biometrics 32:659–667

Dittmar O, Knapp E, Lembcke G (1986) DDR-Buchenertragstafel. IFE-Berichte aus Forschung und Entwicklung Heft 4 1986

Dixon PM (2002) Encyclopedia of environmetrics, vol 3. Wiley, Chichester, pp 1796–1803

Dobson AJ (1990) An introduction to generalized linear models. Chapman and Hall, Statistic Text Series, London

Dohrenbusch A, Bredemeier M, Lamersdorf N (2003) Manipulation of nutrient and water input of a Norway Spruce Ecosystem. Springer, Berlin

Dong L, Zhang L, Li F (2015) A three-steps proportional weighting system of nonlinear biomass equation. For Sci 61:35–45

Donnelly KP (1978) Simulations to determine the variance and edge effect of total nearest-neighbor distance. In: Hodder I (ed) Simulation methods in archaeology. Cambridge University Press, London, pp 91–95

Dragoi M, Ciornei I (2013) A balck-box approach on assessing the opportunity cost of deforestation. Land Use Policy 34:314–320

Dralet M (1820) Traité des forêts des arbres résineux et des terrains adjacent, sur les montagnes de la France. Vieussieux, Toulouse

Drew TJ, Flewelling JW (1979) Stand density management: an alternative approach and its application to Douglas-fir plantations. For Sci 25:518–532

Drexhage M, Colin F (2001) Estimation root system biomass from breast-height diameters. Forestry 74:491–497

Dreyfus P, Bonnet F-R (1995) CAPSIS – Logiciel de simulation de conduites sylvicoles. Rev For Fr XLVII(sp):111–115

Du Plessis M, Zwolinski J (2003) Site and stand analysis for growth prediction of Eucalyptus grandis on the Zululand coastal plain. Southern Hemisphere For J 198

Dunkel K, Elsasser P, Oesten G, Roeder A (1994) Wertschätzung des Waldes aus der Sicht der Waldbesucher – Ergebnisse einer Zielgebietsbefragung im Pfälzerwald. Mitt. d. Forstl. Vers. Anst. Rheinld-Pfalz 27(94):1–72

Düser R (1978) Programmierte Berechnung stehenden Holzes mit dem Taschenrechner nach der Düser-Flori-Formel. Beilage der AFZ, Nr. 36

Dykstra DP (1984) Mathematical programming for natural resource management. McGraw-Hill, New York, 318 p

Ebert W, Langer R (1977) Frankfurter Stadtwald. Published by Forstamt Frankfurt, 64 pp

Edwards M, Scotcher JSB (2000) Significant conservation areas in timber plantations. In: Owen DL (ed) South African forestry handbook 2000, The Southern African Institute of Forestry, vol 2, pp 579–581

Eid T (1993) Models for economical forest management planning in Norway. In: Proceeding of the symposium modelling in forest management planning and managerial economics – a critical investigation. Lithuanian Agricultural Academy, pp 35–43

Eid T (2000) Use of uncertain inventory data in forestry scenario models and consequential incorrect harvest decisions

Eid T, Hobbelstad K (2000) AVVIRK-2000: a large-scale forestry scenario model for long-term investment, income and harvest analyses. Scand J For Res 15(4):472–482

Ek AR, Monserud RA (1974) Dimensional relationships of forest and open-grown stems in Wisconsin. Univ. Wis. Forestry Res. Note 181, 7

Ekö PM, Ågestam E (1994) A comparison of naturally regenerated and planted Scots pine (*Pinus sylvestris* L.) on fertile sites in southern Sweden. For u Landsc Res 1:III–126

Elith J, Leathwick JR (2009) Species distribution models: ecological explanation and prediction across space and time. Annu Rev Ecol Evol Syst 40(1):677

Ellenberg H (1986) Vegetation Mitteleuropas mit den Alpen in ökologischer Sicht. 6. Auflage. Ulmer, Stuttgart

Ellenberg H, Mueller-Dombois D (1967) Tentative physiognomic-ecological classification of plant formations of the Earth [based on a discussion draft of the UNESCO working group on vegetation classification and mapping.] Berichte des Geobotanischen Institutes der Eidg. Techn. Hochschule, Stiftung Rübel, Zürich 37(1965–1966):21–55

Ellenberg H, Weber HE, Düll R, Wirth V, Werner W, Paulissen D (1991) Zeigerwerte von Pflanzen Mitteleuropas. Scripta Geobotanica XVIII. Goltze, Göttingen

Elliott M (2011) Marine science and management means tackling exogenic unmanaged pressures and endogenic managed pressures—a numbered guide. Mar Pollut Bull 62(4):651–655

Ellis EC (2011) Anthropogenic transformation of the terrestrial biosphere. Philos Trans R Soc 369:1010–1035

Emborg J, Christensen M, Heilmann-Clausen J (2000) The structural dynamics of Suserup Skov, a near-natural temperate deciduous forest in Denmark. For Ecol Manage 126(2):173–189

Environmental Systems Research Institute Inc (1990) Understanding GIS, The ARC/INFO Method

Erb K-H, Kastner T, Plutzar C, Bais ALS, Carvalhais N, Fetzel T, Gingrich S, Haberl H, Lauk C, Niedertscheider M, Pongratz J, Thurner M, Luyssaert S (2017) Unexpectedly large impact of forest management and grazing on global vegetation biomass. Nature 553:73–76. (04 January 2018)

Ervasti S, Heikinheimo L, Holopainen V, Kuusela K, Sirén G (1965) The development of Finland's forests in 1964–2000. Silva Fennica 117(2):35 p

Ervasti S, Heikinheimo L, Kuusela K, Mäkinen VO (1970) Forestry and forest industry production alternatives in Finland, 1970–2015. Folia Forestalia 88:65 p

Escribano N, Ariño AH, Galicia D (2016) Biodiversity data obsolescence and land uses changes. PeerJ 4:e2743

Essl F, Lenznerl B, Courchamp F, Dullinger S, Jeschke JM, Kühn I, Leung B, Moser D, Roura-Pascual N, Seebens H (2019) Introducing AlienScenarios: a project to develop scenarios and models of biological invasions for the 21st century. NeoBiota 45:1–17. https://doi.org/10.3897/neobiota.45.33366

Everard DA (2000) Biodiversity. In: Owen DL (ed) South African forestry handbook 2000, The Southern African Institute of Forestry, vol 2, pp 563–566

Eyre FH (1980) Forest cover types of the United States and Canada. Society of American Foresters, Bethesda, Maryland, l48 p (1 map sheet)

Fahrmeir L (2007) Lebensdauer- und Ereignisanalyse [Survival and Event History Analysis]. In: Vorlesungsskript, Institut für Statistik. Ludwig-Maximilians-Universität, München. 141 pp. (in German)

Faith DP (1992) Conservation evaluation and phylogenetic diversity. Biol Conserv 61(1):1–10

Falinski J (1986) Vegetation dynamics in temperate lowland primeval forests: ecological studies in Bialowieza Forest. Dr W Junk/Kluwer Academic Publishers, Dordrecht, 525 pp

Fan Y, Li H, Miguez-Macho G (2013) Global patterns of groundwater table depth. Science 339(6122):940–943

FAO (2015) Global forest resources assessment. Food and Agricultural Organisation of the United Nations. http://www.fao.org/forest-resources-assessment/en/. Accessed 16 June 2017

FAO (2018) The state of the world's forests 2018 – forest pathways to sustainable development. Rome. Licence: CC BY-NC-SA 3.0 IGO

Fehrmann L, Kuhr M, Gadow Kv (2002) Zur Analyse der Grobwurzelsysteme großer Waldbäume an Fichte [*Picea abies* (L.) Karst.] und Buche [*Fagus sylvatica*]. Forstachiv 74(3):96–102

Fenelon JP, Lebart L (1971) Statistique et informatique appliquées. Dunod, Paris

Feng G, Mi X, Yan H, Li FY, Svenning J-C, Ma K (2016) CForBio: a network monitoring Chinese forest biodiversity. Sci Bull. https://doi.org/10.1007/s11434-016-1132-9

Ferris R, Humphrey JW (1999) A review of potential biodiversity indicators for application in British forests. Forestry 72(4):313–328

FIA (2012) Forest inventory and analysis. National core field guide. Volume I: Field data collection procedures for phase 2 plots. Version 6.0.: USDA, Forest Service

Fisher R, The A (1935) Design of experiments, 1st edn. Oliver & Boyd, London

Fisher RA, Corbet SA, Williams CB (1943) The relation between the number of species and the number of individuals in a random sample of an animal population. J Anim Ecol 12:42–58

Fluet-Chouinard E, Lehner B, Rebelo L-M, Papa F, Hamilton SK (2015) Development of a global inundation map at high spatial resolution from topographic downscaling of coarse-scale remote sensing data. Remote Sens Environ 158:348–361

Flury P (1929) Über den Aufbau des Plenterwaldes. Mitteilungen der schweizerischen Centralanstalt für das Versuchswesen, Band 15:305–357

Folch R., Camarasa JM (2000) Encyclopaedia of the Biosphere Vol. 1-11; Gale Group

Fontes L, Tomé, J, Tomé M, Cortiçada A, Faias SP, Soares P, Pacheco CA, Araújo C (2004) Calibration of 3-PG for Eucalyptus globulus plantations in Portugal. In: Borralho NMG, Pereira JS, Marques C, Coutinho J, Madeira M, Tomé M (eds) Eucalyptus in a changing world. Proc. Iufro Conf., Aveiro, 11–15 Oct. (RAIZ, Instituto de Investigação da Floresta e do Papel, Portugal), pp 344–345

Forest Survey of India (2013) India state of forest report 2013. Ministry of Environment and Forests, Dehradun, 252 p

FORPLAN (1986) FORPLAN – an evaluation of a forest planning tool. USDA Gen Techn Rep RM-140, 164pp

Forrester DI (2014) The spatial and temporal dynamics of species interactions in mixed-species forests: from pattern to process. For Ecol Manage 312:282–292

Forrester DI, Vanclay JK, Forrester RI (2011) The balance between facilitation and competition in mixtures of Eucalyptus and Acacia changes asstands develop. Oecologia 166:265–272

Foster DR, Donahue B, Kittredge DB, Motzkin G, Hall B, Turner BL, Chilton E (2008) New England's forest landscape. In: Redman CL, Foster DR (eds) Ecological legacies and conservation patterns shaped by agrarian history, Agrarian landscapes in transition. Oxford University Press, New York

Fourcade Y, Besnard AG, Secondi J (2018) Paintings predict the distribution of species, or the challenge of selecting environmental predictors and evaluation statistics. Glob Ecol Biogeogr 27(2):245–256

Franklin JF, Berg DR, Thornburgh DA, Tappeiner JC (1997) Alternative silvicultural approaches to timber harvesting: variable retention harvest systems. In: Kohn KA, Franklin JF (eds) Creating a forestry for the 21st century: the science of ecosystem management. Island Press, Washington, DC, 111p

Franklin JF, Spies TA, Van Pelt R, Carey AB, Thornburgh DA, Berg DR, Lindenmayer DB, Harmon ME, Keeton WS, Shaw DC, Bible K, Chen JQ (2002) Disturbances and structural development of natural forest ecosystems with silvicultural implications, using Douglas-fir forests as an example. For Ecol Manage 155(1-3):399–423

Franz F (1990) Ernst Assmann's Wirken in München – ein Rückblick zu seinem zehnten Todestag. In Franz and Utschig 1990:1–16

Freese F (1962) Testing accuracy. For Sci 6:139–145

Friedman JH (2002) Stochastic gradient boosting. Comput Stat Data Anal 38(4):367–378

Früh T (1992) C-Bilanz auf der Basis der Pipe-Modell-Theorie, Produktionsmodell. Reports of the For-schungszentrum Waldökosysteme (Göttingen), Band 31:355–367

Fujimori T, Kiyono Y (1986) Dynamics of crown structure and effects of crown control on tree growth in Cryptomeria japonica. In: Fujimori T, Whitehead (eds) Crown and canopy structure in relation to productivity. Published by the Forestry and Forest Products Institute, Ibaraki, pp 243–262. ISBN 4-9900058-1-3

Füldner K (1995) Strukturbeschreibung von Buchen-Edellaubholz-Mischwäldern. Cuvillier Verlag, Göttingen: 146 S

Füldner K, Gadow Kv (1994) How to define a thinning in a mixed deciduous beech forest. In: Proc IUFRO conference in Lousa, Portugal: mixed stands – research plots, measurements and results, models, pp 31–42

Füldner K, Sattler S, Zucchini W, Gadow K v (1996) Modellierung personenabhängiger Auswahlwahrscheinlichkeiten bei der Durchforstung. AFJZ 167(8):159–162

Gadow K (1986) Observations on self-thinning in pine plantations. S Afr J Sci 82(7):364–368

Gadow Kv (1987) Untersuchungen zur Konstruktion von Wuchsmodellen für schnellwüchsige Plantagenbaumarten. Forstl. Forschungsberichte, Universität München Nr. 77, 147 S

Gadow Kv (1988) Thinning rules for pine plantations. In: Proceedings of the resource technology 88 conference, Fort Collins, Colorado, 20-23 Junie 1988: 230-231

Gadow K (1991) The structure and derivation of a rule-based model for pruning decisions. Biometrie u Informatik in Medizin u Biologie 23(3):157–166

Gadow K (1993) Zur Bestandesbeschreibung in der Forsteinrichtung. Forst und Holz 48(21):602–606

Gadow K (1996) Modelling growth in managed forests-realism and limits of lumping. Sci Total Environ 183:167–177

Gadow K (1999) Waldstruktur und Diversität. Allg Forst- u Jagdzeitg 170(7):117–122

Gadow K (2000) Evaluating risk in forest planning models. Silva Fennica 34(2):181–191

Gadow K (ed) (2001) Risk analysis in forest management, Managing forest ecosystems, vol 2. Springer, Dordrecht

Gadow K (2004) Waldstruktur und Wachstum. Universitätsdrucke Göttingen

Gadow K (2013) Messung und Modellforschung – Grundlagen der Forsteinrichtung (Measurement and modeling research – basis for forerst design). Allgemeine Forst und Jagdzeitung 184(7/8):143–158

Gadow Kv (2017) Permanent forest observational plots – assessment and analysis. Report prepared for FAO, October, 2017:. 38pp

Gadow Kv, Bredenkamp BV (1992) Forest management. Academica Press, Pretoria, 151 p

Gadow K, Füldner K (1992) Bestandesbeschreibung in der Forsteinrichtung. Tagungsbericht der Arbeitsgruppe Forsteinrichtung Klieken bei Dessau 15(10):92

Gadow K, Heydecke H (2000) Wachstum und Durchforstung in einem Buchenmischbestand. Forst & Holz 56(3):86–88

Gadow K, Hui GY (1993) Stammzahlentwicklung und potentielle Bestandesdichte bei *Cunninghamia lanceolata*. Centralblatt für das gesamte Forstwesen 110(2):41–48

Gadow Kv, Hui GY (1997) Modellierung forstucher Eingriffe in Plantagenwäldern. Forstarchiv 68:59–63

Gadow Kv, Hui GY (1999) Modelling forest development. Kluwer Academic Publishers, Dordrecht, 213 pp. ISBN 0-7923-5488-5

Gadow Kv, Hui GY (2002) Characterizing forest spatial structure and diversity. In: Proceeding of the SUFOR conference, sustainable forestry in temperate regions, April: 7–9

Gadow K, Hui GY (2007) Can the tree species-area relationship be derived from prior knowledge of the tree species richness? For Stud/Metsanduslikud Uurimused 46:13–22

Gadow Kv, Kleinn C (2005) Forest management, science-based and understandable. In: Peterson CE, Maguire DA (eds) Balancing ecosystem values – innovative experiments for sustainable forestry. US Dept of Agr. Forest Service, Gen. Tech. Rep. PNW-GTR-635: 15-23

Gadow Kv, Kotzé H (2015) Tree survival and maximum density of planted forests – observations from South African spacing studies. For Ecosyst 1:21. https://doi.org/10.1186/s40663-014-0021-4

Gadow K, Meskauskas E (1997) Stichprobenverfahren zur Erfassung von Naturverjüngung. AFZ/Der Wald 52(5):247–248

Gadow Kv, Pukkala T (2008) Designing green landscapes, Vol. 15 of the book series "Managing forest ecosystems". Springer, Dordrecht, 322 p

Gadow Kv, Puumalainen J (2000) Scenario planning for sustainable forest management. In: Gadow Kv, Pukkala T, Tomé M (eds) Sustainable forest management. Vol. 1 in the book series Managing Forest Ecosystems, Kluwer Academic Publishers, 319–356

Gadow K, Schmidt M (1998) Periodische Inventuren und Eingriffsinventuren. Forst u Holz 53(22):667–671

Gadow Kv, Tewari VP (2010) Domesticating forest ecosystems – theory and application. Paper version of a presentation delivered at the International workshop "Use of modern technologies in forestry and advancements in wood science", held on 19th November 2009 at the Institute of Wood Science and Technology in Bangalore, India, 8 pp

Gadow K, Hui GY, Albert M (1998) Das Winkelmass – ein Strukturparameter zur Beschreibung der Individualverteilung in Waldbeständen. Centralbl für das ges Forstw 115:1–10

Gadow K, Kurttila M, Leskinen P, Leskinen L, Nuutinen T, Pukkala T (2007) Designing forested landscapes to provide multiple services. CAB Rev 2(038):1–15

Gadow Kv, Nuutinen T, Kellomäki S (2008) Adaptive design of forested landscapes. In: Gadow K, Pukkala T (eds) Designing green landscapes, Vol. 15 of the Book series "Managing forest ecosystems". Springer, Dordrecht, pp 3–30

Gadow K, Zhang CY, Zhao XH (2009) Science-based forest design. Int J Math Comput For Nat-Resour Sci 1(1):1–12

Gadow Kv, Zhang CY, Wehenkel C, Pommerening A, Corral-Rivas J, Korol M, Myklush S, Hui GY, Kiviste A, Zhao XH (2011) Forest structure and diversity. In: Pukkala T, Gadow K (eds) Continuous cover forestry, Book series managing forest ecosystems, vol 24. Springer, Heidelberg, pp 29–84

Gadow K, Zhang CY, Wehenkel C, Pommerening A, Corral-Rivas J, Korol M, Myklush S, Hui GY, Kiviste A, Zhao XH (2012) Forest structure and diversity. In: Pukkala T, Gadow KV (eds) Continuous cover forestry. Springer, Dordrecht, pp 29–83

Gadow Kv, Zhao XH, Corral Rivas JJ, (2013) Retention strategies for multi-species forests. In: Proceeding of the international symposium for the 50th anniversary of the forestry sector planning in Turkey, 26–28 November 2013, Antalya

Gadow K, Zhang GQ, Durrheim G, Drew D, Seydack A (2016a) Diversity and production in an afromontane forest. For Ecosyst 3:15. (http://forestecosyst.springeropen.com/articles/10.1186/s40663-016-0074-7)

Gadow K, Zhao XH, Tewari VP, Zhang CY, Kumar A, Rivas JJC, Kumar R (2016b) Forest observational studies: an alternative to designed experiments. Eur J For Res 135(3):417–431

Gadow Kv, Hao MH, Kahriman A, Göktürk A, Corral-Rivas J, Ganeshaiah KN (2018) Biodiversity and biodistance among forest communities – new methods based on observational plots from Turkey, Mexico and South Africa – new methods based on observational plots from Turkey, Mexico and South Africa. Presentation at the Beijing Forestry University during a Forest Ecosystems workshop on 18 September, 2018

Gaffrey D (1988) Forstamts- und bestandesindividuelles Sortimentierungsprogramm als Mittel zur Planung, Aushaltung und Simulation. Diplomarbeit, Fachbereich Forstwissenschaft, Universität Göttingen

Gaffrey D (1996) Sortenorientiertes Bestandeswachstums-Simulationsmodell auf der Basis intraspezifischen, konkurrenzbedingungten Einzelbaumwachstums-insbesondere hinsichtlich des Durchmessers- am Beispiel der Douglasie. Diss. Forschungszentrum Waldökosysteme d. Uni. Göttingen, 413 p

Gandomi A, Haider M (2015) Beyond the hype: big data concepts, methods, and analytics. Int J Inf Manag 35(2):137–144

Ganeshaiah KN, Chandrashekara K, Kumar ARV (1997) Avalanche index: a new measure of biodiversity based on biological heterogeneity of the communities. Curr Sci 73:128–133
Ganeshalah KN, Shaanker UR (2000) Measuring biological heterogeneity of forest vegetation types: avalanche index as an estimate of biological diversity. Biodivers Conserv 9:953–963
Ganeshalah KN, Shaanker UR (2003) A decade of diversity. University of Agricultural Sciences Bangalore/India. Internal report, 100 p
García O (1988) Growth modelling—a (re)view. N Z J For Sci 33(3):14–17
Garcia O (1990) Linear programming and related approaches in forest planning. N Z J For Sci 20(3):307–331
Garcia O (1991) Linear programming and related approaches in forest planning. N Z J For Sci 20(3):307–331
García O (1994) The state-space approach in growth modeling. Canadian Journal of Forest Research-Revue Canadienne de Recherche Forestiere 24:1894–1903
García O (2003) Dimensionality reduction in growth models: an example. For Bio Modell Inf Sci 1:1–15
García O (2011) A parsimonious dynamic stand model for interior spruce in British Columbia. For Sci 57(4):265–280
García O (2013) Building a dynamic growth model for trembling aspen in western Canada without age data. Can J For Res 43(3):256–265
García O, Burkhart HE, Amateis RL (2011) A biologically-consistent stand growth model for loblolly pine in the Piedmont physiographic region, USA. For Ecol Manage 262(11):2035–2041
Garland JT, Harvey PH, Ives AR (1992) Procedures for the analysis of comparative data using phylogenetically independent contrasts. Syst Biol 41:18–32
Gartner (2018) Retrieved from http://www.gartner.com/it-glossary/big-data/ on 13th January, 2018
Gasparini P, Tabacchi G (2009) L'Inventario Nazionale delle Foreste e dei serbatoi forestali di Carbonio INFC[page12image168413440]2005. Secondo inventario forestale nazionale italiano. Metodi e risultati. Ministero delle Politiche Agricole, Alimentari e Forestali. 653 (Corpo Forestale dello Stato. Consiglio per la Ricerca e la Sperimentazione in[page12image168419776] Agricoltura, Unità di ricerca per il Monitoraggio e la Pianificazione Forestale, Bologna, Italy, 2011)
Gaston KJ, Blackburn TM (2000) Pattern and process in macroecology. Blackwell Science, Oxford
Gaul HT, Stüber V (1996) Der Eichen-Nelder-Verbandsversuch Göhrde. Forst und Holz 51:70–75
Gause GF (1934) The struggle for existence. The Williams & Wilkins Company, Baltimore
Gehrhardt E (1930) Ertragstafeln für reine und gleichartige Hochwaldbestände von Eiche, Buche, Tanne, Fichte, Kiefer, grüner Douglasie und Lärche. 2. Aufl. Julius Springer
Gehrmann D (1975) Die Bewertung des Windwurfrisikos der Fichte auf verschiedenen Standortstypen. Mitt. d. Hess. Landesforstverwaltung, No 12
Geiger R (1954) Klassifikation der Klimate nach W. Köppen [Classification of climates after W. Köppen]. Landolt-Börnstein – Zahlenwerte und Funktionen aus Physik, Chemie, Astronomie, Geophysik und Technik, alte Serie. Springer, Berlin, 3. pp 603–607
Geldenhuys CJ (2009) Managing forest complexity through application of disturbance-recovery knowledge in development of silvicultural systems and ecological rehabilitation in natural forest systems in Africa. J For Res 15:3–13
Geldenhuys CJ, Bezuidenhout L (2008) Practical guideline for the rehabilitation of forest related streambank vegetation with removal of invader plant stands along the Berg River, Western Cape. Report FW-02/08, FORESTWOOD cc, Pretoria. 39 pp
Geldenhuys CJ, Mucina L (2006) Afrotemperate, subtropical and azonal forests. In: Mucina L, Rutherford MC (eds) The vegetation of South Africa, Lesotho and Swaziland, Strelitzia 16. South African National Biodiversity Institute, Pretoria, pp 585–614
Gerrard DJ (1969) Competition quotient – a new measure of the competition affecting individual forest trees. Mich State Univ Agr Exp Stn Res Bull No 20
Giannetti F, Puletti N, Quatrini V, Travaglini D, Bottalico F, Corona P, Chirici G (2018) Integrating terrestrial and airborne laser scanning for the assessment of single-tree attributes in Mediterranean forest stands. Eur J Rem Sens 51(1):795–807

Gibson CC, McKean MA, Ostrom E (2000) People and forests: communities, institutions, and governance. MIT Press, Cambridge, MA

Gillis MD, Omule AY, Brierley T (2005) Monitoring Canada's forests: the National Forest Inventory. For Chron 81:214–221. https://doi.org/10.5558/tfc81214-2

Gingrich SE (1967) Measuring and evaluating stocking and stand density in Upland hardwood forest in the Central States. For Sci 13:38–53

Gitay H, Rexburg SH, Wilson JB (1991) Species-area relation in a New Zealand tussock grassland, with implications for nature reserve design and for community structure. J Veg Sci 2:113–118

Gleichmar W, Gerold D (1998) Indizes zur Charakterisierung der horizontalen Baumverteilung. Forstwiss Centralbl 117:69–80

González JR, Palahí M, Trasobares A, Pukkala T (2006) A fire risk model for forest stands in Catalonia (north-east of Spain). Ann For Sci 63:169–173

González EMS, González EM, Márquez LMA (2007) Vegetación y Ecoregiones de Durango. CIIDIR-IPN. Plaza y Valdés, S.A. de C.V. México, D.F. 219 p

González-Elizondo MS, González-Elizondo M, Tena-Flores JA, Ruacho-González L, López-Enríquez IL (2012) Vegetación de la sierra madre occidental, México: Una síntesis. Acta botánica mexicana 100:351–403

Goodbody TRH, Coops NC, White JC (2019) Digital aerial photogrammetry for updating area-based forest inventories: a review of opportunities, challenges, and future directions. Curr For Rep. https://doi.org/10.1007/s40725-019-00087-2

Gorska I (1976) Badania archeologiczne w Puszczy Białowieskiej'. [Archaeological investigations in Bialowieza Forest] Archeologia Polski 21(1)

Goward SN, Williams DL, Arvidson T, Rocchio LE, Irons JR, Russell CA, Johnston SS (2017) Landsat's enduring legacy: pioneering global land observations from space. American Society for Photogrammetry and Remote Sensing, Bethesda. (586 p)

Grace J, Berninger F, Nagy L (2002) Impacts of climate change on the tree line. Ann Bot (London) 90(4):537–544

Gracia C, Sabaté S, López B, Sánchez A (2001) Presente y futuro del bosque mediterráneo: balance de carbono, gestión forestal y cambio global. In: Zamora R, Puignaire FI (eds) Aspectos funcionales de los ecosistemas mediterráneos. CSIC-AEET, Granada, pp 351–372

Green PJ, Sibson R (1977) Computing dirichlet tesselations in the plane. Comp J 21(2):168–173

Greenberg JA, Dobrowski SZ, Vanderbilt VC (2009) Limitations on maximum tree density using hyperspatial remote sensing and environmental gradient analysis. Remote Sens Environ 113:94–101

Gregorius H (1974) Genetischer Abstand zwischen Populationen – Zur Konzeption der genetischer Abstandsmessung. Silvae Genetica 23:22–27

Grier CC, Vogt KA, Keyes ML, Edmonds RL (1981) Biomass distribution and above- and below-ground production in young and mature Abies amabilis zone ecosystems of the Washington Cascades. Can J For Res 11:155–167

Griesel F, Gadow Kv (1995) Naturgemäßer – naturnaher – ökologischer Waldbau – Begriffsbestimmungen, Konzepte und Richtlinien, Anwendungsbeispiele. Internal report, Institute of Forsteinrichtung, Univ of Göttingen: 20pp

Griffith DA, Peres-Neto PR (2006) Spatial modelling in ecology: the flexibility of eigenfunction spatial analyses. Ecology 87(10):2603–2613

Griscom BW, Adams J, Ellis PW, Houghton RA, Lomax G, Miteva D A, Schlesinger WH

Grossman DH, Faber-Langendoen D, Weakley AS, Anderson M, Bourgeron P, Crawford R, Goodin K, Landaal S, Metzler K, Patterson KD, Pyne M, Reid M, Sneddon L (1998). International classification of ecological communities: terrestrial vegetation of the United States. Volume I. The National Vegetation Classification System: development, status, and applications. The Nature Conservancy, Arlington, Virginia, USA

Grundner F, Schwappach A (1942) Massentafeln zur Bestimmung des Holzgehaltes stehender Waldbäume und Waldbestände. 9th edn, Berlin

Grut M (1970) Equations for calculating height increment and site index of Pinaster pine in the Cape province. S Afr For J 102:43–50

Guilford JP (1954) Psychometric methods. Mc Graw-Hill, New York

Gulbeyaz O, Bond-Lamberty B, Akyurek Z, West TO (2018) A new approach to evaluate the MODIS annual NPP product (MOD17A3) using forest field data from Turkey. Int J Remote Sens 39(8):2560–2578. https://doi.org/10.1080/01431161.2018.1430913

Guldin JM (1991) Uneven-aged BDq regulation of Sierra Nevada mixed conifers. Western J Appl For 6(2):27–32

Guldin JM (2002) Continuous cover forestry in the United States—experience with southern pines. In: Gadow Kv, Nagel J, Saborowski J (eds) Continuous cover forestry. Vol. 4 in the book series Managing forest ecosystems, Kluwer Academic Publishers, pp 295–307

Gunaratne AMTA, Gunatilleke CVS, Gunatilleke IAUN, Weerasinghe M, Gundersen P, Schmidt IK, Raulund-Rasmussen K (2006) Leaching of nitrate from temperate forests – effects of air pollution and forest management. Environ Rev 14:1–57

Gundersen P (1995) Impacts of nitrogen deposition – scientific background. UN ECE Convention on Long Range Transboundary Air Pollution. Annual Synoptic Report, pp 9–18

Gurjanov M, Gadow K (2005) Analyse der nachhaltigen Nutzung im schlagweisen Betrieb am Beispiel des Forstbetriebes Lissino bei St. Petersburg, Russland. Forstarchiv 76:177–183

Gustafsson L, Baker SC, Bauhus J, Beese WJ, Brodie A, Kouki J, Lindenmayer DB, Lõhmus A, Pastur GM, Messier C, Neyland M, Palik B, Sverdrup-Thygeson A, Volney WJA, Wayne A, Franklin JF (2012) Retention forestry to maintain multifunctional forests: a world perspective. BioScience 62(7):633–645

Hagan MT, Demuth HB, Beale MH, Jesús OD (2014) Neural network design, 2nd edn. Amazon, Orlando

Hakes W (1994) On the predictive power of numerical and Braun-Blanquet classification: an example from beechwoods. J Veg Sci 5:153–160

Halaj J, Grék J, Pánek F, Petráš R, Řehák J (1987) Rastové tabuľky hlavných drevín ČSSR. Príroda, Bratislava

Halaj J, Petráš R, Pánek F, Grék J (1988) Yield tables for spruce, fir, oak and beech. Acta Instituti Forestalis Zvolensis 7, Príroda, Bratislava

Hall RB (1994) Use of the crown competition factor concept to select clones and spacings for short-rotation woody crops. Tree Physiol 14(7-8-9):899–909

Hall DB, Clutter M (2004) Multivariate multilevel nonlinear mixed effects models for timber yield predictions. Biometrics 60:16–24

Hall N, Johnston RD, Chippendale GM (1970) Forest trees of Australia. Forest and Timber Bureau, Canberra, 334 p

Halpern CB, McKenzie D (2001) Disturbance and post-harvest ground conditions in a structural retention experiment. For Ecol Manage 154:215–225

Hampton SE, Strasser CA, Tewksbury JJ, Gram WK, Budden AE, Batcheller AL, Duke CS, Porter JH (2013) Big data and the future of ecology. Front Ecol Environ 11(3):156–162. https://doi.org/10.1890/120103

Hansen GD (1987) Effects of diameter distribution on the growth of northern hardwoods. Can J For Res 17(1)

Hansen MC, Potapov PV, Moore R, Hancher M, Turubanova SA, Tyukavina A, Thau D, Stehman SV, Goetz SJ, Loveland TR, Kommareddy A, Egorov A, Chini L, Justice CO, Townshend JRG (2013) High-resolution global maps of 21st-century forest cover change. Science 342(6160):850–853

Hanson PJ, Weltzin JF (2000) Drought disturbance from climate change: response of United States forests. Sci Total Environ 262:205–220

Hao M, Zhang C, Zhao X, Gadow KV (2018) Functional and phylogenetic diversity determine woody productivity in a temperate forest. Ecol Evol 8:2395–2406

Hao M, Ganeshaiah KN, Zhang C, Zhao X, von Gadow K (2019a) Discriminating among forest communities based on taxonomic, phylogenetic T and trait distances. For Ecol Manage 440(2019):40–47

Hao M, Corral-Rivas JJ, González-Elizondo MS, Ganeshaiah KN, Nava-Miranda MG, Zhang C, Zhao X, von Gadow K (2019b) Assessing biological dissimilarities between five forest communities. For Ecosyst 6:30

Harari YN (2019) 21 lessons for the 21st century. Jonathan Cape, London, 352 p

Härdtle W (1995) On the theoretical concept of the potential natural vegetation and proposals for an up-to-date modification. Folia Geobot 30:263. https://doi.org/10.1007/BF02803708

Harmon ME, Franklin JF, Swanson F, Sollins P, Gregory SV, Lattin JD, Anderson NH, Cline SP, Aumen NG, Sedell JR, Lienkaemper GW, Cromack K Jr, Cummins KW (1986) Ecology of coarse woody debris in temperate ecosystems. In: MacFadey A, Ford ED (eds) Advances in ecological research 15. Academic Press, Orlando, pp 133–302

Harms KE, Condit R, Hubbell SP, Foster RB (2001) Habitat associations of trees and shrubs in a 50-ha neotropical forest plot. J Ecol 89:947–959

Harper JL (1977) Population biology of plants. Academic Press, London

Harrison SP, Bartlein P (2012) Chapter 14 — records from the past, lessons for the future: What the palaeorecord implies about mechanisms of global change. In: Henderson-Sellers A, McGuffie K (eds) The future of the world's climate, 2nd edn. Elsevier, Boston, pp 403–436

Harrison WC, Daniels RF (1988) A new biomathematical model for growth and yield of loblolly pine plantations. In: Ek AR, Shifley SR, Burk TE editors. Forest growth modeling and prediction. Volume I. USDA Forest Service, GTR NC-120. pp 293–304

Harrison S, Yu G, Takahara H, Prentice I (2001) Plant diversity and palaeovegetation in East Asia. Nature 413:129–130

Hart HMJ (1928) Stamtal en Dunning – een Orienteerend Onderzoek naar de Beste Plantwijdte en Dunningswijze voor den Djati. Wageningen, H. Venman & Zonen

Hartig GL (1795) Anweisung zur Taxation der Forste, oder zur Bestimmung des Holzertrags der Wälder. Heyer, Gießen

Hartig T (1847) Vergleichende Untersuchungen über den Ertrag der Rotbuche. Verlag Albert Förstner, Berlin

Hartley MJ (2002) Rationale and methods for conserving biodiversity in plantation forests. For Ecol Manage 155:81–95

Hartmann J, Moosdorf N (2012) The new global lithological map database GLiM: A representation of rock properties at the Earth surface. Geochem Geophys Geosyst 13(12):n/a–n/a

Hartung J (1985) Statistik. Lehr- und Handbuch der angewandten Statistik. R. Oldenbourg Verlag, München Wien, 806 p

Harvey AC (1976) Estimating regression models with multiplicative heteroscedasticity. Econometrica 44:461–465

Hasenauer H (1994) Ein Einzelbaumwachstumssimulator für ungleichaltrige Fichten-, Kiefern- und Buchen-Fichtenmischbestände. Forstl. Schriftenreihe d. Universität f. Bodenkultur, Wien, Bd. 8: 152 S

Hasenauer H, Moser M, Eckmüllner O (1994) Programmbeschreibung. Distanzabhängiger Wachstumssimulator für Mischbestände, MOSES 2.0. Inst. f. Waldwachstumsforschung, Universität f. Bodenkultur, Wien: 36 S

Hasenauer H, Leitgeb E, Sterba H (1996) Der A-Wert nach Johann als Konkurrenzindex für die Abschätzung von Durchforstungseffekten. Forst-u J-Ztg 167:169–174

Hasenauer H, Burkhart HE, Amateis RL (1997) Basal area development in thinned and unthinned loblolly pine plantations. Can J For Res 27:265–271

Hashimoto R (1986) Comparative study of thinning methods in a young Sugi (Cryptomeria japonica) plantation. Changes in canopy structure and light environment. In: Fujimori T, Whitehead (eds) Crown and canopy structure in relation to Productivity. Published by the Forestry and Forest Products Institute, Ibaraki, pp 243–262. ISBN 4-9900058-1-3

Heaton RJ (2000) The silvicuture, nutrition and economics of short rotation coppice in the uplands of mid-Wales. PhD thesis, Cardiff University

Heck CR (1904) Freie Durchforstung. Springer, Berlin

Hector A, Philipson C, Saner P, Chamagne J, Dzulkifli D, O'Brien M et al (2011) The Sabah biodiversity experiment: a long-term test of the role of tree diversity in restoring tropical forest structure and functioning. Philos Trans R Soc B 366:3303–3315

Hedemann O (1939) L'histoire de la foret de Białowieża jusqu'a 1798 (The history of the Forest of Białowieża before 1798). Institut de Recherches des Forets Domaniales Warszawa, Travaux et comptes rendus, Seria A, Nr. 41: 310 pp

Hegyi F (1974) A simulation model for managing jackpine stands. In: Fries J (ed) Growth models for tree and stand simulation. Royal College of Forestry, Stockholm, pp 74–90

Heikkinen J (2006) Assessment of uncertainty in spatially systematic sampling. In: Kangas A, Maltamo M (eds) Forest inventory: methodology and applications, Managing forest ecosystems, vol 10. Springer, Dordrecht, pp 155–176

Heikurainen L, Kuusela K, Linnamies O, Nyyssönen A (1960) Metsiemme hakkuumahdollisuudet. Pitkän ajan tarkastelua. Silva Fennica 110. 52 s. (In Finnish)

Heinonen T, Pukkala T (2007) The use of cellular automaton approach in forest planning. Can J For Res 37:2188–2200

Helliwell R (1993) Newsletter 4. Continuous Cover Forestry Group, UK.

Henderson GS, Hammer RD, Grigal DF, (1990) Can measurable soil properties be in- tegrated into a framework for characterizing forest productivity? In: Gessel SP, Lacate DS, Weetman GF, Powers RF (eds) Sustained productivity of forest soils. Proc. 7th North Am. Forest soils conf. Vancouver, pp 137–154

Hendrick RL (2001) Forest types and classification. In: Evans J (ed) The forest handbook, vol 1. Blackwell Science Ltd, Malden

Hengl T, Walsh MG, Sanderman J, Wheeler I, Harrison SP, Prentice IC (2018a) Global mapping of potential natural vegetation: an assessment of machine learning algorithms for estimating land potential. PeerJ Preprints 6:e26811v1. https://doi.org/10.7287/peerj.preprints.26811v1

Hengl T, Nussbaum M, Wright MN, Heuvelink GB (2018b) Random forest as a generic frame- work for predictive modeling of spatial and spatio-temporal variables. PeerJ Preprints 6:e26693v1

Henne A (1976) Ziele, Zielgewichte und Nutzwertanalyse in der mittelfristigen forstlichen Planung Allg. Forstzeitg. 31. Jg.: 675–678

Henningsen A, Hamann JD (2007) Systemfit: A package for estimating systems of simultaneous equations in R. J Stat Softw 23(4):1–40. http://www.jstatsoft.org/v23/i04/

Henttonen HM, Kangas A (2015) Optimal plot design in a multipurpose forest inventory. For Ecosyst 2:31

Hérault B, Pétronelli P (2017) Paracou research station, Biodiversity plots 1991–2016. CIRAD, French Guiana

Herrick JE, Urama KC, Karl JW, Boos J, Johnson M-VV, Shepherd KD, Hempel J, Bestelmeyer BT, Davies J, Larson Guerra J, Kosnik C, Kimiti DW, Losinyen Ekai A, Muller K, Norfleet L, Ozor N, Reinsch T, Sarukhan J, West LT (2013) The global Land-Potential Knowledge System (LandPKS): Supporting evidence-based, site-specific land use and management through cloud computing, mobile applications, and crowdsourcing. J Soil Water Conserv 68(1):5A–12A

Hertel D (1999) Das Feinwurzelsystem von Rein- und Mischbeständen der Rotbuche: Struktur, Dynamik und interspezifische Konkurrenz. Dissertationes Botanicae 317, 190 p

Hertling UM, Lubke RA (1999) Use of Ammophila arenaria for dune stabilization in South Africa and its current distribution – perceptions and problems. Environ Manag 24:467–482

Hessenmöller D (2002) Modelle zur Wachstums- und Durchforstungssimulation im Göttinger Kalkbuchenwald. Dissertation, University Göttingen, p 163

Hessenmöller D, Gadow K (2001) Beschreibung der Durchmesserverteilung von Buchenbeständen mit Hilfe der bimodalen Weibullfunktion. Allg Forst- u J-Ztg 172(3):46–50

Hevia A, Cao QV, Álvarez-GonzÁlez JG, Ruiz-GonzÁlez AD, von Gadow K (2015) Compatibility of whole-stand and individual-tree models using composite estimators and disaggregation. For Ecol Manage 348:46–56

Hewlett JD, Pienaar L (1973) Design and analysis of the catchment experiment. In: White EH (ed) Proceedings of the Symposium on use of small watersheds in determining effects of forest land use on water quality. University of Kentucky, Lexington, pp 88–106

Hijmans RJ, Elith J (2018) Species distribution modeling with R. Environmental Science and Policy, University of California

Hill M (1973) Diversity and evenness: a unifying notation and its consequences. Ecology 54:427–431

Hille M, Hessenmöller D Möhring B, Gadow Kv (1999) Evaluating alternative silvicultural regimes in a 110-year old beech forest. In: Proc IUFRO conference from theory to practice – gaps and solutions in managerial economics and accounting in forestry. Prag, pp 94–100

Hilmi GF (1957) Theoretical bio-geo physics of forest. --- AN SSSR. Moscow. USSR. [In Russian]

Hines WGS, Hines RJO (1979) The Eberhardt index and the detection of nonrandomness of spatial point distributions. Biometrika 66:73–80

Hinrichs L (2006) Automatic generation of treatment paths for mixed forest stands of spruce (Picea abies) and beech (Fagus sylvatica). In: Proceeding of the IUFRO Conf. Precision Forestry, 5–10 March, 2006; Univ. Stellenbosch, South Africa, pp 261–276

HMSP, Burslem DFRP (2011) Release from root competition promotes tree seedling survival and growth following transplantation into human-induced grasslands in Sri Lanka. For Ecol Manage 262:229–236

Hobbelstad K, Hofstad O (1988) AVVIRK-3, a model for long-term forest planning. Med Norsk Institutt for Skogsforskning 41(35):504–516

Hobbs RJ, Arico S, Aronson J, Baron JS, Bridgewater P, Cramer VA, Epstein PR, Ewel JJ, Klink CA, Lugo AE, Norton D, Ojima D, Richardson DM, Sanderson EW, Valladares F, Vilà M, Zamora R, Zobel M (2006) Novel ecosystems: theoretical and management aspects of the new ecological world order. Glob Ecol Biogeogr 15:1–7

Hodgkins EJ (1960) Forest site classification in the southeast: an evaluation in Southern Forest Soils. In: 8th Am. For. Symp. Proc. 1959. Lousiana State Univ. Press, pp 34-48

Hoen H (1996) Forestry scenario modelling for economic analysis – experiences using the GAYA-JLP model. In: Päivinen R, Roihuvuo L, Siitonen M (eds) Large-scale forestry scenario models – experiences and requirements, European Forest Institute. EFI Proceedings No. 5, pp 79–88

Hoffmann A, Hoffmann S (2005) Irrfahrt Biodiversität – Eine kritische Sicht der europäischen Biodiversitätspolitik. Metropolis Verlag; 401 pp

Hoffmann CW, Usoltsev VA (2001) Modelling root biomass distribution in Pinus sylvestris forests of the Turgai Depression of Kazakhstan. For Ecol Manage 149:103–114

Hofmann M (2005) Pappeln als nachwachsender Rohstoff auf Ackerstandorten – Kulturverfahren, Ökologie und Wachstum unter dem Aspekt der Sortenwahl. Doctoral Dissertation; University of Göttingen

Hoganson HM (1996) Using Dtran for the Minnesota GEIS. In: Päivinen R, Roihuvuo L, Siitonen M (eds) Large-scale forestry scenario models – experiences and requirements, European Forest Institute. EFI Proceedings No. 5, pp 143–152

Hoganson HM, Rose DW (1984) A simulation approach for optimal timber management scheduling. For Sci 30(1):200–238

Hollenstein K (1997) Analyse, Bewertung und Management von Naturrisiken. Hochschulverlag AG der ETH Zürich: 191 p

Holmes MJ, Reed DD (1991) Competition indices for mixed species northern hardwoods. For Sci 37(5):1338–1349

Hosmer DW, Lemeshow S (2000) Applied logistic regression, 2nd edn. Wiley, New York, 373 p

Ho-Tung Lin, Tzeng Yih Lam, Gadow K v, Kershaw JA Jr (2019) Effects of nested plot designs on assessing stand attributes, species diversity and spatial forest structures. For Ecol Manage 457:117658

Hu Y, Zhang Q, Zhang Y, Yan H (2018) A deep convolution neural network method for land cover mapping: a case study of Qinhuangdao, China. Remote Sens (Basel) 2018(10):2053. https://doi.org/10.3390/rs10122053

Huang S, Price D, Titus SJ (2000) Development of ecoregion-based height-diameter models for white spruce in boreal forests. For Ecol Manage 129:125–141

Huang X, Ma M, Wang X, Tang X, Yang H (2018) The uncertainty analysis of the MODIS GPP product in global maize croplands. Front Earth Sci 12(2). https://doi.org/10.1007/s11707-018-0716-x

Hubbell SP (2001) The unified neutral theory of biodiversity and biogeography. Princeton University Press, Princeton\Oxford, 375 p

Hui GY, Albert M (2004) Stichprobensimulationen zur Schätzung nachbarschaftsbezogener Strukturparameter in Waldbeständen. Allg Forst- u Jagdzeitg 175(10):199–209

Hui GY, Gadow Kv (1997) Entwicklung und Erprobung eines Einheitsschaftmodells für die Baumart Cunninghamia lanceolata. Forstw Cbl 116(5):315–321

Hui GY, Kv G (2002) Das Winkelmass: Herleitung des optimalen Standardwinkels. Allgemeine Forst-und Jagdzeitung 173(10):173–177

Hui C, Richardson DM (2017) Invasion dynamics. Oxford University Press, Oxford, 322 p

Hui GY, Sheng WT, Gadow Kv (1994) Study on the yield modelling system of Chinese fir plantation. For Sci 7(4):353–358

Hui GY, Albert M, Gadow KV (1998) Das Umgebungsmaß als Parameter zur Nachbildung von Bestandesstrukturen. Forstwiss Centralbl 117:258–266

Hui GY, Albert M, Chen BW (2003) Reproduktion der Baumverteilung im Bestand unter Verwendung des Strukturparameters Winkelmaß. Allg Forst und Jagdzeitung 174:109–116

Hui GY, Gadow Kv, Hu YB, Xu H (2007) Structure-based forest management. China Forestry Publishing House, Beijing, China, 200 pp. (in Chinese, with English abstract)

Hui C, Richardson DM, Robertson MP, Wilson JRU, Yates CJ (2011a) Macroecology meets invasion ecology: linking the native distributions of Australian acacias to invasiveness. Divers Distrib 17:872–883

Hui G, Zhao X, Zhao Z, Gadow K (2011b) Evaluating tree species spatial diversity based on neighborhood relationships. For Sci 57(4):292–300

Hull RB, Buhyoff CJ, Daniel TC (1984) Measurement of scenic beauty – the law of comparative judgement and scenic beauty estimation procedures. For Sci 30(4):1084–1096

Hulme PE (1996) Herbivory, plant regeneration and coexistence. Ecology 84:609–613

Humphrey J, Stevenson A, Whitfield P, Swailes J, (2002) Life in the deadwood. Forestry Commission: 18 pp

Hundeshagen JC (1826) Die Forstabschätzung auf neuen wissenschaftlichen Grundlagen. H. Laupp, Tübingen

Hurlbert SH (1971) The nonconcept of species diversity: a critique and alternative parameters. Ecology 52:577–586

Husch B, Miller CI, Beers TW (1982) Forest mensuration. Wiley, New York

Hussein KA, Albert M, Gadow K (2001) The crown window- a simple device for measuring tree crowns. Forstw Cbl 119:43–50

Hussein KA, Schmidt M, Kotzé H, Gadow K (2008) Parameter-parsimonious taper functions for describing stem profiles. Scientia Silvae Sinicae 44(6):1–8

Huston MA (1997) Hidden treatments in ecological experiments: re-evaluating the ecosystem function of biodiversity. Oecologia 110:449–460

Hyink DM (1980) Diameter distributions approaches to growth and yield modelling. In: Proceedings of a workshop on forecasting stand dynamic. Lakehead University, Thunder Bay, Ontario, June 24–25

Hyink DM, Moser JW (1983) A generalized framwork for projecting forest yield and stand structure using diameter distributions. For Sci 29(1):85–95

Hynynen J (1993) Self-thinning models for even-aged stands of pinus sylvestris, *Picea abies* and *Betula pendula*. Scand J For Res 8:326–336

Hynynen J (1995) Predicting the growth response to thinning for Scots pine stands using individual-tree growth models. Silva Fennica 29:225–247

Hyytiäinen K, Haigth RG (2011) Optimizing Continuous Cover Forest Management. In: Pukkala T, Gadow Kv (eds.) Continuous Cover Forestry, Managing Forest Ecosystems 23. pages 195-227

Hyytiäinen K, Haight RG (2012) Optimizing continuous cover forest management. In: T Pukkala and Kv Gadow (eds), Continuous Cover Forestry. Springer, pp 195–227

Hyytiäinen K, Hari P, Kokkila T, Mäkelä A, Tahvonen O, Taipale J (2004) Connecting a process-based forest growth model to stand-level economic optimization. Can J For Res 34:2060–2073
Illian J, Penttinen A, Stoyan H, Stoyan D (2008) Statistical analysis and modeling of spatial point patterns. Wiley, Chichester, 534p
Ilvessalo Y (1927a) The forests of Suomi Finland: Results of the general survey of the forests of the country carried out during the years 1921–1924. Communicationes ex Instituto Quaestionum Forestalium Finlandie 11:321–395. (in Finnish with English summary)
Ilvessalo Y (1927b) Suomen metsät. Tulokset vuosina 1921–1924 suoritetusta valtakunnan metsien arvioimisesta. Communicationes ex Instituto Quaestionum Forestalium Finlandiae 11. 421 + 192 p. (In Finnish)
Itô H, Sumida A, Isagi Y, Kamo K (1997) The crown shape of an evergreen Oak, Quercus glauca, in a hardwood community. J For Res 2:85–88
Jaccard P (1900) Contribution au problème de l'immigration post-glaciaire de la flore alpine. Bulletin de la Société Vaudoise des Sciences Naturelles 36:87–130
Jack WH (1967) Single tree sampling in even-aged plantations for survey and experimentation. IUFRO-Proceedings, Section 25, Munich, pp 379–403
Jacobsen C, Rademacher P, Meesenburg H, Meiwes KJ (2002) Gehalte chemischer Elemente in Baumkompartimenten. Literaturstudie und Datensammlung im Auftrag des BMELF: 64p
Jaehne S, Dohrenbusch A (1997) Ein Verfahren zur Beurteilung der Bestandesdiversität. Summary: a method to evaluate forest stand diversity. Forstwissenshaftliches Centralblatt 116:333–345
Jansen JJ (1996) Opbrengsttabel voor de beuk (Fagus sylvatica) in Nederland – een OPTAB – modificatie door middel van empirische simulatie, gebaseerd op proefperken uit Noord-Duitsland en Zuid-Zweden. Hinkeloord report No. 16, Landbouwuniversiteit Wageningen, 36 p
Jensen M, Hofmann G (2002) Entwicklungszyklen des baltischen Buchenwaldes. AFZ/Der Wald 19:1012–1018
Jia W, Sun Y, Pukkala T, Jin X (2020) Improved cellular automaton for stand delineation. Forests 11(1):37. https://doi.org/10.3390/f11010037
Jin X, Pukkala T, Li F (2016) Fine-tuning heuristic methods for combinatorial optimization in forest planning. Eur J For Res 135(4):765–779
Johann K (1982) Der „A-Wert" – ein objektiver Parameter zur Bestimmung der Freistellungsstärke von Zentralbäumen. [The "*A*-thinning index" – an objective parameter for the determination of release intensity of frame trees]. Tagungsbericht der Jahrestagung 1982 der Sektion Ertragskunde im Deutschen Verband Forstlicher Forschungsanstalten in Weibersbrunn, 146–158
Johann K, Pollanschütz J (1981) Standraumregulierung bei Fichte und Buche, Mischwuchsregulierung. Seminarreihe der Forstlichen Bundesversuchsanstalt, Wien, 51 S
Johnson KN, Rose D (1986) Forplan version 2: an overview, Technical report. USDA Forest Service Land Management Planning System Section, Washington, DC
Johnson KN, Scheurman HL (1977) Techniques for prescribing optimal timber harvest investment under different objectives – discussion and synthesis. For Sci Monogr 18
Jones JR (1969) Review and comparison of site evaluation methods. USDA For. Serv, Rocky
Jonsson B, Jacobsson J, Kallur H. (1993) The forest management planning package. Theory and application. Studia For. Suec. 189. 56 s
Jost L (2006) Entropy and diversity. Oikos 113:363–374
Jost L, Chao A, Chazdon RL (2011) Compositional similarity and β (beta) diversity. In: Magurran AE, McGill BJ (eds) Biological diversity: frontiers in measurement and assessment. Oxford University Press, New York, pp 66–84
Juma R, Pukkala T, de Miguel S, Muchiri M (2014) Evaluation of different approaches to individual tree growth and survival modelling using data collected at irregular intervals – a case study for Pinus patula in Kenya. For Ecosyst 1:14
Kafatos FC, Eisner T (2004) Unification in the century of biology. Science 303:1257
Kahle M, Stüber V, Pommerening A (1996) Strukturbeschreibung in Buchen-Edellaubholzbeispielbeständen. Institut für Forsteinrichtung und Ertragskunde, Georg-August-Universität Göttingen, 61 p. (In German)

Kahn M (1994) Modellierung der Höhenentwicklung ausgewählter Baumarten in Abhängigkeit vom Standort. Dissertation, Forstwissenschaftliche Fakultät, Universität München

Kahn M (1995) Quasikausale Modellierung des Standort-Leistungsbezuges als Voraussetzung zum Aufbau flexibler Mischbestands-Modelle. Forstwissenschaftliches Centralblatt 114:175–187

Kajala L (1996) Lapin metsästrategia. Publications 2/1996. Ministry of Acriculture and Forestry.. (In Finnish)

Kallio M, Propol A, Seppälä R (1981) A model for the forest sector. In: Dantzig GB, Dempster MAH, Kallio MJ (eds) Large scale linear programming. Internat. Inst. for Applied Systems Analysis, Laxenburg, pp 1055–1100

Kangas J, Kangas A (2002) Multiple criteria decision support methods in forest management. In: Pukkala T (ed) Multi-objective forest planning, Vol. 6 of the book series Managing forest ecosystems; Kluwer Academic Publishers, 37–70

Kangas AS, Kangas J (2004) Probability, possibility and evidence: approaches to consider risk and uncertainty in forestry decision analysis. For Pol Econ 6:169–188

Kangas A, Maltamo M (2002) Anticipating the variance of predicted stand volume and timber assortments with respect to stand characteristics and field measurements. Silva Fennica 36(4):799–811

Kangas J, Kangas A, Leskinen P, Pykäläinen J (2001) MCDM methods in strategic planning of forestry on state-owned lands in Finland – application and experiences. J Multi-Crit Dec Anal 10:257–271

Kareiva P, Watts S, McDonald R, Boucher T (2007) Domesticated nature: shaping landscapes and ecosystems for human welfare. Science 316(5833):1866–1869

Karger DN, Conrad O, Böhner J, Kawohl T, Kreft H, Soria-Auza RW, Zimmermann NE, Linder HP, Kessler M (2017) Climatologies at high resolution for the earth's land surface areas. Sci Data 4:170122

Kärkkäinen M (1980) Grading of pine sawlog stems. Commun Inst For Fenn 96(5):1–152. (In Finnish with English summary)

Karvonen L (2000) Guidelines for landscape ecological planning. Forestry Publications of Metsähallitus 26. 46 p. + map

Kassier HW (1976) 'n Oorsig van beplanning en beplanningstegnieke in die bosbou met spesiale verwysing na oesreëling. MSc thesis, University of Stellenbosch

Kassier HW (1993) Dynamics of diameter and height distributions in commercial timber plantations, PhD dissertation, Faculty of Forestry, University of Stellenbosch, South Africa

Kato F, Mülder D (1983) Qualitative Gruppendurchforstung der Buche. Allg Forst- u J-Ztg 154:139–145

Kaul ON, Gupta AC, Sharma DC (1975) Preservation plots in India. Indian Forest Bulletin No. 271. FRI, Dehradun

Keenan RJ, Reams GA, Achard F, de Freitas JV, Grainger A, Lindquist E (2015) Dynamics of global forest area: Results from the FAO Global Forest Resources Assessment 2015. For Ecol Manage 352(2015):9–20

Keeney RL, Raiffa H (1976) Decisions with multiple objectives: preferences and value tradeoffs. Wiley

Keeny RL (1982) Decision analysis – an overview. Oper Res 30(2):803–838

Kellomäki S (1986) A model for the relationship between branch number and biomass in Pinus sylvestris crowns and the effect of crown shape and stand density on branch and stem biomass. Scand J For Res 1:455–472

Kellomäki S, Kurttio O (1991) A model for the structural development of a Scots pine crown based on modular growth. For Ecol Manage 43:103–123

Kellomäki S, Strandman H (1995) A model for the structural growth of young Scots pine crowns based on light interception by shoots. Ecol Model 80:237–250

Kellomäki S, Väisänen H (1995) Model computations on the effect of rising temperature on soil moisture and water availability in forest ecosystems dominated by Scots pine in the boreal zone in Finland. Clim Change 32:423–445

Kellomäki S, Väisänen H. Strandman H (1993) FINNFOR: A model for calculating the response of a boreal forest ecosystem to climate change. Univ. of Joensuu. Faculty of Forestry. Research Notes 6

Kellomäki S, Karjalainen T, Väisänen H (1997) More timber from boreal forests under changing climate? For Ecol Manage 94:195–208

Kelly S, Grenyer R, Scotland RW (2014) Phylogenetic trees do not reliably predict feature diversity. Divers Distrib 20:600–612

Kenfack D, Chuyong GB, Condit R, Russo SE, Thomas DW (2014) Demographic variation and habitat specialization of tree species in a diverse tropical forest of Cameroon. For Ecosyst 1:22

Kennel R (1972) Die Buchendurchforstungsversuche in Bayern von 1870 bis 1970. Forschungsberichte der forstl. Forschungsanstalt München Nr. 7.

Kershaw JA Jr, Ducey MJ, Beers TW, Husch B (2016) Forest mensuration, 5th edn. © 2017 Wiley. https://doi.org/10.1002/9781118902028

KFPI (2013) Field guide for the 6th National Forest Inventory and forest health and vitality survey. 108, Korea Forest Promotion Institute, Seoul

Kilkki P (1968) Income-oriented cutting budget. Acta For Fenn 91:54 p

Kilkki P, Siitonen M (1976) Principles of a forestry information system. XVI IUFRO World Congress, Division IV, Proceedings, pp. 154–163

Kimmins JP (1987) Forest ecology. Macmillan Publishing Company, New York

Kimmins JP (1990) Modeling the sustainability of forest production and yield for a changing and uncertain future. For Chron 66:271–280

Kimmins JP, Comeau PG, Kurz W (1990) Modelling the interactions between moisture and nutrients in the control of forest growth. For Ecol Manage 30:361–379

Kinnunen J, Maltamo M, Päivinen R (2007) Standing volume estimates of forests in Russia: how accurate is the published data. Forestry 80(1):53–64

Kint V, Meirvenne MV, Nachtergale L, Geuden G, Lust N (2003) Spatial methods for quantifying forest stand structure development: a comparison between nearest-neighbor indices and variogram analysis. For Sci 49(1):36–49

Kirchhoff B (2003) Erfassung und Beschreibung ausgewählter Waldflächen der Nordmongolei. Master Thesis, University Göttingen: 84 p

Kiviste A, Álvarez JG, Rojo A, González AD (2002) Funciones de crecimiento de aplicación en el ámbito forestal. Comunicaciones I.N.I.A. Serie: Recursos Naturales. No. 4

Kiviste A, Laarmann D, Hordo M, Sims A (2007) Factors influencing tree mortality on the basis of permanent forest sample plot data Paper presented at the International Workshop on disturbance regimes in a changing environment. 03. – 06. October 2007 in Tukums, Latvia

Kiviste A et al (2015) Monitoring and modeling of forest ecosystems: the Estonian Network of Forest Research Plots/Metsaökosüsteemide seire ja modelleerimine metsa kasvukäigu püsiproovitükkide võrgustiku abil. For Stud 62:26–38

Klädtke J, Kohnle U, Kublin E, Ehring A, Pretzsch H, Uhl E, Spellmann H, Weller A (2012) Wachstum und Wertleistung der Douglasie in Abhängigkeit von der Standraumgestaltung. Schweiz Z Forstwes 163(3):96–104

Klein JP, Moeschberger ML (1997) Survival analysis: techniques for censored and truncated data. Springer, New York, 502 pp

Kleinn C, Köhl M (1999) Long term observations and research in forestry. In: Proceedings volume., International IUFRO Symposium held in Costa Rica, February 23–27, 1999

Kleinn C, Stahl G, Fehrmann L, Kangas A (2010) Issues in forest inventories as an input to planning and decision processes. Allg Forst- u JZtg 181(11/12):205–210

Kleinschmit D, Mansourian S, Wildburger C, Purret A (eds) (2016) Illegal logging and related timber trade – dimensions, drivers, impacts and responses. A global scienti c rapid response assessment report. IUFRO World Series Volume 35.Vienna. 148 p

Klier G (1969) Beitrag zu den Baumabstandsmethoden in der Forstwirtschaft. Arch. Forstwes., Berlin 18 Jg. Heft 12, S. 1257–1264

Knoebel BR, Burkhart HE (1991) A bivariate distribution approach to modeling forest diameter distributions at two points of time. Biometrics 47:241–253

Knoebel BR, Burkhart HE, Beck DE (1986) A growth and yield model for thinned stands of yellow-poplar. Suppl For Sci 32(2)

Knoke T, Seifert T (2008) Integrating selected ecological effects of mixed European beech–Norway spruce stands in bioeconomic modelling. Ecol Model 210:487–498

Koch F (2005) Entwicklung der Waldbaurichtlinien in der DDR. Unpublished contribution to the Seminar Steuerung und Analyse der Waldentwicklung. Institut für Waldinventur und Waldwachstum Univerity Göttingen

Köhl M, Green E (1991) Empirische Bayes-Schätzer zur Datenanalyse in Forstinventuren. Forstw Cbl 110:393–399

Köhl M, Magnussen S, Marchetti M (2006) Sampling methods, remote sensing and GIS multiresource forest inventory. Springer, Berlin

Köhler A (1951) Vorratsermittlung in Buchenbeständen nach Stammdurchmesser und Baumabstand. Allgemeine Forst und Jagdzeitung (AFJZ) 69–74

Kohnle U, Hanewinkel M, Klädtke J (2011) Extracting environmentally driven growth trends from diameter increment series based on a multiplicative decomposition model. Can J For Res 41(8):1577–1589

Kolbert E (2014) The sixth extinction; an unnatural history. Bloomsbury, London

Koleff P, Gaston KJ, Lennon JJ (2003) Measuring beta diversity for presence-absence data. J Anim Ecol 72(3):367–382

Komiteamietintö (1995) 5. Metsä 2000-ohjelman tarkistustoimikunnan mietintö. Ministry of Acriculture and Forestry. (In Finnish)

Konohira Y, Amano M (1986) Methods for the forests allowable cut calculation in Japan. In: Carcea F (ed) Determination of the forests; allowable cut in various countries of the world. IUFRO, Bucharets, pp 149–157

Koop H (1981) Vegetatiesstructuur en dynamiek van twee natuurlijke bossen: het Neuenburger en Hasbrucher Urwald, Verslagen van landbouwkundige onderzoekingen 904 PUDOW Wageningen

Koop H (1989) Forest dynamics. Springer

Köppen W (1884) Translated by Volken, E.; Brönnimann, S. "Die Wärmezonen der Erde, nach der Dauer der heissen, gemässigten und kalten Zeit und nach der Wirkung der Wärme auf die organische Welt betrachtet" [The thermal zones of the earth according to the duration of hot, moderate and cold periods and to the impact of heat on the organic world)]. Meteorologische Zeitschrift (published 2011). 20 (3): 351–360. Bibcode:2011MetZe..20..351K. doi:10.1127/0941-2948/2011/105 – via

Korol M, Gadow K (2003) Ein Einheitsschaftmodell für die Baumart Fichte (A generalized taper model for Spruce). Forstw Cbl 122:175–182

Korpel S (1992) Ergebnisse der Urwaldforschung für die Waldwirtschaft im Buchen-Ökosystem. Allgemeine Forstzeitschrift 47(21):1148–1152

Košir Z (1966) Beitrag zur Erforschung der Urwaldstruktur reiner Buchenwälder. In: Gesellschaftsmorphologie, Ber. Int. Symp. Int. Ver. Veg.-Kunde, Rinteln: 4.–7

Kotar M (1993) Bestimmung des Verteilungsmusters von Bäumen im Wald. Tagung der Sektion Forstliche Biometrie und Informatik der DVFFA in Freising 1993:15

Kotze H (2008) FORSAT-Forest-level – Users Manual – Version 1.2. Unpublished Research report, Komatiland Forests. 52 pp

Kouba J (1989) The theory of an estimate of the development of calamities and of management of the process of forest adjustment to normal forest. Lesnictvi 35(10):925–944

Kouba J (2002) Das Leben des Waldes und seine Lebensunsicherheit [Forest life and its temporal uncertainty]. Forstwiss Centralbl 121:211–228 (in German, with English summary)

Koutika L-S, Richardson DM (2019) Acacia mangium Willd: benefits and threats associated with its increasing use around the world. For Ecosyst 6(2). https://doi.org/10.1186/s40663-019-0159-1

Kozak A (1988) A variable exponent taper equation. Can J For Res 18:1363–1368

Kozak A (2004) My last words on taper functions. For Chron 80(4):507–515

Kraft G (1884) Beiträge zur Lehre von den Durchforstungen, Schlagstellungen und Lichtungshieben. Klindworth's Verlag, Hannover

Krajicek JE, Brinkman KA, Gingrich SF (1961) Crown competition. A measure of density. For Sci 7:35–42

Kramer H (1988) Waldwachstumslehre. Verlag Paul Parey, Hamburg/Berlin, 374 p

Kramer H, Akça A (1995) Leitfaden zur Waldmesslehre. Sauerländer, J D, Frankfurt a. Main, 298p

Kramer PJ, Kozlowski TT (1979) Physiology of woody plants. Academic press, New York, 811 p

Krebs CJ (1999) Ecological methodology, 2nd edn. Addison Wesley Longman, Inc, Menlo Park, 620p

Krengel U (2005) Einführung in die Wahrscheinlichkeitstheorie und Statistik: Für Studium, Berufspraxis und Lehramt, 8th edn. Vieweg, Wiesbaden

Kreutzer K (1993) Changes in the role of nitrogen in Central European forests. In: Huettl RF, Mueller-Dombois D (eds) Forest decline in the Atlantic and Pacific region. Springer, Berlin/Heidelberg, pp 82–96

Kreutzer K, Weis T (1998) The Höglwald field experiments – aims, concept and basic data. Plant and Soil 199(1):1–10

Kristöfel F (2017) Das Forstliche Versuchswesen in Österreich. In: Ledermann T (ed) Dauerversuche – Wertvolles Erbe oder Schwere Last? BFW-Berichte 153/2017. Bundesforschungszentrum für Wald, Vienna, pp 57–68

Kroeger W (1992) Grundzüge der Sicherheit technischer Systeme. Lecture notes, ETHZ Zurich

Kuehl RO (1994) Statistical principles of research design and analysis. Duxbury Press, Belmont, 686 p

Kuhn M (2008) Building predictive models in R using the caret package. J Stat Softw 28(1):1–26

Kuhn M, Johnson K (2013) Applied predictive modeling. Springer

Kuhn M, Weston S, Keefer C, Coulter N, Quinlan R (2017) Cubist: rule-and instance-based regression modeling. R package version 0.2.2

Kuijper DPJ, Jędrzejewska B, Brzeziecki B, Churski M, Jędrzejewski W, Żybura H (2010) Fluctuating ungulate density shapes tree requirement in natural stands of the Białowieża Primeval Forest, Poland. J Veg Sci 21:1082–1098

Kuiper LC, Coutts MP (1992) Spatial disposition and extension of the structural root system of Douglas-fir. For Ecol Manage 47:111–125

Kumar BM, Nair PKR (2006) Tropical homegardens: a time-tested example of sustainable agroforestry. Springer, 370 pp

Kurth A (1976) Lecture notes, Ertragskunde. Eidgen. Techn. Hochsch., Zürich

Kurth W (1994) Morphological models of plant growth: possibilities and ecological relevance. Ecol Modell Int J Ecol Modell Syst Ecol 75/76:299–308. Verlag Elsevier

Kurth W (1999) Die Simulation der Baumarchitektur mit Wachstumsgrammatiken. Habil., 327 pp. ISBN 3-932089-37-5 (In German)

Kurth H, Gerold D, Dittrich K (1987) Reale Waldentwicklung und Zielwald – Grundlagen nachhaltiger Systemregelung des Waldes. Wiss. Zeitschr. d. TU Dresden: 121–137

Kurttila M (2001) The spatial structure of forests in the optimization calculations of forest planning – a landscape ecological perspective. For Ecol Manage 142:127–140

Kurttila M, Pukkala T (2003) Combining holding-level economic goals with spatial landscape-level goals in the planning of multiple ownership forestry. Landsc Ecol 18:529–541

Kurttila M, Pukkala T, Loikkanen J (2002) The performance of alternative spatial objective types in forest planning calculations: a case for flying squirrel and moose. For Ecol Manage 166:245–260

Kurz WA, Beukema SJ, Apps MJ (1996) Estimation root biomass and dynamics for the carbon budget of the Canadian forest sector. Can J For Res 26:1973–1979

Lacointe A (2000) Carbon allocation among tree organs: a review of basic processes and representation in functional-structural tree models. Ann For Sci 57:521–533

Lal JB (1989) India's forests: myth and reality. Natraj Publishers, New Delhi, 304 p

Lämås T, Eriksson LO (2003) Analysis and planning systems for multi-resource, sustainable forestry—The Heureka research programme at SLU. Can J For Res 33:500–508

Lance GN, Williams WT (1967) Mixed-data classificatory programs I – Agglomerative systems. Aust Comput J 1(1):15–20

Lange M, Türke M, Pašalić E, Boch S, Hessenmöller D, Müller J, Prati D, Socher SA, Fischer M, Weisser WW, Gossner MM (2014) Effects of forest management on ground-dwelling beetles (Coleoptera; Carabidae, Staphylinidae) in Central Europe are mainly mediated by changes in forest structure. For Ecol Manage 329:166–176

Langsæter A (1941) Om tynning I enaldret granog furuskog [Thinning of even-aged spruce and pine forests]. Medd Norsk Skogfor 8:131–216

Lappi J (1992) JLP – a linear programming package for management planning. The Finnish Forest Res. Inst., Res. Paper 414: 131 pp.

Lappi J (1997) A longitudinal analysis of height/diameter curves. For Sci 43(4):555–570

Lary DJ, Alavi AH, Gandomi AH, Walker AL (2016) Machine learning in geosciences and remote sensing. Geosci Front 7(1):3–10. Special Issue: Progress of Machine Learning in Geosciences

Lässig R (1991) Zum Wachstum von Fichtensolitären [Picea abies (L.) Karst.] in Südwestdeutschland. Dissertation, Univ. Freiburg, 136 S. (In German)

Laughton FS (1937) The silviculture of the indigenous forests of the Union of South Africa with special reference to the forests of the Knysna region. Sci. Bull. 157, Forestry Series 7, Government Printer, Pretoria: 169 p

Laurén A, Finér L, Koivusalo H, Kokkonen T, Karvonen T, Kellomäki S, Mannerkoski H, Ahtiainen M (2005) Water and nitrogen processes along a typical water flowpath and streamwater exports from a forested catchment and changes after clearcutting: a modelling study. Hydrol Earth Syst Sci 9(2):1–17

Lausch A, Heurich M, Fahse L (2013) Spatio-temporal infestation patterns of *Ips typographus* (L.) in the Bavarian Forest National Park, Germany. Ecol Indic 31:73–81

Law R, Watkinson AR (1989) Competition. In: Cherrret JM (ed) Ecological concepts. Blackwell, Oxford, pp 243–284

Law R, Illian J, Burslem DFRP, Gratzer G, Gunatilleke CVS, Gunatilleke IAUN (2009) Ecological information from spatial patterns of plants: insights from point process theory. J Ecol 97:616–628

Lawton JH (1999) Are there general laws in ecology? Oikos 84:177–192

Le Goff N, Ottorini J-M (2001) Root biomass and biomass increment in a beech (Fagus sylvatica L.) stand in North-East France. Ann For Sci 58:1–13

Le Maitre DC (1999) Adjustments of WR90 naturalized flows to allow for streamflow reductions based on the CSIRs streamflow reduction model. Report for the Department of Water Affairs and Forestry, Pretoria: South Africa

Leak WB (1964) An expression of diameter distribution of unbalanced, uneven-aged stands and forests. For Sci 10:39–50

Leak WB (1965) The J-shaped probability distribution. For Sci 11:405–409

Lee DT (1980) Two-dimensional Voronoi diagram in the LP-metric. J ACM 27(4):604–618

Lee WK (1993) Wachstums- und Ertragsmodelle für Pinus densiflora in der Kangwon-Provinz, Korea. Diss., Forstwiss. Fachbereich, Univ. Göttingen. Cuvillier Verlag, 178 p

Lee WK, Gadow K (1997) Iterative Bestimmung der Konkurrenzbäume in Pinus densiflora Beständen. AFJZ 168(3/4):41–44

Lee WK, Gadow K, Chung D-J, Lee J-L (2003) Individual tree dbh-growth in Pinus densiflora and Quercus variabilis mixed stands in central Korea. Ecol Model 176:187–200

Legendre P, Cáceres MD (2013) Beta diversity as the variance of community data: dissimilarity coefficients and partitioning. Ecol Lett 16(8):951–963

Legendre P, Gallagher ED (2001) Ecologically meaningful transformations for ordination of species data. Oecologia 129:271–280

Legendre P, Legendre L (1998) Numerical ecology, 2nd edn. Elsevier, Amsterdam

Legendre P, Legendre L (2012) Numerical ecology, 3rd edn. Elsevier, Amsterdam

Leibundgut H (1978) Die Waldpflege. Paul Haupt, Bern/Stuttgart

Leinster T, Cobbold CA (2012) Measuring diversity: the importance of species similarity. Ecology 93(3):477–489

Leisch F (2004) FlexMix: a general framework for finite mixture models and latent class regression in R. J Stat Softw 11(8):2004. (http://www.jstatsoft.org/v11/i08/)

LeMay V, Kozak A, Muhairwe K, Kozak RA (1993) Factors affecting the performance of Kozak's (1988) variable-exponent taper function. Wood u Wiant 1993:34–53

Lemm R (1991) Ein dynamisches Forstbetriebs-Simulationsmodell. Dissertation, ETH Zürich: 235 p

Lennon JJ, Koleff P, Greenwood JJD, Gaston KJ (2001) The geographical structure of British bird distributions: diversity, spatial turnover and scale. J Anim Ecol 70:966–979

Leong M, Roderick GK (2015) Remote sensing captures varying temporal patterns of vegetation between human-altered and natural landscapes. PeerJ 3:e1141

Lepě J, Stursa J (1989) Species-area curve, life history strategies, and succession: a field test of relationships. Vegetatio 83:249–257

Leskinen P (2001) Statistical methods for measuring preferences. University of Joensuu, Publications in social sciences no. 48: 111 p.

Leskinen P, Kangas J (1998) Analysing uncertainties of interval judgment data in multiple-criteria evaluation of forest plans. Silva Fennica 32:363–372

Lessard VC (2001) Updating Indiana annual forest inventory and analysis plot data using eastern broadleaf forest diameter growth models. In: Reams GA, McRoberts RE, Van Deusen PC (eds) Proceedings of the second annual forest inventory and analysis symposium; 2000 October 17–18; Salt Lake City, UT. Gen. Tech. Rep. SRS-47. U.S. Department of Agriculture, Forest Service, Southern Research Station, Asheville, NC pp 66–69

Leuchner M, Hertel C, Rötzer T, Seifert T, Weigt R, Werner H, Menzel A (2012) Solar radiation as a driver for growth and competition in forest stands. In: Matyssek R, Schnyder H, Ernst D, Munch JC, Oßwald W, Pretzsch H (eds) Growth and defence in plants: resource allocation at multiple scales, vol 220, Ecol Stud., pp 175–191

Leuschner A (1990) Forest regulation, harvest scheduling and planning techniques. John Wiley, 281 p

Leuschner C (2002) Lebensraum Baumkrone – Schatzkiste der Biodiversität. Georgia Augusta, Wissenschaftsmagazin der Georg-August-Universität Göttingen; No. 1: 18–25

Levavasseur G, Vrac M, Roche D, Paillard D (2012) Statistical modelling of a new global potential vegetation distribution. Environ Res Lett 7(4):044019

Levis C, Souza PF, Schietti J, Emilio T, JLPdV P et al (2012) Historical human footprint on modern tree species composition in the Purus-Madeira Interfluve, Central Amazonia. PLoS One 7(11):e48559. https://doi.org/10.1371/journal.pone.0048559

Lewandowski A, Gadow Kv (1997) Ein heuristischer Ansatz zur Reproduktion von Waldbeständen [A heuristic method for reproducing forest stands]. Allg Forst- u J -Zeitung 168:170–174

Lewandowski A, Pommerening A (1997) Zur Beschreibung der Waldstruktur – Erwartete und beobachtete Artendurchmischung. Forstwiss Cbl 116:129–139

Li J, Roy DP (2017) A global analysis of sentinel-2A, sentinel-2B and Landsat-8 data revisit intervals and implications for terrestrial monitoring. Remote Sens (Basel) 9(9):902

Li X, Chen W, Cheng X, Wang L (2016) A comparison of machine learning algorithms for mapping of complex surface-mined and agricultural landscapes using ziyuan-3 stereo satellite imagery. Remote Sens (Basel) 8(6):514

Li X, Farooqi TJA, Jiang C, Liu S, Sun OJ (2019) Spatiotemporal variations in productivity and water use efficiency across a temperate forest landscape of Northeast China. For Ecosyst 6:22. https://doi.org/10.1186/s40663-019-0179-x

Liang JJ, Crowther TW, Picard N, Wiser S, Zhou M, Alberti G, Schulze E-D, McGuire AD, Bozzato F, Pretzsch H, de Miguel S, Paquette A, Hérault B, Scherer-Lorenzen M, Barrett CB, Glick HB, Hengeveld GM, Nabuurs G-J, Pfautsch S, Viana H, Vibrans AC, Ammer C, Schall P, Verbyla D, Tchebakova N, Fischer M, Watson JV, Chen HYH, Lei X, Schelhaas M-J, Lu H, Gianelle D, Parfenova EI, Salas C, Lee E, Lee B, Kim HS, Bruelheide H, Coomes DA, Piotto D, Sunderland T, Schmid B, Gourlet-Fleury S, Sonké B, Tavani R, Zhu J, Brandl S, Vayreda J, Kitahara F, Searle EB, Neldner VJ, Ngugi MR, Baraloto C, Frizzera L, Bałazy R, Oleksyn J, Zawiła-Niedźwiecki T, Bouriaud O, Bussotti F, Finér L, Jaroszewicz B, Jucker T, Valladares F, Jagodzinski AM, Peri PL, Gonmadje C, Marthy W, O'Brien T, Martin EH, Marshall AR, Rovero F, Bitariho R, Niklaus PA, Alvarez-Loayza P, Chamuya N, Valencia R, Mortier F,

Wortel V, Engone-Obiang NL, Ferreira LV, Odeke DE, Vasquez RM, Lewis SL, Reich PB (2016a) Positive biodiversity-productivity relationship predominant in global forests. Science 354. https://doi.org/10.1126/science.aaf8957

Liang X, Kankare V, Hyyppä J, Wang Y, Kukko A, Haggrén H, Yu X, Kaartinen H, Jaakkola A, Guan F, Holopainen M, Vastaranta M (2016b) Terrestrial laser scanning in forest inventories. ISPRS J Photogramm Remote Sens 115:63–77. https://doi.org/10.1016/j.isprsjprs.2016.01.006

Liao W, Nogami K (1999) Prediction of near-view scenic beauty in artificial stands of Hinoki (Chamaecyparis obtusa S. et Z.). J For Res 4:93–98

Liesebach M, Wuehlisch G, Muhs HJ (1999) Aspen for short-rotation coppice plantations on agricultural sites in Germany: effects of spacing and rotation time on growth and biomass production of aspen progenies. For Eco Manag 121:25–39

Lindenmayer DB, Fischer J, Felton A, Crane M, Michael D, Macgregor C, Montague-Drake R, Manning A, Hobbs RJ (2008) Novel ecosystems resulting from landscape transformation create dilemmas for modern conservation practice. Conserv Lett 1:129–135

Lindner M (2000) Developing adaptive forest management strategies to cope with climate change. Tree Physiol 20:299–307

Lindner M, Maroschek M, Netherer S, Kremer A, Barbati A, Garcia-Gonzalo J, Seidl R, Delzon S, Corona P, Kolström M, Lexer MJ, Marchetti M (2010) Climate change impacts, adaptive capacity, and vulnerability of European forest ecosystems. For Ecol Manage 259:698–709. https://doi.org/10.1016/j.foreco.2009.09.023

Liu C, Zhang L, Davis CJ, Solomon DS, Gove JH (2002) A finite mixture model for characterizing the diameter distributions of mixed-species forest stands. For Sci 48(4):653–661

Ljung GM, Box GEP (1978) On a measure of lack of fit in time series models. Biometrika 65:297–303

Lloyd J, Taylor JA (1994) On the temperature dependence of soil respiration. Funct Ecol 8:315–323

Lockow K.-W (2004) Weißerle im Norddeutschen Tiefland.. Landesforstanstalt Eberswalde

Loetsch F, Haller KE (1973) Forest inventory, Volume I. Munich, BLV Verlagsgesellschaft mbH

Longuetaud F, Seifert T, Leban JM, Pretzsch H (2008) Long-term dynamics of crowns in sessile oak stands by means of spatial analysis. For Ecol Manage 255:2007–2019

López Sánchez CA, Gorgoso Varela J, Castedo Dorado F, Rojo Alboreca A, Rodríguez Soalleiro R, Álvarez González JG, Sánchez Rodríguez F (2003) A height-diameter models for Pinus radiata D. Don in Galicia (Northwest Spain). Ann For Sci 60:237–245

López-Serrano PM, López-Sánchez CA, Díaz-Varela RA, Corral-Rivas JJ, Solís-Moreno R, Vargas-Larreta B, Álvarez-González JG (2015) Estimating biomass of mixed and uneven-aged forests using spectral data and a hybrid model combining regression trees and linear models. iForest. (early view). 10.3832/ifor1504-008 [online 2015-09-21]

Lorenz M, Varjo J, Bahamondez C, Brack C, Clarke M, Gillis M, Hirvonen H, Kleinn C, Riebau A, Sase H, Totsuka T (2005) Forest assessment for changing information needs. Forests in the global balance – changing paradigms. IUFRO World Series. Volume 17. pp 139–150

Lorimer CG (1983) Test of age-independent competition indices for individual trees in natural hardwood stands. For Ecol Manage 6:343–360

Losos JB (2008) Phylogenetic niche conservatism, phylogenetic signal and the relationship between phylogenetic relatedness and ecological similarity among species. Ecol Lett 11:995–1003

Lotwick HW, Silverman BW (1982) Methods for analysing spatial processes of several types of points. J R Stat Soc Ser B 44:406–413

Low T (2012) Australian acacias: weeds or useful trees? Biol Invasions 14:2217–2227

Lowrance WW (1980) The nature of risk. In: Schwing RC, Albers WA (eds) Societal risk assessment. How safe is safe enough? Plenum Press, New York, pp 5–14

Lozupone C, Knight R (2005) UniFrac: a new phylogenetic method for comparing microbial communities. Appl Environ Microbiol 71:8228–8235

Lubke RA (2004) Vegetation dynamics and succession on sand dunes of the eastern coasts of Africa. In: Martínez ML, Psuty NP (eds) Ecological studies, Coastal dunes, ecology and conservation, vol 171. Springer, Berlin/Heidelberg, pp 67–84

Luhmann N (1995) Social Systems. Stanford University Press, Stanford

Lujan-Soto JE, Corral-Rivas JJ, Aguirre-Calderón OA, Gadow K (2015) Grouping forest tree species on the Sierra Madre Occidental, Mexico. Allgemeine Forst und Jagdzeitung 186(3-4):63–71

Lundström A, Söderberg U (1996) Outline of the HUGIN system for long-term forecasts of timber yields and possible cut in Sweden. In: Päivinen R, Roihuvuo L, Siitonen M (eds) Large-scale forestry scenario models: experiences and requirements. Proceedings of the international seminar and summer school, June 15–22, 1995, Joensuu, Finland. EFI-Proceedings No. 5, pp 63–77

Lüpke Nv, Hardtke A, Lück M, Hessenmöller D, Ammer Ch, Schulze ED (2011) Bestandesvorrat, Baumartenvielfalt und Struktur kleinparzellierter Privatwälder im Hainich Forstarchiv 82:203-215

Luyssaert S, Schulze ED, Börner A, Knohl A, Hessenmöller D, Law BE, Ciais P, Grace J (2008) Oldgrowth forests as global carbon sinks. Nature, Vol 455, 11 September 2008. https://doi.org/10.1038/nature07276

Luzopone C, Knight R (2005) UniFrac: a new phylogenetic method for comparing microbial communities. Appl Environ Microbiol 71(12):8228–8235

Ma KP (2015) Biodiversity monitoring in China: from CForBio to Sino BON. Biodivers Sci 23:1–2. (In Chinese)

Mabvurira D, Miina J (2002) Individual-tree growth and mortality models for Eucalyptus grandis (Hill) Maiden plantations in Zimbabwe. For Ecol Manage 161(1–3):231–245

MacLean CD, Bolsinger CL (1973) Estimating dunning's site index from plant indicators. USDA For. Serv., Pacific Northwest Research Station, Portland, OR, Res. Note PNW-197, 10 pp

Madrigal Collazo A, Alvarez Gonzalez JG, Rodriguez Soalleiro R, Rojo Alboreca A (1999) Tablas de producción para los montes españoles. Fundación Conde del Valle de Salazar, Madrid, 253 p

Maguire DA, Hann DW (1989) The relationship between cross crown dimensions and sapwood area at crown base in Douglas-fir. Can J For Res 19(5):557–565

Maguire DA, Mainwaring D, Halpern CB (2006) Stand dynamics after variable retention harvesting in mature Douglas-fir forests of western North America. Allgemeine Forst und_Jagdzeitung 177:120–131

Magurran AE (2004) Measuring biological diversity. Blackwell, Oxford

Mailly D, Turbis S, Pothier D (2003) Predicting basal area increment in a spatially explicit, individual tree model: a test of competition measures with black spruce. Can J For Res 33(3):435–443

Mäkelä A, Landsberg J, Ek AR, Burk TE, Ter-Mikaelian M, Agren GI, Oliver CD, Puttonen P (2000) Process-based models for forest ecosystem management: current state of the art and challenges for practical implementation. Tree Physiol 20:289–298

Mali P (1972) Managing by objectives – an operating guide to faster and more profitable results. Wiley

Malone T, Liang J, Packee EC (2009) Cooperative Alaska forest inventory. 42 (Gen. Tech. Rep. PNW-GTR-785, USDA Forest Service, Pacific Northwest Research Station, Portland, OR)

Mandallaz D (2008) Sampling techniques for forest inventories. Chapman Hall/CRC, Boca Raton

Manley B (1996) Application of the IFS/FOLPI forest estate modelling system to large-scale forestry scenario modelling in New Zealand. In: Päivinen R, Roihuvuo L, Siitonen M (eds) EFI Proceedings No. 5 (pp 133–141). Proceedings of the international seminar and summer school: large-scale forestry scenario models: experiences and requirements, Joensuu, Finland, 15–22 June1995. Joensuu: European Forest Institute

Map (2015) http://www.ctfs.si.edu/look/images/CTFS-Map-Big.jpg. Retrieved on 24 May, 2015

Marazzi CA (1985) A value analysis method for the evaluation of telecommunication systems bid proposals. IEEE Trans Eng Manag EM-32(2):55–62

March JG, Simon HA (1958) Organisations. Wiley, New York

Mårell A, Leitgeb E (2004) European long-term research for sustainable forestry. www.enfors.org, Appendix MC4-8a

Margalef R (1957) Information theory in ecology. Genet Syst 3:36–71

Marques P (1991) Evaluating site quality of even-aged stands in northern Portugal using direct and indirect methods. For Ecol Manage 41:193–204

Marris E (2009) Ragamuffin earth. Nature l460, 2163:450–453

Marris E (2013) Perspective: coming of age in a trash forest. In: Hobbs et al. Wiley-Blackwell, Chapter 29: 245

Marsh EK (1957) Some preliminary results from O'Connor's correlated curve trend (CCT) experiments on thinning and espacements and their practical significance. Commonwealth For. Conf., 1957

Martin GL, Ek AR (1984) A comparison of competition measures and growth models for predicting plantation red pine diameter and height growth. For Sci 30(3):731–743

Martínez A (1999) Silvicultura práctica en renovales puros y mixtos, y bosques originales remanentes del tipo forestal Roble-Raulí-Coigüe. In: Donoso C, Lara A (eds) Silvicultura de los bosques nativos de Chile, pp 145-176

Mason LR, Hutchings SS (1967) Estimating foliage yields on Utah juniper from measurements of crown diameter. J Range Manage 20(3):161–166

Mason B, Kerr G, Pommerening A, Edwards C, Hale S, Ireland D, Moore R (2005) Continuous cover forestry in British conifer forests. In: Forest Research, 2005: Annual report and accounts 2003–2004:38–53

Matala J, Hynynen J, Miina J, Ojansuu R, Peltola H, Sievänen R, Väisänen H, Kellomäki S (2003) Comparison of a physiological model and a statistical model for prediction of growth and yield in boreal forests. Ecol Model 161:95–116

Matérn B (1960) Spatial variation. Meddelanden från statens skogsforskningsinstitut 49(5):1–44

Matèrn B (1986) Spatial variation, Lecture notes in statistics, 2nd edn. Springer, Berlin, p 151 S

Matney TG, Sullivan AD (1982) Compatible stand and stock tables for thinned and unthinned loblolly pine stands. For Sci 28:161–171

Matney TG, Belli KL, Farrar RM Jr (1990) Parameter free diameter-distribution recovery. In: Division 4 Proceedings of the IUFRO World Congress Meeting. School of Forestry and Wildlife Resources, Virginia Polytechnic Institute and State University. Blacksburg, Va. pp 95–107

Matsue K, Iwagami S, Yamamoto T, Shibayama Z (1999) Modelling crown form of Hinoki (Chamaecyparis obtusa) based on branch measurement. J For Plann 5:13–18

Matsumura N (1988) Entwicklung und Anwendungsmöglichkeiten eines einzelbaumorientierten Simulationsmodells für Douglasie. Docvtoral Diss. Georg-August-Universität Göttingen

Matthews JD (1963) Some applications of genetics and physiology in thinning. Forestry 36:172–180

Matthews E, Payne R, Rohweder M, Murray S (2000) Forest ecosystems: pilot analysis of global ecosystems. World Resource Institute, Washinton, DC

Matyssek R (2003) Kosten und Nutzen von Raumbesetzung und –ausbeute. AFZ/Der Wald 17:862–863

Mawson JC, Thomas JW, Degraff RM (1976) Program HTVOL. The determination of tree crown volume by layers. USDA For Serv Res Pap NE-354

Max TA, Burkhart HE (1976) Segmented polynomial regression applied to taper equations. For Sci 22(33):283–289

May RM (1975) Patterns of species abundance and diversity. In: Cody ML, Diamond JM (eds) Ecology and evolution of communities. Belknap Harvard University, Cambridge, pp 81–120

Mayer KE, Laudenslayer WF (1988) A guide to wildlife habitats of California. California Department of Forestry and Fire Protection, 166 pp

McCartney MP, Yawson DK, Magagula TF, Seshoka J (2004) Hydrology and water resources development in the Olifants River Catchment, Working Paper 76. International Water Management Institute (IWMI), Colombo

McComb WC, Spies TA, Emmingham WH (1993) Douglas-fir forests: managing for timber and mature forest habitat. J For 91(12):31–42

McCullagh P, Nelder JA (1989) Generalized Linear Models, 2nd edition. Chapman and Hall, London

McDill ME, Amateis RL (1992) Measuring forest site quality using the parameters of a dimensionally compatible height growth function. For Sci 38:409–429

McElhinny C, Gibbons P, Brack C, Bauhus J (2005) Forest and woodland stand structural complexity: its definition and measurement. For Ecol Manage 218:1–24

McGinley P (2012) Decision analysis software survey. OR/MS Today 39

McIntire EJB, Fajardo A (2009) Beyond description: the active and effective way to infer processes from spatial patterns. Ecology 90:46–56

McRoberts RE (2003) Compensating for missing plot observations in forest inventory estimation. Can J For Res 33:1990–1997

McRoberts RE, Wendt DG, Nelson MD, Hansen MH (2002) Using a land cover classification based on satellite imagery to improve the precision of forest inventory area estimates. Remote Sens Environ 81:36–44

McRoberts RE, Bechtold WAB, Patterson PL, Scott CT, Reams GA (2005) The enhanced forest inventory and analysis programme of the USDA forest service: historical perspective and announcement of statistical documentation. J For 103:304–308

McRoberts RE, Tomppo EO, Næsset E (2007) Advances and emerging issues in national forest inventories. Scand J For Res 25:368–381

Meffe GK (1996) Conserving genetic diversity in natural systems. Szaro & Johnston 1996:41–57

Mehtätalo L (2004) A longitudinal height-diameter model for Norway spruce in Finland. Can J For Res 34:131–140

Meixner O, Haas R (2002) Computergestützte Entscheidungsfindung. Redline Wirtschaft bei Ueberreuter, Frankfurt/Wien, 262 p

Menezes LM, Bunn DW, Taylor JW (2000) Review of guidelines for the use of combined forecasts. Eur J Oper Res 120:190–204

Meng XY (1991) The significance of taper equations and stand diameter structures in developing merchantable volume tables. J Beijing For Univ 13(2):14–20

MERA (1968) Metsätalouden rahoitusohjelma III. Julk. Metsätiedotustoimisto Helsinki. 64 p

Meskimen G, Franklin EC (1978) Spacing Eucalyptus grandis in southern Florida: a question of merchantable versus total volume. Southern J Appl For 1:3–5

Messier C, Nikinmaa E (2000) Effects of light availability and sapling size on the growth, biomass allocation, and crown morphology of understory sugar maple, yellow birch, and beech. Ecosci 7:345–356

Messier C, Bauhus J, Doyon F, Maure F, Sousa-Silva R, Nolet P, Mina M, Aquilue N, Fortin M-J, Puettmann K (2019) The functional complex network approach to foster forest resilience to global changes. For Ecosyst 6:21. https://doi.org/10.1186/s40663-019-0166-2

Metsätilastollinen vuosikirja (2006) Skogsstatistisk årsbok. Finnish Statistical Yearbook of Forestry. SVT Maa-, metsä- ja kalatalous 2006. Metsäntutkimuslaitos. 438 p

Meyer HA (1933) Eine mathematisch-statistische Untersuchung über den Aufbau des Plenterwaldes. Schweiz Zeitschr Forstwes 84(88–103):124–131

Michaelis KA (1907) Gute Bestandespflege mit Starkholzzucht, eine der wichtigsten Aufgaben unserer Zeit. Neudamm

Midgley DC, Pitman WV, Middleton BJ (1994) Surface water resources of South Africa, Volumes I, II, III, IV, V and VI, Reports No's. 298/1.1/94, 298/2.1/94, 298/3.1/94, 298/4.1/94, 298/5.1/94 and 298/6.1/94. Pretoria, South Africa: Water Research Commission

Mikulka B (1955) Versuche zur zahlenmäßigen Erfassung der Qualität von Waldbeständen. Mitt. der Schw. Anst. f. d. forstl. Vers. Wes. 23. Bd. 1

Ministry of Agriculture and Forestry (1999) Finland's National Forest Programme 2010. Publications 2/1999. 37 p

Mischerlich G (1975) Waldwachstum und Umwelt. Volume 3: Boden, Luft und Produktion. J.D. Sauerländer's Verlag, Frankfurt am Main

Mitchel SJ, Beese WJ (2002) The retention system: reconciling variable retention with the principles of silvicul- tural systems. For Chronicle 78(3):397–403

Mitchell KJ (1975) Dynamics and simulated yield of Douglas-fir. For Sci Monogr 17–39

Mitchell FJG, Cole E (1998) Reconstruction of long-term successional dynamics of temperate woodland in Bialowieza Forest, Poland. J Ecol 86:1042–1059

Mitlöhner R (1998) Pflanzeninterne Potentiale als Indikatoren fü den tropischen Standort. University Göttingen, Dissertation

Mitsch WJ (2012) What is ecological engineering? Ecol Eng 45:5–12

Mitscherlich G (1952) Der Tannen-Fichten-(Buchen)-Plenterwald. Heft 8 d. Schriftenreihe d. Bad. Forstl. Vers. Anst., Freiburg: 42 p

Mitscherlich G (1961) Untersuchungen in Plenterwäldern des Schwarzwaldes. AFJZ 132:61–73 + 85–95

Mitscherlich G (1970) Wald, wachstum und Umwelt. J. D. Sauerländer's Verlag, Frankfurt am Main, 142 pp. ISBN 3-7939-0110-6. (In German)

Mitscherlich G (1975). Wald, Wachstum und Umwelt - Boden, Luft und Produktion. J.D. Sauerländer's Verlag.

Mitscherlich G, Ganssen R (1951) Die Ergebnisse zweier Buchendurchforstungsversuche in höheren Lagen des Schwarzwaldes. Allg Forst- u J-Ztg 123:1–15

Mlinsek D (1994) Der naturnahe Waldbau – sein kognitiver Weg – eine Herausforderung. Der Dauerwald 10:35–43

Moeur M (1981) Crown width and foliage weight of northern Rocky mountain conifers. USDA For. Serv. Res. Pap. INT.-283

Moeur M (1993) Crown width and foliage weight of northern Rocky Mountain conifers, USDA, Forest Service, Intermountain Forest and Range Experiment Station, Ogden, Utah, Research Paper INT-283, p 14

Mohren GMJ, Van Gerwen CP, Spitters CJT (1984) Simulation of primary production in even-aged stands of Douglas-fir. For Ecol Manage 9:27–49

Møller J, Waagepetersen RP (2007) Modern statistics for spatial point processes. Scand J Stat 34:643–684

Monod J (1950) La technique de culture continue, the´orie et applications. Ann l'Institut Pasteur 79:390–410

Monserud RA (1976) Simulation of forest tree mortality. For Sci 22:438–444

Monserud RA (2002) Large-scale management experiments in the moist maritime forests of the Pacific Northwest. Landsc Urban Plan 59(3):159–180

Monserud RA, Sterba H (1996) A basal area increment model for individual trees growing in even- and uneven-aged forest stands in Austria. For Ecol Manage 80(1–3):57–80

Monserud RA, Sterba H (1999) Modeling individual tree mortality for Austrian forest species. For Ecol Manage 113:109–123

Montero G, Martinez F, Alía R, Candela JA, Ruiz-Peinado R, Cañellas I, Mutke S, Calama R, Gutierrez M, Pavon J, Alonso A, Del Rio M, Bachiller A (2004) El Pino Piñonero (Pinus pinea L.) en Andalucía. Consejería de Medio Ambiente, Junta de Andalucía: 261 p

Montgomery DC, Johnson LA, Gardiner JS (1990) Forecasting and time series analysis, 2nd edn. McGraw-Hill, Inc., New York/San Francisco, 381 pp. ISBN 0-07-042858-1

Moosmayer H-U, Schöpfer W (1972) Beziehungen zwischen Standortsfaktoren und Wuchsleistung der Fichte. AFJZ 143(10):203–215

Moravie MA, Durand M, Houllier F (1999) Ecological meaning and predictive ability of social status, vigour and competition in a tropical rain forest (India). For Ecol Manage 177:221–240

Motz K, Sterba H, Pommerening A (2010) Sampling measures of tree diversity. For Ecol Manage 260:1985–1996

Muchiri MN, Pukkala T, Miina J (2002) Optimising the management of maize – Grevillea robusta fields in Kenya. Agr Syst 56:13–25

Muhairwe CK (2006) Revised measurement manual for PGP's in Native Forests. (Forests NSW internal publication)

Mühlenberg M, Hondong H, Dulamsuren CH, Gadow Kv (2000) Large-scale biodiversity research in the southern Taiga, Northern Mongolia. In: Szaro RC, Peterson CE, Gadow Kv, Kräuchi N (eds) Creat- ing a legacy for sustainable science-based forest management: lessons learned from field experiments. Forest Snow and Landscape Research. 78 (1). Swiss Federal Research Institute WSL, Birmensdorf, Switzerland, pp 93–118

Mukul SA, Saha N (2017) Conservation benefits of tropical multifunctional land-uses in and around a forest protected area of Bangladesh. Landarzt 6:2

Murray DM, Gadow K (1991) Prioritizing mountain catchment areas. J Environ Manage 32(4):357–366

Murray DM, Gadow K (1993) A flexible yield model for regional timber forecasting. Southern J Appl For 17(2):112–115

Nabuurs GJ, Päivinen R (eds) (1996) Large scale forestry scenario models – a compilation and review. Eur For Insitute Working Papers 10. Joensuu, Finland. 174 pp

Nabuurs GJ, Päivinen R, Sallnäs O, Kupka I (1997) A large scale forestry scenario model as a planning tool for European forests. In: Moiseev NA, von Gadow K, Krott M (eds) Planning and decision making for forest management in the market economy. IUFRO conference, Pushkino, Moscow, 25-29 September 1996. Cuvillier Verlag, Göttingen, pp 89–102

Nagel J (1994) Ein Einzelbaumwachstumsmodell für Roteichenbestände. Forst und Holz. 49. Jahrgang Nr 3, 69–75

Nagel J (1996) Anwendungsprogramm zur Bestandesbewertung und zur Prognose der Bestandesentwicklung. Forst u Holz 51(3):76–78

Nagel J, Albert M, Schmidt M (2002) Das waldbauliche Prognose- und Entscheidungsmodell BWINPro 6.1. Forst und Holz 57(15/16):486–493

Nagel J, Spellmann H, Pretzsch H (2012) Zum Informationspotenzial langfristiger forstlicher Versuchsflächen und periodischer Waldinventuren für die waldwachstumskundliche Forschung. Allg Forst und Jagdzeitung 183(5/6):111–116

Nagelkerke NJD (1991) A note on a general definition of the coefficient of determination. Biometrika 78:691–692

Nagumo H, Shiraishi N, Tanaka M, (1981) Computer programming for the construction of a sugi local yield-table - program TUSYCS. In: Nishizawa M (ed) Proceedings of forest resource inventory, growth models, management planning and remote sensing, XVII IUFRO World Congress. Kyoto, Japan, pp 103–114

Nance W, Grissom JE, Smith WR (1988) A new competition index based on weighted and constrained area potentially available. In: Ek AR, Shifley SR, Burke TE (eds) Forest growth modelling and prediction, vol NC-120, USDA For Serv Gen Techn. Rep., pp 134–142

Nänni UW (1970) Trees, water and perspective. S Afr For J 75:9–17

Nebel G, Kvist LP, Vanclay JK, Christensen H, Luis Freitas L, Ruiz J (2001) Structure and floristic composition of flood plain forests in the Peruvian Amazon. For Ecol Manage 150:27–57

Nelder JA (1962) New kinds of systematic designs for spacing experiments. Biometrics 18:283–307

Nepal SK, Somers GL (1992) A generalized approach to stand table projections. For Sci 38:120–133

Neumann M, Zhao M, Kindermann G, Hasenauer H (2015) Comparing MODIS net primary production estimates with terrestrial national forest inventory data in Austria. Remote Sens (Basel) 2015(7):3878–3906. https://doi.org/10.3390/rs70403878

Newbold P, Granger CWJ (1974) Experience with forecasting univariate time series and the combination of forecasts. J R Stat Soc Ser A 137:131–165

Newbold T, Hudson LN, Hill SLL et al (2016) Global effects of land use on local terrestrial biodiversity. Nature 520:45–50

Newnham RM (1976) Computer may cut costs $1-2/cord. Pulp Pap Can 77(9):73–79

Newnham RM (1991) LOGPLAN II – a model for planning logging and regeneration activities. For. Can. Petawawa Nat. For. Inst. Inf. Rep. PI-X-102

Ni R, Baiketuerhan Y, Zhang C, Zhao XH, Gadow K (2014) Analysing structural diversity in two temperate forests in northeastern China. For Ecol Manage 316:139–148

Nielsen CNN (1995) Detailed instructions for root architecture assessments with the Rootarch-method. Arboretum, Royal Vet. and Agriculture University, Int. Rep. No. 7, Hoersholm (DK), 12 p

Niinemets Ü, Valladares F (2006) Tolerance to shade, drought, and waterlogging of temperate northern hemisphere trees and shrubs. Ecol Monogr 76(4):521–547

Niklasson M, Zin ME, Zielonka T, Feijen M, Korczyk AF, Churski M, Samojlik T, Jedrzejewska B, Gutowski JM, Brzeziecki B (2010) A 350-year tree-ring fire record from Białowieza Primeval Forest, Poland: implications for Central European lowland fire history. J Ecol 98:1319–1329

Nilson A (2006) Modeling dependence between the number of trees and mean tree diameter of stands, stand density and stand sparsity. In: Cieszewski CJ, Strub M (eds), Second international conference on forest measurement and quantitative methods and management & the

2004 Southern Mensurationists Meeting 15–18 June 2004 Hot Springs, Arkansas USA: 74–94. University of Georgia, Athens

Nord-Larsen T (2006) Modeling individual-tree growth from data with highly irregular measurement intervals. For Sci 52:198–208

North Karelian Forestry Programme 2006–2010 (2006) Forestry Centre Pohjois-Karjala. 101 p. + appendices

Noss RF (1990) Indicators of monitoring biodiversity: a hierarchical approach. Conserv Biol 4:355–364

Nothdurft A, Schmidt M (2010) Kalibrierfähige Modelle für simultane Prognosen von Durchmesser und Höhe bei Waldinventuren – integrierte Nutzung von Langzeitbeobachtungen auf Versuchsflächen. DVFFA – Sektion Ertragskunde, Jahrestagung 2010, pp 125–130

Nothdurft A, Saborowski J, Nuske RS, Stoyan D (2010) Density estimation based on k-tree sampling and point pattern reconstruction. Can J For Res 40:953–967

Nuutinen T, Hirvelä H (2006) Hakkuumahdollisuudet Suomessa valtakunnan metsien 10. inventoinnin perusteella. Metsätieteen aikakauskirja 1B(2006):223–237

Nuutinen T, Salminen O (1999) Hakkuumahdollisuusarviot. In: Reunala A, Halko L, Marila M (eds) Kansallinen metsäohjelma 2010 – Taustaraportti. Maa- ja metsätalousministeriön julkaisuja 6(1999):124–131

Nuutinen T, Hirvelä H, Horne P, Härkönen K, Naskali A, Ovaskainen V, Salminen O, Seppälä H, Siitonen M, Toropainen M (1996) Metsänsuojelun taloudelliset ja työllisyysvaikutukset. Metsäntutkimuslaitos. Loppuraportti Metsänsuojelun ja työllisyyden rahoitustoimikunnalle. Mimeographed. 19 p. + appendices

Nuutinen T, Hirvelä H, Hynynen J, Härkönen K, Hökkä H, Korhonen KT, Salminen O (2000) The role of peatlands in Finnish wood production – an analysis based on large-scale forest scenario modelling. Silva Fennica 34(2):131–153

Nuutinen T, Hirvelä H, Salminen O (2005a) Valtakunnan metsien 9. inventointiin perustuvat hakkuumahdollisuusarviot vuosille 2003-2032 Lapin metsäkeskuksen alueella. Metsätieteen aikakauskirja 2B(2005):289–305

Nuutinen T, Hirvelä H, Salminen O (2005b) Etelä-Suomen metsien kehitys – vuosille 2001–2005 tehtyjen alueellisten metsäohjelmien vaikutusanalyysi. Metlan työraportteja 12:22 p

Nuutinen T, Hirvelä H, Salminen O (2005c) Alueelliset hakkuumahdollisuudet Suomessa. Metlan työraportteja 13:73 p

Nuutinen T, Matala J, Hirvelä H, Härkönen K, Peltola H, Väisänen H, Kellomäki S (2006) Regionally optimized forest management under changing climate. Clim Change 79(3-4):315–333

Nyland RD (2002) Silviculture. Concepts and applications, 2nd edn. McGraw-Hill, New York

Nyland RD (2016) Silviculture: concepts and applications. Waveland Press, Long Grove

O'Connell BM et al (2014) The forest inventory and analysis database: database description and user guide version 6.0.1 for Phase 2. 748 (U.S. Department of Agriculture, Forest Service., Washington, D.C., 2014)

O'Connor AJ (1935) Forest research with special reference to planting distances and thinning. Govt. Printer, Pretoria

O'Connor AJ (1960) Thinning research. J South Afr For Assoc 34:65–88

Ogden NH, Wilson JRU, Richardson DM, Hui C, Davies SJ, Kumschick S, Le Roux JJ, Measey J, Saul W-C, Pulliam JRC (2019) Emerging infectious diseases and biological invasions: a call for a one health collaboration in science and management. R Soc Open Sci 6:181577. https://doi.org/10.1098/rsos.181577

O'Hara KL, Gersonde RF (2004) Stocking control concepts in uneven-aged silviculture. Forestry 77(2):131–143

O'Hehir J (2001) Growth and yield models for South Australian Radiata Pine Plantations incorporating Fertilising and thinning. PhD dissertation, Institute of Land and Food Resources, University of Melbourne

Öhman K (2000) Creating continuous areas of old forest in long-term forest planning. Can J For Res 30:1817–1823

Öhman K, Eriksson O (1999) Creating continuous area of old forest in long-term forest planning. Can J For 30:1917–1823

Oksanen J, Guillaume BF, Kindt R, Legendre P, Minchin PR, O'Hara RB, Simpson GL, Solymos P, Stevens MHH, Wagner H (2013) Vegan: community ecology package. R package version 2.0-9

Oliver CD, Larson BC (1996) Forest stand dynamics. Wiley, New York

Olson DM, Dinerstein, E, Wikramanayake, ED, Burgess ND, Powell GVN, Underwood EC, D'amico JA, Itoua I, Strand HE, Morrison JC, Loucks CJ, Allnutt TF, Ricketts TH, Kura Y, Lamoreux JF, Wettengel WW, Hedao P, Kassem KR (2001) Terrestrial ecoregions of the world: a new map of life on earth. BioScience 51(11), (1 November). https://doi.org/10.1641/0006-3568(2001)051[0933:TEOTWA]2.0.CO;2

Ontario Ministry of Natural Resources (2000) Ontario forest accord. Queen's Printer for Ontario. (http://www.mnr.gov.on.ca/mnr/oll/ofaab/accord.html, as seen on Sept 27, 2000)

Oohata S-I (1986) Some quantitative analysis of tree form. Further extensions to the pipe model theory. In: Fujimori T, Whitehead (eds) Crown and canopy structure in relation to productivity. Published by the Forestry and Forest Products Institute, Ibaraki, Japan, pp. 243-262. ISBN 4-9900058-1-3.

Oppili P (2013) Plantations to turn into natural forests. The Hindu, September 5, 2013

Oren R, Ellsworth DS, Johnsen KH, Phillips N, Ewers BE, Maier C, Schäfer KRV, McCarthy GH, Hendrey G, McNulty SG, Katul GG (2001) Soil fertility limits carbon sequestration by forest ecosystems in a CO_2-enriched atmosphere. Nature 411:469–472

Ormerod DW (1973) A simple bole model. For Chronicle 49:136–138

Otto H-J (1994) Die Verwirklichung naturgemäßer Waldwirtschaft in den Niedersächsischen Landesforsten – Chancen und Probleme. Der Dauerwald 10:3–23

Otto HJ (2002) Antriebskräfte natürlicher Buchenwalddynamik in Europa. Forst u Holz 57(21):649–653

Ottorini JM (1991) Growth and development of individual Douglas Fir stands for applications to simulation in silviculture. Annales des Sciences Forestieres 48

Ough K, Murphy A (1998) 'Understorey islands: A method of protecting understorey flora during clearfelling operations', VSP Internal Report 29, Department of Natural Resources and Environment, Victoria

Overbeck M, Schmidt M (2012) Modelling infestation risk of Norway spruce by Ips typographus (L.) in the Lower Saxon Harz Mountains (Germany). For Ecol Manage 266 (2012) 115–125

Packalen P, Pukkala T, Pascual A (2019). Combining spatial and economic criteria in tree-level harvest planning. Manuscript submitted to Forest Ecosystems

Packalen P, Pukkala T, Pascual A (2020) Combining spatial and economic criteria in tree-level harvest planning. For Eco 7:18

Palik B, Mitchell RJ, Pecot S, Battaglia M, Pu M (2003) Spatial distribution of overstory retention influences resources and growth of longleaf pineseedlings. Ecol Appl 13:674–686

Pan Y, Birdsey RA, Fang J, Houghton R, Kauppi PE, Kurz WA, Phillips OL, Shvidenko A, Lewis SL, Canadell JG, Ciais P, Jackson RB, Pacala SW, David McGuire A, Piao S, Rautiainen A, Sitch S, Hayes D (2011) A large and persistent carbon sink in the world's forests. Science 333(6045):988–993

Paquette A, Messier C (2011) The effect of biodiversity on tree productivity: from temperate to boreal forests. Glob Ecol Biogeogr 20:170–180. https://doi.org/10.1111/j.1466-8238.2010.00592.x

Paquette A, Joly S, Messier C (2015) Explaining forest productivity using tree functional traits and phylogenetic information: two sides of the same coin over evolutionary scale? Ecol Evol 5:1774–1783

Pardé J, Bouchon J (1990) Dendrométrie. Ecole Nationale du Genie Rural des Eaux et des Forêts, Nancy

Parmesan C (2006) Ecological and evolutionary responses to recent climate change. Annu Rev Ecol Evol Syst:637–669

Parresol BR (1999) Assessing tree and stand biomass: a review with examples and critical comparisons. Forest Science 45:573–593

Parrott D, Brinks J, Lhotka J (2012) Designing Nelder wheel plots for tree density experiments. New For 43(2):245

Paterson SS (1956) The forest area of the world and its potential productivity. Goteburg University Press, Goteburg

Paterson SS (1959) A study based on productivity regions, of some basic factors in Swedish forest productivity. Festskrift Till O. Jonasson. Goteborg, pp 159–168,. (In Swedish).

Paterson SS (1962) Introduction to Phytochorology of Norden. Medd Stat Skogsforskininginstitut 50:1–145

Paulsen JC (1795) Praktische Anweisung zum Forstwesen.. Detmold

Pearson SM, Turner MG, Gardner RH, O'Neill RV (1996) An organism-based perspective of habitat fragmentation. Szaro & Johnston 1996:41–57

Peìrez-Harguindeguy N, Díaz S, Garnier E, Lavorel S, Poorter H, Jaureguiberry P (2013) New handbook for standardised measurement of plant functional traits worldwide. Aust J Bot 61:167–234

Pelz DR, Kohnle U (2013) Forest observational studies in Germany. In: Zhao XiuHai, Zhang ChunYu, Gadow, Kv (eds) Forest observational studies. Proceedings of an international workshop at Beijing Forestry University, 20/21 September 2012:109–124

Penttinen A, Stoyan D, Henttonen HM (1992) Marked point process in forest statistics. For Sci 38:806–824

Pesonen M (1995) Non-industrial private landowners' choices of timber management strategies and potential allowable cut – case of Pohjois-Savo. Acta Forestalia Fennica 247: 31 pp

Petchey OL, Gaston KJ (2006) Functional diversity: back to basics and looking forward. Ecol Lett 9:741–758

Peters R (1997) Beech forests, Geobotany, vol 24. Kluwer Academic Publishers, Dordrecht, 187 p

Peterson CE, Monserud RA (2002) Compatibility between wood production and other values and uses on forested lands – a problem analysis. Gen. Tech. Rep. PNW-GTR-564. Portland, OR: US Dept of Agriculture, Forest Service, Pacific Northwest Research Station, 51 p

Peyron J-L (1993) Présentation illustreé d'une méthode de planification de la gestion forestière et de détermination de l'effort de régéneration. Rev For Fr XLV(1):59–73

Philip MS (1994) Measuring trees and forests. Wallingford, Cab International. 310 S

Phillips JFV (1931) Forest succession and ecology in the Knysna Region. In: Schulze E-D, Caldwell MM (eds) Memoirs of the botanical survey of South Africa 14. The Government Printer, Pretoria

Pielou EC (1961) Segregation and symmetry in two-species populations as studied by nearest neighbour relations. J Ecol 49:255–269

Pielou EC (1977) Mathematical ecology. Wiley, New York, 384 p

Pienaar LV (1989) A stand table projection approach to yield prediction in plantations. South-Afr For J 149:44–47

Pienaar LV, Harrison WM (1988) A stand table projection approach to yield prediction in unthinned even-aged stands. For Sci 34:804–808

Pienaar V, Shiver B (1981) Survival functions for site-prepared slash pine plantations in the flatwoods of Georgia and Northern Florida. South J App For 5:59–62

Pienaar LV, Turnbull KJ (1973) The Chapman-Richards generalization of von Bertalanffy's growth model for basal area growth and yield in evenaged stands. For Sci 19(1):2–22

Pindyck RS, Rubinfeld DL (1991) Econometric models and economic forecasts. McGraw-Hill, INC., New York/Hamburg/London, 596 pp. ISBN 0-07-100866-7

Pippuri I, Maltamo M, Packalen P, Mäkitalo J (2013) Predicting species-specific basal areas in urban forests using airborne laser scanning and existing stand register data. Eur J For Res 132:999–1012

Pitkänen TP, Raumonen P, Kangas A (2019) Measuring stem diameters with TLS in boreal forests by complementary fitting procedure. ISPRS J Photogramm Remote Sens 147:294–306

Pnevmaticos SM, Mann SH (1972) Dynamic programming in tree bucking. For Prod J 22(2):26–30

Pommerening A (1997) Eine Analyse neuer Ansätze zur Bestandesinventur in strukturreichen Wäldern [An analysis of new approaches towards stand inventory in structure-rich forests].

Ph.D. dissertation, Faculty of Forestry and Forest Ecology, University of Göttingen, Cuvillier Verlag Göttingen: 187 pp

Pommerening A (2001) Continuous cover forestry – chance and challenge for forest science. Institute of Chartered Foresteres (ICF) news 1/2001, special feature "Alternatives to clearfelling", pp 8–10

Pommerening A (2002) Approaches to quantifying forest structures. Forestry 75(3):305–324

Pommerening A (2006) Evaluating structural indices by reversing forest structural analysis. For Ecol Manage 224:266–277

Pommerening A (2008) Analysing and modeling spatial woodland structure. Habilitationsschrift (D.Sc. dissertation), University of Natural Resources and Applied Life Sciences, Vienna, Austria

Pommerening A, Grabarnik P (2019) Individual-based methods in forest ecology and management. Springer Nature, 411 p

Pommerening A, Stoyan D (2006) Edge-correction needs in estimating indices of spatial forest structure. Can J For Res 36:1723–1739

Pommerening A, Stoyan D (2008) Reconstructing spatial tree point patterns from nearest neighbor summary statistics measured in small subwindows. Can J For Res 38:1110–1122

Poschmann C, Riebenstahl C, Schmidt-Kallert E (1998) Umweltplanung und –bewertung. Justus Perthes Verlag Gotha GmbH, Gotha

Pressler R (1865) Zur Verständigung über den Reinertragswaldbau und dessen Betriebsideal. AFJZ 36, pp 41–55 and pp 173–191 and pp 261–271

Preston FW (1962) The canonical distribution of commonness and rarity: Part I. Ecology 43:185–215

Pretzsch H (1992a) Konzeption und Konstruktion von Wuchsmodellen für Rein- und Mischbestände, vol 115. Schriftenreihe d. Forstw. Fak. University München, p 332

Pretzsch H (1992b) Modellierung der Kronenkonkurrenz von Fichte und Buche in Rein-und Mischbeständen. Allg Forst-u Jagd Zeitung 163(11-12):203–213. (In German)

Pretzsch H (1993) Analyse und Reproduktion räumlicher Bestandesstrukturen – Versuche mit dem Strukturgenerator STRUGEN. Schriften aus der Forstl. Fak. d. Univ. Göttingen u. d. Nieders. Forstl. Vers. Anst., Band 114, J.D. Sauerländer's Verlag

Pretzsch H (1995) Perspektiven einer modellorientierten Waldwachstumsforschung. Forstwissenschaft-liches Centralblatt 114:188–209

Pretzsch H (1997) Analysis and modeling of spatial stand structures. Methodological considerations based on mixed beech-Larch stands in Lower Saxony. For Ecol Manage 97:237–253

Pretzsch H (2001) Modellierung des Waldwachstums. Parey Buchverlag Berlin, 341 S

Pretzsch H (2002) Grundlagen der Waldwachstumsforschung. [The priciples of forest growth science.]. Parey Buchverlag, Berlin

Pretzsch H (2003) Diversität und Produktivität von Wäldern (Diversity and productivity of forests). Allg Forst u Jagdztg 174(5/6):88–97

Pretzsch H (2005) Stand density and growth of Norway spruce (Picea abies (L.) Karst.) and European beech (Fagus sylvatica L.): evidence from long-term experimental plots. Eur J For Res 124:193–205. https://doi.org/10.1007/s10342-005-0068-4

Pretzsch H (2009) Forest dynamics, growth and yield. Springer, Berlin, p 664

Pretzsch H, Biber P (2016) Tree species mixing can increase maximum stand density. Can J For Res 46:1179–1193

Pretzsch H, Kahn M (1998) Forschungsvorhaben „ Konzeption und Konstruktion von Wuchs- und Prognosemodellen für Mischbestände in Bayern": Abschlußbericht Projekt W28 Teil 2. Konzeption und Konstruktion des Wuchsmodells SILVA 2.2 – Methodische Grundlagen. Lehrstuhl für Waldwachstumskunde der Ludwig-Maximilians-Universität München, Freising, 279 p

Pretzsch H, Biber P, Ďurský J, Gadow Kv, Hasenauer H, Kändler G, Kenk G, Kublin E, Nagel J, Pukkala T, Skovsgaard JP, Sodtke R, Sterba H (2002) Recommendations for standardized documentation and further development of forest growth simulators. Forstwissenschaftliches Centralblatt 121(3):138–151

Pretzsch H, Dieler J, Seifert T, Rötzer T (2012) Climate effects on productivity and resource use efficiency of Norway spruce (Picea abies [L.] Karst.) and European beech (Fagus sylvatica [L.]) in stands with different spatial mixing patterns. Trees 26:1343–1360

Pretzsch H, Biber P, Gadow Kv (2015) Ernst assmann: a German pioneer in forest production ecology and quantitative silviculture. Eur J Forest Res. 10.1007/s10342-015-0872-4. Published online 24. March 2015.

Pretzsch H, Forrester DI, Bauhus J (2017) Mixed species forests. Springer 653 p

Preussner K (1974) Aufstellung und experimentelle Überprüfung mathematischer Modelle für die Entwicklung der Durchmesserverteilung von Fichtenbeständen. Doctoral Dissertation. Univ. Tharandt: 123p

Prodan M (1966) Holzmeßlehre. J.D. Sauerländer's Verlag, Frankfurt a. M, 644 p

Puettmann KJ, Wilson SMG, Baker SC, Donoso PJ, Drössler L, Amente G, Harvey BD, Knoke T, Lu Y, Nocentini S, Putz FE, Yoshida T, Bauhus J (2015) Silvicultural alternatives to conventional even-aged forest management – what limits global adoption. For Ecosyst 2:8. https://doi.org/10.1186/s40663-015-0031-x

Pukkala T (1989) Prediction of tree diameter and height in a Scots pine stand as a function of the spatial pattern of trees. Silva Fenn 23:83–99

Pukkala T (2002a) Multi-objective forest planning, Vol. 6 of the book series Managing forest ecosystems. Kluwer Academic Publishers, 264 p.

Pukkala T (2002b) Measuring non-wood forest outputs in numerical forest planning. In: Pukkala T (ed) Multi-objective forest planning. Kluwer Academic Publishers, Dordrecht, pp 173–207

Pukkala T (2004) Dealing with ecological objectives in the Monsu planning system. Silva Lusitana, Special issue (2004): 1–15

Pukkala T (2008) Integrating multiple services in the numerical analysis of landscape design. In: Gadow Kv, Pukkala T (eds) Designing green landscapes. Vol 15 of the Springer book series managing forest ecosystems, pp 137–168

Pukkala T (2019) Using ALS raster data in forest planning. J For Res. https://doi.org/10.1007/s11676-019-00937-6

Pukkala T (2019a) Optimized cellular automaton for stand delineation. J For Res 30(1):107–119

Pukkala T (2019b) Using ALS raster data in forest planning. J For Res 30(5):1581–1593

Pukkala T, Gadow Kv (eds) (2012) Continuous cover forestry, Book series managing forest ecosystems Vol 24, © Springer

Pukkala T, Kangas J (1993) A heuristic optimization method for forest planning and decision making. Scand J For Res 8:560–570

Pukkala T, Nuutinen T, Kangas J (1995) Integrating the amenity of forest area into numerical forest planning. Landsc Urban Plan 132:185–195

Pukkala T, Heinonen T, Kurttila M (2008) An application of the reduced cost approach to spatial forest planning. For Sci 55(1):13–22

Pukkala T, Packalen P, Heinonen T (2014) Dynamic treatment units in forest management planning. Manag For Ecosyst 33:373–392

Pukkala T, Lähde E, Laiho O (2016) Which trees should be removed in thinning treatments? For Ecosyst 2(23):2–12

Pullan W, Bhadeshia H (eds) (2000) Structure in science and art. Cambridge University Press, Cambridge

Purdy J (2015) After nature – a politics for the anthropocene. Harvard University Press, Cambridge

Puumalainen J (1996) Die Beta-Funktion und ihre analytische Parameterbestimmung für die Darstellung von Durchmesserverteilungen. Arbeitspapier 15–96, Inst. f. Forsteinrichtung, Univ. Göttingen, p 12

Puumalainen J (1998) Optimal cross-cutting and sensitivity analysis for various log dimension constraints by using dynamic programming approach. Scand J For Res 13:74–82

Puumalainen J, Gadow K, Korhonen K (1998) Flexible Ergebnisberechnung einer Eingriffsinventur. [Flexible calculation of a thinning inventory]. Forstarchiv 69:145–150

Qian H, Jin Y (2016) An updated megaphylogeny of plants, a tool for generating plant phylogenies and an analysis of phylogenetic community structure. J Plant Ecol 9:233–239

Qin J, Cao QV (2006) Using disaggregation to link individual-tree and whole-stand growth models. Can J For Res 36:953–960

Rademacher P (2002) Ermittlung der Ernährungssituation, der Biomasseproduktion und der Nährelementakkumulation mit Hilfe von Inventurverfahren sowie Quantifizierung der Entzugsgrößen auf Umtriebsebene in forstlich genutzten Beständen. Habilitationsschrift, Univ. Göttingen

Raffa KF, Aukema BH, Bentz BJ, Carroll AL, Hicke JA, Turner MG, Romme WH (2008) Cross-scale drivers of natural disturbances prone to anthropogenic amplification: the dynamics of bark beetle eruptions. BioScience 58(6):501–517. https://doi.org/10.1641/B580607

Rahman MS, Akter S, Al-Amin M (2015) Forest and agro-ecosystem productivity in Bangladesh: a climate vegetation productivity approach. For Sci Technol. https://doi.org/10.1080/21580103.2014.957358

Raia AP (1974) Managing by objectives. Scott, Foreman & Co., Illinois

Randall JH, Gadow K (1990) An evaluation of the use of "Gentan" probabilities in forestry planning. South Afr For J 154:27–35

Ranneby B, Cruse T, Hägglund B, Jonasson H, Swärd J (1987) Designing a new national forest survey for Sweden. Studia Forestalia Suecica 177

Rao CR (1982) Diversity and dissimilarity coefficients: a unified approach. Theor Popul Biol 21:24–43

Rautiainen O (1998) Modelling the yield and growth of uneven-aged Shorea robusta stands. Ph.D thesis, University of Joensuu

Rautiainen O (1999) Growth dynamics and management of Shorea robusta forests in Southern Nepal. Diss. Faculty of Fortestry, University of Joensuu

Raven PH, Evert RF, Eichhorn SE (1987) Biology of plants. Worth Publishers Inc., New York, 685 p

R-Core Team (2012) R: A language and environment for statisticalcomputing. R Foundation for Statistical Computing, version 2.15.1, Vienna, Austria, http://www.R-project.org

Reddy SC Jha CS Diwakar PG Dadhwal VK (2015) Nationwide classification of forest types of India using remote sensing and GIS. Environ Monit Assess 187(12). https://doi.org/10.1007/s10661-015-4990-8. https://www.researchgate.net/publication/285020645_Nationwide_classification_of_forest_types_of_India_using_remote_sensing_and_GIS/figures?lo=1

Reed DD, Green EJ (1984) Compatible stem taper and volume ratio equations. For Sci 30(4):977–990

Reif A, Schulze ED, Ewald J, Rothe A (2013) Waldkalkung – Bodenschutz contra Naturschutz. Waldökologie, Landschaftsforschung und Naturschutz

Reineke LH (1933) Perfecting a stand density index for even-aged forests. J Agric Res 46(7):627–638

Reininger H (1987) Zielstärken-Nutzung oder die Plenterung des Altersklassenwaldes. Österreichischer Agrarverlag, Wien

Remote sensing: The R package sperrorest. In 2012 IEEE International geoscience and remote sensing symposium, pp 5372–5375

Rennie PJ (1963) Methods of assessing site capability. Commonw For Rev 42:306–317

Reyes-Hernandez V, Comeau PG, Bokalo M (2013) Static and dynamic maximum size-density relationships for mixed trembling aspen and white spruce stands in western Canada. For Ecol Manage 289:300–311

Reyes-Hernandez VJ, Comeau PG (2015) The influence of stocking and stand composition on productivity of boreal trembling aspen-white spruce stands. Forests 2015(6):4573–4587. https://doi.org/10.3390/f6124387

Reynolds KM, Schmoldt DL (2006) Computer-aided decision-making. In: Shao GF, Reynolds KM (eds) Computer applications in sustainable forest management, Kluwer/Springer book series "Managing forest ecosystems", vol 11, pp 143–169

Richards FJ (1959) A flexible growth function for empirical use. J Exp Bot 10:290–300

Richards AE, Forrester DI, Bauhus J, Scherer-Lorenzen M (2010) The influence of mixed tree plantations on the nutrition of individual species: a review. Tree Physiol 30(9):1192–1208

Richardson DM, Macdonald IAW, Hoffmann JH, Henderson L (1997) Alien plant invasions. In: Cowling RM, Richardson DM, Pierce SM (eds) Vegetation of Southern Africa. Cambridge University Press, Cambridge, pp 535–570

Richardson DM, Pyšek P, Carlton JT (2011a) A compendium of essential concepts and terminology in invasion ecology. In: Fifty years of invasion ecology – The legacy of Charles Elton. Wiley-Blackwell, Oxford, pp 409–420

Richardson DM, Carruthers J, Hui C, Impson FAC, Miller JT, Robertson MP, Rouget M, Le Roux JJ, Wilson JRU (2011b) Human-mediated introductions of Australian acacias – a global experiment in biogeography. Divers Distrib 17:771–787

Richardson DM, Hui C, Nuñez MA, Pauchard A (2014) Tree invasions – patterns, processes, challenges and opportunities. Biol Invasions 16:473–481. https://doi.org/10.1007/s10530-013-0606-9

Richter J (1994) Neue Aspekte der Fichtendurchforstung. AFZ 49(12):632–637

Ricotta C, Burrascano S (2009) Testing for differences in beta diversity with asymmetric dissimilarities. Ecol Indic 9:719–724

Riemer T, Gadow K, Sloboda B (1995) Ein Modell zur Beschreibung von Baumschäften. Allg Forst- und J Ztg 166(7):144–147

Ripley BD (1977) Modelling spatial patterns. J R Stat Soc. Series B (Methodological), 172–212.

Ritchie MW, Hann DW (1997) Implications of disaggregation in forest growth and yield modeling. For Sci 43(2):223–233

Roberge JM, Mönkkönen M, Toivanen T, Kotiaho J (2013) Retention forestry and biodiversity conservation: a parallel with agroforestry. Nat Conserv 4(0):29. https://doi.org/10.3897/natureconservation.4.5116

Rodriguez L (1996) A microcomputer program for solving forest scheduling problems with heuristic approaches. In: Päivinen R, Roihuvuo L, Siitonen M (eds) Large-scale forestry scenario models – experiences and requirements, European Forest Institute. EFI Proceedings No. 5, pp 153–166

Rodriguez Soalleiro RJ, Alvarez Gonzalez JG, Vega Alonso G (1994) Pineiro do Pais – Modelo dinamico de crecemento de masas regulares de Pinus pinaster Aiton en Galicia. Xunta de Galicia:40 S

Rois Diaz M (2000) Analyse eines Buchen-Edellaubholz-Mischbestandes im Göttinger Wald. Master thesis, Faculty of Forestry and Forest Ecology, Georg-August-University, Göttingen

Rojo A, Montero G (1996) El Pino Silvestre en la Sierra de Guadarrama. Ministerio de Agriculture, Pesca y Alimentacion, Madrid, 293 S

Rojo A, Ulises D-A, Rodríguez-Soalleiro R, Gadow Kv (2005) Modelling silvicultural and economic alternatives for Scots pine (Pinus sylvestris L.) plantations in Northwestern Spain. Forestry 78(3) 17pp

Römisch K (1983) Ein mathematisches Modell zur Simulation von Wachstum und Durchforstung gleichaltriger Reinbestände. TU Dresden, Sektion Forstwirtschaft, Tharandt, Diss. A

Ross RT (1934) Optimum orders for the presentation of pairs in the method of paired comparisons. J Educ Psychol 25:375–382

Ross E (2017) Amazon rainforest has been shaped by an ancient hunger for fruits and nuts. Nature, https://doi.org/10.1038/nature.2017.21576

Rothe A, Binkley D (2001) Nutritional interactions in mixed species forests: a synthesis. Can J For Res 31:1855–1870

Rothe A, Brandt S, Hurler R (1999) Waldbewirtschaftung und Nitratbelastung des Grundwassers. AFZ/Der Wald 10:531–533

Rottmann M (1986) Wind- und Sturmschäden im Wald.. Frankfurt

Rötzer T, Seifert T, Gayler S, Priesack E, Pretzsch H (2012) Effects of stress and defence allocation on tree growth: Simulation results at the tree and stand level. In: Matyssek R, Schnyder H, Ernst D, Munch JC, Oßwald W, Pretzsch H (eds) Growth and defence in plants: resource allocation at multiple scales, vol 220, Ecol Stud., pp 401–432

Ruggiero LF, Jones LC, Aubry KB (1991) Plant and animal habitats associations in Douglas-fir forests of the Pacific Northwest: an overview. In: Ruggiero LF, Aubry KB, Carey AB, Huff HM

(tecn cords.) Wildlife and Vegetation of Unmanage Douglas-fir forest. USDA Forest Service, GTR-PNW-285. pp 447–462
Ruppert K (1972) 600 Jahre Frankfurter Stadtwald. Published by Forstamt Frankfurt
Rusanen K, Finér L, Antikainen M, Korkka-Niemi K, Backman B, Britschgi R (2004) The effect of forest cutting on the quality of groundwater in large aquifers in Finland. Boreal Environ Res 9:253–261
Saaty TL (1980) The analytic hierarchy process. McGraw-Hill
Saaty TL, Alexander JM (1981) Thinking with models, Pergamon Press
Saborowski J (1982) Entwicklung biometrischer Modelle zur Sortimentenprognose. Dissertation, Universität, Göttingen
Sachsse H (1991) Kerntypen der Rotbuche. Forstarchiv 62:238–242
Sagarin R, Pauchard A (2012) Observation and ecology. Island Press, 213 pp
Sahney S, Benton MJ, Ferry PA (2010) Links between global taxonomic diversity, ecological diversity and the expansion of vertebrates on land. Biol Lett 6(4):544–547
Sakai AK, Allendorf FW, Holt JS, Lodge DM, Molofsky J, With KA, Baughman S, Cabin RJ, Cohen JE, Ell-strand NC, McCauley DE, O'Neil P, Parker IM, Thompson JN, Weller SG (2001) The population biology of invasive species. Annu Rev Ecol Syst 32:305–332
Sakamoto Y, Ishiguro M, Kitagawa G (1986) Akaike information criterion statistics. In: KTK scientific publisher. D Reidel publishing company, Tokyo, 290 pp. ISBN 90-277-2253-6
Salas C, LeMay V, Núñez P, Pacheco P, Espinosa A (2006) Spatial patterns in an old-growth Nothofagus obliqua forest in south-central Chile. For Ecol Manage 231:38–46
Sallnäs O (1996) The IIASA-model for analysis of harvest potentials. EFI Proceedings 5. Seminar and summer school on 'Large scale forestry scenario models' held in Joensuu, Finland. 15-22 June 1995. European Forest Institute pp 19–30
Salminen S, Salminen O (1998) Metsävarat Keskisessä Suomessa 1988–92 sekä koko Etelä-Suomessa 1986–92. Summary: Forest resources in Middle Finland, 1988–92, and in South Finland, 1986–92. Metsäntutkimuslaitoksen tiedonantoja 710. 137 s
Sánchez Orois S, Kotzé H, Corral Rivas JJ (2003) Ein kompatibles Modell für die Fortschreibung von Durchmesserverteilungen. Vortrag anlässlich der Jahrestagung der Sektion Ertragskunde des Deutschen Verbandes Forstlicher Forschungsanstalten in Torgau am 2. – 4. Juni 2003
Sánchez S, Rodríguez R (2002) Modelling the growth and management of uneven-aged maritime pine-broadleved species forests in Galicia (NW Spain). Scand J For Res 17:538–547
Sanderson EW, Jaiteh M, Levy MA, Redfrod KH, Wannebo AV, Woolmer G (2002) The human footprint and the last of the wild. BioScience 52:891–904
Sands PJ, Battaglia M, Mummery D (2000) Application of process-based models to forest management: experience with ProMod, a simple plantation productivity model. Tree Physiol 20:383–392
SAS Institute Inc. (1989) SAS/STAT User's Guide. Version 6, 4th edn., Vol 2. SAS Institute Inc., Cary, NC. 846p
Savin EN, Milyutin LI, Krasnoshhekov JN, Korotkov IA, Suncov AV, Dugarzhav C, Cogoo Z, Dorzhsuren C, Zhamjansurjen S, Gombosuren N (1988) Forests of the Mongolian People's Republic (Larch forests of the Eastern Khentey). Soviet-Mongolian Expedition. Nauka, Siberian Department: Novosibirsk, Russia. p. 176. (In Russian)
Sayer J, Campbell B (2004) The science of sustainable development. Cambridge University Press, Cambridge
Schädelin W (1942) Die Auslesedurchforstung als Erziehungsbetrieb höchster Wertleistung. 3. Aufl., Bern
Schama S (1995) Landscape and memory. Harper Collins, London, 652 pp
Scherer-Lorenzen M, Schulze ED, Don A, Schumacher J, Weller E (2007) Exploring the functional significance of forest diversity: a new long-term experiment with temperate tree species (BIOTREE). PerspPlant Ecol Evol Syst 9:53–70
Schimper AFW, von Faber FC (1935) Pflanzengeographie auf physiologischer Grundlage, 3rd edn. Fischer, Jena

Schmidt W (2000) Biotische Mannigfaltigkeit – Indikator für intakte Wälder? Beispiel aus Studien über Bodenvegetation in Naturwaldreservaten. NUA 4:108–127

Schmidt M (2001) Prognosemodelle für ausgewählte Holzqualitätsmerkmale wichtiger Baumarten. Dissertation, Institut für Waldinventur und Waldwachstum, Universität Göttingen: 302 p

Schmidt M, Gadow K, Höfle H (2005) Vorkommen und Ausprägung von fakultativen Kerntypen bei Rotbuche auf niedersächsischen Kalk- und Rötstandorten. Allg Forst und Jagdzeitung 176(11/12):200–211

Schmidt M, Hanewinkel M, Kändler G, Kublin E, Kohnle U (2010) An inventory-based approach for modeling single- tree storm damage — experiences with the winter storm of 1999 in southwestern Germany. Can J For Res 40:1636–1652

Schmidt M, Kiviste A, Gadow K (2011) A spatially explicit height-diameter model for scots pine in Estonia. Eur J For Res 130:303–315. https://doi.org/10.1007/s10342-010-0434-8)

Schmoldt DL, Kangas J, Mendoza GA, Pesonen M (eds) (2001) The analytic hierarchy process in natural resource and environmental decision making. Vol 5 of the book series Managing forest ecosystems, Kluwer Academic Publishers, 305 p

Schober R (1953) Ertragstafeln und Forsteinrichtung. Forstwiss Cbl 72:1–13

Schober R (1972) Die Rotbuche 1971. J. D. Sauerländer's Verlag, Frankfurt am Main, 333 S

Schober R (1975) Ertragstafeln wichtiger Baumarten. J.D. Sauerländer's Verlag

Schober R (1978) Ertragstafeln wichtiger Baumarten. Sauerländer's Verlag

Schober R (1991) Eclaircies par le haut et arbres d'avenir. Rev For Fr XLIII:385–401

Schölch M, Rodenkirchen H, Risse F-J (2014) ANW-Exkursion durch Mischwälder im Nordosten und mittleren Westen der USA – Teil II Waldbauexperimente und Urwälder. Der Dauerwald 50:45–53

Scholes RJ, Biggs R (2005) A biodiversity intactness index. Nature 434(3):45–49

Schönau APG (1988) A re-appraisal of a site evaluation study in Black Wattle with special reference to soil factors. In: Gadow K (ed) Research contributions to plantation forestry. Faculty of Forestry, Stellenbosch, pp 83–102

Schönau APG, Aldworth WJK (1991) Site evaluation in black wattle with special reference to soil factors. S Afr For J 156:35–43

Schönhar S (1997) Die Waldbodenvegetation als Standortsweiser. AFJZ 164:173–180

Schreuder HT, Gregoire TG, Wood GB (1993) Sampling methods for multiresource forest inventory. Wiley, New York

Schröder FG (1998a) Lehrbuch der Pflanzengeographie. Quelle & Meyer Verlag, Wiesbaden, 457 p

Schröder J (1998b) Beschreibung von Bestandesstrukturen im Knysna-Wald, Südafrika. Diplomarbeit, Fak. f. Forstwiss. u. Waldökologie, Univ. Göttingen

Schröder J, Gadow K (1999) Testing a new competition index for maritime pine in North-Western Spain. Can J For Res 29:280–283

Schröder J, Rodríguez R, Vega G (2002) An age-independent basal area increment model for maritime pine trees in Northwestern Spain. For Ecol Manage 157:55–64

Schuck A, Päivinen R, Hytönend T, Pajari B (2002) Compilation of forestry terms and definitions. European Forest Institute, Joensuu

Schulze ED (2017) Effects of forest management on biodiversity in temperate deciduous forests: an overview based on Central European beech forests. J Nat Conserv 43:213–226

Schulze ED, Beck E, Müller-Hohenstein K (2005) Plant ecology. Springer, Berlin

Schulze E-D, Bouriaud L, Bussler H, Gossner M, Walentowski H, Hessenmöller D, Bouriaud O, Kv G (2014) Opinion paper: forest management and biodiversity. Web Ecol 14(1):3

Schulze ED, Craven D, Durso AM, Reif J, Guderle M, Kroiher F, Hennig P, Weiserbs A, Schall P, Ammer C, Eisenhauer N (2019) Positive association between forest management, environmental change, and forest bird abundance. For Ecosyst 6:3

Schütz J-Ph (1989) Der Plenterbetrieb. Unterlage zur Vorlesung Waldbau III, ETH Zürich, 54pp

Schütz J-P (1994) Geschichtlicher Hergang und aktuelle Bedeutung der Plenterung in Europa. AFJZ 165(5–6):106–114

Schütz J-P (2002) Silvicultural tools to develop irregular and diverse forest structures. Forestry 75:329–338

Schütz J-P (2006) Modeling the demographic sustainability of pure beech plenter forests in Eastern Germany. Ann For Sci 63:93–100

Schütz J-Ph, Pukkala T, Donoso P, Gadow Kv (2012) Historical emergence and current application of CCF. In: Pukkala T, Gadow K.v (eds) Continuous cover forestry. book series managing forest ecosystems. vol 24, © Springer, pp 1–28

Schwappach A (1890) Wachstum und Ertrag normaler Fichtenbestände. Julius Springer Verlag: 100 p

Schwappach A (1911) Die Rotbuche. Neumann Verlag, Neudamm

Schweizerisches Landesforstinventar (1988) Berichte Nr. 1986: Ergebnisse der Aufnahme 1982–1986. Eidgenössische Anstalt für das forstliche Versuchswesen. Birmensdorf

Schwichtenberg G, Sánchez Orois S (2005) Optimising management paths for an Acacia mangium plantation in Colombia. Forstarchiv 76:215–222

Scott DF, Lesch W (1997) Streamflow responses to afforestation with Eucalyptus grandis and Pinus patula and to felling in the Mokobulaan experimental catchments, Mpumalanga Province, South Africa. J Hydrol 199:360–377

Scott DF, Smith RE (1997) Preliminary empirical models to predict reductions in total and low flows resulting from afforestation. Water SA 23:135–140

Scott DF, Le Maitre DC, van Wyk E (1999) The effects of forestry on low flows: further studies on low flow in wet and dry cycles and in relation to catchments of increasing size. Contract report No. ENV/S-C 99059, CSIR, Pretoria, South Africa

Seastedt TR, Hobbs RJ, Suding KN (2008) Management of novel ecosystems: are novel approaches required? Front Ecol Environ 6:547–553

Seber GAF, Wilde CJ (2003) Nonlinear regression. Wiley, Hoboken

Seckendorf A v (1881) Das Forstliche Versuchswesen, insbesondere dessen Zweck und wirthschaftliche Bedeutung. Faesy & Frick, Vienna, 132p

Sedjo RA (1995) Ecosystem management – an uncharted path for public forests. Resources, (Resources for the Future) No. 121: 10–20

Seebach C (1846) Ertragsuntersuchungen im Buchenhochwalde. Krit Blätter f Forst- u Jagdwiss 23:74–88

Seifert T (2003) Integration von Holzqualität und Holzsortierung in behandlungssensitive Waldwachstumsmodelle. Technical University Munich, Dissertation, p 314

Seifert T, Breibeck J, Seifert S, Biber P (2010) Resin pocket occurrence in Norway spruce depending on tree and climate variables. For Ecol Manage 260:302–312

Seifert T, du Toit B, Jooste GH, Seibold S, Durrheim G, Seydack A (2011) Towards an individual tree forest growth model for natural forests in South Africa: testing competition indices to model individual tree diameter growth in afro-temperate forests. In: Proceedings of the fifth national national forest and woodlands symposium, Richards Bay South Africa, Richards

Seifert T, Seifert S, Seydack A, Durrheim G, Gadow Kv (2014) Competition effects in an afrotemperate forest. For Ecosyst 1:13

Seydack AHW (2002) Continuous cover forestry systems in tropical and subtropical forests: current state and future perspectives. In Gadow Kv, Nagel J, Saborowski J (eds) Continuous cover forestry: assessment, analyses, scenarios; Kluwer Academic Publishers, pp 309–334

Seydack AHW (2012) Regulation of timber yield sustainability for tropical and subtropical moist forests: ecosilvicultural paradigms and economic constraints. In: Pukkala T, Gadow Kv (eds) Continuous Cover Forestry

Seydack AHW, Vermeulen WJ, Heyns HE, Durrheim GP, Vermeulen C, Willems D, Ferguson MA, Huisamen J, Roth J (1995) An unconventional approach to timber yield regulation for multi-aged, multispecies forests. II. Application to a South African forest. For Ecol Manage 77:155–168

Seydack AHW, Durrheim G, Louw JH (2011) Spatiotemporally interactive growth dynamics in selected South African forests: Edaphoclimatic environment, crowding and climate effects. For Ecol Manage 261:1152–1169

Seydack AHW, Durrheim G, Louw JH (2012) Forest structure in selected South African forests: edaphoclimatic environment, phase and disturbance. Eur J For Res 131:261–281

Shafer GN (1988) A site growth model for *Pinus eliottii* in the Southern Cape. S Afr For J 146:12–17

Shafer GN (1989) Site indicator species for predicting productivity of pine plantations in the Southern Cape. S Afr For J 148:7–17

Shannon CE, Weaver W (1949) The mathematical theory of communication. Univ. of Illinois Press, Urbana

Sharma M, Zhang S (2004) Height–diameter models using stand characteristics for Pinus banksiana and Picea mariana. Scand J Fort Res 19: 442–451

Sharma RP, Zdenek V, Vacek S, Kucera M (2019) A nonlinear mixed-effects height-to-diameter ratio model for several tree species based on czech national forest inventory data. Forests 10(1):70. https://doi.org/10.3390/f10010070

Shaughnessy GL (1980) Historical ecology of alien woody plants in the vicinity of Cape Town, South Africa. PhD dissertation, University of Cape Town. 421 pp

Shimatani K, Kubota Y (2004) Quantitative assessment of multispecies spatial pattern with high species diversity. Ecol Res 19:149–163

Shiver BD (1988) Sample sizes and estimation methods for the Weibull distribution for unthinned slash pine plantation diameter distributions. For Sci 34(3):809–814

Shoch D, Siikamäki JV, Smith P, Woodbury P, Zganjar C, Blackman A, Campari J, Conant RT, Delgado C, Elias P, Gopalakrishna T, Hamsik MR, Herrero M, Kiesecker J, Landis E, Laestadius L, Leavitt SM, Minnemeyer S, Polasky S, Potapov P, Putz FE, Sanderman J, Silvius M, Wollenberg E, Fargione J (2017) Natural climate solutions. Proc Natl Acad Sci 114(44):11645–11650

Shortt JS, Burkhart HE (1996) A comparison of loblolly pine plantation growth and yield models for inventory updating. South J Appl For 20(1):15–22

Shugart HH (1984) A theory of forest dynamics: the ecological implication of forest successsion models. Springer, New York, p 278 S

Shumway RH (1988) Applied statistical time series analysis. Prentice-Hall, Englewood Cliffs, 379 pp. ISBN 0-13-041500-6

Shvidenko A, Venevsky S, Raille G, Nilsson S (1995) A system for evaluation of growth and mortality in Russian forests. Water Air Soil Pollut 82:333–348

Shvidenko AZ, Schepaschenko DG, Nilson S, Buluy YI (2006) Tables and models of growth and productivity of forests of major forests forming species of Northern Eurasia. Ministry of Natural Resources, Federal Agency of Forest Management, International Institute for Applied Systems Analysis, Vienna, 803 p

Siegel A, Etzkorn I (2013) Simple: conquering the crisis of complexity. Grand Central Publishing, 256 pp

Siitonen M (1983) A long term forestry planning system based on data from the Finnish national forest inventory. Univ of Helsinki, Dept of For Mens & Mgmt Res Note 17:195–207

Siitonen M (1990) Suomen metsävarat 1990 ja metsien kehitysmahdollisuudet 1990–2030. Selvitys Metsä 2000 -ohjelman tarkastustoimikunnalle. Moniste, 56 s

Siitonen M, Härkönen K, Hirvelä H, Jämsä J, Kilpeläinen H, Salminen O, Teuri M (1996) MELA Handbook 1996 Edition. Metsäntutkimuslaitoksen tiedonantoja 622. The Finnish Forest Research Institute, Research Papers 622. 452 p

Silva-Flores R, Pérez-Verdín G, Wehenkel C (2014) Patterns of tree species diversity in relation to climatic factors on the Sierra Madre Occidental, Mexico. PLoS One 9(8):e105034

Simpson EH (1949) Measurement of diversity. Nature 163:688

Sinoquet H, Le Roux X, Adam B, Ameglio T, Daudet FA (2001) RATP: a model for simulating the spatial distribution of radiation absorption, transpiration and photosynthesis within canopies: application to an isolated tree crown. Plant Cell Environ 24:395–406

Sivananthawerl T (2001) Growth and site conditions of Eucalyptus grandis in Sri Lanka. Doctoral Diss., Georg-August-University, Göttingen

Skovsgaard JP (1997) Tyndingsfri drift af sitkagran. The Research Series, Forskningcentret for Skov & Landskab, Nr 19: 525 pp

Slik JWF et al (2018) Phylogenetic classification of the world's tropical forests. Proc Natl Acad Sci U S A 115(8):1837-1842. doi: https://doi.org/10.1073/pnas.1714977115.. Epub 2018 Feb 5.

Sloboda B (1984) Bestandesindividuelles biometrisches Schaftformmodell zur Darstellung und zum Vergleich von Formigkeit und Sortimentausbeute sowie Inventur. Tagungsbericht d. Sektion Ertragskunde, Neustadt

Smaltschinski T (1998) Charakterisierung von Baumverteilungen. Forstwiss Centralbl 117:355–361

Smith RE, Scott DF (1992) The effect of afforestation on low flows in various regions of South Africa. Water SA 18:185–194

Socolar JB, Gilroy JJ, Kunin WE, Edwards DP (2016) How should beta-diversity inform biodiversity conservation? Trends Ecol Evol 31(1):67–80

Solberg B (ed) (1996) Long-term trends and prospects in wood supply and demand for wood and implications for sustainable forest management. EFI research report No. 6, 106 p

Song M, Hu Q, Tian Y, Ouyang H (2012) Seasonal patterns of root and shoot interactions in an alpine meadow on the Tibetan Plateau. J Plant Ecol 5:182–190

Sørensen TA (1948) A method of etablishing groups of equal amplitude in plant sociology based on similarity of species content, and its application to analysis of the vegetation on Danish commons. Biologiske Skrifter Kongelige Danske Videnskabernes Selskab 5:1–34

Spanos KA, Feest A (2007) A review of the assessment of biodiversity in forest ecosystems. Manag Environ Qual 18(4):475–486

Späth H (1973) Spline-Algorithmen zur Konstruktion glatter Kurven und Flächen. Oldenbourg, München

Speidel G (1955) Die Wertklasse als Gütemaßstab in der Forsteinrichtung. Forstarchiv:217–224

Speidel G (1971) Planung im Forstbetrieb. Paul Parey, Hamburg

Spellmann H, Wagner S, Nagel J, Guericke M, Griese F (1996) In der Tradition stehend, neue Wege beschreitend. Forst u Holz 51(11):363–368

Spiecker M (1994) Wachstum und Erziehung wertvoller Waldkirschen. Mitt. d. Forstl. Vers. u. Forschungsanstalt Baden-Württemberg, No 181

Spies TA (1997) Forest stand structure, composition, and function. In: Creating a forestry for the 21st century: the science of ecosystem management. Oxford University Press, Oxford

Sprugel DG (1991) Disturbance, equilibrium, and environmental variability – what is natural vegetation in a changing environment. Biol Conserv 58:1–18

Srivastava VK, Giles DEA (1987) Seemingly unrelated regression equations models. Marcel Dekker Inc., New York

Stadt KJ, Huston C, Coates KD, Feng Z, Dale MRT, Lieffers WJ (2007) Evaluation of competition and light estimation indices for predicting diameter growth in mature boreal mixed forests. Ann For Sci 64(5):477–490

Stape JL, Binkley O (2010) Insights from full-rotation Nelder spacing trials with Eucalyptus in Sao Paulo, Brazil. South For 72(2):91–98

Staudhammer CL, LeMay VM (2001) Introduction and evaluation of possible indices of stand structural diversity. Can J For Res 31:1105–1115

Staupendahl K (1997) Ein neues Stichprobenverfahren zur Erfassung und Beschreibung von Naturverjüngung. Vortrag anlässlich der Jahrestagung der Sektion Forstliche Biometrie und Informatik des Deutschen Verbandes Forstlicher Versuchsanstalten am 24. 9. 97 in Freiburg

Staupendahl K (2011) Modellierung der Überlebenswahrscheinlichkeit von Waldbeständen mithilfe der neu parametrisierten Weibull-Funktion [Modelling the survival probability of forest stands using the reparametrised Weibull function]. Forstarchiv 82:10–19. (in German, with English summary)

Staupendahl K, Gadow K (2008) Eingriffsinventuren und Dynamisches Betriebswerk – Instrumente der Operativen Planung im Forstbetrieb. Forstarchiv 79:16–27

Staupendahl K, Möhring B (2011) Integrating natural risks into silvicultural decision models: a survival function approach. For Pol Econ (accepted)

Staupendahl K, Puumalainen J (2000) Modellierung des Einflusses von Durchforstungen auf die Durchmesserverteilung von gleichaltrigen Fichtenreinbeständen. Centralblatt für das gesamte Forstwesen, 116. Jahrgang, Heft 4:249–262

Staupendahl K, Zucchini W (2006) Estimating the spatial distribution in forest stands by counting small angles between nearest neighbors. Allg Forst- u J-Ztg 177(8/9):160–168

Staupendahl K, Zucchini W (2011) Schätzung von Überlebensfunktionen der Hauptbaumarten auf der Basis von Zeitreihendaten der Rheinland-Pfälzischen Waldzustandserhebung [Estimating survival functions for the main tree species based on time series data from the forest condition survey in Rheinland-Pfalz, Germany]. Allg. Forst- Jagdztg. 182(5/6). (in German, with summaries in English and French)

Steffen W (2015) Planetary boundaries: Guiding human development on a changing planet. Science 347:6223

Steffen W, Crutzen PJ, McNeill JR (2007) The Anthropocene: are humans now overwhelming the great forces of nature. AMBIO J Hum Environ 36:614–621

Steffen W, Leinfelder R, Zalasiewicz J, Waters CN, Williams M, Summerhayes C, Barnosky AD, Cearreta A, Crutzen P, Edgeworth M, Ellis EC, Fairchild IJ, Galuszka A, Grinevald J, Haywood A, do Sul JI, Jeandel C, McNeill JR, Odada E, Oreskes N, Revkin A, Richter DB, Syvitski J, Vidas D, Wagreich M, Wing SL, Wolfe AP, Schellnhuber HJ (2016) 4:324–345. https://doi.org/10.1002/2016EF000379

Stehle TC (1980) Waaisandbestuurplan vir die Alexandriakusstaatsbos 1981/2 tot 1990/1. Internal report, Department of Forestry, South Africa

Steingass F (1996) Beschreibung der Schaftprofile von Douglasien. Diplomarbeit, Fachbereich Forstwissenschaft, Universität Göttingen

Steinmeyer A, Gadow K (1994) Saaty's AHP – dargestellt am Beispiel der Waldbiotopkartierung. Centralblatt f d ges Forstwesen 112(1):53–65

Steneker GA, Jarvis JM (1963) A preliminary study to assess competition in a white spruce – trembling aspen stand. For Chron 39:334–336

Stephenson W, Williams WT, Cook SD (1972) Computer analyses of Petersen's original data on bottom communities. Ecol Monogr 42(4):387–415

Sterba H (1975) Assmanns Theorie der Grundflächenhaltung und die "Competition-Density-Rule" der Japaner Kira, Ando und Tadaki. Centralblatt für das gesamte Forstwesen 92(1):46–62

Sterba H (1981) Natürlicher Bestockungsgrad und Reinecke's SDI. Centralblatt für das gesamte Forstwesen 98(2):101–116

Sterba H (1990) Validitätsprüfung des Einzelbaumsimulators für Fichten-Kiefern-Mischbestände. Tagungs-bericht. Dt. Verb. Forstl. Forschungsanstalten – Sektion Ertragskunde, pp 101–117

Sterba H (1991) Forstliche Ertragslehre. Lecture notes (Heft 4), Vienna, p 159

Sterba H, Monserud RA (1997) Applicability of the forest stand growth simulator Prognaus for the Austrian part of the Bohemian Massif. Ecol Model 98:23–34

Sterba H, Zingg A (2006) Abstandsabhängige und abstandsunabhängige Bestandesstrukturbeschreibung. Allg Forst- u J-Ztg 177(8/9):169–176

Stöcker G (2002) Analyse und Vergleich von Bestandesstrukturen naturnaher Fichtenwälder mit Lorenz-Funktionen und Gini-Koeffizienten. Centralblatt für das gesamte Forstwesen 119:12–39

Stöhr FK (1963) Erweiterungsmöglichkeiten der Winkelzählprobe. Dissertation, Albert-Ludwig-University, Freiburg/Br. 83 S

Stoyan D, Penttinen A (2000) Recent applications of point process methods in forestry statistics. Stat Sci 15:61–78

Stoyan D, Stoyan S (1994) Fractals, random shapes and point fields. Methods of geometrical statistics. Wiley, Chichester, 394p

Struck G (1999) Effekte von Pflanzverband, Kulturdüngung und Schlagräumung am Beispiel junger Traubeneichenbestände. Dissertation, Fak. f. Forstwiss. u. Waldökologie, University of Göttingen, 196 p

Sturm K (1995) Möglichkeiten, Probleme und Defizite der Integration naturschutzrelevanter Daten im Rahmen der Forsteinrichtung. Schr f Landschaftspfl u Natursch 43:231–245

Sunseri T (2012) Exploiting the Urwald: German post-colonial forestry in Poland and central Africa, 1900–1960. Past Present 214:305–342

Susmel L (1980) Normalizzazione delle foreste alpini – basi ecosistemiche – equilibrio – modelli colturali – produttivita. Liviane editrice, Padova, 437 p

Suzuki T (1971) Forest Transition as a stochastic process. Mitt FBVA Wien Heft 91:137–150

Suzuki SN, Kachi N, Suzuki J-I (2008) Development of a local size-hierarchy causes regular spacing of trees in an even-aged *Abies* forest: analyses using spatial autocorrelation and the mark correlation function. Ann Bot 102:435–441

Swart DH, Reyneke PG (1998) The role of driftsands at Waenhuiskrans, South Africa. J Coast Res Special Issue, No. 3

Swenson NG (2011) Phylogenetic beta diversity metrics, trait evolution and inferring the functional beta diversity of communities. PLoS One 6:e21264

Swenson NG (2014) Functional and phylogenetic ecology in R. Springer, New York

Swenson NG, Stegen JC, Davies SJ, Erickson DL, Forero-Montana J, Hurlbert AH, Kress WJ, Thompson J, Uriarte M, Wright SJ, Zimmerman JK (2012) Temporal turnover in the composition of tropical tree communities: functional determinism and phylogenetic stochasticity. Ecology 93:490–499

Szaro RC, Johnston DW (1996) Biodiversity in managed landscapes. Oxford University Press, New York. 778 p

Szaro RC, Peterson CE, Gadow K (2006) Operational experiments for sustainably managing forests. Allg Forst- u J-Ztg 177(6/7):98–104

Szczukowski S, Tworkowski J, Klasa A, Stolarski M (2002) Productivity and chemical composition of wood tissues of short rotation willow coppice cultivated on arable land. University of Warmia and Mazury, Olsztyn

Tabaku V (1999) Struktur von Buchen-Urwäldern in Albanien im Vergleich mit deutschen BuchenNaturwaldreservarten und -Wirtschaftswäldern. Dissertation. Forstl. Fak. Universtiy of Göttingen. Cuvillier Verlag, Göttingen

Talbot MMB, Bate GC (1991) The structure of vegetation in bush pockets of trangressive coastal dune fields. S Afr J Bot 57:156–160

Talents LA, Lovett JC, Hall JB, Hamilton AC (2005) Phylogenetic diversity of forest trees in the Usambara mountains of Tanzania: correlations with altitude. Bot J Linn Soc 149(2):217–228

Tang S, Zhang H, Xu H (2000) Study on establishing and estimating a compatible biomass model. Scientia Silvae Sinicae 36(Suppl. 1):19–27. (In Chinese with English abstract)

Tang ZY, Wang ZH, Zheng CY et al (2006) Biodiversity in China's mountains. Front Ecol Environ 4:347–352

Temesgen H, Gadow K (2003) Generalized height-diameter models – an application for major tree species in complex stands of interior British Columbia. Eur J For Res 123(1):45–51

Terauds A, Cooper J, Chown SL, Ryan P (2010) Marion & Prince Edward – Africa's Southern Islands. SUN Press, Stellenbosch, p 350. ISBN 978-1-920338-42-8

Tewari VP, Arya R (2005) Degradation of arid rangelands in Thar Desert, India: a review. Arid Land Res Manag 19(1):1–12

Tewari VP, Gadow Kv (2008) Modelling potential density limiting survival, stand density and basal area growth for pure even-aged *Dalbergia sissoo* stands in a hot arid region of India. For Trees Livelihoods 18:133–150

Tewari VP, Gadow Kv (2010) Domesticated ecosystems – analysis and design – an example involving Indira Gandhi Canal Plantations in Rajasthan, India. Poster presented at the IUFRO World Congress in Seoul/Korea; August 2010

Tewari VP, Singh M (2006) Tree-crop interaction in the Thar desert of Rajasthan, India. Sècheresse 17(1-2):326–332

Tewari VP, Álvarez-González JG, García O (2014a) Developing a dynamic growth model for teak plantations in India. For Ecosyst 1:9

Tewari VP, Sukumar R, Kumar R, Gadow K (2014b) Forest observational studies in India – past developments and future prospects. For Ecol Manage 316:32–46

Tewari VP, Alvarez-Gonzalez JG, Gadow Kv (2014c) Dynamic base-age invariant site index models for Tectona grandis in Peninsular India. Southern Forests 76(1):21–27

The Forest 2000 Programme (1986) Guidelines for developing Finnish forestry and the forest industries. Silva Fennica 20(1):35–44

Thomasius H (1988) Stabilität natürlicher und künstlicher Waldökosysteme sowie deren Beeinflußbarkeit durch forstliche Maßnahmen. AFZ 43(1037–1043):1064–1068

Thomasius HO (1976) Anwendungsbeispiel eines Verfahrens zur Berechnung ertragskundlich optimaler Bestandesdichtewerte und Diskussion dieses Verfahrens. Als Manuskr. gedr. Techn. Univ. Dresden, Preprint Nr. 21-16-76

Thomasius HO (1978) Ableitung eines Verfahrens zur Berechnung der ertragskundlich optimalen Bestandesdichte. Beitr. Forstwirtsch 12, 83

Thomasius HO, Thomasius H (1976) Anwendungsbeispiel zu einem Verfahren der Berechnung ertragskundlich optimaler Bestandesdichtewerte und Diskussion dieses Verfahrens. Informationen TU Dresden, Sektion Forstwirtschaft, Bereich Waldbau und Forstschutz, Tharandt, pp 21–76

Thomasius HO, Thomasius H (1978) Ableitung eines Verfahrens zur Berechnung der ertragskundlich optimalen Bestandesdichte. Beiträge f. d. Forstwirtschaft 12(79)

Thurstone LL (1927) A law of comparative judgement. Psychol Rev 34:273–286

Tilman D, Lehman CL, Thomson KT (1997) Plant diversity and ecosystem productivity: theoretical considerations. Proc Natl Acad Sci U S A 94:1857–1861

Tinley KL (1985) Coastal dunes of South Africa. South African National Scientific Programmes Report, No 109. CSIR, Pretoria

Tjørve E (2003) Shapes and functions of species-area curves: a review of possible models. J Biogeogr 30:827–835

Tomé M, Burkhart HE (1989) Distance-dependent competition measures for predicting growth of individual trees. For Sci 35(3):816–831

Tomé M, Falcao A, Amaro A (1997) Globulus v 1.0.0: A regionalised growth model for eucalypt plantations in Portugal. In: Ortega A, Gezan S (eds) Proceedings of the IUFRO Conference: Modelling growth of fast-grown tree species. Valdivia, Chile, 5–7 September 1997

Tomppo E (2004) Resource assessment: Forest resources. In: Burley J, Evans J, Youngquist JA (eds) Encyclopedia of forest sciences. Amsterdam, Elsevier, pp 965–973

Tomppo E (2006) The Finnish National forest inventory. In: Kangas A, Maltamo M (eds) Forest inventory, methodology and applications. Springer, Dordrecht, pp 179–194

Tomppo E (2009) Establishing methodological basis and building partnership for Brazil's National Forest Inventory. A mission report for the Food and Agriculture Organization of the United Nations (FAO). FAO, Rome

Tomppo E, Czaplewski RL (2002) Potential for a remote-sensing-aided forest resource survey for the whole globe. Unasylva 53(210):16–18

Tomppo E, Gschwantner T, Lawrence M, McRoberts RE (eds) (2010) National forest inventories. Pathways for common reporting, vol XXVI. Springer, 612 p

Tomppo E, Malimbwi R, Katila M, Mäkisara K, Henttonen HM, Chamuya N, Zahabu E, Otieno J (2014) A sampling design for a large area forest inventory: case Tanzania. Can J For Res 44:931–948

Torgerson WS (1958) Theory and methods of scaling. Wiley, New York

Torquato S (2002) Random heterogeneous materials. Microstructure and macroscopic properties. Interdisciplinary applied mathematics 16. Springer, New York

Torres Rojo JM (2000) Sostenibiladad del volumen de cosecha calculado con el método de ordenación de montes. Madera y Bosques 6(2):57–72

Torres-Rojo JM (2014) Exploring volume growth-density of mixed multiaged stands in northern Mexico. Agrociencia 48:447–4613

Torres-Rojo JM, Magaña-Torres OS (2006) Management of Mexican community forests with timber production objectives. Allgemeine Forst und Jagdzeitung 177(3/4):63–71

Trasobares A, Pukkala T (2005) Optimising the management of uneven-aged Pinus sylvestris L. and Pinus nigra Arn. Mixed stands in Catalonia, north-east Spain. Ann For Sci 61:747–758

Tress B, Tress G (2001) Capitalising on multiplicity: a trans- disciplinary systems approach to landscape research. Landsc Urban Plann 57:143–157

Trincado GF (1996) Modellierung der Schaftform von Fichten (Picea abies Karst.) und Buchen (Fagus Sylvatica L.). MSc thesis. University of Göttingen, 50 p

Trincado G, Gadow K (1996) Zur Sortimentschätzung stehender Laubbäume. Clbl ges Forstw 1:27–38

Trincado GV, Kiviste A, Gadow K (2003a) Preliminary site index models for native Roble (Nothofagus obliqua) and Raulí (N. alpina) in Chile. N Z J For Sci 32(3):322–333

Trincado G, Quezada R, Gadow Kv (2003b) A comparison of two stand table projection methods for young Eucalyptus nitens (Maiden) plantations in Chile. For Ecol Manage 6193:1–9

Trumbore S, Brando P, Hartmann H (2015) Forest health and global change. Science 349(6250):814–818. https://doi.org/10.1126/science.aac6759

Turner J, Thompson CH, Turvey ND, Hopmans P, Ryan PJ (1990) A soil technical classification for *Pinus radiata* (D. Don) plantations. I. Development. Aust J Soil Res 28:797–811

Turvey, N. D., Booth, T. H. u. Ryan, P. J., 1990: A soil technical classification for *Pinus radiata* (D. Don) plantations. II. A basis for estimation of crop yield. Aust J Soil Res 28: 813-824.

Ulrich B (1987) Stabilität, Elastizität und Resilienz von Waldökosystemen unter dem Einfluß saurer. Depositionen Forstarchiv 58:232–239

Ulrich B (1993) Prozeßhierarchie in Waldökosystemen. Biologie in Unserer Zeit 23(5):322–329

Ung C-H (1993) Growth and branchiness of Picea mariana (Mill.) BSP. In: Canadian Forest Service Modelling Working Group. Proceedings of the seventh annual meeting and workshop, p 9. ISBN 0-662-21104-9

Ung C-H, Raulier F, Oullet D, Dhôte J-F (1997) L'indice de compétition interindividuelle de Schütz. Can J For Res 27:521–526

United Nations (1992) Internationally agreed glossary of basic terms related to desaster management. United Nations Department of Humanitarian Affairs, Geneva

United Nations Department of Humanitarian Affairs (1992) Internationally agreed glossary of basic terms related to disaster management. Geneva. http://reliefweb.int/sites/reliefweb.int/files/resources/004DFD3E15B69A67C1256C4C006225C2-dha-glossary-1992.pdf

Upton G, Fingleton B (1989) Spatial data analysis by example. Vol. 2: Categorical and directional data. Wiley, Chichester, 416 p

Upton G, Fingleton B (1990) Spatial data analysis by example. Vol. 1: Point pattern and quantitative data, 2nd edn. Wiley, Chichester, 410 p

UN-CCD (2000) United Nations Convention to Combat Desertification. https://www.unccd.int

Valladares F, Niinemets Ü (2008) Shade tolerance, a key plant feature of complex nature and consequences. Annu Rev Ecol Evol Syst 39:237–257

Valsta LT (1986) Mänty-rauduskoivusekametsikön hakkuuohjelman optimointi. Summary: Optimizing thinnings and rotation for mixed, even-aged pine-birch stands. Folia Forestalia 666:23 S

Van Daalen JC (1993) The value of crown position and form as growth indicators in mixed evergreen forests. S Afr For J 165:29–35

Van Laar A (1981) The response of Pinus radiata to initial spacing. S Afr For J 121:52–63

Van Laar A, Akça A (1997) Forest mensuration. Cuvillier Verlag Göttingen, 418 pp. ISBN 3-89588-874-5

Van Laar A, Akça A (2007) Forest mensuration. Vol 13 of the Springer Book Series managing forest ecosystems, 418 p

Van Laar A, Jansen JJ (2001) Models for diameter distributions in even-aged Douglas Fir stands in the Netherlands. In: Akca A, Hoffmann B, Schumann K, Staupendahl K (eds) Waldinventur, Waldwachstum und Forstplanung. Festschrift K. v. Gadow, Zohab Publisher, Göttingen, pp 185–194

Van Lill WS, Kruger FJ, Van Wyk DB (1980) The effect of afforestation wilh Eucalyptus grandis Hill exMaiden and Pinus patula Schlecht. et Cham. on streamflow from experimental catchments at Mokobulaan' Transvaal. J Hydrol 48:107–118

Van Wyk G (1985) Genetic improvement of Eucalyptus grandis. In: Paper read at the meeting of Eucalypt related research, NTRI, Pretoria on 8 August 1985

Vanclay JK (1988) A stand growth model for cypress pine. In Leech JW, McMurtrie RE, West PW, Spencer RD, Spencer BM (eds) Modelling Trees, Stands and Forests. Proc. Workshop, Aug. 1985, Univ. of Melbourne. School of Forestry, Univ. of Melbourne, Bulletin No. 5, pp 310–332

Vanclay JK (1992) Assessing site productivity in tropical moist forests: a review. For Ecol Manage 54:257–287

Vanclay JK (1994) Modelling forest growth – applications to mixed tropical forests. CAB International, Wallingford, 312 pp. ISBN 0 85198913 6

Vanclay JK (1995) Growth models for tropical forests: a synthesis of models and methods. For Sci 41:7–42

Vanclay JK, Henry NB (1988) Assessing site productivity of indigenous cypress pine forest in southern Queensland. Commonw For Rev 67:53–64

Vargas-Larreta B, Castedo-Dorado F, Álvarez González JG, Barrio-Anta M, Cruz-Cobos F (2009) A generalized height–diameter model with random coefficients for uneven-aged stands in El Salto, Durango (Mexico). Forestry 82:445–462

Vauhkonen J, Tokola T, Packalen P, Maltamo M (2009) Identification of single-tree attributes using airborne laser scanning-based height, intensity and alpha shape metrics. For Sci 55:37–47

Vera FWM (2000) Grazing ecology and forest history. CAB International, Wallingford, 506 pp. http://lib.du.ac.ir/documents/10157/60360/Grazing+Ecology+and+Forest+History+-+(2000).pdf

Verheyen K, Ceunen K, Ampoorter E, Baeten L, Bosman B, Branquart E, Carnol M, De Wandeler H, Grégoire JC, Lhoir P, Muys B, Setiawan NN, Vanhelle-Mont M, Ponette Q (2013) Assessment of the functional role of tree diversity, the example of the multi-site FORBIO-experiment. Plant Ecol Evol 146:1–10

Verheyen K, Vanhellemont M, Auge H, Baeten L, Baraloto C, Barsoum N, Bilodeau-Gauthier S, Bruelheide H, Castagneyrol H, Godbold D, Haase J, Hector A, Jactel H, Koricheva J, Loreau M, Mereu S, Messier C, Muys B, Nolet P, Paquette A, Parker J, Perring M, Ponette Q, Potvin C, Reich P, Smith A, Weih M, Scherer-Lorenzen M (2015) Contributions of a global network of tree diversity experiments to sustainable forest plantations. Ambio. [Epub ahead of print]

Verma M, Negandhi D, Wahal AK, Kumar R (2013) Revision of rates of NPV applicable for different class/category of forests. Indian Institute of Forest Management. Bhopal, India

Versfeld DB (1994) The forest industry and management for water conservation. In: Southern African forestry handbook 1994. South African Institute of Forestry, pp 657–674

Vidal RVV (ed) (1993) Applied simulated annealing. Springer, Berlin/Heidelberg

Vinatier F, Tixier P, Duyck P-F, Lescourret F (2011) Factors and mechanisms explaining spatial heterogeneity: a review of methods for insect populations. Method Ecol Evol 2(1):11–22

Vincke P (1992) Multicriteria decision-aid. Wiley, New York

Vining J, Merrick MS, Price EA (2008) The distinction between humans and nature: human perceptions of connectedness to nature and elements of the natural and unnatural. Hum Ecol Rev 15(1):1–11

Virgilietti P, Buongiorno J (1997) Modelling forest growth with management data – a matrix approach for the Italian Alps. Silva Fennica 31(1):27–42

Volz K-R (1995) Zur ordnungspolitischen Diskussion über die nachhaltige Nutzung der Zentralressource Wald. Forst u Holz 50(6):163–170

Vose D (2008) Risk analysis: a quantitative guide, 3rd edn. Wiley

von Schütz JE (1757) Ökonomisches Bedenken von dem Nachteil deren Gehaue in den sogenanten Schwartzen Hölzer. In: Ökonomisches Bedenken über allerhand in die Hauswirtschaft einschlagende Sachen, vol 1. Stössel, Chemnitz, pp 1–18

Von Winterfeldt D, Edwards W (1986) Decision analysis and behavioral research. Cambridge University Press, Cambridge

Vose JM, Allen HL (1988) Leaf area, stemwood growth and nutrition relationships in loblolly pine. For Sci 34(3):547–563

Waggener TR (1969) Some economic implications of sustained yield as a forest regulation model. Contribution No. 6, University of Washington, 22 pp

Waldherr M (1997) Risikoverluste und Erntealter. AFZ/Der Wald, pp 206–207

Walotek P, Murach D, Gadow Kv (2006) Die Korbweide (Salix viminalis) – Ertragschätzung und Nutzungsmodell. Contribution to the German meeting of yield scientists (Ertragskunde-Tagung) on 29/30 May 2006

Wang Y, Woodcock CE, Buermann W, Stenberg P, Voipio P, Smolander H, Hame T, Tian Y, Hu J, Knyazikhin Y, Myneni RB (2004) Evaluation of the MODIS LAI algorithm at a coniferous forest site in Finland. Remote Sens Environ 2004(91):114–127

Wang J, Zhang C, Xia F, Zhao X, Wu L, Kv G (2011) Biomass structure and allometry of Abies nephrolepis (Maxim) in Northeast China. Silva Fenn 45(2):211–226

Wang J, Wu L, Zhao X, Fan J, Zhang C, Gadow K (2013a) Influence of ground flora on Fraxinus mandshurica seedling growth on abandoned land and beneath forest canopy. Eur J For Res 132(2):313–324

Wang J, Shen J, Wu Y, Tu C, Soininen J, Stegen JC, He J, Liu X, Zhang L, Zhang E (2013b) Phylogenetic beta diversity in bacterial assemblages across ecosystems: deterministic versus stochastic processes. ISME J 7:1310–1321

Wang J, Zhang C, Zhao X, Gadow Kv (2014) Reproductive allocation of two dioecious Rhamnus species in temperate forests of Northeast China. iForest 7:25–32. http://www.sisef.it/iforest/contents/?id=ifor1020-007

Wapner P (2014) The changing nature of nature: environmental politics in the Anthropocene. Glob Environ Pol 14:36–54

Ware GO, Clutter JL (1971) A mathematical programming system for the management of industrial forests. For Sci 17:428–445

Webb LJ (1959) A physiognomic classification of Australian rainforests. J Ecol 47:551–570

Webb CO (2000) Exploring the phylogenetic structure of ecological communities: an example for rain forest trees. Am Nat 156:145–155

Webb CO, Ackerly DD, Kembel SW (2008) Phylocom: software for the analysis of phylogenetic community structure and trait evolution. Bioinformatics 24:2098–2100

Wehenkel C, Corral-Rivas JJ, Hernandez-Díaz JC, Gadow Kv (2011) Estimating balanced structure areas in multi-species forests on the Sierra Madre Occidental, Mexico. Ann For Sci 68. https://doi.org/10.1007/s13595-011-0027-9

Wehenkel C, Corral-Rivas JJ, Gadow KV (2014) Quantifying differences between ecosystems with particular reference to selection forests in Durango/Mexico. For Ecol Manage 316:117–124

Weibull W (1951) A statistical distribution of wide applicability. J Appl Mech 18:293–297

Weise W (1880) Ertragstafeln für die Kiefer. Julius Springer Verlag, Berlin

WeiSheng Z, Tomppo E, Healey SP, Gadow K (2015) The national forest inventory in China: history – results – international context. For Ecosyst 2:23

Weiskittel AR, Hann DW, Kershaw JA (2011) Forest growth and yield modeling. Wiley, New York, p 415. ISBN-13: 9780470665008

Weiss G (2001) Administrating risk – a social science perspective on natural hazard prevention based on an Austrian case study. In: Gadow Kv (ed) 2001: risk analysis in forest management. Vol 2 in the book series managing forest ecosystems. Kluwer Academic Publishers, New York, pp 171–200

Wenk G (1996) Durchmesserverteilungen im Buchenplenterwald (als Bestandteil des Forschungsprojektes „Sortenorientiertes Wachstumsmodell"). Tagungsband des Deutschen Verbandes Forstlicher Versuchs-anstalten. Sektion Ertragskunde, Mai 1996 in Neresheim

Wenk G, Antanaitis V, Smelko S (1990) Waldertragslehre. Deutscher Landwirtschaftsverlag, Berlin

West PW (1982) Comparison of stand density measures in even-aged regrowth eucalypt forest of southern Tasmania. Can J For Res 13:22–31

Westphal C, Tremer N, Oheimb Gv, Hansen J, Gadow K, Härdtle W (2006) Is the reverse J-shaped diameter distribution universally applicable in European virgin beech forests? For Ecol Manag 223:75–83

White CM (1960) Multiple goals in the theory of the firm. In: Boulding KE, Spivey WA (eds) Linear programming and the theory of the firm. New York

White F (1978) The afromontane region. In: Werger MJA (ed) Biogeography and ecology of Southern Africa. Dr. W. Junk, The Hague, pp 464–513

Whittaker RH (1972) Evolution and measurement of species diversity. Int Ass Plant Tax 21:213–251
Wiant HV, Ramirez MA, Barnard JE (1975) Predicting oak site index by species composition in West Virginia. J For 73:666–667
Wicht CL (1949) Forestry and water supplies in South Africa. Department of Forestry Bulletin No. 33, Pretoria
Wiecko E (1972) Puszcza Białowieska. Panstwowe Wydawnictwo Naukowe, Warsaw. 193 pp
Wiedemann E (1931) Die Rotbuche. Mitteilung aus Forstwirtschaft und Forstwissenschaft, 96 S
Wiedemann E (1949) Ertragstafeln der wichtigsten Holzarten. M.& H. Schaper, Hannover
Wiegard C, Netzker D, Gadow K (1997) Die Erdstückmethode der Wertinventur. Forstarchiv 68:144–148
Wiens JJ, Graham CH (2005) Niche conservatism: integrating evolution, ecology, and conservation biology. Annu Rev Ecol Evol Syst 36:519–539
Williams CB (1964) Patterns in the balance of nature. Academic, London
Williams MR (1995) An extreme-value function model of the species incidence and species-area relationship. Ecology 76:2607–2616
Wilson JB (1988) Shoot competition and root competition. J Appl Ecol 25:279–296
Windhager M (1999) Evaluierung von vier verschiedenen Waldwachstumssimulatoren. Dissertation, Universität für Bodenkultur Wien, 217 S
Winston PH (1977) Artificial intelligence. Addison-Wesley, Reading
Wirth C, Schumacher J, Schulze E-D (2004) Generic biomass functions for Norway spruce in Central Europe – a meta-analysis approach toward prediction and uncertainty estimation. Tree Physiol 24:121–139
Wood SN (2006) Generalized additive models: an introduction. R. Chapman & Hall/CRC, Boca Raton, p 391
Woollons RC (1998) Even-aged stand mortality estimation through a two-step regression process. For Ecol Manage 105:189–195
Woollons R, Hayward W (1985) Revision of a growth and yield model for radiata pine in New Zealand. For Ecol Manage 11:191–202
Wright EF, Canham CD, Coates KD (2000) Effects of suppression and release on sapling growth for 11 tree species of northern, interior British Columbia. Can J For Res 30:1571–1580
Wulder MA, Hilker T, White JC, Coops NC, Masek JG, Pflugmacher D, Crevier Y (2015) Virtual constellations for global terrestrial monitoring. Remote Sens Environ 170:62–76
Wulder MA, Loveland TR, Roy DP, Crawford CJ, Masek JG, Woodcock CE, Allen RG, Anderson MC, Belward AS, Cohen WB, Dwyer J, Erb A, Gao F, Griffiths P, Helder D, Hermosilla T, Hipple JD, Patrick H, Joseph Hughes M, Huntington J, Johnson DM, Kennedy R, Kilic A, Li Z, Lymburner L, McCorkel J, Pahlevan N, Scambos TA, Schaaf C, Schott JR, Sheng Y, Storey J, Vermote E, Vogelmann J, White JC, Wynne RH, Zhu Z (2019) Current status of landsat program, science, and applications. Remote Sens Environ 225:127–147
Wykoff WR (1990) A basal area increment model for individual conifers in the northern Rocky mountains. For Sci 36(4):1077–1104
Wykoff WR, Crookston NL, Stage AR (1982) User's guide to the stand prognosis model. USDA Forersty Service General Technical Report No. INT–133, p 112
Xia FC (2007) Study on the plant biodiversity and spatial patterns of broad-leaved Korean Pine forest in Changbai Mountain. Doctoral dissertation, Beijing Forestry University, China, 198 pp
Xu W, Luo W, Zhang C, Zhao X, von Gadow K, Zhang Z (2019) Biodiversity-ecosystem functioning relationships of overstorey versus understorey trees in an old-growth temperate forest. Annals of Forest Science
Yamakura T, Kanzake M, Itoh A, Ohkubo T, Ogino K, Chai EOK, Lee HS, Ashton PS (1995) Topography of a large-scale research plot established within a tropical rain forest at Lambir, Sarawak. Tropics 5:41–56
Yan Y, Zhang C, Wang Y, Zhao X, Gadow K (2015) Drivers of seedling survival in a temperate forest and their relative importance at three stages of succession. Ecol Evol:1–13. https://doi.org/10.1002/ece3.1688

Yang YC, Feng FL (1989) The application of parameter prediction models and paramter recovery methods to the analysis of tree growth and stand structure in Taiwan. Q J Chin For 22(1):53–71

Yoda K, Kira T, Ogawa H, Hozumi K (1963) Self thinning in overcrowded pure stands under cultivated and natural conditions. J BiolOsaka City Univ 14:107–132

Yoshida T, Kamitani T (1998) Effects of crown release on basal area growth rates of some broad-leaved tree species with different shade tolerance. J Jpn For Res 3:181–184

Yoshimoto A (2001) Risk analysis in the context of timber harvest planning. In: Gadow Kv (ed) Risk analysis in forest management. Vol 2 of the book series managing forest ecosystems. Kluwer Academic Publishers, Dordrecht, pp 113–148

Yue C, Kohnle U, Hein S (2008) Combining tree- and stand-level models. A new approach to growth prediction. For Sci 54(5):553–566

Zarnoch SJ, Feduccia DP, Baldwin VC, Dell TR (1991) Growth and yield predictions for thinned and unthinned slash pine plantations on cutover sites in the West Gulf region. Res. Pap. SO-264. New Orleans, LA: U.S. Department of Agriculture, Forest Service, Southern Forest Experiment Station. 32 p

Zangemeister C (1976) Nutzwertanalyse in der Systemtechnik. Wittemannsche Buchhandlung, München

Zanne AE, Tank DC, Cornwell WK, Eastman JM, Smith SA, FitzJohn RG, McGlinn DJ, O'Meara BC, Moles AT, Reich PB, Royer DL, Soltis DE, Stevens PF, Westoby M, Wright IJ, Aarssen L, Bertin RI, Calaminus A, Govaerts R, Hemmings F, Leishman MR, Oleksyn J, Soltis PS, Swenson NG, Warman L, Beaulieu JM (2014) Three keys to the radiation of angiosperms into freezing environments. Nature 506:89–92

Zeide B (1987) Analysis of the 3/2 power law of self thinning. For Sci 33:517–537

Zeide B (1991) Self-thinning and stand density. Forest Science 37:517–523

Zeide B (1995) Tolerance and self-tolerance of trees. For Ecol Manage 13:149–166

Zeide B (2004) Optimal stand density: A solution. Can J For Res 34:846–854

Zeide B (2005) How to measure stand density. Trees 19:1–14

Zeide B, Gresham CA (1991) Fractal dimensions of tree crowns in three loblolly pine plantations of coastal South Carolina. Can J For Res 21(8):1208–1212

Zeide B, Pfeifer P (1991) A method for estimation of fractal dimension of tree crowns. For Sci 37(5):1253–1265

Zeleny M (1983) Multiple criteria decision making. McGraw-Hill, New York

Zellner A (1962) An efficient method of estimating seemingly unrelated regression equations and tests for aggregation bias. J Am Stat Assoc 57:348–368

Zhang X, Lei Y (2010) A linkage among whole-stand model, individual-tree model and diameter-distribution model. J For Sci 56:600–608

Zhang S, Amateis RL, Burkhart HE (1997a) Constraining individual tree diameter increment and survival models for loblolly pine plantations. For Sci 43:414–423

Zhang S, Burkhart HE, Amateis RL (1997b) The influence of thinning on tree height and diameter relationships in loblolly pine plantations. South J Appl For 21:199–205

Zhang L, Kong J, Zeng X, Ren J (2008) Plant species identification based on neural network. In: 2008 fourth international conference on natural computation, Jinan, 2008, pp 90–94. https://doi.org/10.1109/ICNC.2008.253

Zhang CY, Zhao XH, Gadow K (2009) Gender, neighboring competition and habitat effects on the stem growth of dioecious Fraxinus mandshurica trees in a northern temperate forest. Ann For Sci 66:812–821

Zhang X, Lei Y, Cao QV (2010a) Compatibility of stand basal area predictions based on forecast combination. For Sci 56(6):552–557

Zhang CY, Petráš R, Zhao X, Gadow K (2010b) Estimating beech growth and survival. A study based on longterm experiments in Slovakia. Allgemeine Forst- und Jagdzeitung 181(3/4):44–53

Zhang CY, Zhao XH, Gadow K (2010c) Partitioning temperate plant community structure at different scales. Acta Oecologica 36(2010):306–313

Zhang X, Lei Y, Cao QV, Chen X, Liu X (2011) Improving tree survival prediction with forecast combination and disaggregation. Can J For Res 41:1928–1935

Zhang CY, Zhao Y, Zhao X, Gadow K (2012) Species-habitat associations in a northern temperate forest in China. Silva Fennica 46(4):501–519

Zhang C, Wei Y, Zhao X, Gadow Kv (2013a) Spatial characteristics of tree diameter distributions in a temperate old-growth forest. PLoS One 8:e58983

Zhang CY, Zhao XH, Gadow Kv (2013b) An analysis of three selective harvest events based on large forest observational studies in North Eastern China. For Ecol Manage 316:100–109

Zhang C, Zhao X, Gadow K (2014) Analysing selective harvest events in three large forest observational studies in north eastern China. For Ecol Manage 316:100–109

Zhang CY, Zhao XH, Gadow K (2015) Maximum density patterns in two natural forests: an analysis based on large observational field studies in China. For Ecol Manage 346:98–105

Zhao M, Running SW, Nemani RR (2006) Sensitivity of moderate resolution imaging spectroradiometer (MODIS) terrestrial primary production to the accuracy of meteorological reanalyses. J Geophys Res 111:1–13

Zhao XH, Zhang CY, Gadow Kv (2012) Forest observational networks. In: Proceedings of the IUFRO-EFI-ICFFI conference ecosystem design for multiple services-with an emphasis on Eurasian Boreal Forests, St. Petersburg Forest Technical University, St. Petersburg/Russia, 9–11 November 2011, pp 34–47

Zhao XH, Corral-Rivas JJ, Zhang CY, Temesgen H, Gadow K (2014) Forest observational studies-an essential infrastructure for sustainable use of natural resources. For Ecosyst 2014:1–8

Zhe Z, Wulder MA, Roy DP, Woodcock CE, Hansen MC, Radeloff VC, Healey SP, Schaaf C, Hostert P, Strobl P, Pekel J-F, Lymburner L, Pahlevan N, Scambos TA (2019) Benefits of the free and open landsat data policy. Remote Sens Environ. https://doi.org/10.1016/j.rse.2019.02.016

Zimmermann HJ (1991) Fuzzy set theory and its applications, 2nd edn. Kluwer Academic Publishers

Zucchini W, Gadow K (1995) Two indices of agreement among foresters selecting trees for thinning. Forest Landsc Res 1:199–206

Zucchini W, Schmidt M, Gadow K (2001) A model for the diameter-height distribution in an uneven-aged beech forest and a method to assess the fit of such models. Silva Fennica 35(2):168–183

Zunino CA, Ferrando MT (1997) Modelación del crecimiento y rendimiento de plantaciones de Eucalyptus en Chile. Una primera etapa. In: Ortega A, Gezan S (eds) Proceedings of the IUFRO conference modelling growth of fast-grown tree species. Valdivia (Chile), 5–7 September 1997, pp 155–164

Index

© Springer Nature Switzerland AG 2021

K. von Gadow et al., *Sustaining Forest Ecosystems*, Managing Forest Ecosystems 37, https://doi.org/10.1007/978-3-030-58714-7